List of participants

Alarcon Ricardo
Arizona State University
Department of Physics and Astronomy
85287-1504 Tempe, AZ
USA

Armstrong David
College of William and Mary
P.O. Box 8795
Physics Department
23187-8795 Williamsburg, VA
USA

Aulenbacher Kurt
Institut für Kernphysik
J.J. Becherweg, 45
55099 Mainz
Germany

Baunack Sebastian
Institut für Kernphysik
J.J. Becherweg, 45
55099 Mainz
Germany

Bijker Roelof
ICN-UNAM, AP 70-543
04510 Mexico, DF
Mexico

Bodek Arie
University of Rochester
Department of Physics and Astronomy
14627 Rochester, NY
USA

Bouchiat Marie-Anne
Ecole Normal Supérieure
24, Rue Lhomont
75231 Paris
France

Bouchiat Claude
LPT Ecole Normal Supérieure
24, Rue Lhomont
75231 Paris Cedex 05
France

Bowman David
Los Alamos National Laboratory
Los Alamos
87545 New Mexico, NM
USA

Budker Dmitry
UC Berkley
366, LeConte Hall
94720-7300 Berkley, CA
USA

Capozza Luigi
Institut für Kernphysik
J.J. Becherweg, 45
55099 Mainz
Germany

Chardonnet Christian
Université Paris
Laboratoire de Physique des Lasers
13-99, Avenue J.-B. Clement
93430 Villetaneuse
France

de Jager Kees
Jefferson Lab
12000, Jefferson Avenue
MS 12H2, 23606 Newport News, VA
USA

Decowski Piotr
Smith College
Clark Science Center
01063 Northampton, MA
USA

Derevianko Andrei
University of Nevada
Physics Department
Mail Stop 220
89503 Reno, NV
USA

Desplanques Bertrand
Laboratoire de Physique Subatomique
et de Cosmologie
53, Avenue des Martyrs
38026 Grenoble
France

Diefenbach Joergen
Institut für Kernphysik
J.J. Becherweg, 45
55099 Mainz
Germany

Dubnicka Stanislav
Institut of Physics, SAS
Dubravska cesta
84511 Bratislava
Slovakia

Dubnickova Anna
Comenius University
Department of Theoretical Physics
Faculty of Mathematics, Physics and Informatics
Mlynskadolina, 84248 Bratislava
Slovakia

El-Yakoubi Marouan
IPN Orsay
Université de Paris Sud
Bât. 100M, 91406 Orsay
France

Erler Jens
IF-UNAM Cd. Universitaria (Coyoacan)
04510 Mexico City
Mexico

Gläser Boris
Institut für Kernphysik
J.J. Becherweg, 45
55099 Mainz
Germany

Goeke Klaus
Ruhr-University of Bochum
Institute for Theoretical Physics
Ruhr-University
Universitäts-Str., 150
44801 Bochum
Germany

Gwinner Gerald
University of Manitoba
Department of Physics and Astronomy
R3T 2N2 Winnipeg
Canada

Holstein Barry
University of Massachusetts
Department of Physics - LGRT
01003 Amherst, MA
USA

Hyun Chang Ho
Institute of Basic Science
Sungkyunkwan University
Cheon Cheon Dong
440-746 Suwon
Republic of Korea

Imai Yoshio
Institut für Kernphysik
J.J. Becherweg, 45
55099 Mainz
Germany

Jiang Xiaodong
Rutgers University
12000, Jefferson Avenue
MS 12H, 23606 Newport News, VA
USA

Kaufman Lisa
University of Massachusetts
1126 Lederle Graduate Research Tower (LGRT)
01003-9337 Amherst, MA
USA

Kowalski Stanley
MIT
77, Mass Avenue, Bldg 26-427
02139 Cambridge, MA
USA

Kox Serge
LPSC
53, Avenue des Martyrs
38026 Grenoble
France

Kumar Krishna
University of Massachusetts
Department of Physics
01003 Amherst, MA
USA

Leader Elliot
Imperial College London
Prince Consort Road
SW7 2BW London
UK

Lewis Randy
University of Regina
Department of Physics
S4S 0A2 Regina
Canada

Lhuillier David
CEA - Saclay
DAPNIA/SPhN, Bât. 703
91191 Gif-sur-Yvette
France

Lintz Michel
Laboratoire Kastler Brossel, ENS
24, Rue Lhomont
75231 Paris
France

Londergan John
Indiana University, Nuclear Theory Center
2401, Sampson Lane
47401 Bloomington, Indiana
USA

Maas Frank
Institut de Physique Nucléaire
15, Rue Georges Clemenceau
91406 Orsay Cedex
France

Mack David
TJNAF
12000, Jefferson Avenue
23606 Newport News, VA
USA

From Parity Violation to Hadronic Structure and more...
PAVI 2006

Società Italiana
di Fisica

Springer

Proceedings of

The 3rd Workshop

From Parity Violation to Hadronic Structure and more ...

PAVI 2006

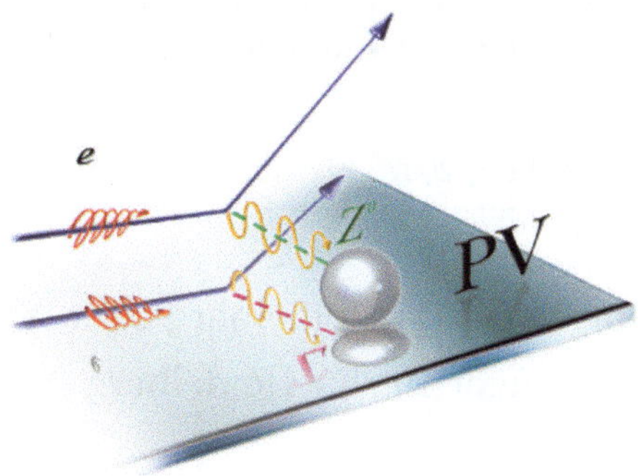

May 16–20, 2006

Milos, Greece

edited by

K. de Jager, S. Kox, D. Lhuillier,
F. Maas, S. Page, C. Papanicolas,
S. Stiliaris and J. Van de Wiele

Dr K. de Jager
Jefferson Lab
12000, Jefferson Avenue
Newport News, VA 23606
USA

Dr S. Kox
LPSC Grenoble
53, Avenue des Martyrs
38026 Grenoble Cedex
France

Dr D. Lhuillier
Service de Physique Nucléaire
CEA/Saclay Bâtiment 703
DSM/DAPNIA/SphN
91191 Gif-sur-Yvette Cedex
France

Dr F. Maas
IPN Orsay, IN2P3/CNRS
15, Rue George Clemenceau
91406 Orsay Cedex
France

Prof. Dr S. Page
Department of Physics and Astronomy
University of Manitoba
Winnipeg, MB
Canada R3T 2N2

Prof. Dr C. Papanicolas
National and Kapodistrian
University of Athens
Physics Department, Ilissia Campus
15771 Athens, Greece

Prof. Dr S. Stiliaris
National and Kapodistrian
University of Athens
Physics Department, Ilissia Campus
15771 Athens
Greece

Dr J. Van de Wiele
IPN Orsay, IN2P3/CNRS
15, Rue George Clemenceau
91406 Orsay Cedex
France

A selection of the articles in this book originally appeared in the journal
The European Physical Journal A – Hadrons and Nuclei
Volume 32, Number 4
ISSN 1434-6001

ISBN-10 3-642-44632-9 Springer Berlin Heidelberg New York
ISBN-13 978-3-642-44632-0 Springer Berlin Heidelberg New York

Springer is a part of Springer Science+Business Media
springer.com

Softcover re-print of the Hardcover 1st edition 2007

Typesetting and Cover design: SIF Production Office, Bologna, Italy
Printing and Binding: Tipografia Compositori, Bologna, Italy

Printed on acid-free paper SPIN: 12112210 – 5 4 3 2 1 0

International Workshop

From Parity Violation to Hadronic Structure and more... (Part III)

16-20 May 2006

Milos Conference Center George Eliopoulos, Milos Island, Greece

TOPICS

- **Overview of the experimental situation**
- **Strangeness in the nucleon: theory and experiment**
- **Standard Model tests, including:**
  - **Weinberg angle measurements through neutrino scattering and Deep Inelastic Scattering**
  - **Parity Violation in Atomic Physics**
- **Hadronic Parity Violation**
- **Probing two-photon exchange effects**
- **Electro-weak radiative corrections involving hadronic structure**
- **Charge symmetry breaking in the nucleon wavefunction**
- **Technical developments associated with PV experiments**
- **Other related physics**

International Advisory Committee

D.H. Beck (UIUC)
M.-A. Bouchiat (LKB Paris)
B. Desplanques (LPSC Grenoble)
D. von Harrach (Univ. of Mainz)
C. Horowitz (Indiana University)
S. Kowalski (MIT)
A. Magnon (DAPNIA Saclay)
W. Marciano (BNL)
U. Meissner (Univ. of Bonn)
W. van Oers (Univ. of Manitoba)
E. De Sanctis (INFN Frascati)
M.J. Ramsey-Musolf (Caltech)
D. Ryckbosch (Gent)
P.A. Souder (Univ. of Syracuse)
A. Thomas (JLab)
M. Vanderhaeghen (College of William and Mary)

Organizing Committee

Kees de Jager (JLab)
Serge Kox (LPSC Grenoble)
David Lhuillier (DAPNIA Saclay)
Frank Maas (IPN Orsay)
Costas N. Papanicolas (Univ. of Athens & IASA)
Shelley Page (Univ. of Manitoba)
Efstathios Stiliaris (Univ. of Athens & IASA, chair)
Jacques Van de Wiele (IPN Orsay)

Organized by the:

Institute of Accelerating Systems and Applications (IASA)
P.O. Box 17214, Athens, 10024, Greece
Tel: +30 210 7257533, Fax: +30 210 7295069
E-mail: pavi06@iasa.gr
Conference Secretary : Mrs Maria Tzamtzi

Conference Web Page: http://www.iasa.gr/pavi06

Marciano William J.
Brookhaven National Laboratory
Physics Department, 510A
11973-5000 Upton, NY
USA

Marmonier Carole
Photonis Avenue
Roger Roncier, BP520
19106 Brive-la-Gaillarde
France

McKeown Robert
California Institute of Technology
1200, E. California Blvd.
91125 Pasadena, CA
USA

Michaels Robert
Jefferson Lab
12000, Jefferson Avenue
23606 Newport News, VA
USA

Moffit Bryan
College of William and Mary
P.O. Box 8795
Physics Department
23187-8795 Williamsburg, VA
USA

Opper Allena
The George Washington University
725, 21rst Str.
NW 20052 Washington, DC
USA

Pacetti Simone
INFN, Laboratori Nazionali di Frascati
Via E. Fermi, 40
00044 Frascati
Italy

Page Shelley
University of Manitoba
Department of Physics and Astronomy
R3T 2N2 Winnipeg
Canada

Parsa Zohreh
Brookhaven National Laboratory
Physics Department, 510A
11973-5001 Upton, NY
USA

Paschke Kent
University of Massachusetts, Amherst
TJNAF
12000, Jefferson Avenue
23188 Newport News, VA
USA

Pasquini Barbara
University of Pavia
Via Bassi, 6
27100 Pavia
Italy

Piekarewicz Jorge
Florida State University
Department of Physics
32306-4350 Tallahassee, FL
USA

Procureur Sebastien
CEA - Saclay
Orme des Merisiers
91190 Gif-sur-Yvette
France

Ramsey-Musolf Michael
CalTech
P.O. Box 254
91102 Pasadena, CA
USA

Real Jean-Sebastian
LPSC CNRS/IN2P3
53, Avenue des Martyrs
38026 Grenoble
France

Reimer Paul
Argonne National Laboratory
9700, S. Cass Avenue
60439 Argonne, IL
USA

Riska Dan-Olof
Helsinki Institute of Physics
P.O. Box 64
FIN-00014, University of Helsinki
Finland

Sanguinetti Stefano
Dipartimento di Fisica "E. Fermi"
Largo B. Pontecorvo, 3
56127 Pisa
Italy

Schäfer Andreas
University of Regensburg
93049 Regensburg
Germany

Schiavilla Rocco
JLab/ODU and University of Pisa
Via Buonarrotti, 2
56127 Pisa
Italy

Schindler Matthias
Institut für Kernphysik
J.J. Becherweg, 45
55099 Mainz
Germany

Snow William
Indiana University/IUCF
2401, Sampson Lane
47408 Bloomington, IN
USA

Souder Paul
Syracuse University
201 Physics Building
13244 Syracuse, NY
USA

Stilliaris Stathis
IASA
P.O. Box 17214
10024 Athens
Greece

Toublan Dominique
University of Maryland
082, Regents Drive
20742 College Park
USA

Tsigutkin Konstantin
UC Berkley
Physics Department
94015 Berkley, CA
USA

Vanderhaeghen Marc
College of William and Mary/JLab
12000, Jefferson Avenue
23606 Newport News, VA
USA

von Harrach Dietrich
Institut für Kernphysik
J.J. Becherweg, 45
55128 Mainz
Germany

Young Ross
Jefferson Lab
12000, Jefferson Avenue
MS 12H2, 23606 Newport News, VA
USA

Zanotti James
University of Edinburgh
School of Physics, JCMB
King's Buildings
Mayfield Road
EH9 3JZ Edinburgh
UK

Zheng Xiaochao
Massachusetts Institute of Technology (MIT)
MS 12C, 12/F357
12000, Jefferson Avenue
23606 Newport News, VA
USA

Sponsors

We gratefully acknowledge the sponsorship of the following Laboratories, Institutions and Companies:

Laboratoire de Recherche sur les Lois Fondamentales de l'Univers

Centre National de la Recherche Scientifique

Hadron Physics I3 (I3HP)

Institut National de Physique Nucleaire et de Physique des Particules

Thomas Jefferson National Accelerator Facility

PHOTONIS Group

S & B Industrial Minerals S.A.

Greek Ministry of National Education and Religious Affairs

Contents

Contributions marked with an asterisk have appeared in Eur. Phys. J. A, Vol. **32**, no. 4 (2007).

■ Preface

■ Introduction

■ Strangeness in the Nucleon and Related Topics: Theory

■ Strangeness in the Nucleon and Related Topics: Experiment

■ Two-Photon Exchange

■ Parity Violation with Hadronic Probes

■ Parity Violation in Atomic Physics

Preface

The third workshop *From Parity Violation to Hadronic Structure and more ...* was held from May 16 to May 20, 2006, at the George Eliopoulos conference center on the Greek island of Milos. It is part of a series that started in Mainz in 2002 and was followed by a second workshop in Grenoble in 2004.

While originally initiated by the extraction of the strangeness contribution to the electromagnetic form factors of the nucleon, we have continuously broadened the focus of this workshop series to the application of parity violation using hadronic probes and to parity violation experiments in atomic physics. Meanwhile there have been many exciting new proposals for using parity violation in other areas like in the search for new physics beyond the standard model or in exploring hadron structure. There are also close connections to the open question on the size of the two-photon exchange amplitude.

Fifty years after the 1956 proposal of Lee and Yang to test the hypothesis of violation of parity symmetry in the weak interaction, the many applications of parity violation in very different experiments are way beyond the scope of what Lee and Yang could have imagined. For the physics topics discussed during this workshop, the application of parity violation has become a standard workhorse that allows to extract many physics topics in different experiments. The small effect which is used to cleanly separate the physics issues discussed here, makes the continuous technical development of new experimental tools necessary. All this has been discussed during the workshop on the beautiful island of Milos in the Greek Cyclades where the association with the ancient Greek myths made this workshop a special event.

Our thanks for making this meeting so special go to Maria Tzamtzi and Christina Grentzelou who organised all the non-scientific aspects of the workshop. Whithout their tireless effort on organising air planes, hotels, ferries, meals, excursions and much more the conference would not have been such a great success.

The whole success of the workshop had not been possible without the support of many involved people and sponsors. Finally, the work and advises of the International Advisory Committee, which has helped us to set up the program and select such an excellent group of presentations, is gratefully acknowledged.

With all the new proposals, with the continued data taking of ongoing experiments we expect with great enthusiasm many new developments and results and are looking forward to the fourth workshop in this series *PAVI 2008*.

Kees de Jager
Serge Kox
David Lhuillier
Frank Maas
Shelley Page
Costas Papanicolas
Stathis Stiliaris
Jacques Van de Wiele

The Editors

Hadron structure on the lattice and in the continuum

A. Schäfer[a]

Institute for Theoretical Physics, University of Regensburg, 93040 Regensburg, Germany

Abstract. Hadron structure physics has in recent years reached a level of precision which allows for a change of perspective. Model-based arguments are often quite unreliable. However, meanwhile they can be more and more replaced by controlled and systematic QCD approaches. The story of the strange electric form factor, which provided much of the motivation for the PAVI Conference series provides a typical example to illustrate this statement. However, high-precision theory is technically very challenging and progress is, therefore, unpleasantly slow. This fact and the present status in general is illustrated by a few typical examples.

PACS. 12.38.-t Quantum chromodynamics – 12.38.Bx Perturbative calculations – 12.38.Gc Lattice QCD calculations

1 Strangeness in the nucleon

The amount of $\bar{s}s$ pairs present in the nucleon ground state and its contribution to, *e.g.*, the nucleon spin or the electric form factors has been the topic of extensive theoretical discussions and very substantial experimental efforts. Early experimental results by the SMC Collaboration were interpreted as signs of a large negative longitudional strangeness polarization Δs [1,2] and early theoretical models [3] suggested large strangeness form factor contributions for the nucleon. Since then, both claims have faded away, mainly due to much improved experimental results. Today it looks as if Δs, G_M^s and G_E^s would all be very close to zero. For Δs this is illustrated by some recent HERMES results [4] in fig. 1 and for G_M^s, G_E^s this was discussed in great detail in the excellent talks by Armstrong, Baunack, Bijker, Real, Young, etc. at this conference.

The upshot for Δs is that while the SMC value [1] $\Delta s = -0.10 \pm 0.03 \pm 0.04$ suggested a surprisingly large strangeness contribution the HERMES value [4] $\Delta s = 0.028 \pm 0.033 \pm 0.009$ (in the range $0.023 \leq x \leq 0.6$) is compatible with zero.

While there is generally speaking no compelling support for large strange quark contributions to the nucleon structure from QCD, there exist interesting model-based arguments in their favor ("intrinsic strangeness", see, *e.g.*, [5]). The basic reason why the theoretical modeling for strangeness is especially tricky is that its phenomenolgical properties are neither characteristic for light quarks (which would imply substantial contributions) nor heavy quarks (which are basically absent from the nucleon structure at lowish scales). Therefore, indeed, the strange sector of the nucleon can very well hold surprises. Note in this context that recent preliminary HERMES data suggest a kaon Sivers asymmetry which is substantially larger

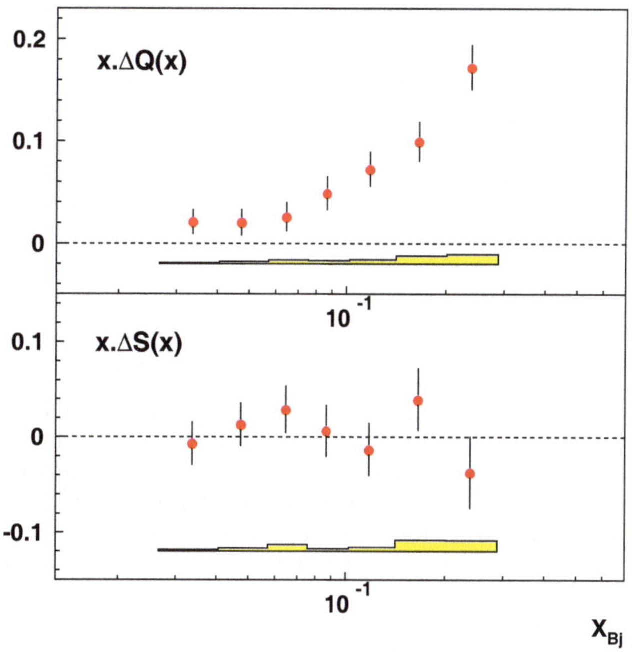

Fig. 1. Strange and non-strange quark helicity distributions, $\Delta S(x) \equiv \Delta s(x) + \Delta\bar{s}(x)$ and $\Delta Q(x) \equiv \Delta u(x) + \Delta\bar{u}(x) + \Delta d(x) + \Delta\bar{d}(x)$, at $\langle Q^2 \rangle = 2.5\,\mathrm{GeV}^2$, as a function of Bjorken x. The error bars are statistical, and the bands at the bottom represent the systematic uncertainties, see hep-ex/0601006.

[a] e-mail: andreas.schaefer@physik.uni-regensburg.de

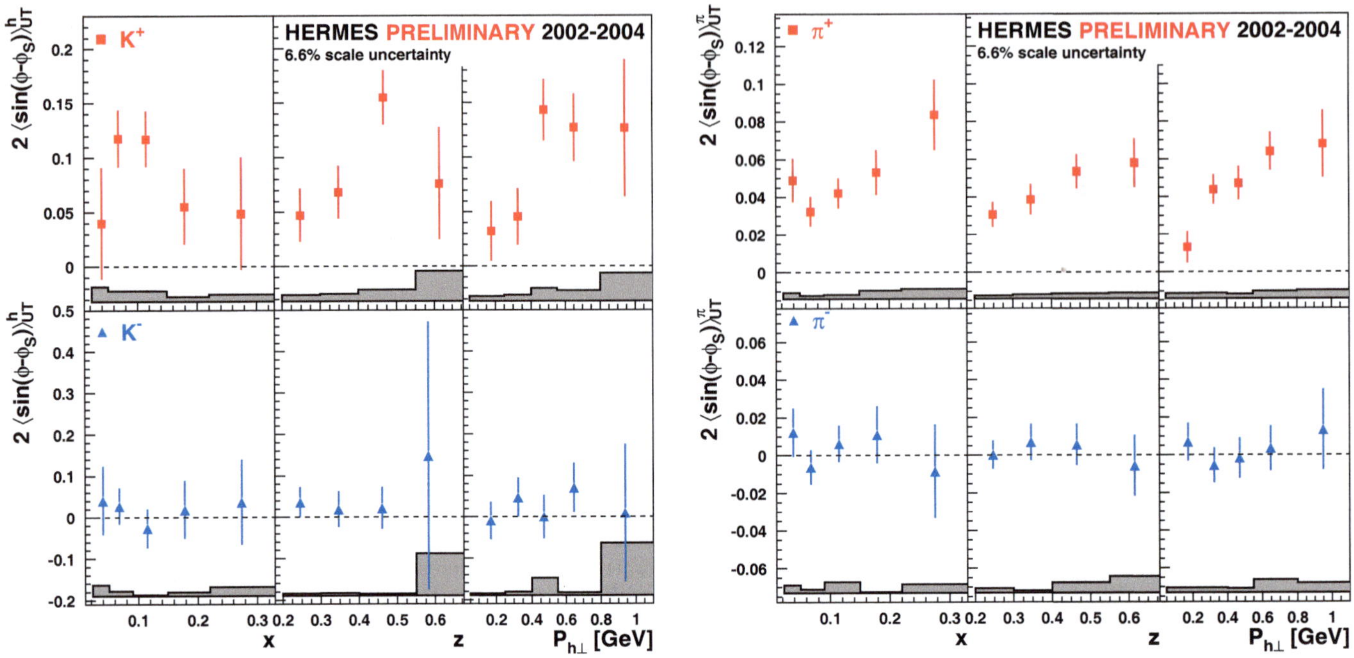

Fig. 2. Sivers asymmetries for kaons (left) and pions (right); note the difference of scales; *preliminary data*; for the definitions and interpretation, see [6]. Basically, the Sivers asymmetry is related to the internal orbital angular-momentum structure of the nucleon and thus a large $\bar{s}$ contribution might be related to a large orbital angular-momentum contribution of $\bar{s}$. However, the theoretical status is still very unsettled.

than that of pions, see fig. 2, a fact which could hardly be explained without a substantial $\bar{s}$ contribution.

Let us finish this introduction with some generic remarks about the comparison of model predictions and high-precision QCD results. If one calculates quantities like Δs, which are no direct observables or protected by conservation laws, one has to specify exactly in which renormalization/factorization scheme they are defined. As

$$\Delta s(x,Q^2,LO) \neq \Delta s(x,Q^2,\overline{MS},NLO,\mu_r) \neq \Delta s(x,Q^2,\overline{MS},NNLO,\mu_r),\dots \quad (1)$$

it is often very difficult to connect statements from models to QCD as it remains unclear which Δs is meant. This is only simple in LO, which is often more or less explicitly assumed, but we know for sure that for most processes and experimental kinematics LO is insufficient.

Also lattice calculations face this problem. They correspond to all order perturbative calculations and thus their results cannot be identified directly with, *e.g.*, the Δs extracted from semi-inclusive polarized lepton-nucleon scattering by an NLO analysis. Furthermore lattice calculation provide only moments of, *e.g.*, distribution functions, as these are given by matrix elements of local operators:

$$\int_0^1 x^2 \Delta s(x,Q^2)\mathrm{d}x|_{N^\infty LO,\overline{MS},\mu_r} \Leftrightarrow \langle PS|\bar{s}(0)\gamma_\mu\gamma_5 D_\nu D_\lambda s(0)|PS\rangle. \quad (2)$$

In this case, however, because perturbative QCD and lattice QCD are both completely defined field-theoretical approaches the relationship can, at least in principle, be worked out exactly (up to limits due to, *e.g.*, the non-convergence of perturbative expansions and lattice artefacts).

Thus, in hadron phenomenology one has the difficult choice between

- *models*, which in general require only limited computational effort but usually give unreliable results, as illustrated by the strangeness-related phenomenology of nucleon structure;
- *QCD*, which requires enormous effort (factorization proofs, $LO \rightarrow NLO \rightarrow NNLO$ perturbation theory, higher twist analysis, lattice calculations) and often needs decades to finally converge onto some firm results.

In the long run and with increasing experimental precision there is no real alternative to the second option, *i.e.* QCD calculations. I shall illustrate how difficult these get in the next sections. I will argue that hadron phenomenology has matured into a field of very high technical standards which first of all requires diligence and patience.

2 GPDs: a unified picture of hadron structure

Many aspects of hadron phenomenology and relationships between different observables, which ten or twenty years ago could only be discussed within specific models, are now clearly established within QCD. Generalized Parton Distributions (GPDs) make this fact transparent by providing a unified parametrization such that the different

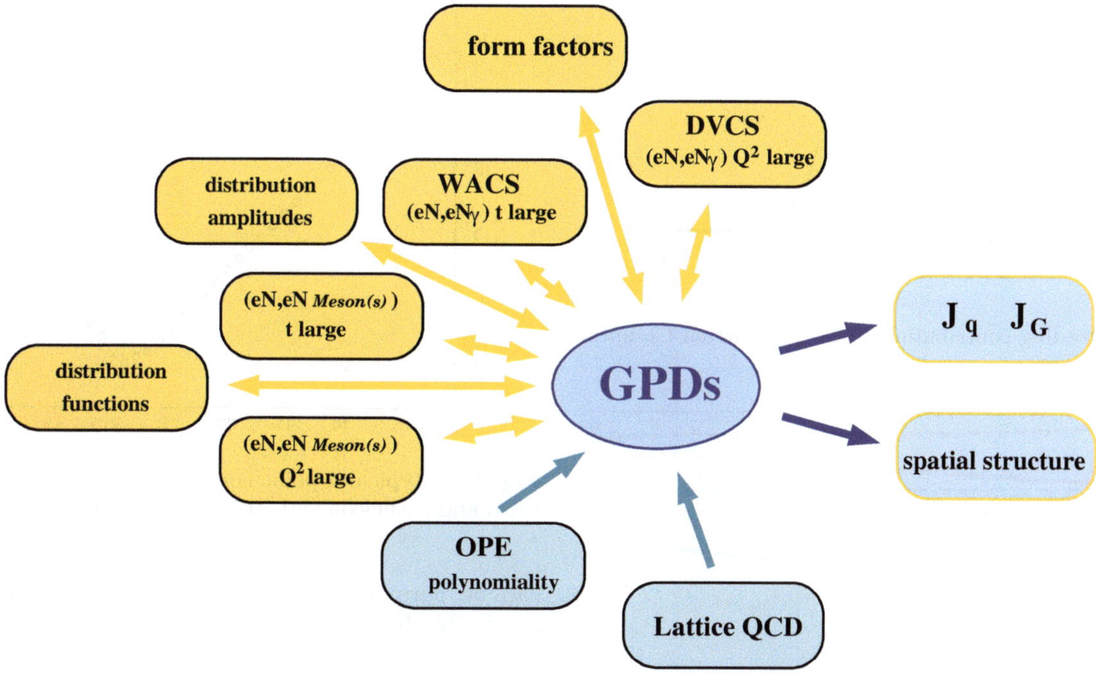

Fig. 3. GPDs combine the information from many experiments in a common parametrization. The Operator Product Expansion (OPE) allows one to establish many links, often even very surprising ones. For example the same operators occur in the hadronic energy momentum tensor and in Deeply Virtual Compton Scattering (DVCS) off hadrons, allowing one to relate the total angular momentum to DVCS and other reactions.

facts can be rephrased as constraints for these parameters. While presently the experimental and lattice data is still somewhat insufficient to do so, in the forseeable future global fits of GPD parametrizations will effectively establish connections between most inclusive, semi-inclusive and exclusive reaction channels. The powerful tool of the operator product expansion allows then to extract from this combined information aspects which are not directly accessible, most notably the transverse structure of fast hadrons and the total angular-momentum contributions to, *e.g.*, the proton spin, see fig. 3.

There exist eight different types of GPDs denoted, *e.g.*, for a quark of flavor q by H_q, E_q, $\tilde{H}_q$, $\tilde{E}_q$, H_{Tq}, E_{Tq}, $\tilde{H}_{Tq}$, and $\tilde{E}_{Tq}$. Each of these depends on three kinematic variables, namely x, $\xi = -Q^2/2P\cdot q$ and $t = \Delta^2$ for the typical reaction

$$h(P_1) + \Gamma^*(q_1) \rightarrow h(P_2) + \Gamma(q_2)$$

with some hadron h and with $\Delta_\mu = q_{2\mu} - q_{1\mu}$, $P_\mu = (P_{1\mu} + P_{2\mu})/2$. Two of the quark GPDs in a nucleon are, *e.g.*, defined by the correlator

$$\int \frac{\mathrm{d}z^-}{2\pi} e^{ix\bar{P}^+z^-} \langle P_2|\,\bar{q}(-\tfrac{1}{2}z)\,\gamma^+ q(\tfrac{1}{2}z)\,|P_1\rangle\Big|_{z^+=0,\,z_\perp=0} = \frac{1}{P^+}\Big[H_q(x,\xi,t)\,\bar{N}(P_2)\gamma^+N(P_1) + E_q(x,\xi,t)\,\bar{N}(P_2)\frac{i\sigma^{+\alpha}\Delta_\alpha}{2M}N(P_1)\Big]. \tag{3}$$

To illustrate the usefulness of GPDs let us come back to the Sivers asymmetries plotted in fig. 2. GPDs encode also information on the transverse distribution of partons in the impact parameter plane. More precisely,

$$H_q(x,0,\mathbf{b}_\perp^2) = \frac{1}{(2\pi)^2}\int \mathrm{d}^2\boldsymbol{\Delta}_\perp\, \mathrm{e}^{\mathrm{i}\mathbf{b}_\perp\boldsymbol{\Delta}_\perp} H_q(x,0,\boldsymbol{\Delta}_\perp^2) \tag{4}$$

can be interpreted as the probabilty to find a quark with longitudinal momentum fraction x in some distance $\mathbf{b}_\perp$ from the center of a nucleon. When calculating moments of GPDs on the lattice one finds, *e.g.*, that for transverse nucleon polarization the quarks show a non-zero transverse shift in the direction orthogonal to the nucleon spin. This shift is related to the orbital angular momentum of the parton and there do exist models which link it to the Sivers asymmetry. Thus, *e.g.*, a strong Sivers asymmetry for strange quarks would be related to a strong transverse shift of the strange quarks, which would in turn be related to large orbital angular momentum. If so, the observation in fig. 2 would be highly relevant for the discussion about Δs and G_m^s, etc. The story of GPDs is a very good example for how difficult and technical QCD calculations have to be to really match the experimental needs. The formalism allows one to extract valuable information from experiment which otherwise would be lost. In view of the enormous costs of producing the experimental data, it is clear that one needs a matching theory effort. Thus since the groundbreaking work of D. Müller, X. Ji, A. Radyushkin *et al.* [7] a large community worked for 10 years to reach the present status, which finally allows one to start to fit GPDs to experimental and lattice data. For three excellent reviews, see [8].

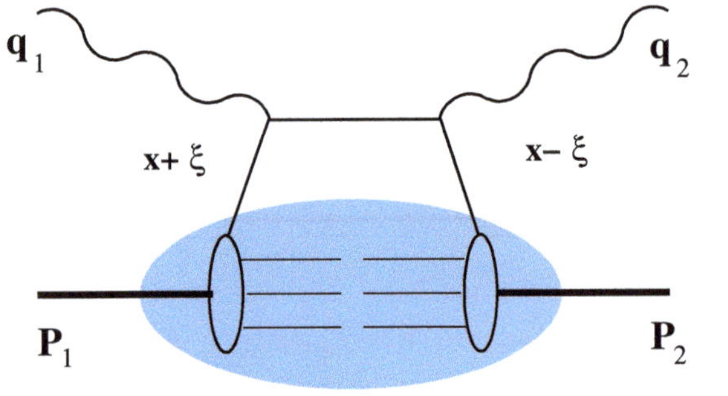

Fig. 4. The leading contribution to Deeply Virtual Compton Scattering (DVCS).

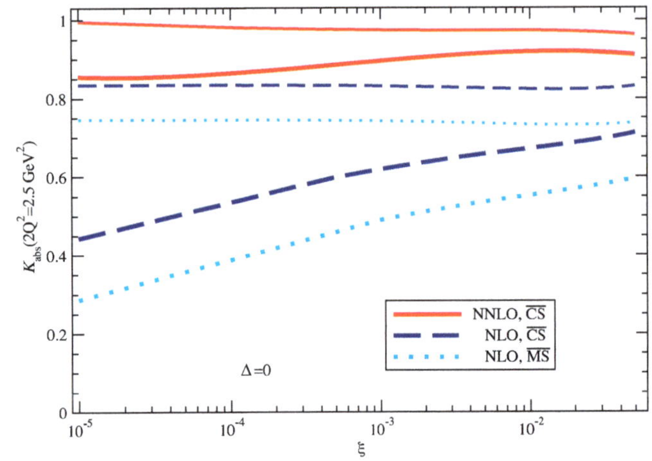

Fig. 5. The ratio $K^P_{\rm abs}$ from eq. (8). The corrections for $LO \to NLO$ (dashed and dotted lines) are obviously much larger than for $NLO \to NNLO$ (solid line). The dashed and dotted curves correspond to different regularization schemes. The thick and thin lines correspond to different model GPDs.

3 QCD perturbation theory at NNLO

Hadron physics is in need of both perturbative and non-perturbative QCD calculations. While non-perturbative techniques like lattice QCD allow one, *e.g.*, to calculate moments of GPDs which ultimately should result in well-justified GPD parametrizations, one depends on perturbative QCD to relate these to actual experiments. Thus, a balanced effort in both areas is crucial. (It is unfortunate that presently perturbative QCD is insufficiently supported, such that in some time from now the technical knowledge for such calculations might get lost.) The state of the art was for many years defined by next-to-leading order (NLO) calculations. Presently, it is moving on to $NNLO$ and even higher-order calculations, which are, however, technically so demanding that ever fewer groups worldwide are able to perform them [9,10].

As an example let me sketch a recent $NNLO$ analysis of DVCS (fig. 4) by Dieter Müller and collaborators [11, 12]. For $|x| \leq |\xi|$ generalized parton distributions are similar to distribution amplitudes for which one knows that Gegenbauer polynomials are eigenfunctions under evolution. Therefore, it was a natural idea to expand GPDs in terms of Gegenbauer polynomials. DVCS cross-sections

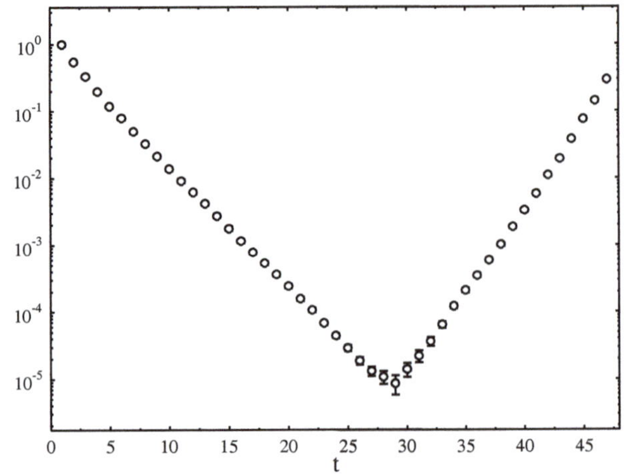

Fig. 6. A typical 2-point correlator with a nucleon source at $t = 0$ and a nucleon sink at the t-value plotted.

can be parametrized in terms of Compton form factors $\mathcal{F}$, $\tilde{\mathcal{F}}$, $\mathcal{H}$, $\tilde{\mathcal{H}}$. *E.g.*, the singlet Compton form factor $\mathcal{H}^S$ has a quark-singlet piece and a gluon piece, which mix under evolution and therefore are best combined to a 2-component vector

$$\mathcal{H}^S(\xi, \Delta^2, \mathcal{Q}^2) = 2\sum_{j=0}^{\infty} \xi^{-j-1} \boldsymbol{C}_j(\mathcal{Q}^2/\mu^2, \alpha_s(\mu)) \times \boldsymbol{H}_j(\Delta^2, \mathcal{Q}^2, \xi), \tag{5}$$

$$\boldsymbol{C}_j(\mathcal{Q}^2/\mu^2, \alpha_s(\mu)) = \frac{2^{j+1}\Gamma(j+5/2)}{\Gamma(j+4)\Gamma(3/2)}\frac{1}{2} \times \int_{-1}^{1} \mathrm{d}x\, \boldsymbol{C}(x, \mathcal{Q}^2/\mu^2, \alpha_s(\mu)|\xi = 1) \times \begin{pmatrix} (j+3)[1-x^2]C_j^{3/2}(x) & 0 \\ 0 & 3[1-x^2]^2 C_{j-1}^{5/2}(x) \end{pmatrix}(x), \tag{6}$$

$$\boldsymbol{H}_j(\eta) = \frac{\Gamma(3/2)\Gamma(j+1)}{2^j\Gamma(j+3/2)}\frac{1}{2}\int_{-1}^{1} \mathrm{d}x\, \eta^{j-1} \times \begin{pmatrix} \eta\, C_j^{3/2}(x/\eta) & 0 \\ 0 & (3/j)\, C_{j-1}^{5/2}(x/\eta) \end{pmatrix} \boldsymbol{H}(x, \eta), \tag{7}$$

where $\boldsymbol{C}$ is the vector of the coefficient functions and $\boldsymbol{H}$ the vector of the singlet GPDs H^q and H^G. One could then expect that the evolution equations could be written in terms of the $\boldsymbol{C}_j$ and $\boldsymbol{H}_j$. Unluckily, however, this expansion converges only in the unphysical region $\xi \geq 1$. Dieter Müller found a different expansion leading to convergent results. This allowed him to solve the problem of $NNLO$ evolution for DVCS, making use of the results derived in a truly impressive effort by Moch, Vermaseren and Vogt [9].

The detailed presentation of theoretical results and model studies by D. Müller, K. Kumericki, K. Passek-Kumericki *et al.* would require more space than what is available here. The bottom line is that at $Q^2 \sim 2\,\mathrm{GeV}^2$ the corrections for the step $NLO \to NNLO$ are far smaller

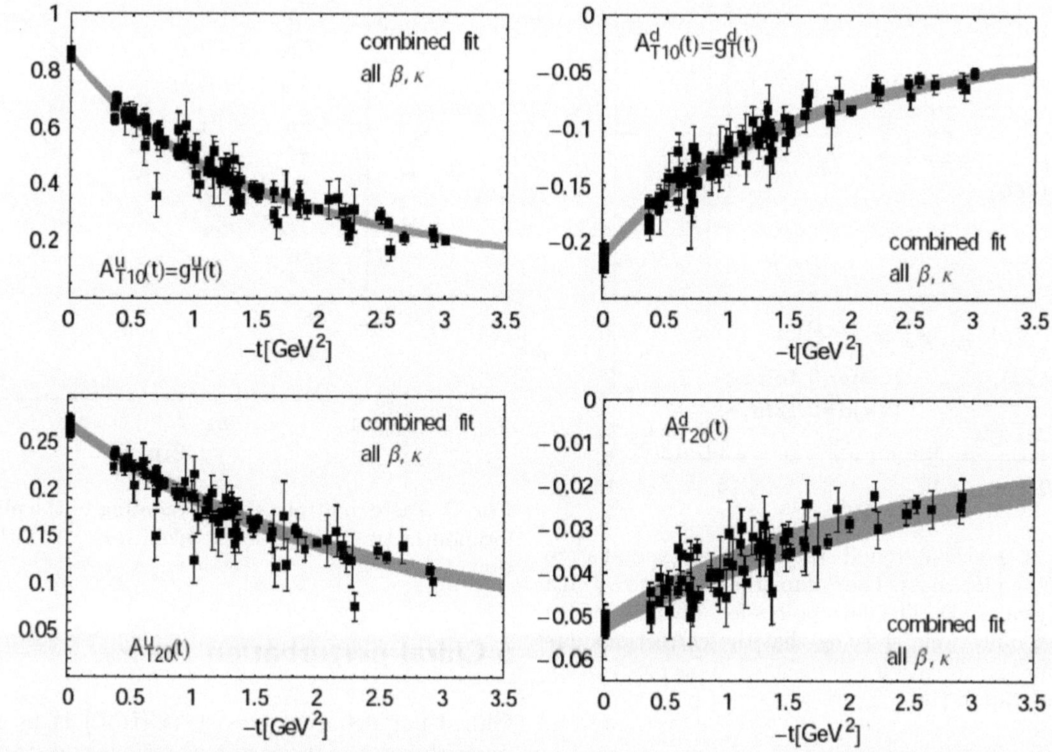

Fig. 7. The generalized form factors A_{T10} and A_{T20}, see eq. (11), together with dipole fits, extrapolated to the physical pion mass. This is a result from the QCDSF Collaboration, see [13].

than those for $LO \to NLO$, *i.e.* the perturbative expansion seems to converge. In fig. 5 this is shown for the relative change in magnitude

$$K^P_{\text{abs}} = \frac{\left|\mathcal{H}^{N^P LO}\right|}{\left|\mathcal{H}^{N^{P-1} LO}\right|}. \quad (8)$$

Note that the $NNLO$ corrections can still be of order 20% (depending on the model GPDs used). Therefore at these Q^2 values an $NNLO$ calculation is really needed to reach an interesting precision. Obviously, the systematic development $LO \to NLO \to NNLO$ over the last 20 years is a perfect example for the hard and long work needed for modern hadron theory.

4 Lattice QCD: QCD beyond perturbation theory

Lattice QCD has been used to calculate so many different aspects of hadron phenomenology that it is completely impossible to give even a very sketchy overview. In view of the fact that James Zanotti gave a review from the QCDSF perspective I prefer to discuss here only the basic ideas and give just a few examples.

By analytic continuation of time to imaginary values (Euclidean time) one can map problems from quantum field theory onto corresponding problems in statistical physics or thermodynamics. The latter can be solved numerically with Monte Carlo techniques. To do so, space and time have to be discretized, and one has to limit oneself to a finite space-time volume, such that the number of variables (the values for all fields at each space-time point) becomes finite. Different theory groups differ primarily in the precise manner in which this discretization is done.

To extract information on specific hadrons, *e.g.* the nucleon, one inserts a combination of quark fields with the correct quantum numbers on two time slices with different Euclidean times, *e.g.* for a proton

$$B_\alpha(t,\boldsymbol{p}) = \sum_{\boldsymbol{x}} e^{i\boldsymbol{p}\cdot\boldsymbol{x}} \epsilon_{ijk} u^i_\alpha(x)\, u^j_\beta(x)(C^{-1}\gamma_5)_{\beta\gamma} d^k_\gamma(x) \quad (9)$$

with the charge conjugation matrix $C = i\gamma^2\gamma^4 = C^{-1}$. Obviously such a simple interpolating current cannot generate the correct multi-particle wave function of the nucleon. Rather it yields a superposition of all states with the same quantum numbers. However, under (Euclidean !) time evolution all higher states are suppressed and the correct wave function is filtered out:

$$\begin{aligned} |B\rangle &\sim c_0|N\rangle + c_1|N'\rangle + c_2|N\pi\rangle + \ldots \\ &\Rightarrow c_0 e^{-E_N t}|N\rangle + c_1 e^{-E_{N'} t}|N'\rangle \\ &\quad + c_2 e^{-E_{N\pi} t}|N\pi\rangle + \ldots \end{aligned} \quad (10)$$

Due to periodicity of the lattice the resulting 2-point function looks as shown in fig. 6 (for baryons in the right

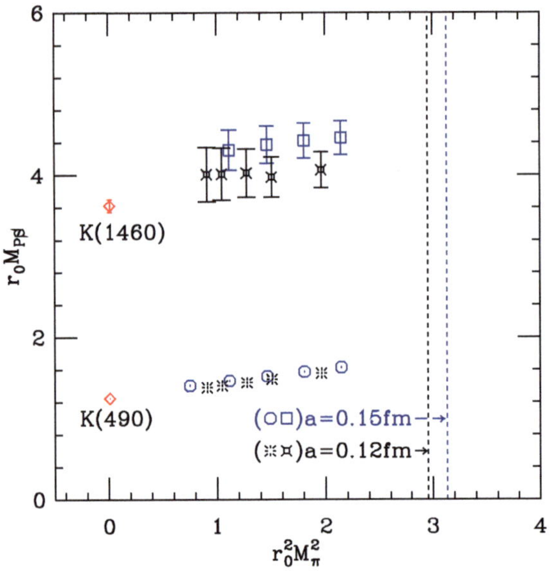

Fig. 8. Masses of the kaon ground state and first excited state as a function of the pion mass. The Sommer parameter $r_0 = 0.5$ fm sets the physical scale. The data points for different lattice spacings a agree quite well, showing that the corrections due to the continuum extrapolation will be small. This plot is from the BGR Collaboration [14].

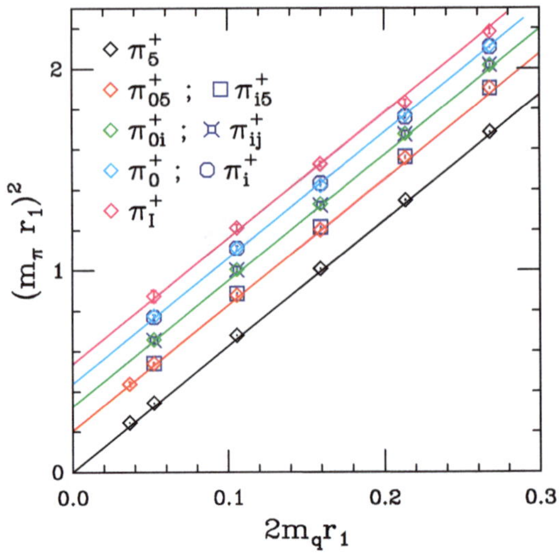

Fig. 9. Taste multiplets corresponding to the pion in staggered fermion lattice calculations, taken from the MILC Collaboration [15].

part one gets the state of opposite parity, here an N^*-resonance). The slope of this correlator gives the mass of the lowest-lying hadron and by inserting different operators on time slices where one has a clean exponential behavior one can calculate various 3-point functions, which give information on different properties of the hadron wave function. As an example fig. 7 shows some results for the moments of the tensor GPD H_T,

$$A_{T10}(t) = \int_{-1}^{1} \mathrm{d}x\ H_T(x,\xi,t)\,,$$
$$A_{T20}(t) = \int_{-1}^{1} \mathrm{d}x\ x\ H_T(x,\xi,t)\,. \qquad (11)$$

Simulations with different lattice spacings correspond to different β's, simulations with different quark/pion masses correspond to different κ's. One of the informations contained in this specific GPD is, *e.g.*, whether the transverse spatial structure of quark distribution functions depends on their polarization. The 2-point functions can also be used to obtain masses of hadron resonances (by various methods), see fig. 8, and to calculate quantitites like decay constants. The enormous computational effort needed for lattice QCD is well known to everybody. Less well known is the similarly large and diverse theoretical work (algorithms, OPE, renormalization, ...) needed for such calculations. Actually, the required effort is such that lattice QCD is dominated by large collaborations of people specialized on only a few of these issues.

5 Chiral perturbation theory

Chiral perturbation theory (CHPT) is by now an indispensible part of modern theoretical hadron physics. (Here I mean by CHPT any comprehensive and consistent description in terms of hadronic degrees of freedom, neglecting all the controversies between different approaches.) Its importance becomes visible in many places, *e.g.* if one wants to combine a hard-QCD part with some soft hadronic interaction.

In recent years CHPT was also strongly advocated to improve the reliability of extrapolations of lattice data to the chiral limit. While the strength of this argument is somewhat debatable for many lattice formulations, it is undisputed for simulations with staggered fermions. Staggered fermions have some problematic features: one has always quadruplets of fermions and correspondingly a total of 5 pion-like states instead of one, see fig. 9. To get single quarks the fourth-root trick has to be used, which could make the sea quarks unphysical. One faces a violation of locality and unitarity, propagators with unphysical double-poles, etc. All of these diseases are hopefully cured in the limit $\lim_{m_\pi \to 0} \lim_{a\to 0}$ which, however, is different from the limit $\lim_{a\to 0} \lim_{m_\pi\to 0}$. To treat this subtle limiting process correctly requires an analytic formulation which can only be (and is) provided by CHPT. Disregarding the fundamental discussion about the validity of the fourth-root trick, it is clear that if one wants to use staggered fermions to calculate hadron structure correlators, which is very attractive as they are computationally cheap, one has to control the analytic properties for small a and m_π and thus definitely needs an appropriate CHPT analysis. A comprehensive description of the status of the field, its successes and problems was given by S. Sharpe at LATTICE 2006 [16].

To illustrate the potential benefits of a combination of lattice calculations and CHPT let us finally show a fit

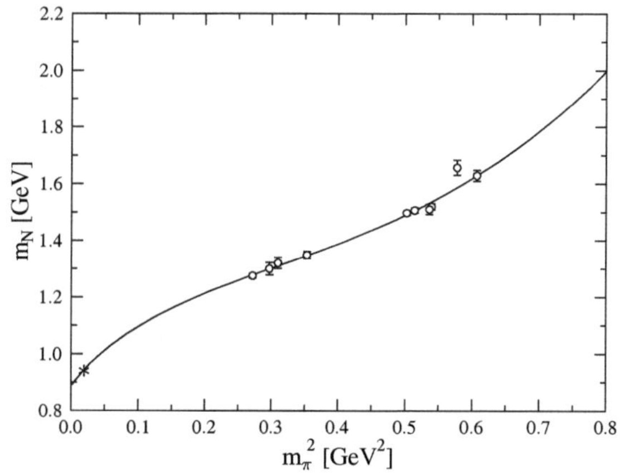

Fig. 10. The nucleon mass as a function of m_π^2 plus a fit from the SSE variant of CHPT [17]. The data are from the QCDSF/UKQCD, CP-PACS and JLQCD Collaborations.

of lattice data for the nucleon mass by CHPT, see [17]. Here improved Wilson fermions were used, which are far less problematic, and thus a different form of CHPT. The combined dependence on m_π and lattice size L was derived using and extending earlier work by Procura, Hemmert and Weise [18]. Figure 10 shows the fit for the m_π-dependence including lattice data from other groups. Even though the data are at best bordering the region in m_π where one would really trust CHPT, the resulting smooth dependence looks very convincing. Furthermore, the L-dependence turns out to be described perfectly.

6 Conclusions

We have presented three different, technically highly developed QCD techniques, namely perturbation theory, lattice QCD and CHPT. All of these progress steadily but also slowly. The typical time scale for higher-order perturbative QCD calculations, careful lattice studies, or a higher-order CHPT analysis, is years and typically one is very happy if one can increase the previous precision by, *e.g.*, a factor of 2. So, there is still a very long way to go, but it seems clear that for many questions phenomenological model building will be replaced by full fledged QCD calculations in the end.

Most of the results presented are due to the HERMES, QCDSF, and BGR Collaborations. I thank all of their members for a long and fruitful collaboration. This work was supported in parts by DFG, BMBF, GSI (ORSCH), EU (I3HP), and HGF (VH-NG-004).

References

1. Spin Muon Collaboration (SMC) (D. Adams *et al.*), Phys. Lett. B **329**, 399 (1994) [arXiv:hep-ph/9404270].
2. See the review: M. Anselmino, A. Efremov, E. Leader, Phys. Rep. **261**, 1 (1995); **281**, 399 (1997)(E); arXiv:hep-ph/9501369; E. Leader, this conference.
3. R.L. Jaffe, Phys. Lett. B **229**, 275 (1989); N.W. Park, H. Weigel, Nucl. Phys. A **451**, 453 (1992); H.W. Hammer, U.G. Meissner, D. Drechsel, Phys. Lett. B **367**, 323 (1996); H.-W. Hammer, M.J. Ramsey-Musolf, Phys. Rev. C **60**, 045204 (1999).
4. HERMES Collaboration (A. Airapetian *et al.*), Phys. Rev. D **71**, 012003 (2005) [arXiv:hep-ex/0407032].
5. J.R. Ellis, A. Kotzinian, D.V. Naumov, Eur. Phys. J. C **25**, 603 (2002) [arXiv:hep-ph/0204206].
6. HERMES Collaboration (A. Hillenbrand), *Transverse spin physics at HERMES*, arXiv:hep-ex/0606019 and references therein.
7. D. Müller, D. Robaschik, B. Geyer, F.M. Dittes, J. Horejsi, Fortschr. Phys. **42**, 101 (1994) [arXiv:hep-ph/9812448]; X.D. Ji, Phys. Rev. Lett. **78**, 610 (1997) [arXiv:hep-ph/9603249]; A.V. Radyushkin, Phys. Lett. B **380**, 417 (1996) [arXiv:hep-ph/9604317].
8. M. Diehl, Phys. Rep. **388**, 41 (2003) [arXiv:hep-ph/0307382]; A.V. Belitsky, A.V. Radyushkin, Phys. Rep. **418**, 1 (2005) [arXiv:hep-ph/0504030]; K. Goeke, M.V. Polyakov, M. Vanderhaeghen, Prog. Part. Nucl. Phys. **47**, 401 (2001) [arXiv:hep-ph/0106012].
9. A. Vogt, S. Moch, J.A.M. Vermaseren, Nucl. Phys. B **691**, 129 (2004) [arXiv:hep-ph/0404111]; S. Moch, J.A.M. Vermaseren, A. Vogt, Nucl. Phys. B **688**, 101 (2004) [arXiv:hep-ph/0403192] and the publications citing these papers as given by hep-SPIRES.
10. B. Melic, D. Müller, K. Passek-Kumericki, Phys. Rev. D **68**, 014013 (2003) [arXiv:hep-ph/0212346].
11. D. Müller, Phys. Lett. B **634**, 227 (2006) [arXiv:hep-ph/0510109].
12. K. Kumericki, D. Müller, K. Passek-Kumericki, A. Schäfer, arXiv:hep-ph/0605237.
13. QCDSF Collaboration (M. Göckeler *et al.*), Phys. Lett. B **627**, 113 (2005) [arXiv:hep-lat/0507001].
14. T. Burch *et al.*, Phys. Rev. D **73**, 094505 (2006) [arXiv:hep-lat/0601026].
15. MILC Collaboration (C. Aubin *et al.*), Phys. Rev. D **70**, 114501 (2004) [arXiv:hep-lat/0407028].
16. See the contribution of S. Sharpe at `http://www.physics.arizona.edu/lattice06/`; see also the review S.R. Sharpe, hep-lat/0607016.
17. QCDSF-UKQCD Collaboration (A. Ali Khan *et al.*), Nucl. Phys. B **689**, 175 (2004) [arXiv:hep-lat/0312030].
18. M. Procura, T.R. Hemmert, W. Weise, Phys. Rev. D **69**, 034505 (2004) [arXiv:hep-lat/0309020].

Future directions in parity violation

From quarks to the cosmos

M.J. Ramsey-Musolf[a]

California Institute of Technology, Pasadena, CA 91125, USA and
University of Wisconsin-Madison, Madison, WI 53706, USA

Abstract. I discuss the prospects for future studies of parity-violating (PV) interactions at low energies and the insights they might provide about open questions in the standard model as well as physics that lies beyond it. I cover four types of parity-violating observables: PV electron scattering; PV hadronic interactions; PV correlations in weak decays and searches for the permanent electric dipole moments of quantum systems.

PACS. 11.30.Er Charge conjugation, parity, time reversal, and other discrete symmetries – 14.60.Cd Electrons (including positrons) – 24.80.+y Nuclear tests of fundamental interactions and symmetries

1 Introduction

This year we mark the fiftieth anniversary of the discovery of parity violation (PV) in the β-decay of polarized ^{60}Co [1] and the decay of polarized muons [2], confirming the prediction of Lee and Yang [3]. The subsequent history of the field is remarkably rich and diverse, as studies of PV interactions have provided key insights about both the nature of the fundamental weak interaction of leptons and quarks as well the internal structure of strongly interacting systems such as the proton and nuclei. As the interest in this series of PAVI meetings amply demonstrates, the study of PV effects remains an engaging topic that will undoubtedly demand our efforts for many years to come. In this paper, I will give my perspective on how the field may develop over the coming decade, bearing in mind that there are likely to be surprises and new ideas that we cannot foresee at present. In doing so, I will endeavor to give a representative reference to the relevant literature (without providing a full-fledged review), and I apologize in advance to anyone whose work I inadvertently omitted.

In thinking about PV, I would like to put the field in the broader context of fundamental symmetries and their relation to the history of the cosmos. Broadly speaking, I break the particle physics of the Universe into three eras: 1) the present era, during which the broken symmetries of the standard model (SM) provide a remarkably successful framework for explaining a host of microscopic and astrophysical phenomena; 2) the era running from the Big Bang to the moment when electroweak symmetry was broken; and 3) the brief period of electroweak symmetry breaking (EWSB). Remarkably, studies of PV interactions at low energies has something to say about the microphysics of each of these eras, as I hope to convince you below. More extensive discussions of PV in the context of low-energy weak interaction studies can be found in refs. [4–7].

[a] e-mail: `mjrm@caltech.edu`

2 PV: the standard model era

It would take an entire talk to summarize the successes of the SM, so I will focus on what I call "unfinished business" of the theory: the simplest part of the SM to write down is the hardest to solve. Explaining how QCD gives rise to the non-perturbative structure of hadrons and nuclei and governs their interactions is this unfinished business, and it remains a central quest for nuclear physicists. In the past decade, PV electron scattering (PVES) has proven to be an important tool in pursuing this quest. Nearly two decades ago, polarized deep inelastic scattering started to create a "spin crisis" suggested that the naive quark model picture of the proton was flawed, as the quarks carried perhaps $\sim$ 30% of the proton's spin (for a review, see *e.g.* ref. [8]). There were also indications that strange quarks, which do not appear in the quark model picture, contributed as much as 10% of the proton spin and were polarized anti-parallel to it. At the same time, people began to wonder whether strange quarks could also play an important role in the nucleon's electromagnetic structure, and early computations by Jaffe [9] and others suggested that a large strange quark contribution was not unreasonable (for recent reviews of the theoretical literature, see, *e.g.*, refs. [10–13]).

As first suggested by Kaplan and Manohar [14] and followed by with experimental ideas by McKeown [15], Beck [16] and others, PVES became the probe of choice to look for these strange quark contributions. As a result of the efforts of the SAMPLE [12], PV A4 [17], HAPPEX [18], and G0 [19] experiments, we now have limits on the size of the strange quark effects. The results are being summarized by other speakers, but I like to characterize them in terms of the isoscalar form factors as these are the only ones to which strange quarks contribute. In the case of the isoscalar magnetic moment, for example, we now know that strange quarks are responsible for at most $(20 \pm 15)\%$. The strange quark contributions to the nucleon electric form factors are similarly bounded to be a small fraction of the total. At the outset of this program of PV studies, various theoretical expectations suggested the possibility of considerably larger effects. Obtaining these experimental results that now rule out many of those early expectations represents a substantial achievement for the field.

The successful use of PVES to probe nucleon strangeness has depended on a close interplay of experiment and theory. Obtaining robust extractions of bounds on the strange vector form factors from the measured PV asymmetries required a careful delineation of various contributions and sources of theoretical uncertainty [20], including hadronic contributions to electroweak radiative corrections [21,22], nuclear wavefunction and many-body effects [23–26], and isospin breaking in the nucleon [27, 28]. Interpreting the results in QCD has inspired recent work using lattice QCD methods [29–32]. Calculating the strange quark effects requires evaluation of "disconnected" operator insertions on the lattice —a highly non-trivial technical challenge. Given the tight bounds on the strange form factors, obtaining robust results from direct lattice computations stands as an important future milestone for the lattice community.

2.1 PV DIS: beyond the parton model?

In essence, PVES gave us a clear look into the low-energy "internal landscape" of the proton beyond the quark model picture. In the future, I think it might play the same role as a way to look beyond the parton model description that has been enormously successful in explaining deep inelastic and Drell-Yan processes. As discussed in detail at the present meeting, there exists considerable interest now in revisiting the PV deep inelastic scattering (DIS) experiment that was carried out at SLAC in the 1970s and that provided the smoking gun for the SM picture of the weak neutral current interaction [33]. In addition to using a more precise version of that experiment to test the electroweak part of the SM and look for new physics, it could also shed new light on the structure of the proton.

One aspect that particularly intrigues me is the possibility to look for higher twist (HT) effects. Such effects generate corrections to parton model expectations that go as powers of $\Lambda^2_{\rm had}/Q^2$ and that are highly suppressed at the energies where parton distribution functions are typically determined. At lower energies, however, their effects should become more apparent. In this respect, there exists a puzzle emerging from inclusive, inelastic electroproduction data from Jefferson Lab. As detailed extensively in ref. [34], the expected HT contributions to structure functions do not behave as expected. Plotting $2x_B F_1(x_B, Q^2)$ *vs.* x, for example, one sees that the data fluctuates about the curves derived from parton distribution functions, that characterize the leading twist contribution. The lower the Q^2, the larger the fluctuations. One can clearly see as well that the location of the peaks corresponds to various well-known resonances, indicating departures from incoherent scattering from uncorrelated partons. One can then convert the structure function data into moments and plot the moments as a function of Q^2. After subtracting out the known elastic contribution, one finds remarkable agreement with parton model predictions for the lowest moments.

The puzzle is: What happened to the parton correlations that are naively responsible for the resonance structure in the structure functions and that ought to show up as important HT corrections to the moments? To be more concrete, consider the momentum sum rules that relate the structure function moments to matrix elements of local operators:

$$M_j^{(n)}(Q^2) = \int \mathrm{d}x_B\, x_B^{n-j}\, F_j(x_B, Q^2) \propto \sum_{n,j} \tilde{C}^{(n)}_{j,k}(Q^2, \mu^2, g)\, A^n_k(\mu), \tag{1}$$

where j labels the structure function; the $A_k^{(n)}(\mu)$ are forward matrix elements operators $\mathcal{O}_k^{(n)}$ having spin n and being labeled by index "k"; μ is the renormalization scale; and the $\tilde{C}^{(n)}_{j,k}$ are Wilson coefficients whose Q^2-dependence can be computed in perturbation theory. The twist of an operator is defined as its dimension $d_{\mathcal{O}}$ minus its spin: $\tau = d_{\mathcal{O}} - n$. For twist-two operators, the Wilson coefficients carry only a logarithmic dependence on Q^2 (summed to all orders using the renormalization group). For twist-four operators, the $\tilde{C}$ carry an additional $1/Q^2$ power dependence. The corresponding operator matrix element introduces an additional $\Lambda^2_{\rm had}$ in the numerator (with a coefficient determined by the details of the operator matrix element), leading to the $\Lambda^2_{\rm had}/Q^2$ power correction to the twist-two logarithmic Q^2-dependence. Successively higher orders in the twist expansion imply higher-order power corrections of this type. Generically, one expects that for $Q^2 \sim 1\,\mathrm{GeV}^2$, these power corrections should become quite important. The puzzle is that in the low-Q^2 data for the lowest moments, they do not seem to be there in a significant way.

It is interesting to note that the $\tau = 4$ operators generally probe correlations between quarks and gluons in the nucleon. Illustrative examples include [35]

$$\begin{aligned} \Delta \cdot Q_n^{1(k,\ell)} &= g\,\bar{q}_R\, \not{\Delta}\, \overleftarrow{d}^{\ell}\, \overrightarrow{d}^{k} q_R \bar{q}_R\, \not{\Delta}\, \overrightarrow{d}^{n-2-k-\ell} q_R \\ \Delta \cdot Q_n^{8(k)} &= i\bar{q}\, \overleftarrow{d}^{k} \Delta_\alpha \gamma_\beta G^{\alpha\beta}\, \overrightarrow{d}^{n-1-k} q\,, \end{aligned} \tag{2}$$

where Δ_μ is a light-like vector contracted with all free operator indices, $G^{\alpha\beta}$ is the gluon field strength tensor, and $d = i\Delta \cdot D$ with D_μ being the gauge covariant derivative. From the examples in eq. (2) one can see that twist-four operators probe quark-quark correlations ($\Delta \cdot Q_n^{1(k,\ell)}$) or quark-gluon correlations ($\Delta \cdot Q_n^{8(k)}$). Thus, probing the effects of such operators is a way to look beyond the parton model picture of the nucleon in which scattering occurs incoherently from quasifree (uncorrelated) constituents.

Simple curve fitting to power corrections suggests that contributions from a given order in the twist expansion beyond twist two may be large, but that cancellations occur between successive orders (see, *e.g.*, ref. [36]). This may be plausible, but is not yet convincing, since we do not know the non-power law Q^2-evolution of HT contributions to the structure functions (*e.g.*, the logarithmic dependence of the $\tilde{C}$'s). It seems equally possible that the leading twist/parton model description fits the lowest moments because the matrix elements of the relevant HT operators are suppressed rather than because of a conspiracy of cancellations. Ideally, one would like to have access to a different combination of HT operators than those entering parity-conserving scattering to help sort out among these possibilities. In this sense, PV DIS may provide us with a novel tool for looking beyond the parton model description of the nucleon that has worked so well for inelastic electroproduction, just as PVES allowed us to look beyond the quark model (for strange quarks) description of the elastic EM nucleon properties.

To carry the parallel a bit further, theoretical computations of the nucleon strange magnetic moment (μ_s) based on dispersion theory suggested that the portion of μ_s arising from the kaon cloud (resonating into the ϕ) could be comparable in magnitude to the isoscalar magnetic moment, but that its effects in electromagnetic interactions are hidden by a conspiracy of cancellations [9,37]. Without looking directly for μ_s using PVES, we would never have been able to test this scenario. The results from the PVES experiments now imply that the picture is more complicated and that it will take additional theoretical efforts to understand the dynamics of $s\bar{s}$ pairs in the vector channel. I suspect that we may be in a similar situation with HT effects and that PV DIS could provide important information that complements what we know from parity-conserving electroproduction.

In exploring this possibility, I believe there is considerable theoretical work to be completed. The experimentalists are far ahead of theorists in developing the PV DIS program, but I am confident that we can lay out the framework for a systematic and interpretable program just as we did over a decade ago for elastic and quasielastic PVES. In doing so, we should address several questions: What is the non-power law Q^2-dependence of the twist-four Wilson coefficients? What information on HT operators from parity-conserving electroproduction can be used as input for the PV case? Which twist-four operators are the PV asymmetries most sensitive to and which kind of quark-quark or quark-gluon correlations do they probe? What would we expect for the corresponding matrix elements in QCD? I hope that by the time of the next PAVI meeting in two years we will begin to have some answers to these questions.

2.2 Hadronic PV: four-quark operators in the nuclear domain

The purely hadronic weak interaction (HWI) has been a topic of considerable scrutiny over the years, and yet it remains enigmatic. What makes the HWI both interesting and challenging is that it involves a complex interplay of weak and strong interactions. The effects of perturbative QCD can be computed fairly reliably, as they lead to the running of four-quark operators from the electroweak scale down to hadronic scales. The real difficulties arise in understanding the low-energy matrix elements of those operators. Short of first principles, lattice QCD computations, theorists have relied on a number of methods over the years to try and compute these matrix elements: chiral perturbation theory, quark models, large-N_C symmetry, $SU(3)$ symmetry, etc. Unfortunately, the data brings all these efforts up short.

As recently discussed in ref. [38], the strangeness changing ($\Delta S = 1$) sector presents a number of puzzles that remain to be explained. The reason for the $\Delta I = 1/2$ rules is, perhaps, the most well known. In addition, the non-leptonic decays of hyperons have defied a description within the framework of chiral perturbation theory (ChPT), which has been very successful in describing strong and electromagnetic interactions of hadrons. In the case of the $\Delta S = 1$ HWI, one can adequately characterize either the S- or P-wave non-leptonic decays with ChPT, but not both. As suggested in refs. [39,40], one may need to include a host of baryon resonances to arrive at a satisfying description. An explanation of the PV radiative decays such as $\Sigma^+ \to p\gamma$ is even more elusive. In this case, one measures a PV asymmetry $\alpha_{BB'}$ of the outgoing photon relative to the polarization of the decaying hyperon. This asymmetry arises from an interference of the parity-conserving $M1$ and parity-violating $E1$ amplitudes. In the pure $SU(3)$ limit, the latter must vanish [41]. Thus, one would expect $\alpha_{BB'} \sim m_s/\Lambda_{\rm had}$, characteristic of the size of $SU(3)$-breaking. Experimentally, however, one finds asymmetries four to five times larger in magnitude.

Apparently, the symmetries of low-energy QCD are not terribly helpful when applied to the $\Delta S = 1$ HWI. What we do not know is whether this failure reflects the presence of the strange quark, with its problematic mass (not too large, not too small), or more fundamental dynamics associated with four-quark matrix elements in hadrons. One way to find out is to study the $\Delta S = 0$ HWI, for which the effects of the strange quark ought to be relatively unimportant. The best way for doing so is to study PV effects in hadronic and nuclear systems. Since one would like to get at the effects of the weak interaction, and since the strong and electromagnetic interactions are many orders of magnitude stronger than the weak interaction in $\Delta S = 0$ processes, one must rely on PV observables to filter out the latter from the former.

Unfortunately, the state of our understanding of PV effects in hadronic and nuclear systems remains rather poor, despite years of dedicated experimental and theoretical effort [38]. The problem is two-fold. First, in order to see the effects of the PV $\Delta S = 0$ HWI, experimentalists had to rely on processes in nuclei, such as PV γ-decays analogous to the hyperon radiative decays. Due to accidental near degeneracies of opposite-parity states in nuclei such as ^{18}F, the effects of the PV HWI could be enhanced by several orders of magnitude over the nominal $\lesssim 10^{-7}$ scale. While such enhancements made the measurement of PV effects eminently more feasible, the use of nuclei also complicated their interpretation. Second, for many years the theoretical framework used to perform this interpretation relied on a picture of PV meson-exchange interactions, wherein the exchanged meson experienced a PV interaction with one nucleon and a parity-conserving strong interaction with the other. To make such a framework practical, one needed to truncate the tower of exchanged mesons to the lightest few that could saturate all the quantum numbers of the low-energy PV NN interaction. What the effective vertices in such a picture mean in terms of the underlying HWI is not clear —especially when one folds in the approximations needed to perform computations with this meson-exchange model in nuclei.

Two new developments have changed the prospects for this field, and I am now optimistic that substantial progress can be achieved in the future. First, experimental advances have opened the way to performing measurements of $\sim 10^{-7}$ PV observables in few-body systems, thereby circumventing the need to contend with nuclear many-body issues. A comprehensive program of few-body PV measurements is now being planned at the Spallation Neutron Source, National Institute of Standards and Technology, and elsewhere that will bring about a sea change in the experimental information we have available to us. Second, the framework for interpreting these measurements has been reformulated using the ideas of effective field theory (EFT) [42,38]. The use of EFT allows us to systematically classify the terms in the PV NN interaction according to powers of a small momentum or mass, labeled by "Q". The lowest-order interaction —which occurs at $\mathcal{O}(Q^{-1})$— is purely long range and arises from pion exchange, as in the old meson-exchange picture. There is one low-energy constant (LEC) associated with this interaction, h_π^1 (or f_π in past literature). The next set of interactions arise at next-to-next-to leading order (NNLO), or $\mathcal{O}(Q)$ and comprise three classes of terms: 1) a medium-range term in the potential associated with two pion exchange; 2) short-range contributions to the PV potential that in the meson exchange picture were parameterized by heavy meson exchange; and 3) a long-range pion-exchange meson exchange current (MEC). The medium-range term in the potential introduces no new LECs beyond the h_π^1 entering the LO potential, but it is a qualitatively new effect that had not been systematically included in the old meson exchange picture. The short-range terms introduce five additional LECs that correspond to the five independent S-P amplitudes that characterize the PV NN interaction at very low energies ($E \ll m_\pi$): ρ_t, λ_t, and $\lambda_s^{1,2,3}$. The long-range MEC introduces one new constant, $\bar{C}_\pi$, that also does not appear in the older framework. Thus, to $\mathcal{O}(Q)$, one has seven LECs that must be determined from experiment.

The goal of the program at the SNS, NIST, and other facilities is to determine these seven constants from PV experiments with few-body systems. The advantage of the few-body systems is that one can perform precise, *ab initio* few-body computations using the lowest-order PV NN potential and extract the LECs without encountering many-body nuclear-physics uncertainties. A successful completion of this program therefore requires completion of a comprehensive set of few-body computations. For recent work along these lines, see refs. [38,43]. At the same time, one would like to compute the LECs from first principles in QCD using the lattice so that —at the end of the day— one can confront the experimental values with theoretical expectations. At present, one has estimates for the LECs based on naive dimensional analysis and on a correspondence with the well-known "DDH" quark model/$SU(6)_w$ ranges for the PV meson-nucleon vertices in the old framework (for a summary of these estimates, see ref. [38]). Clearly, one would like to go beyond these model estimates to *bona fide* QCD computations, and to that end, initial work using chiral perturbation theory —needed for extrapolations of lattice computations to the physical light quark domain— has been performed for h_π^1 [44,45]. One physical insight derived from these studies is that the role of quark-quark correlations in the four-quark weak matrix elements that generate h_π^1 could be more important than implied by the DDH quark model/$SU(6)_w$ computation. The results of such QCD computations, when compared with experimental results, could also help us find out whether the $\Delta S = 0$ suffers from the same departures from QCD symmetry-based expectations as one finds in the $\Delta S = 1$ sector, or whether the latter are just a reflection of the problematic presence of the strange quark.

An additional follow-up to the few-body hadronic PV program would be to revisit the nuclear PV observables. It would be interesting to test whether the NNLO PV NN interaction, determined by the few-body program, can explain the results of nuclear PV experiments when it is used in many-body computations. If so, the results could have consequences for the interpretation of other nuclear weak-interaction observables that are sensitive to the effects of four-quark operators. One such observable of particular interest these days is neutrinoless double-beta decay ($0\nu\beta\beta$). It is well known that if one finds a non-zero signal in $0\nu\beta\beta$, one has smoking gun evidence for the Majorana nature of the neutrino. Ideally, one would also like to use a non-zero rate to determine the absolute scale of neutrino mass, since for light Majorana neutrino exchange, the rate is proportional to the square of the effective mass. It is entirely possible, however, that the exchange of some heavy Majorana particle, such as a neutralino in supersymmetry or a heavy right-handed Majorana neutrino, could contribute to the rate at the same level as the exchange of the light Majorana neutrino. In order to separate the two contributions, one would like to be able to compute the effects of the former, which entail calculating nuclear matrix elements of four-quark operators. The corresponding nuclear operators can be classified according to Q-counting as with the PV NN interaction. What we do not know is whether or not the leading terms in this classification suffice to explain four-quark matrix elements in nuclei. Using the NNLO PV NN interaction to confront nuclear PV data would provide us with a test. If successful, one would then have some confidence that the lowest-order terms in the EFT for $0\nu\beta\beta$ give a reasonable description of heavy-particle exchange effects in that process.

One alternate probe of the $\Delta S = 0$ HWI that remains interesting involves PVES for the $N \to \Delta$ interaction. In contrast to the situation for elastic PVES, the $N \to \Delta$ asymmetry need not vanish at $Q^2 = 0$ —a consequence of Siegert's theorem well known in electromagnetic physics. In the real photon limit, the asymmetry has the simple form [46]

$$A_{PV}^{N\to\Delta} = -2\frac{d_\Delta}{C_3^V}\frac{m_N}{\Lambda_{\rm had}} + \cdots \qquad (3)$$

where d_Δ is a LEC that characterizes the PV $\gamma N\Delta$ electric dipole interaction. It is the $\Delta S = 0$ analog of the $\Delta S = 1$ $E1$ amplitude responsible for the PV hyperon radiative decays discussed above. It appears realistic that a useful determination of d_Δ could be performed with either the G0 or Q-Weak apparatus. The results of such measurements could shed additional light on the hyperon radiative decay puzzles. If, for example, parity-mixing with nucleon resonances is responsible for the large $\alpha_{BB'}$ discussed above, and if the same kind of dynamics apply to the $\Delta S = 0$ HWI, then one would expect $A_{PV}^{N\to\Delta}$ to be quite large —of order 10^{-6}. On the other hand, if the dynamics are substantially different in the two cases, then naive dimensional analysis would suggest a considerably smaller $N \to \Delta$ asymmetry. Either way, the results would be interesting, and I think it is important to keep the prospects for such a measurement on the radar screen for the future.

3 PV: beyond the standard model

The search for physics beyond the standard model (BSM) clearly lies at the forefront of particle physics as well as at the intersection of nuclear physics with particle physics and cosmology. This search is motivated by a number of open questions involving the microphysics of the early Universe that the SM cannot solve. These questions include: Why is there more matter than antimatter in the Universe? Were all the forces of nature unified into a single force at the end of the Big Bang, and if so, why don't we see this unification in the running of the SM gauge couplings? What is the origin of neutrino mass? Why is electric charge quantized? Why is the electroweak scale so low or, in terms of more familiar quantities, why is the Fermi constant so large?

Looking ahead to the start of the LHC era, we are hoping to uncover direct evidence for one or more of the possible extensions of the SM that would address some of these puzzles, including supersymmetry, extended gauge symmetries and the associated gauge bosons, extra dimensions and the corresponding modes that would appear in four-dimensional spacetime, *etc.* At the same time as we move into the next energy frontier, experiments that push the precision frontier are poised to provide important, complementary information. Most of these experiments involve processes at low energy and many entail the study of PV observables. I will discuss the future directions in two classes: weak decays and PV electron scattering.

3.1 Weak decays: PV correlations

The study of various correlations in weak decays has been a topic of considerable interest for decades, and recent experimental advances are moving the field into a new era. For concreteness, let me consider the weak decays of the muon and neutron. In the case of polarized muons, the spatial distribution and energy dependence of the outgoing positron or electron are characterized by the Michel parameters that enter the partial rate [47,48]

$$\begin{aligned} d\Gamma = &\frac{G_\mu^2 m_\mu^5}{192\pi^3}\frac{d\Omega}{4\pi}x^2\, dx \left\{ \frac{1+h(x)}{1+4\eta(m_e/m_\mu)} \right. \\ &\times \left[12(1-x) + \frac{4}{3}\rho(8x-6) + 24\frac{m_e}{m_\mu}\frac{(1-x)}{x}\eta\right] \\ &\left. \pm P_\mu\, \xi \cos\theta \left[4(1-x) + \frac{4}{3}\delta(8x-6) + \frac{\alpha}{2\pi}\frac{g(x)}{x^2}\right] \right\}, \end{aligned} \qquad (4)$$

where $x = |\boldsymbol{p}_e|/|\boldsymbol{p}_e|_{\rm max}$, $\theta = \cos^{-1}(\hat{p}_e \cdot \hat{s}_\mu)$, P_μ is the $\mu^\pm$ polarization, and $h(x)$ and $g(x)$ are momentum-dependent radiative corrections. The quantities ρ, η, δ, and ξ are Michel parameters (there are an additional five that do not concern us here). Note that ξ and δ are associated with the anisotropic part of the charged lepton spectrum that reflects PV in the underlying weak interaction. In the SM, one has $\rho = \delta = 3/4$, $P_\mu\xi = 1$, and $\eta = 0$. Deviations from these values would reflect the presence of non $(V-A)\times(V-A)$ interactions. Recently, the TWIST Collaboration has completed new determinations of ρ, δ, and $P_\mu\xi$, reducing the uncertainty by a factor of two or more over previous determinations [49–51].

The effects of non $(V-A)\times(V-A)$ interactions on these parameters can be described by an effective low-energy Lagrangian

$$\mathcal{L}^{\mu\text{-decay}} = -\frac{4G_\mu}{\sqrt{2}}\sum_\gamma g_{\epsilon\mu}^\gamma \;\bar{e}_\epsilon \Gamma^\gamma \nu_e\, \bar{\nu}_\mu \Gamma_\gamma \mu_\mu\,, \qquad (5)$$

where the sum runs over Dirac matrices $\Gamma^\gamma = 1$ (S), γ^α (V), and $\sigma^{\alpha\beta}$ (T) and the subscripts ϵ and μ denote the chirality (R, L) of the final-state lepton and muon, respectively[1]. The SM gives $g^V_{LL} = 1$ with all other $g^\gamma_{\epsilon\mu} = 0$. For example, deviations of δ and $P_\mu\xi$ from their SM values are given by

$$\delta - \frac{3}{4} = \frac{9}{4}\left[|g^V_{RL}|^2 + 2|g^T_{RL}|^2 + \mathrm{Re}\, g^T_{RL} g^{S*}_{RL} - (L \leftrightarrow R)\right],$$
$$1 - \xi\frac{\delta}{\rho} = 2||g^V_{RR}|^2 + \frac{1}{2}|g^S_{RR}|^2 + \frac{1}{2}|g^S_{RR}|^2 + \frac{1}{2}|g^S_{LR} - 2g^T_{LR}|^2 . \qquad (6)$$

Various BSM scenarios can lead to non-zero values for the $g^\gamma_{\epsilon\mu}$ appearing in these PV Michel parameters. For example, a non-zero value of $g^V_{LR,RL}$ can occur in left-right symmetric models when the left-handed (LH) and right-handed (RH) W-bosons mix [52]. Similarly, supersymmetric loop graphs can give rise to a non-vanishing g^S_{RR} when the superpartners of LH and RH charged leptons mix [53]. The latter effects are too small to generate an observable deviation of δ from the SM value, since this effect arises at one-loop level and since g^S_{RR} enters δ quadratically. On the other hand, this parameter interferes linearly with g^V_{LL} in the quantity η, which is assumed to be zero when the value of the Fermi constant is extracted from the muon lifetime, τ_μ. If the L-R superpartner mixing is large, the correction to τ_μ could be large enough to affect the value of G_μ to be determined from the next generation of muon lifetime experiments currently underway at PSI.

It turns out that the scale of the mass of light neutrinos has significant implications for the values of some of the $g^\gamma_{\epsilon\mu}$. In particular, for $\epsilon\mu = LR$ or RL —corresponding to neutrinos of opposite chirality— radiative corrections involving the associated operators can generate contributions to the neutrino mass matrix. In the absence of "unnatural" large cancellations between these radiative contributions and tree level neutrino mass terms, these radiative contributions should not be large compared to the scale of entries in the neutrino mass matrix. Based on such "naturalness" considerations, one would expect the $|g^\gamma_{RL,LR}|$ to be bounded from above by the scale of m_ν. These implications were recently noted in ref. [54] and followed up in several studies, considering both weak decays [55] and neutrino magnetic moments [56–58].

In analyzing the neutrino mass naturalness implications, one has to be careful to employ a set of gauge invariant operators (the four-fermion operators in eq. (5) are not) while taking into account the flavor structure of all gauge invariant operators that can contribute to muon decay [55]. As a result, one finds that the scale of m_ν implies rigorous bounds on the $|g^V_{LR,LR}|$ ($\lesssim 10^{-4}$) and constrains some, but not all, of the gauge invariant four-fermion operators that can give rise to scalar and tensor interactions involving neutrinos of opposite chirality. Thus, if future measurements of these PV Michel parameters yielded non-zero deviations from the SM, one would likely conclude that the source is one of the gauge invariant four-fermion operators

$$\epsilon^{ij}\bar{L}^A_i \ell^C_R \bar{L}^B_j \nu^D_R \qquad (7)$$

with generation indices $A = B = 1$, $C = 2$ or $A = B = 2$, $C = 1$, as these operators cannot contribute to the neutrino mass matrix through radiative corrections[2].

An analogous situation pertains in the semileptonic sector. For weak decays of systems containing light quarks, one can write down an effective Lagrangian analogous to eq. (5):

$$\mathcal{L}^{\beta\text{-decay}} = -\frac{4G_\mu}{\sqrt{2}} \sum_{\gamma,\,\epsilon,\,\delta} a^\gamma_{\epsilon\delta}\ \bar{e}_\epsilon \Gamma^\gamma \nu_e\, \bar{u}\Gamma_\gamma d_\delta, \qquad (8)$$

where the $a^\gamma_{\epsilon\delta}$ play the role of the $g^\gamma_{\epsilon\mu}$ in this sector. There exist several equivalent representations of the low-energy effective semileptonic interaction [59,60], but I prefer the form in eq. (8) because of its similarity to the muon decay effective Lagrangian. In the standard model, one has $a^V_{LL} = V_{ud}$ at tree level. Beyond tree level, one must correct this expression for the difference between radiative corrections to the muon and β-decay amplitudes because we have normalized $\mathcal{L}^{\beta\text{-decay}}$ in terms of the muon decay Fermi constant.

As members of this community well know, obtaining a reliable determination of a^V_{LL} is of considerable interest since the value of V_{ud} is needed to test the unitarity of the CKM matrix. This subject has been discussed extensively elsewhere [7,59,61], so I will not dwell on the details here. It is worth noting, however, that in addition to the using of super allowed Fermi transitions nuclei to determine a^V_{LL}, considerable effort is being devoted to obtaining this parameter from a combination of the neutron lifetime (τ_n) and one or more of its PV decay correlations that appear in the partial rate [62]:

$$\mathrm{d}\Gamma \propto \mathcal{N}(E_e)\mathrm{d}\Omega_e \mathrm{d}\Omega_\nu \mathrm{d}E_e \left\{1 + a\frac{\boldsymbol{p}_e\cdot\boldsymbol{p}_\nu}{E_e E_\nu} + b\frac{\Gamma m_e}{E_e} + \langle\boldsymbol{J}\rangle\cdot\left[A\frac{\boldsymbol{p}_e}{E_e} + B\frac{\boldsymbol{p}_\nu}{E_\nu} + D\frac{\boldsymbol{p}_e\times\boldsymbol{p}_\nu}{E_e E_\nu}\right] + \boldsymbol{\sigma}\cdot\left[N\langle\boldsymbol{J}\rangle + G\frac{\boldsymbol{p}_e}{E_e} + Q'\hat{p}_e\hat{p}_e\cdot\langle\boldsymbol{J}\rangle + R\langle\boldsymbol{J}\rangle\times\frac{\boldsymbol{p}_e}{E_e}\right]\right\}, \qquad (9)$$

where $\mathcal{N}(E_e) = p_e E_e (E_0 - E_e)^2$; E_e (E_ν), $\boldsymbol{p}_e$ $(\boldsymbol{p}_\nu)$, and $\boldsymbol{\sigma}$ are the β (neutrino) energy, momentum, and polarization, respectively; $\boldsymbol{J}$ is the polarization of the decaying nucleus; and $\Gamma = \sqrt{1-(Z\alpha)^2}$. Since the strong interaction renormalizes the vector and axial vector components of the $V-A$ quark currents in eq. (8), and since the neutron decay amplitude depends on both components, one requires both τ_n and the ratio of axial vector to vector hadronic couplings, $\lambda = g_A/g_V$. The latter can be obtained, for example, from the correlation coefficients a, A,

[1] The use of the subscript "μ" to denote both the chirality of the muon and the flavor of the corresponding neutrino is an unfortunate historical convention.

[2] Here, L is an $SU(2)_L$ doublet, ℓ_R is an $SU(2)_L$ charged lepton singlet, and i, j are generation indices.

or B:

$$a=\frac{1-\lambda^2}{1+3\lambda^2},\quad A=-2\frac{\lambda(1+\lambda)}{1+3\lambda^2},\quad B=2\frac{\lambda(\lambda-1)}{1+3\lambda^2},\quad (10)$$

where both A and B are associated with PV correlations. To date, the most precise determinations of the A parameter have been obtained from neutron decay experiments at ILL, while a new measurement of A is underway using ultracold neutrons at LANSCE. Future neutron decay experiments at the SNS will provide a comprehensive set of correlation measurements that could lead to significantly lower systematic uncertainties in the value of λ that is achievable with any single measurement alone.

Measurements of these correlation coefficients can also probe non-$(V-A)\times(V-A)$ interactions that enter eq. (8). For example, supersymmetric box graph corrections can generate the scalar and tensor interactions parameterized by a^S_{RR}, a^S_{RL}, and a^T_{RL} in the presence of mixing between LH and RH scalar fermions [53]. The effect of such interactions would show up most strongly in terms in eq. (9) that depend on m_e/E_e, such as the "Fierz interference" term proportional to b as well as the PV neutrino correlation parameter, B [7]:

$$B^{\rm SUSY}_{\rm box}=-2\left(\frac{\Gamma m_e}{E_e}\right)\frac{\lambda}{1+3\lambda^2}\,\mathrm{Re}\Bigg\{4\lambda\left(\frac{g_T}{g_A}\right)\left(\frac{a^T_{RL}}{a^V_{LL}}\right)^* + \left[2\left(\frac{g_T}{g_A}\right)\left(\frac{a^T_{RL}}{a^V_{LL}}\right)^* - \left(\frac{g_S}{g_V}\right)\left(\frac{a^S_{RL}+a^S_{RR}}{a^V_{LL}}\right)^*\right]\Bigg\},$$

where g_S and g_T are scalar and tensor form factors. If the L-R superpartner mixing is near maximal and the superpartner masses are not too heavy, then it is possible that $|B|$ can be as large as $\sim 10^{-3}$, while future correlation experiments could probe the energy-dependent part of B at the few $\times 10^{-4}$ level. Observation of such large effects could be problematic for supersymmetry, as it would imply either fine tuning in order to obtain proper electroweak symmetry-breaking and/or superheavy Higgs scalars that could not be observed at the LHC[3].

3.2 PV electron scattering

Future measurements of PV electron scattering asymmetries at Jefferson Lab will provide similarly interesting probes of new physics. For elastic scattering from a target f the asymmetry has the general form [20]

$$A_{PV}=\frac{G_\mu}{4\sqrt{2}\pi\alpha}Q^2\left[Q^f_W+F(\theta,Q^2)\right],\quad (11)$$

where $F(\theta,Q^2)$ is a form factor term and where the weak charge Q^f_W is given by

$$Q^f_W=\hat\rho_{NC}(0)\left[2I^f_3-4Q_f\hat\kappa(0,\mu)\hat s^2(\mu)\right]+\hat\lambda^f_V+\left(-1+4\hat s^2\right)\hat\lambda^e_A+\mathrm{box}.\quad (12)$$

[3] There would still be one light SM-like Higgs boson.

Here, I^f_3 is the third component of the target's weak isospin, $\hat s^2\equiv\sin^2\hat\theta_W(\mu)$ gives the square of the sine of the weak mixing angle in the $\overline{MS}$ scheme, $\hat\rho_{NC}$ and $\hat\kappa$ denote a universal set of radiative corrections, $\hat\lambda^f_{V,A}$ are the vertex plus external leg corrections, and "+box" indicate box graph corrections. In the case of SUSY radiative corrections to the weak charges of the electron and proton —targets of interest to the future Jefferson Lab program— the dominant effects enter through $\hat\kappa$ [63]. For the Q^p_W, these corrections could be as large as $\sim 4\%$ while for Q^e_W the relative effect could be twice as large. In both cases, the effect of the SUSY corrections always decreases the effective weak mixing angle, given by

$$\sin^2\hat\theta(Q^2)^{\rm eff}=\hat\kappa(Q^2,\mu)\hat s^2(\mu)\,,\quad (13)$$

thereby leading to a relative *increase* in the magnitude of the weak charges.

It is worth noting that the central value for $\sin^2\hat\theta(0)^{\rm eff}$ obtained from the SLAC E158 Møller experiment [64] —though in agreement with the SM prediction at better than the 2σ level— is slightly larger than the SM prediction. If the results of the Q-Weak measurement or future Jefferson Lab Møller experiment were to agree with this central value but with smaller error bars, the largest SUSY corrections would be disfavored. This could be particularly interesting in light of the value for the muon anomalous magnetic moment, which favors the same SUSY parameter space that would give the largest increases in the weak charges. Thus, one could imagine a situation where the results for $g_\mu-2$ and the PVES measurements could lead to some tension within SUSY.

One way such a situation might be avoided is to allow for so-called "R parity violating" (RPV) effects in SUSY [65]. These interactions are entirely supersymmetric but entail the violation of lepton (L) or baryon number (B). The L-violating interactions are particularly interesting for PVES, where they generate tree level effects. In the case of the weak charge of the electron, for example, one has [66]

$$\frac{\delta Q^e_W}{Q^e_W}\approx-45\left(\frac{100\,\mathrm{GeV}}{\tilde m}\right)^2|\lambda_{12k}|^2,\quad (14)$$

where $\tilde m$ is the mass of the exchanged scalar lepton and λ_{12k} is the relevant L-violating coupling. Note that this effect does not enter here directly into the PV ee amplitude, but rather indirectly through the normalization of the amplitude to the muon decay Fermi constant. The quantity λ_{12k} is, in fact, an RPV coupling that enters that process.

For $\tilde m=1\,\mathrm{TeV}$, a 5% measurement of Q^e_W would be sensitive to $\lambda_{12k}\sim 0.3$. This sensitivity is within a factor of two what has been achieved to date in direct searches for L-violation with $\mu\to e\gamma$ experiments —a rather remarkable statement of the power for PVES as a probe of new physics. It is also interesting to observe that the effect of RPV interactions is always to *decrease* the magnitude of Q^e_W from its SM value, in contrast to the situation with SUSY loop effects. The E158 results hint in this direction, and the results of a future experiment with substantially smaller error bars could provide stronger indications for RPV effects. Such a result could be significant, because the presence of RPV implies that superpartners can ultimately decay to SM particles, so the lightest supersymmetric particle could not have lived long enough to form the cold dark matter of the Universe. In addition, loop effects with the RPV interactions generate a Majorana mass for the neutrino. Thus, if one had conclusive evidence for RPV in SUSY, one would know that neutrinos are Majorana particles even if future $0\nu\beta\beta$ experiments yielded null results.

Measurements of the weak charges can provide similarly interesting and complementary probes of other BSM physics scenarios, such as the presence of new light Z' bosons, doubly charged Higgs scalars, leptoquarks, *etc.* (for recent discussions, see refs. [67,68]).

4 PV: electroweak symmetry breaking and the origin of matter

An important task for nuclear physics is to explain why there exists any baryonic matter in the Universe at all. If, in fact, the Universe was matter-antimatter symmetric at the end of the inflationary epoch, then there would have to have been some dynamics in the particle physics of the evolving cosmos to create a matter-antimatter asymmetry. Four decades ago, Sakharov enumerated the necessary ingredients in those dynamics [69]: 1) a violation of baryon number; 2) the presence of both C and CP-violation; and 3) a departure from thermal equilibrium at some juncture[4]. In principle, the SM contains all three ingredients. Baryon number violation arises through anomalous processes called "sphaleron transitions" between different electroweak vacua. The C-violation needed arises through the axial vector couplings of gauge bosons to fermions, while the CP-violation appears via the complex phase in the CKM matrix. Finally, when the Universe cooled through the electroweak temperature and the Higgs got its vacuum expectation value, one could have seen a departure from thermal equilibrium. It is well known, however, that neither the CP-violation implied by observations in the kaon and B-meson systems, nor the electroweak phase transition (EWPT) in the SM, are sufficiently strong to explain the observed baryon asymmetry, characterized here by the baryon to photon entropy density ratio[5]:

$$Y_B \equiv \frac{n_B}{s} = \begin{cases} (7.3 \pm 2.5) \times 10^{-11}, & \text{BBN} \\ (9.2 \pm 1.1) \times 10^{-11}, & \text{WMAP} \end{cases} \tag{15}$$

where "BBN" and "WMAP" indicate values derived from Big Bang nucleosynthesis [70] and the cosmic microwave background [71], respectively.

[4] The latter assumes that CPT is an exact symmetry. If CPT was violated, then a baryon asymmetry could have been created during equilibrium dynamics.

[5] The mass of the Higgs boson is too heavy to allow for a strong first-order EWPT in the SM.

Clearly, it is up to BSM physics to account for the small, but anthropically crucial, value of Y_B. A number of BSM scenarios that may provide such an explanation have been proposed and discussed extensively in the literature. Here, I wish to focus on the possibility that new physics during the era of electroweak symmetry breaking was responsible. In SUSY, for example, there exist abundant sources of new CP-violation whose effects are not *a priori* suppressed as in the SM, and the SUSY Higgs sector may provide for the requisite strong first-order EWPT. It is timely to think about these and other electroweak scenarios because both the LHC and precision measurements will be probing physics at this scale. If nothing else, we have some chance of testing and either ruling out electroweak scale baryogenesis (EWB) or providing evidence for its viability. In this respect, PV will play a key role.

The most powerful probes of new CP-violation present during the EWPT are searches for the permanent electric dipole moments of various elementary particles, atoms, and nuclei. As discussed elsewhere [4,7,72], we are on the brink of a revolution in EDM searches, as experimentalists expect to push the sensitivity of these measurements to BSM CP-violation by several orders of magnitude during the LHC era. These measurements will not be sensitive enough to observe EDMs predicted in the electroweak sector of the SM, but they could uncover the effects of BSM CP-violation. Thus, if new electroweak scale physics is discovered at the LHC; if LHC and future linear collider studies of the Higgs sector show that a strong first order EWPT is viable with such new physics; and if the EDM searches yield non-zero results, the possibility that the baryon asymmetry was produced at the electroweak symmetry-breaking era would be quite compelling.

The basic idea of an EDM search is quite simple. One peforms a Larmour precession measurement with a sample of, *e.g.*, polarized neutrons in a configuration of magnetic and electric fields. The component of the precision frequency due to the EDM, $d\boldsymbol{J}$, is given by

$$\nu^{\mathrm{EDM}} = -\frac{d\boldsymbol{J}\cdot\boldsymbol{E}}{h}\,. \tag{16}$$

Note that this effect is odd under both time reversal and parity. In contrast, the Larmour frequency due to the interaction of the magnetic field and magnetic moment are P- and T-even. To separate the two effects, one exploits the PV aspect of ν^{EDM} and looks for a change in the precession frequency upon reversal of the direction of $\boldsymbol{E}$. This reversal amounts to performing a parity transformation since $\boldsymbol{E}$ changes sign under such a transformation but $\boldsymbol{J}$ does not.

Searches for the EDM of the electron, neutron, muon, and neutral atoms have been pursued for many years, with increasingly stringent upper bounds on the magnitudes of EDMs being achieved in each case. For recent reviews of EDM searches, see, *e.g.*, refs. [4,7,72]. These null results can have significant implications for the viability of EWB. In a simple supergravity scenario in supersymmetric baryogenesis, Y_B can depend on two phases [73–76]:

$$Y_B \equiv \frac{\rho_B}{s_\gamma} = F_1 \sin\phi_\mu + F_2 \sin(\phi_\mu + \phi_A)\,, \tag{17}$$

where ϕ_μ and ϕ_A are the CP-violating phases associated with the supersymmetric "μ" term and SUSY-breaking tri-scalar "A" terms, respectively. The coefficients $F_{1,2}$ depend on the other parameters of the SUSY model and on the detailed transport dynamics during the electroweak phase transition. The latter entails a detailed competition between CP-violating asymmetries in the scattering of superpartners from the spacetime varying Higgs vacuum expectation values and CP-conserving interactions that like to minimize free energy by causing these asymmetries to relax to zero.

It turns out that for this scenario, the most important term is the F_1-term. To obtain the observed baryon asymmetry, the corresponding value of $\sin\phi_\mu$ has to be greater in magnitude than about 0.2. Present limits on the EDM of the electron, for example, imply that $|\sin\phi_\mu|$ is considerably smaller, if the mass of the selectron is below about a TeV. In this case, d_e is dominated by its one-loop contribution and this scenario for EWB is close to being ruled out. On the other hand, if the selectron mass is larger than a few TeV, the d_e becomes two-loop dominated and larger CP-violating phases are —consistent with EWB— allowed by present EDM limits. Taking the minimal value for $|\sin\phi_\mu|$ allowed by the observed baryon asymmetry and the two-loop EDM, one would expect a value for $|d_e|$ of order $10^{-28}\,e\,\mathrm{cm}$ or larger if this scenario for EWB is to remain viable [77]. A similar conclusion applies to the neutron EDM. New experiments being carried out for both the electron and neutron could reach this level of sensitivity within the next several years —making the coming period particularly interesting for the interplay of EDMs and cosmology.

5 Conclusions

It seems clear to me that the coming decade will be a period of intense interest in the studies of PV interactions at low and intermediate energies. The field has come a long way from its inception 50 years ago and blossomed into a remarkably rich and diverse area of physics. Recent experimental advances, together with new theoretical developments have put the field on the cusp of a new era. Refining this tool to probe both BSM physics as well as the structure and dynamics of hadrons should engage our efforts for many years to come.

I would like to thank B. Holstein, B. Desplanques, R. McKeown, K. Kumar, P. Souder, C.-P. Liu, and C. Keppel for useful conversations. This work was supported under U.S. Department of Energy contract DE-FG02-05ER41361 and National Science Foundation Award PHY-05556741.

References

1. C.S. Wu, E. Ambler, R.W. Hayward, D.D. Hoppes, R.P. Hudson, Phys. Rev. **105**, 1413 (1957).
2. R.L. Garwin, L.M. Lederman, M. Weinrich, Phys. Rev. **105**, 1415 (1957).
3. T.D. Lee, C.N. Yang, Phys. Rev. **104**, 254 (1956).
4. J. Erler, M.J. Ramsey-Musolf, Prog. Part. Nucl. Phys. **54**, 351 (2005) [arXiv:hep-ph/0404291].
5. M.J. Ramsey-Musolf, AIP Conf. Proc. **842**, 661 (2006) [arXiv:hep-ph/0603023].
6. M.J. Ramsey-Musolf, arXiv:nucl-th/0608035.
7. M.J. Ramsey-Musolf, S. Su, arXiv:hep-ph/0612057.
8. B.W. Filippone, X.D. Ji, Adv. Nucl. Phys. **26**, 1 (2001) [arXiv:hep-ph/0101224].
9. R.L. Jaffe, Phys. Lett. B **229**, 275 (1989).
10. M.J. Ramsey-Musolf, Eur. Phys. J. A **24**, S2, 197 (2005) [arXiv:nucl-th/0501023].
11. E.J. Beise, M.L. Pitt, D.T. Spayde, Prog. Part. Nucl. Phys. **54**, 289 (2005) [arXiv:nucl-ex/0412054].
12. SAMPLE Collaboration (D.T. Spayde *et al.*), Phys. Lett. B **583**, 79 (2004) [arXiv:nucl-ex/0312016].
13. D.H. Beck, B.R. Holstein, Int. J. Mod. Phys. E **10**, 1 (2001) [arXiv:hep-ph/0102053].
14. D.B. Kaplan, A. Manohar, Nucl. Phys. B **310**, 527 (1988).
15. R.D. Mckeown, Phys. Lett. B **219**, 140 (1989).
16. D.H. Beck, Phys. Rev. D **39**, 3248 (1989).
17. A4 Collaboration (F.E. Maas), Prog. Part. Nucl. Phys. **55**, 320 (2005).
18. HAPPEX Collaboration (A. Acha *et al.*), arXiv:nucl-ex/0609002.
19. G0 Collaboration (D.S. Armstrong *et al.*), Phys. Rev. Lett. **95**, 092001 (2005) [arXiv:nucl-ex/0506021].
20. M.J. Musolf, T.W. Donnelly, J. Dubach, S.J. Pollock, S. Kowalski, E.J. Beise, Phys. Rep. **239**, 1 (1994).
21. S.L. Zhu, S.J. Puglia, B.R. Holstein, M.J. Ramsey-Musolf, Phys. Rev. D **62**, 033008 (2000) [arXiv:hep-ph/0002252].
22. M.J. Musolf, B.R. Holstein, Phys. Lett. B **242**, 461 (1990).
23. M.J. Musolf, R. Schiavilla, T.W. Donnelly, Phys. Rev. C **50**, 2173 (1994) [arXiv:nucl-th/9405012].
24. E. Hadjmichael, G.I. Poulis, T.W. Donnelly, Phys. Rev. C **45**, 2666 (1992).
25. R. Schiavilla, J. Carlson, M.W. Paris, Phys. Rev. C **67**, 032501 (2003) [arXiv:nucl-th/0212038].
26. C.P. Liu, G. Prezeau, M.J. Ramsey-Musolf, Phys. Rev. C **67**, 035501 (2003) [arXiv:nucl-th/0212041].
27. G.A. Miller, Phys. Rev. C **57**, 1492 (1998) [arXiv:nucl-th/9711036].
28. B. Kubis, R. Lewis, Phys. Rev. C **74**, 015204 (2006) [arXiv:nucl-th/0605006].
29. S.J. Dong, K.F. Liu, A.G. Williams, Phys. Rev. D **58**, 074504 (1998) [arXiv:hep-ph/9712483].
30. R. Lewis, W. Wilcox, R.M. Woloshyn, Phys. Rev. D **67**, 013003 (2003) [arXiv:hep-ph/0210064].
31. D.B. Leinweber *et al.*, Phys. Rev. Lett. **94**, 212001 (2005) [arXiv:hep-lat/0406002].
32. D.B. Leinweber *et al.*, Phys. Rev. Lett. **97**, 022001 (2006) [arXiv:hep-lat/0601025].
33. C.Y. Prescott *et al.*, Phys. Lett. B **77**, 347 (1978); **84**, 524 (1979).
34. W. Melnitchouk, R. Ent, C. Keppel, Phys. Rep. **406**, 127 (2005) [arXiv:hep-ph/0501217].
35. R.L. Jaffe, M. Soldate, Phys. Rev. D **26**, 49 (1982).
36. CLAS Collaboration (M. Osipenko *et al.*), Phys. Rev. D **67**, 092001 (2003) [arXiv:hep-ph/0301204].
37. H.W. Hammer, M.J. Ramsey-Musolf, Phys. Rev. C **60**, 045204 (1999); **62**, 049902 (2000)(E) [arXiv:hep-ph/9903367].
38. M.J. Ramsey-Musolf, S.A. Page, arXiv:hep-ph/0601127.

39. B. Borasoy, B.R. Holstein, Phys. Rev. D **59**, 094025 (1999) [arXiv:hep-ph/9902351].
40. B. Borasoy, B.R. Holstein, Phys. Rev. D **59**, 054019 (1999) [arXiv:hep-ph/9902431].
41. Y. Hara, Phys. Rev. **134**, B701 (1964).
42. S.L. Zhu, C.M. Maekawa, B.R. Holstein, M.J. Ramsey-Musolf, U. van Kolck, Nucl. Phys. A **748**, 435 (2005) [arXiv:nucl-th/0407087].
43. C.P. Liu, arXiv:nucl-th/0609078.
44. S.L. Zhu, S.J. Puglia, B.R. Holstein, M.J. Ramsey-Musolf, Phys. Rev. D **63**, 033006 (2001) [arXiv:hep-ph/0005281].
45. S.R. Beane, M.J. Savage, Nucl. Phys. B **636**, 291 (2002) [arXiv:hep-lat/0203028].
46. S.L. Zhu, C.M. Maekawa, B.R. Holstein, M.J. Ramsey-Musolf, Phys. Rev. Lett. **87**, 201802 (2001) [arXiv:hep-ph/0106216].
47. L. Michel, Proc. R. Phys. Soc. London, Ser. A **63**, 514 (1950).
48. C. Bouchiat, L. Michel, Phys. Rev. **106**, 170 (1957).
49. TWIST Collaboration (J.R. Musser), Phys. Rev. Lett. **94**, 101805 (2005) [arXiv:hep-ex/0409063].
50. TWIST Collaboration (A. Gaponenko *et al.*), Phys. Rev. D **71**, 071101 (2005) [arXiv:hep-ex/0410045].
51. B. Jamieson *et al.*, Phys. Rev. D **74**, 072007 (2006).
52. P. Herczeg, *Beta Decay And Muon Decay Beyond The Standard Model*, in *Precision Tests of the Standard Electroweak Model*, edited by P. Langacker (World Scientific, 1995).
53. S. Profumo, M.J. Ramsey-Musolf, S. Tulin, arXiv:hep-ph/0608064.
54. G. Prezeau, A. Kurylov, Phys. Rev. Lett. **95**, 101802 (2005) [arXiv:hep-ph/0409193].
55. R.J. Erwin, J. Kile, M.J. Ramsey-Musolf, P. Wang, arXiv:hep-ph/0602240.
56. N.F. Bell, V. Cirigliano, M.J. Ramsey-Musolf, P. Vogel, M.B. Wise, Phys. Rev. Lett. **95**, 151802 (2005) [arXiv:hep-ph/0504134].
57. S. Davidson, M. Gorbahn, A. Santamaria, Phys. Lett. B **626**, 151 (2005) [arXiv:hep-ph/0506085].
58. N.F. Bell, M. Gorchtein, M.J. Ramsey-Musolf, P. Vogel, P. Wang, Phys. Lett. B **642**, 377 (2006) [arXiv:hep-ph/0606248].
59. N. Severijns, M. Beck, O. Naviliat-Cuncic, arXiv:nucl-ex/0605029.
60. P. Herczeg, Prog. Part. Nucl. Phys. **46**, 413 (2001).
61. E. Blucher *et al.*, arXiv:hep-ph/0512039.
62. J.D. Jackson, S.B. Treiman, H.W. Wyld, Phys. Rev. **106**, 517 (1957).
63. A. Kurylov, M.J. Ramsey-Musolf, S. Su, Phys. Rev. D **68**, 035008 (2003) [arXiv:hep-ph/0303026].
64. SLAC E158 Collaboration (P.L. Anthony *et al.*), Phys. Rev. Lett. **95**, 081601 (2005) [arXiv:hep-ex/0504049].
65. V.D. Barger, G.F. Giudice, T. Han, Phys. Rev. D **40**, 2987 (1989).
66. M.J. Ramsey-Musolf, Phys. Rev. C **60**, 015501 (1999) [arXiv:hep-ph/9903264].
67. J. Erler, A. Kurylov, M.J. Ramsey-Musolf, Phys. Rev. D **68**, 016006 (2003) [arXiv:hep-ph/0302149].
68. A. Czarnecki, W.J. Marciano, Int. J. Mod. Phys. A **15**, 2365 (2000) [arXiv:hep-ph/0003049].
69. A.D. Sakharov, Pisma Zh. Eksp. Teor. Fiz. **5**, 32 (1967) (JETP Lett. **5**, 24 (1967)).
70. Particle Data Group (S. Eidelman *et al.*), Phys. Lett. B **592**, 1 (2004).
71. WMAP Collaboration (D.N. Spergel *et al.*), Astrophys. J. Suppl. **148**, 175 (2003) [arXiv:astro-ph/0302209].
72. M. Pospelov, A. Ritz, Annals Phys. **318**, 119 (2005) [arXiv:hep-ph/0504231].
73. P. Huet, A.E. Nelson, Phys. Rev. D **53**, 4578 (1996) [arXiv:hep-ph/9506477].
74. M. Carena, M. Quiros, M. Seco, C.E.M. Wagner, Nucl. Phys. B **650**, 24 (2003) [arXiv:hep-ph/0208043].
75. C. Lee, V. Cirigliano, M.J. Ramsey-Musolf, Phys. Rev. D **71**, 075010 (2005) [arXiv:hep-ph/0412354].
76. T. Konstandin, T. Prokopec, M.G. Schmidt, M. Seco, Nucl. Phys. B **738**, 1 (2006) [arXiv:hep-ph/0505103].
77. V. Cirigliano, S. Profumo, M.J. Ramsey-Musolf, JHEP **0607**, 002 (2006) [arXiv:hep-ph/0603246].

Quark model analysis of strangeness form factors of the proton

D.O. Riska[a]

Helsinki Institute of Physics, 00014 University of Helsinki, Finland

Abstract. The present empirical information on the proton strangeness form factors suggests that the *uuds* subsystem of the $uuds\bar{s}$ configuration has the flavor spin symmetry $[4]_{FS}[22]_F[22]_S$, mixed orbital symmetry $[31]_X$ and that the $\bar{s}$ is in the ground state. This $uuds\bar{s}$ configuration yields the empirical signs for both the form factors G_E^s and G_M^s. Transition matrix elements between the *uud* and $uuds\bar{s}$ components are significant.

PACS. 12.39.-x Phenomenological quark models – 13.40.Gp Electromagnetic form factors – 14.20.Dh Protons and neutrons

1 Introduction

The present experimental situation is that the strangeness magnetic form factor $G_M^s(q^2)$ of the proton is positive [1–4], while the strangeness electric form factor $G_E^s(q^2)$ [4,5] and the strangeness axial form factor [6] are negative. The positive sign of strangeness magnetic moment indicates that long-range K meson loop fluctuations are suppressed [7]. Phenomenological K and K^* loop calculations and recent extrapolated quenched lattice calculations yield small but still negative values [8,9].

Here it is noted that the $uuds\bar{s}$ configuration with one orbitally excited quark and the $\bar{s}$ in the ground state, and which is expected to have the lowest energy, leads to the same signs as the empirical form factors.

2 Strange meson loops

A kaon loop fluctuation of the proton yields a negative value for the strangeness magnetic moment. Because the kaon is a pseudoscalar, the K-hyperon loops that involve spin flip give twice the contribution of loops without spin flip. Therefore, the loops in which the hyperon spin is opposite to that of the proton determine the sign of the loop contribution to the magnetic moment. As then the sign of both the charge and the spin of the strange quark in the hyperon is opposite to that of the proton, its net contribution to the magnetic moment is positive. When the charge $-e/3$ is factored out, the net "strangeness" magnetic moment becomes negative. Because of the spin flip, the kaon in the loop has $\ell_z = 1$, and its strange antiquark also yields a negative contribution, once the charge is factored out [10]. Positive loop contributions to the strangeness magnetic moment only arise from shorter-range loops with KK^* transition couplings, but even these are expected to be very small [11]. These observations carry over to most, if not all, model calculations of the strangeness magnetic moment [12].

If the strange meson loops are evaluated with phenomenological couplings at the hadron level, the results are unrealistically large, unless strongly cut-off at rather small momenta [13]. At the quark level, the loops contributions are smaller as a consequence of the smaller coupling strengths (*e.g.*, $\mu_s \sim -0.06\,\mu_N$ [8]).

Quenched lattice calculations have mostly given negative values for μ_s, the exception being ref. [14]. The most recent quenched lattice calculation with a chiral extrapolation to low mass gave $\mu_s = -0.046 \pm 0,019\,\mu_N$ [9].

3 qqqq$\bar{s}$ configurations

The *uuds* subsystem of the $uuds\bar{s}$ component of the proton has to be completely antisymmetric. This requires that either all the quarks are in the spatially symmetric ground state with a mixed symmetry spin and flavor configuration or that one of the quarks is in the P-state so that the spatial state has mixed symmetry, while the spin-flavor configuration is completely symmetric. In the former case the $\bar{s}$ is in the P-state and in the latter case it is in the ground state [15].

In the case of the spatially symmetric *uuds* configurations the $\bar{s}$ is in the P-state. All of these configurations, with exception of the energetically unfavored spin $S = 2$ state, lead to negative values for the strangeness magnetic moment (table 1) [16]. For the $S = 2$ state, moreover, the strangeness contribution to the proton spin Δ_s would

[a] e-mail: riska@mappi.helsinki.fi

Table 1. The flavor and spin configurations of the *uuds* states with all quarks in the ground state [17] and the corresponding operator matrix elements [16].

uuds ground state	$\Delta_s(P_{s\bar{s}})$	$\mu_s(\frac{m_p}{m_s}P_{s\bar{s}})$
$[31]_{FS}[211]_F[22]_S$	−1/3	−1/3
$[31]_{FS}[211]_F[31]_S$	85/216	−95/216
$[31]_{FS}[22]_F[31]_S$	5/12	−5/12
$[31]_{FS}[31]_F[22]_S$	−1/3	−1/3
$[31]_{FS}[31]_F[31]_S$	281/648	−259/648
$[31]_{FS}[31]_F[4]_S$	1/6	7/6

Table 2. The flavor and spin configurations of the *uuds* states in the first orbitally excited state [17] with total angular momentum $J = 1$ [16].

uuds P-state	$\Delta_s(P_{s\bar{s}})$	$\mu_s(\frac{m_p}{m_s}P_{s\bar{s}})$
$[4]_{FS}[22]_F[22]_S$	−1/3	1/2
$[4]_{FS}[31]_F[31]_S$	−2/27	73/108
$[31]_{FS}[211]_F[22]_S$	−1/3	37/72
$[31]_{FS}[211]_F[31]_S$	−23/108	707/1296
$[31]_{FS}[22]_F[31]_S$	−1/6	7/12
$[31]_{FS}[31]_F[22]_S$	−1/3	37/72
$[22]_{FS}[211]_F[31]_S$	−7/36	5/9
$[31]_{FS}[31]_F[31]_S$	−43/324	2371/3888
$[22]_{FS}[22]_F[22]_S$	−1/3	1/2
$[211]_{FS}[211]_F[22]_S$	−1/3	35/72
$[31]_{FS}[31]_F[4]_S$	1/6	497/648
$[211]_{FS}[211]_F[31]_S$	−43/324	811/1296
$[22]_{FS}[31]_F[31]_S$	−7/36	251/324
$[22]_{FS}[22]_F[4]_S$	1/6	3/4
$[211]_{FS}[22]_F[31]_S$	−1/6	7/12
$[211]_{FS}[211]_F[4]_S$	1/6	157/216
$[211]_{FS}[31]_F[22]_S$	−1/3	35/72
$[211]_{FS}[31]_F[31]_S$	−23/108	227/432

be positive, which is experimentally contraindicated. For the long-range K meson loop fluctuation the $\bar{s}$ is in the P-state, which corresponds to the configurations with the *uuds* subsystem in the ground state.

The other alternative is that the spatial configuration of the *uuds* subsystem has the mixed symmetry $[31]_X$ while the spin flavor state is completely symmetric $[4]_{FS}$. These configurations are listed in table 2. All of these configurations in contrast lead to positive values for the strangeness magnetic moment, and, with exception again for the energetically unfavored spin $S = 2$ state, at the same time lead to negative values for Δ_s [16]. These results show that a negative strangeness magnetic moment implies that the $uuds\bar{s}$ configuration is such that one of the quarks is in the P-state and the $\bar{s}$ is in the ground state. As such configurations are incompatible with the long-range K meson hyperon loop fluctuation, those have to be suppressed, and the experimental signature arises from a short-range structure. This is illustrated by the quark model calculation of the strangeness form factors below.

The *uuds* configurations in tables 1 and 2 are ordered according to increasing symmetry of the flavor and spin state. Under the assumption that the hyperfine interaction between the quarks is spin and/or isospin dependent, this ordering corresponds to increasing energy [15]. The configurations on top thus are expected to have the lowest energy and therefore to be more likely to have a significant probability in the proton.

4 The strangeness form factors

If the flavor-spin symmetry of the *uuds* system is the mixed symmetry $[4]_{FS}[22]_F[22]_S$, which is likely to have the lowest energy, the corresponding strangeness form factors in the harmonic-oscillator quark model are [18]:

$$G_E^s(q^2) = -\frac{q^2}{24\omega^2}\frac{e^{-q^2/4\omega^2}}{\sqrt{1+q^2/4m_s^2}}P_{s\bar{s}}\,, \tag{1}$$

$$G_M^s(q^2) = \frac{m_p}{2m_s}\frac{e^{-q^2/4\omega^2}}{\sqrt{1+q^2/4m_s^2}}P_{s\bar{s}}\,, \tag{2}$$

$$G_A^s(q^2) = -\frac{1}{3}e^{-q^2/4\omega^2}P_{s\bar{s}}\,. \tag{3}$$

Here, $P_{s\bar{s}}$ is the probability of the $uuds\bar{s}$ component in the proton, and m_p and m_s are the proton and strange quark masses, respectively. The parameter ω is the oscillator parameter for the $uuds\bar{s}$ system.

Besides these "diagonal" matrix elements between the $uuds\bar{s}$ states, "non-diagonal" matrix elements between the *uud* and $uuds\bar{s}$ components of the proton also have to be taken into account. These depend both on the explicit wave function model and on the model for the $s\bar{s}-\gamma$ vertices. If these vertices are taken to have the forms $\bar{v}(p')\gamma_\mu u(p)$ and $\bar{v}(p')\gamma_\mu\gamma_5 u(p)$, and no account is taken of the interaction between the annihilating $s\bar{s}$ pair and the proton, these transition matrix elements lead to the following form factor contributions:

$$G_E^s(q^2) = -\delta\, C_{35}\frac{\sqrt{3}}{6}\frac{q^2}{m_s\omega}\frac{e^{-q^2/4\omega^2}}{\sqrt{1+q^2/4m_s^2}}\sqrt{P_{s\bar{s}}P_{uud}}\,, \tag{4}$$

$$G_M^s(q^2) = \delta\, C_{35}\frac{2\sqrt{3}}{3}\frac{m_p}{\omega}\frac{e^{-q^2/4\omega^2}}{\sqrt{1+q^2/4m_s^2}}\sqrt{P_{s\bar{s}}P_{uud}}\,, \tag{5}$$

$$G_A^s(q^2) = \delta\, C_{35}\frac{\sqrt{3}}{6}\frac{q^2}{m_s\omega}e^{-q^2/4\omega^2}\sqrt{P_{s\bar{s}}P_{uud}}\,. \tag{6}$$

Here, P_{uud} is the probability of the *uud* component of the proton. The factor C_{35} is the overlap integral of the wave function of the *uud* and the corresponding component of

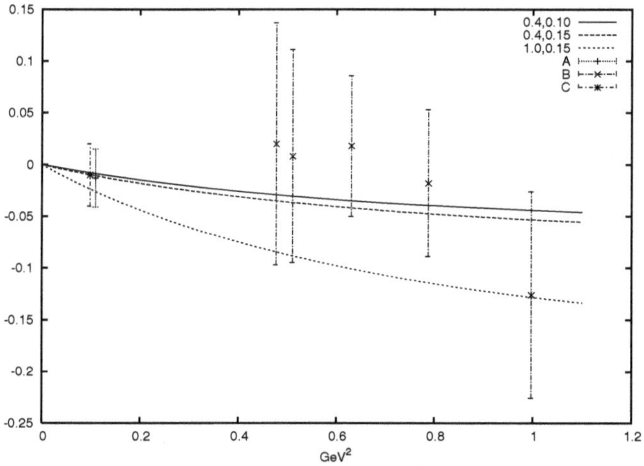

Fig. 1. The strangeness electric form factor for $C_{35} = 0.4$ and $C_{35} = 1.0$ (first number in the curves). The second value is the value of $P_{s\bar{s}}$. The data points are from [4,19] ("A"), [20] ("B") and [5] ("C") [18].

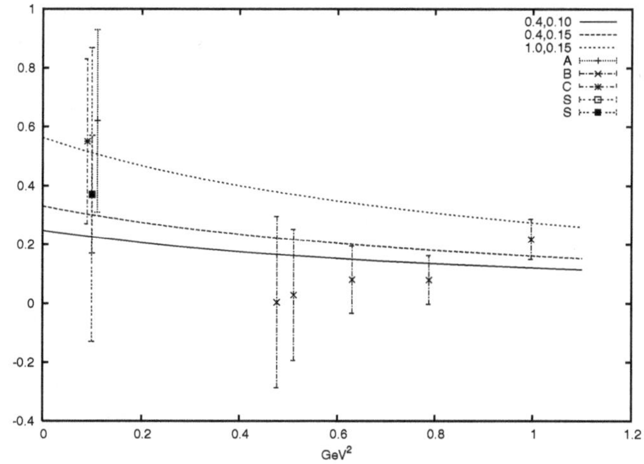

Fig. 2. The strangeness magnetic form factor for $C_{35} = 0.4$ and $C_{35} = 1.0$ (notation as in fig. 1). The data points are from [1] ("S"), [4,19] "A", [20] ("B") and [5] ("C").

the $uuds\bar{s}$ configuration. In the oscillator model, this factor is

$$C_{35} = \left(\frac{2\omega\omega_3}{\omega^2 + \omega_3^2}\right)^{9/2}. \tag{7}$$

Here, ω_3 is the oscillator constant for the uud component of the proton. In the case of compact $uuds\bar{s}$ wave function, for which $\omega \sim 2\omega_3$, the value for C_{35} is $C_{35} \sim 0.4$. The model parameters are the oscillator parameter ω, the probability $P_{s\bar{s}}$ of the $uuds\bar{s}$ component (here, $P_{uud} = 1 - P_{s\bar{s}}$) and the phase factor δ in the non-diagonal contribution. The constituent mass of the strange quark will be taken to be $400\,\mathrm{MeV}/c^2$. A good description of the empirical form factors is obtained with $\delta = +1$.

The empirical values for the strangeness form factors given in refs. [4,19,20] indicate that they all fall slowly with momentum transfer up to $q^2 = 1\,\mathrm{GeV}^2$. This slow falloff indicates that the wave function of the strangeness component is compact relative to the proton radius. This may be described by taking ω as large as 1 GeV, which corresponds to a matter radius function radius of $1/\omega \simeq 0.2\,\mathrm{fm}$. The calculated strangeness electric form factor is shown in fig. 1. A much smaller value for ω, the non-diagonal contribution (4), would give rise to a too large value of the strangeness radius. These results were obtained with the overlap factor C_{35} (7) taken to be 0.4 and 1.0, respectively. The values of the probability $P_{s\bar{s}}$ were taken to be 0.1 and 0.15 as indicated in the curves. The calculated strangeness radius is positive, as the s quark is preferentially in the P-state and the $\bar{s}$ is in the S-state. Therefore, the charge distribution of the strange component is positive at short distance and negative at longer distances.

The calculated values for G_M^s obtained with the same parameter values are shown in fig. 2. The best description of the data is obtained by taking the probability of the $uuds\bar{s}$ component to be $P_{s\bar{s}}$ in the range 10–15% and the value of the phase factor δ in the non-diagonal contribution (4) to be positive ($\delta = +1$). Here again the slow falloff with q^2 is notable.

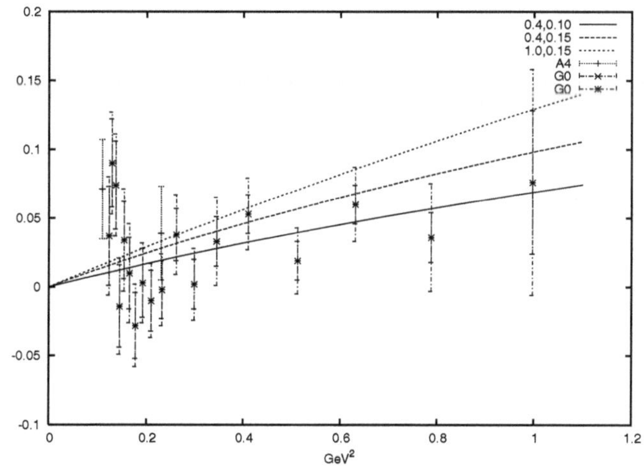

Fig. 3. The strangeness form factor combination $G_E^s + \eta G_M^s$ for 3 values of $P_{s\bar{s}}$ for $C_{35} = 0.4$ and $C_{35} = 1.0$ (notation as in fig. 1) The data points are from [3,21] ("A4") and [4] ("G0").

The calculated combination $G_E^s(q^2) + \eta G_M^s(q^2)$ of form factors with this parameterization is compared to the results of the A4 [3,21] and G0 experiments [4] in fig. 3. In this case, the overall features of the empirical values are best reproduced with $C_{35} = 0.4$ and $P_{s\bar{s}} = 0.10$.

5 Discussion

This quark model analysis of the strangeness form factors shows that the empirical signs and momentum dependence can be understood if $\bar{s}$ remains in the ground state and the $uuds$ system is in the P-state. This configuration is opposite to those which correspond to the $K\Lambda$ loop fluctuations.

The $uuds\bar{s}$ configuration, for which the $uuds$ subsystem has the mixed flavor-spin symmetry $[4]_{FS}[22]_F[22]_S$, is likely to have a very low energy, if the hyperfine interaction between the quarks is spin and flavor dependent, and thus is the most likely configuration to form a notable component of the proton. This configuration leads both to a positive value for μ_s and to negative values for G_E^s. These features correspond rather well to the empirical situation of the world data in 2005, with $G_E^s(0) = -0.12 \pm 0.29$, $G_M^s(0) = 0.62 \pm 0.32$ and also to the world data of 2006 with $G_E^s(0) = -0.006 \pm 0.016$, $G_M^s(0) = 0.28 \pm 0.20$ [22,23]. More definite conclusions have to await full convergence of the different experiments.

Informative correspondence and discussion with C.S. An and B.S. Zou is gratefully acknowledged.

References

1. D.T. Spayde *et al.*, Phys. Lett. B **583**, 79 (2004).
2. K.A. Aniol *et al.*, Phys. Rev. C **69**, 065501 (2004).
3. F.E. Maas *et al.*, Phys. Rev. Lett. **94**, 152001 (2005).
4. D.S. Armstrong *et al.*, Phys. Rev. Lett. **95**, 092001 (2005).
5. K.A. Aniol *et al.*, Phys. Lett. B **635**, 275 (2006).
6. S.F. Pate, Phys. Rev. Lett. **92**, 082002 (2004).
7. M.J. Musolf, M. Burkardt, Z. Phys. C **61**, 433 (1994).
8. L. Hannelius, D.O. Riska, Phys. Rev. C **62**, 045204 (2000).
9. D.B. Leinweber *et al.*, Phys. Rev. Lett. **94**, 212001 (2005).
10. L. Hannelius, D.O. Riska, L.Ya. Glozman, Nucl. Phys. A **665**, 353 (2000).
11. P. Geiger, N. Isgur, Phys. Rev. **55**, 299 (1997).
12. D.H. Beck, R.D. McKeown, Annu. Rev. Nucl. Part. Sci. **51**, 189 (2001).
13. H. Forkel, M. Nielsen, F. Navarra, Phys. Rev. C **61**, 055206 (2000).
14. R. Lewis, W. Wilcox, R.M. Woloshyn, Phys. Rev. D **67**, 013003 (2003).
15. B.S. Zou, D.O. Riska, Phys. Rev. Lett. **95**, 072001 (2005).
16. C.S. An, B.S. Zou, D.O. Riska, Phys. Rev. C **73**, 035207 (2006).
17. C. Helminen, D.O. Riska, Nucl. Phys. A **699**, 624 (2002).
18. D.O. Riska, B.S. Zou, Phys. Lett. B **636**, 265 (2006).
19. D.H. Beck, *Particles and Nuclei International Conference, PANIC 2005*, `www.panic05.lanl.gov`.
20. S.F. Pate, G. MacLachlan, D. McKee, V. Papavassilou, hep-ex/0512032.
21. F.E. Maas *et al.*, Phys. Rev. Lett. **93**, 022002 (2004).
22. D.S. Armstrong, K. Carter, CERN Cour., **45**, 8 (2005).
23. D.S. Armstrong, this conference; CERN Cour. **46**, 5 (2006).

Strange nucleon form factors: Solitonic approach to G_M^s, G_E^s, $\tilde{G}_A^p$ and $\tilde{G}_A^n$ and comparison with world data

K. Goeke[1,a], H.-C. Kim[2], A. Silva[3,4], and D. Urbano[3,4]

[1] Institut für Theoretische Physik II, Ruhr-Universität Bochum, D-44780 Bochum, Germany
[2] Department of Physics and Nuclear Physics and Radiation Technology Institute (NuRI), Pusan National University, 609-735 Busan, Republic of Korea
[3] Departamento de Física and Centro de Física Computacional, Universidade de Coimbra, P-3000 Coimbra, Portugal
[4] Faculdade de Engenharia da Universidade do Porto, R. Dr. Roberto Frias s/n, P-4200-465 Porto, Portugal

Paper reprinted from Eur. Phys. J. A with permission

Abstract. We summarize the results of the chiral quark-soliton model (χQSM) concerning basically all form factors necessary to interpret the present data of the parity-violating electron scattering experiments SAMPLE, HAPPEX, A4 and G0. The results particularly focus on the recently measured asymmetries and the detailed data for various combinations of G_M^s, G_E^s, $\tilde{G}_A^p$ and $\tilde{G}_A^n$ at $Q^2 = 0.1\,\text{GeV}^2$. The calculations yield positive strange magnetic and electric form factors and a negative axial vector one, all being rather small. The results are very close to the combined experimental world data from parity-violating electron scattering and elastic νp and $\bar{\nu} p$ scattering.

PACS. 12.40.-y Other models for strong interactions – 14.20.Dh Protons and neutrons

1 Introduction

The strange quark contribution to the distributions of charge and magnetization in the nucleon has been a very important issue well over decades, since it provides a vital clue in understanding the structure of the nucleon and in particular in probing the quark sea. There are some indications of about 4% contribution to the momentum sum rule of deep inelastic lepton scattering, of roughly 15% to the spin of the nucleon extracted from polarized deep inelastic scattering, or of up to 30% contribution to the mass of the nucleon, where all these numbers show rather large uncertainties. Recently, the strangeness content of the nucleon has been studied particularly intensively since parity-violating electron scattering (PVES) has demonstrated to provide an essential tool for probing the sea of $s\bar{s}$ pairs in the vector channel [1]. In fact, various PVES experiments have been already conducted from which the strange vector form factors can be extracted [2–12]. The results from the SAMPLE, HAPPEX, PVA4, and G0 Collaborations have shown evidence for a non-vanishing strange quark contribution to the structure of the nucleon. In particular, evidence was found that the strange magnetic moment of the proton is positive [11], suggesting that the strange quarks reduce the proton's magnetic moment. This is an unexpected and surprising finding, since a majority of theoretical studies favors a negative value. One of the models, which yield a positive strange magnetic moment of the proton, is the chiral quark soliton model (χQSM). It will be used in the present paper to investigate the form factors G_M^s, G_E^s, $\tilde{G}_A^p$ and $\tilde{G}_A^n$ and to compare them with world data.

Using the χQSM the present authors have recently investigated the set of six electromagnetic form factors ($G_{E,M}^{u,d,s}$) and three axial-vector ones ($G_A^{u,d,s}$) [13–16]. The results show a good agreement with the data of the SAMPLE, HAPPEX, A4 and G0 experiments. This includes parity-violating asymmetries (PVA) which have been measured by the G0 experiment over a range of momentum transfers in the forward direction [12]. We even predicted the PVAs of the future G0 experiment at backward angles [17]. In the present contribution we perform more detailed comparison including the most recent data of the HAPPEX experiment on He-4 and the results of the PVAs combined with elastic νp and $\bar{\nu} p$ scattering.

2 Chiral quark soliton model

The χQSM has been used several times to calculate strange properties of the nucleon and of hyperons. It is

[a] e-mail: Klaus.Goeke@tp2.ruhr-uni-bochum.de

an effective relativistic quark theory based on the instanton degrees of freedom of the QCD vacuum and has been derived from QCD in the large-N_c limit. In the end it turns out to be the simplest possible quark theory which allows for spontaneously broken chiral symmetry. It results in an effective chiral action for valence and sea quarks both moving in a static self-consistent Goldstone background field [18–20]. For this model it is absolutely natural to have strange quark contributions to the nucleon. The χQSM has very successfully been applied to mass splittings of hyperons, to electromagnetic and axial-vector form factors [18] of the baryon octet and decuplet and to forward and generalized parton distributions of the nucleon. With one set of four parameters, unchanged for years, it reproduces all appropriate observables of light baryons with an accuracy of (10–30)%. This parameter set consists of an effective current mass for up- and down-quarks, a cut-off parameter in the relativistic proper-time regularization scheme, and a quark-pion coupling constant corresponding to a constituent mass. These parameters are fitted to the pion decay constant, the pion mass, and baryonic properties as proton charge radius and delta-nucleon mass splitting. In addition we assume an effective current strange quark mass of 180 MeV. A numerical iteration procedure yields then the self-consistent mean field whose lowest states get occupied until baryon number $B = 1$ is reached. The resulting solitonic state is semiclassically rotated in space and iso-space in order to project on proper spin and hypercharge quantum numbers.

As far as strange form factors are concerned the formalism used in the present investigation can be found in the paper of Silva *et al.* [17] and references therein. We just mention for clarity some relations between form factors, which are often differently denoted in the literature. We refer to the papers of Musolf *et al.* [21], of Alberico *et al.* [22] and of Maas *et al.* [7]. Altogether we have:

$$\tilde{G}_A^p = -(1+R_A^1)G_A^{(3)}(Q^2) + R_A^0 + G_A^s, \quad (1)$$

$$\tilde{G}_A^n = -3(1+R_A^1)G_A^{(3)}(Q^2) + R_A^0 + G_A^s, \quad (2)$$

$$\tilde{G}_A^p = G_A^e = -\frac{1}{2}G_A^{NC} \quad (3)$$

with the values for the electro-weak radiative corrections [21]:

$$R_A^1 = -0.41 \pm 0.24, \qquad R_A^0 = 0.06 \pm 0.14. \quad (4)$$

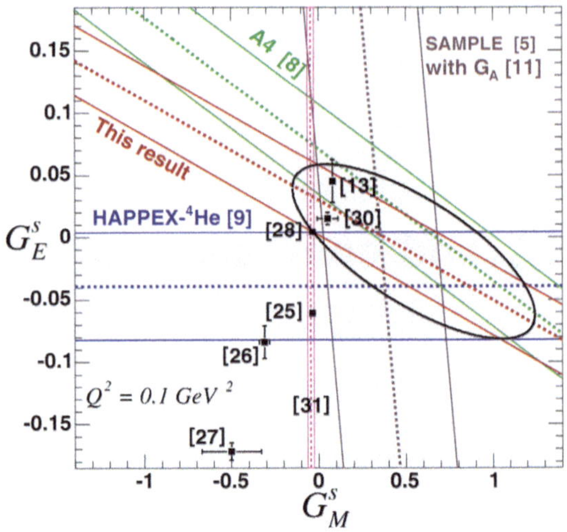

Fig. 1. The world data on the strange form factors G_M^s and G_E^s at $Q^2 = 0.1\,\text{GeV}^2$ including the HAPPEX data on He-4 of 2004. The figure is taken from ref. [11]. The numbers indicate the references of theoretical calculations. The χQSM is given by [13].

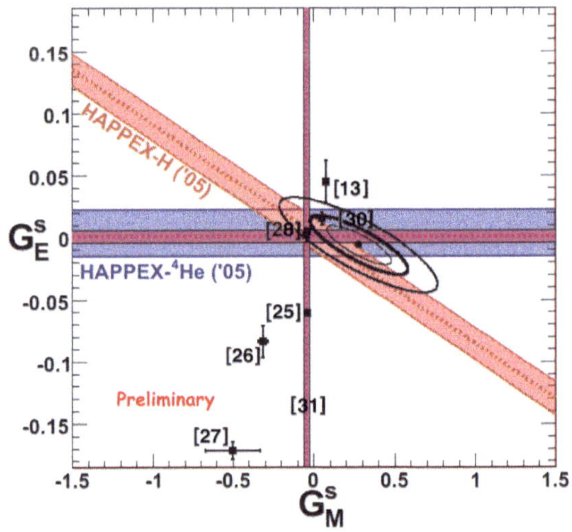

Fig. 2. The world data on the strange form factors G_M^s and G_E^s at $Q^2 = 0.1\,\text{GeV}^2$ including the HAPPEX data on He-4 of 2004 and of 2005 (preliminary). The numbers indicate the references of theoretical calculations. The figure is taken from ref. [23]. The χQSM is given by [13].

3 Electron and neutrino scattering data

The experimental situation is by far the best at $Q^2 = 0.1\,\text{GeV}^2$, where in addition to the usual linear combinations of electric and magnetic form factors the measurements of parity violation on He-4 allowed an extraction of G_E^s. The experimental results of the HAPPEX Collaboration are $G_E^s = -0.038 \pm 0.042 \pm 0.010$ measured at $Q^2 = 0.091\,\text{GeV}^2$ [10] and, more recently, $G_E^s = -0.002 \pm 0.017$ at $Q^2 = 0.1\,\text{GeV}^2$ [23].

Also the combined data $G_E^s = -0.006 \pm 0.016$ [23] are consistent with zero. Experimental evidence from the SAMPLE and HAPPEX Collaborations gives a positive value of the strange magnetic form factor G_M^s at $Q^2 = 0.1\,\text{GeV}^2$ of $G_M^s = 0.37 \pm 0.20 \pm 0.26 \pm 0.07$ [24], $G_M^s = 0.55 \pm 0.28$ [11] and $G_M^s = 0.12 \pm 0.24$ [23], respectively. The overall comparison of the χQSM calculation [13] with the world data and with other model calculations [25–29], and lattice gauge calculations [30,31] is given in fig. 1. If one adds the preliminary 2005 data of the HAPPEX-He-4 experiment the regions of confidence get smaller and one obtains fig. 2. Only those theoretical calculations are selected which are somehow in the vicinity of

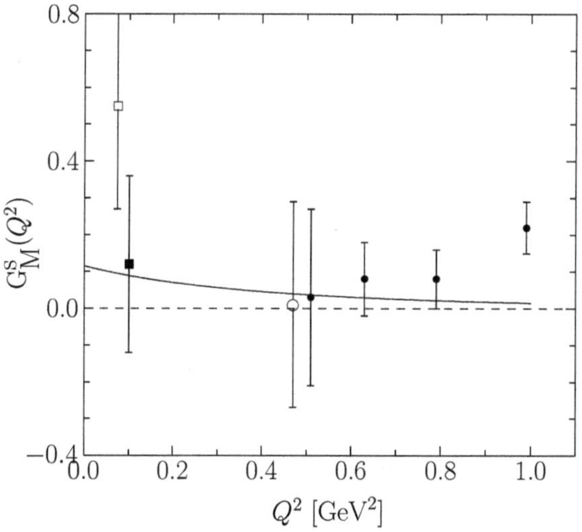

Fig. 3. The strange magnetic form factor $G^s_M(Q^2)$ of the nucleon: The χQSM is compared with the analysis of Pate *et al.* [36] involving simultaneously parity-violating *ep* data and data from ν and $\bar{\nu}$ scattering. The open circle is from a combination of HAPPEx and E734 data, while the closed circles are from a combinaton of G0 and E734 data. The open square is from ref. [11] and involve parity-violating *ep* data only, similarly as the closed square from refs. [23] taken from [29].

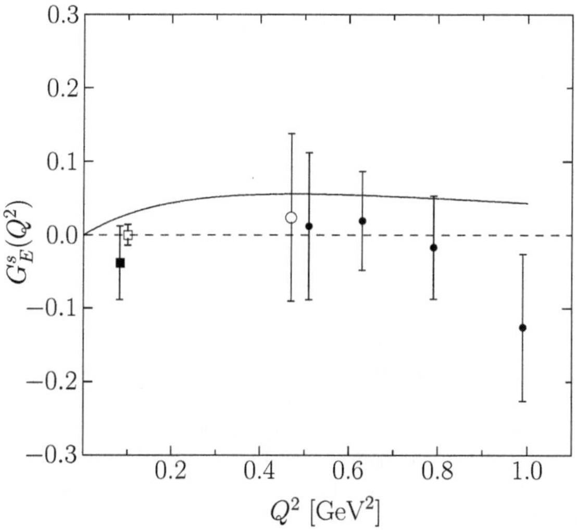

Fig. 4. The strange electric form factor $G^s_E(Q^2)$ of the nucleon: The χQSM is compared with the analysis of Pate *et al.* [36]. The open circle is from a combination of HAPPEX and E734 data, while the closed circles are from a combinaton of G0 and E734 data. The open square is from ref. [10] and involve parity-violating *ep* data only, similarly as the closed square from refs. [23] taken from [29].

the experimental data. Actually they do not give a consistent picture. This is also true for lattice-QCD calculations (LQCD). For example those of Lewis *et al.* [30] advocate a positive magnetic strange moment, whereas the recent results of Leinweber *et al.* [31–33], indicate a negative one.

It is interesting to combine the data from PV electron scattering with the data from elastic νp and $\bar{\nu} p$ scattering off protons [34], which provides independent information

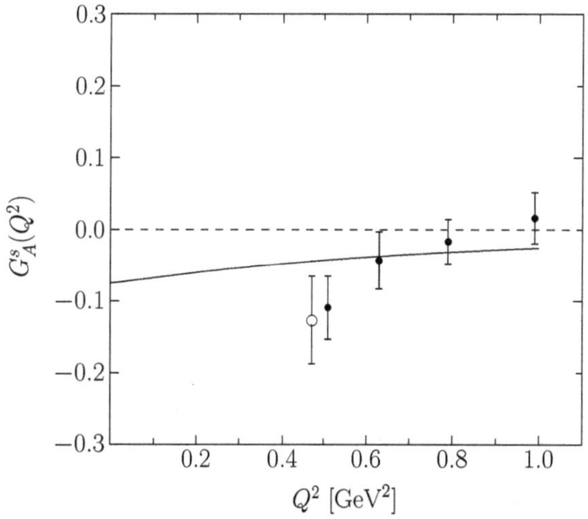

Fig. 5. The strange axial vector form factor $G^s_A(Q^2)$ of the nucleon: The χQSM is compared with the analysis of Pate *et al.* [36]. The open circle is from a combination of HAPPEX and E734 data, while the closed circles are from a combination of G0 and E734 data.

on the strange form factors [35]. A comparison with the results of such an extraction [36] can be seen in figs. 3, 4 and 5. Apparently the χQSM is compatible with more or less all data available up to $Q^2 = 1\,\text{GeV}^2$, which is the range where the χQSM can provide form factors. It is not excluded that the experiments favor a negative $G^s_E(Q^2)$ whereas the χQSM yields a positive one.

It is also interesting to compare the present calculations with the analysis of Young *et al.* in ref. [37]. These authors use systematic expansions of all the unknown form factors to simultaneously analyze the current data sets and extract the values at $Q^2 = 1\,\text{GeV}^2$, indepen-

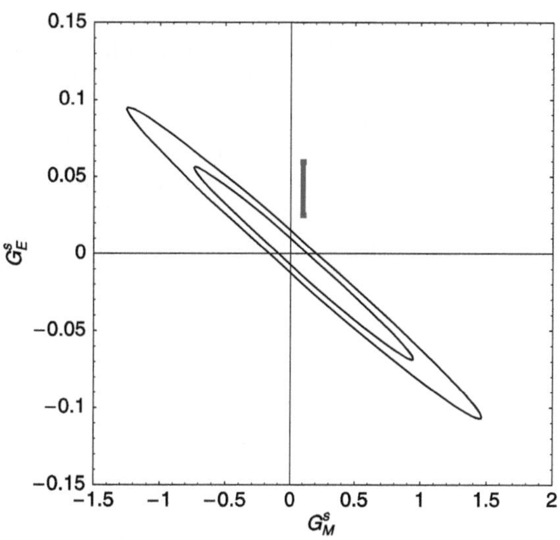

Fig. 6. The contours display the 68% and 95% confidence intervals for the joint determination of G^s_M and G^s_E at $Q^2 = 0.1\,\text{GeV}^2$. The result of the χQSM is indicated. The ellipses originate from a theory-independent combined fit to all parity-violating data by Young *et al.* [37].

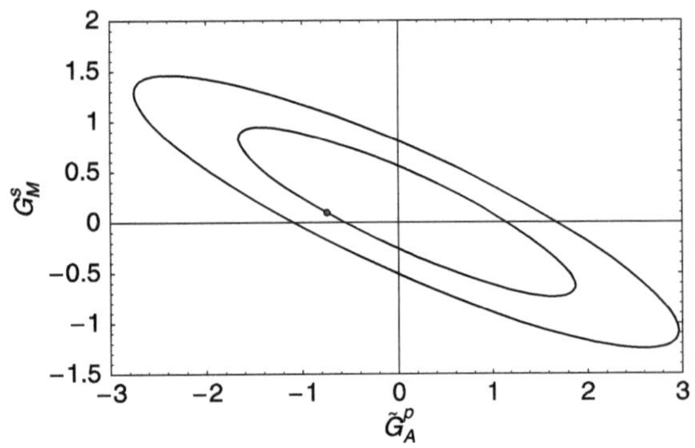

Fig. 7. The contours display the 68% and 95% confidence intervals for the joint determination of G_M^s and $\tilde{G}_A^p$ at $Q^2 = 0.1\,\mathrm{GeV}^2$. The result of the χQSM is indicated. The ellipses originate from a theory-independent combined fit to all parity-violating data by Young *et al.* [37].

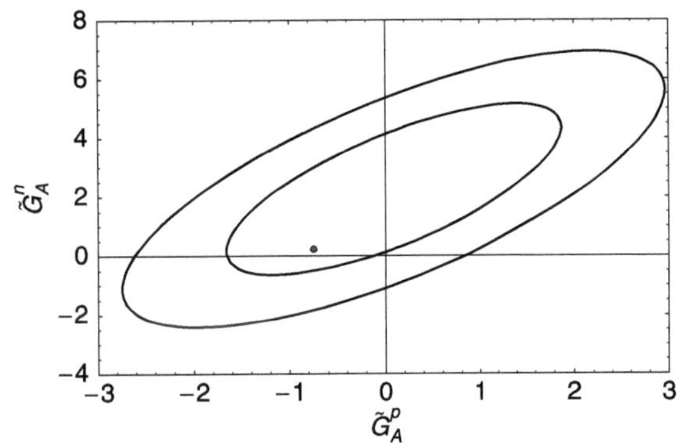

Fig. 8. The contours display the 68% and 95% confidence intervals for the joint determination of $\tilde{G}_A^n$ and $\tilde{G}_A^p$ at $Q^2 = 0.1\,\mathrm{GeV}^2$. The result of the χQSM is indicated. The ellipses originate from a theory-independent combined fit to all parity-violating data by Young *et al.* [37].

dent of theoretical input, except assuming the constraint of charge symmetry. Figure 6 shows this analysis for G_M^s and G_E^s, Figure 7 for G_M^s and $\tilde{G}_A^p$ and Figure 8 for $\tilde{G}_A^n$ and $\tilde{G}_A^p$. The error bar of the χQSM-result in fig. 6 is caused by a systematic error of the model in case of a purely strange observable. It originates from the inability of the χQSM to describe simultaneously mesonic tails with different Yukawa masses [13]. For quantities, which are not purely strange, this systematic error is usually negligible.

4 Summary

In the present theoretical work, we have investigated various form factors which are relevant for the analysis of parity-violating electron scattering experiments SAMPLE, HAPPEX, A4 and G0 and the scattering of ν and $\bar{\nu}$ scattering off nucleons. These form factors are G_M^s, G_E^s, $\tilde{G}_A^n$ and $\tilde{G}_A^p$. We used for the study the electromagnetic and strange vector and axial vector form factors calculated in the chiral quark soliton model, yielding both small but positive magnetic and electric strange form factors, see refs. [13–16]. All these χQSM form factors were obtained with one fixed set of four model parameters, which has been adjusted several years ago to basic mesonic and baryonic observables. As seen already in a previous paper of the present authors [17] the parity-violating asymmetries obtained in the present work are in a good agreement with the experimental data, which implies that the present model produces reasonable form factors of many different quantum numbers. We also predicted in that paper the parity-violating asymmetries for the future G0 experiment at backward angles. In the present paper we demonstrated that our theoretical numbers reproduce also form factors from a combined analysis of parity-violating electron scattering and ν and $\bar{\nu}$ scattering. Altogether, comparing the results of the χQSM with the overall world data one observes a remarkable agreement. It seems that the chiral quark soliton model, which has been applied over several years with an accuracy of (10–30)% to many observables of baryons in the octet and decuplet, is formulated in terms of proper effective degrees of freedom. The model is in fact the simplest quark model which describes spontaneous breaking of chiral symmetry. It is based on the $N \to \infty$ expansion of the QCD and appears to describe the properties of the light baryons reasonably well. Perhaps one can learn from the present comparison of the χQSM with experiment that the degree of freedom of spontaneous breaking of chiral symmetry governs not only the up and down sector of the nucleon but also its strange quark content.

The authors are grateful to Frank Maas for useful comments and discussions. AS acknowledges partial financial support from Portugese Praxis XXI/BD/15681/98. The work has also been supported by the Korean-German grant of the Deutsche Forschungsgemeinschaft and KOSEF (F01-2004-000-00102-0). The work is partially supported by the Transregio-Sonderforschungsbereich Bonn-Bochum-Giessen, by the Verbundforschung of the Federal Ministry for Education and Research, by the COSY-Jülich project, and by the EU integrated infrastructure initiative "Hadron Physics Project" under contract No. RII3-CT-2004-506078. The work of HCK is also supported by Korea Research Foundation (Grant No. KRF-2003-070-C00015).

References

1. D.B. Kaplan, A. Manohar, Nucl. Phys. B **310**, 527 (1988).
2. SAMPLE Collaboration (B. Mueller *et al.*), Phys. Rev. Lett. **78**, 3824 (1997).
3. SAMPLE Collaboration (D.T. Spayde *et al.*), Phys. Rev. Lett. **84**, 1106 (2000); Phys. Lett. B **583**, 79 (2004).
4. SAMPLE Collaboration (R. Hasty *et al.*), Science **290**, 2117 (2000).
5. HAPPEX Collaboration (K.A. Aniol *et al.*), Phys. Lett. B **509**, 211 (2001).
6. F.E. Maas *et al.*, Phys. Rev. Lett. **94**, 152001 (2005).
7. A4 Collaboration (F.E. Maas *et al.*), Phys. Rev. Lett. **93**, 022002 (2004).
8. A4 Collaboration (F.E. Maas), Eur. Phys. J. A **17**, 339 (2003).

9. HAPPEX Collaboration (K.A. Aniol *et al.*), Phys. Rev. C **69**, 065501 (2004).
10. HAPPEX Collaboration (K.A. Aniol *et al.*), Phys. Rev. Lett. 96, 022003 (2006).
11. HAPPEX Collaboration (K.A. Aniol *et al.*), Phys. Lett. B **635**, 275 (2006).
12. G0 Collaboration (D.S. Armstrong *et al.*), Phys. Rev. Lett. **95**, 092001 (2005).
13. A. Silva, H.C. Kim, K. Goeke, Phys. Rev. D **65**, 014016 (2002); **66**, 039902 (2002)(E).
14. A. Silva, H.-Ch. Kim, K. Goeke, Eur. Phys. J. A **22**, 481 (2004).
15. A. Silva, PhD Thesis, Ruhr-Universität Bochum, unpublished (2004).
16. A. Silva, H.-Ch. Kim, D. Urbano, K. Goeke, Phys. Rev. D **72**, 094011 (2005).
17. A. Silva, H.Ch. Kim, D. Urbano, K. Goeke, Phys. Rev. D **74**, 054011 (2006).
18. C.V. Christov *et al.*, Prog. Part. Nucl. Phys. **37**, 91 (1996).
19. R. Alkofer, H. Reinhardt, H. Weigel, Phys. Rep. **265**, 139 (1996).
20. M. Wakamatsu, H. Yoshiki, Nucl. Phys. A **524**, 561 (1991).
21. M.J. Musolf, T.W. Donnelly, J. Dubach, S.J. Pollock, S. Kowalski, E.J. Beise, Phys. Rep. **239**, 1 (1994).
22. W.M. Alberico, S.M. Bilenky, C. Maieron, Phys. Rep. **358**, 227 (2002).
23. HAPPEX Collaboration (K. Paschke, D.S. Armstrong, K. De Jager), talk by D.S. Armstrong, this conference, `http://kalypso.iasa.gr/pavi06/`.
24. SAMPLE Collaboration (D.T. Spayde *et al.*), Phys. Lett. B **583**, 79 (2004).
25. N.W. Park, H. Weigel, Nucl. Phys. A **541**, 453 (1992).
26. H.W. Hammer, U.G. Meissner, D. Drechsel, Phys. Lett. B **367**, 323 (1996).
27. H.W. Hammer, M.J. Ramsey-Musolf, Phys. Rev. C **60**, 045204 (1999); **62**, 049902 (2000)(E).
28. V.E. Lyubovitskij, P. Wang, T. Gutsche, A. Faessler, Phys. Rev. C **66**, 055204 (2002).
29. R. Bijker, arXiv:nucl-th/0607058.
30. R. Lewis, W. Wilcox, R.M. Woloshyn, Phys. Rev. D **67**, 013003 (2003).
31. D.B. Leinweber *et al.*, Phys. Rev. Lett. **97**, 022001 (2006).
32. D.B. Leinweber *et al.*, Phys. Rev. Lett. **94**, 212001 (2005).
33. D.B. Leinweber, S. Boinepalli, A.W. Thomas, A.G. Williams, R.D. Young, J.B. Zhang, J.M. Zanotti, Eur. Phys. J. A **24**, S2, 79 (2005).
34. L.A. Ahrens *et al.*, Phys. Rev. D **35**, 785 (1987).
35. S.F. Pate, Phys. Rev. Lett. **92**, 082002 (2004).
36. S.F. Pate, G.A. MacLachlan, D.W. McKee, V. Papavassiliou, arXiv:hep-ex/0512032.
37. R.D. Young, J. Roche, R.D. Carlini, A.W. Thomas, arXiv:nucl-ex/0604010.

Implications of the JLab proton polarization data for the behavior of strange nucleon form factors

S. Dubnička[1] and A.Z. Dubničková[2,a]

[1] Institute of Physics, Slovak Academy of Sciences, Bratislava, Slovak Republic
[2] Department of Theoretical Physics, Comenius University, Bratislava, Slovak Republic

Paper reprinted from Eur. Phys. J. A with permission

Abstract. A special eight-resonance unitary and analytic model of nucleon electromagnetic structure is used to analyze first the classical proton form factor data obtained by the Rosenbluth technique, and then also the contradictory JLab proton polarization data on the ratio $\mu_p G_{Ep}(Q^2)/G_{Mp}(Q^2)$, with the aim to investigate the implications of the latter for the behavior of strange nucleon form factors.

PACS. 12.40.Vv Vector-meson dominance – 13.40.Gp Electromagnetic form factors – 14.20.Dh Protons and neutrons

Owing to the quark structure of the proton and neutron, one does not know the explicit form of the nucleon matrix element of the electromagnetic (EM) current $J_\mu^{EM} = 2/3\bar{u}\gamma_\mu u - 1/3\bar{d}\gamma_\mu d - 1/3\bar{s}\gamma_\mu s$, and as a result, two independent scalar functions, called nucleon EM form factors (FFs), of the squared four-momentum $t = -Q^2$ transferred by the exchanged virtual photon are introduced for the proton and neutron, respectively. Very natural is an introduction of Dirac $F_{1N}(t)$ and Pauli $F_{2N}(t)$ FFs:

$$\langle N|J_\mu^{EM}|N\rangle = \bar{u}(p')\left\{\gamma_\mu F_{1N}(t) + i\frac{\sigma_{\mu\nu}q_\nu}{2m_N}F_{2N}(t)\right\}u(p). \tag{1}$$

The most suitable form factors for extracting experimental information are Sachs electric $G_{EN}(t)$ and magnetic $G_{MN}(t)$ FFs:

$$G_{EN}(t) = F_{1N}(t) + \frac{t}{4m_N^2}F_{2N}(t),$$
$$G_{MN}(t) = F_{1N}(t) + F_{2N}(t) \tag{2}$$

giving in the Breit frame the charge and magnetization distributions within the nucleon, respectively. However, for a construction of various models of nucleon EM structure, the isoscalar and isovector Dirac and Pauli FFs, to be defined by the matrix elements

$$\langle p'|J_\mu^{I=0}|p\rangle = \bar{u}(p')\left[\gamma_\mu F_1^{I=0}(t) + i\frac{\sigma_{\mu\nu}q^\nu}{2m_N}F_2^{I=0}(t)\right]u(p) \tag{3}$$

of the isoscalar EM current $J_\mu^{I=0} = 1/6(\bar{u}\gamma_\mu u + \bar{d}\gamma_\mu d) - 1/3\bar{s}\gamma_\mu s$ and

$$\langle p'|J_\mu^{I=1}|p\rangle = \bar{u}(p')\left[\gamma_\mu F_1^{I=1}(t) + i\frac{\sigma_{\mu\nu}q^\nu}{2m_N}F_2^{I=1}(t)\right]u(p) \tag{4}$$

of the isovector EM current $J_\mu^{I=1} = 1/2(\bar{u}\gamma_\mu u - \bar{d}\gamma_\mu d)$, are the most appropriate.

One of the important tasks of modern hadronic physics is to clarify the role of hidden flavors in the structure of the nucleon. The contribution of the strange quarks is of special interest as their mass is within the range of the mass scale of QCD ($m_s \approx \Lambda_{QCD}$), so the dynamic creation of sea strange-quark-antiquark pairs should dominate in comparison with heavier c, b and t quark-antiquark pair creation.

The momentum dependence of the nucleon matrix element of the strange-quark vector current $J_\mu^s = \bar{s}\gamma_\mu s$ is contained in the Dirac $F_1^s(t)$ and Pauli $F_2^s(t)$ strange nucleon FFs:

$$\langle p'|\bar{s}\gamma_\mu s|p\rangle = \bar{u}(p')\left[\gamma_\mu F_1^s(t) + i\frac{\sigma_{\mu\nu}q^\nu}{2m_N}F_2^s(t)\right]u(p) \tag{5}$$

or in the strange electric $G_E^s(t)$ and strange magnetic $G_M^s(t)$ FFs:

$$G_E^s(t) = F_1^s(t) + \frac{t}{4m_N^2}F_2^s(t), \qquad G_M^s(t) = F_1^s(t) + F_2^s(t), \tag{6}$$

which, as a consequence of the isospin zero value of the strange quark, contribute only to the behavior of the isoscalar nucleon FFs and not to the isovector ones.

[a] e-mail: `dubnickova@fmph.uniba.sk`

Recent measurements [1–3] of recoil polarization in elastic scattering of polarized electrons on unpolarized protons at JLab have been used to extract data on the ratio $\mu_p G_{Ep}(Q^2)/G_{Mp}(Q^2)$ for $0.49\,\mathrm{GeV}^2 \leq Q^2 \leq 5.54\,\mathrm{GeV}^2$, which disagree with Rosenbluth extractions from cross-section measurements. Taking into account the dominance of $G_{Mp}(t)$ in the unpolarized cross-section, we believe that the behavior of $G_{Ep}(t)$ is responsible for this discrepancy. As a result, there are two sets of experimental data on nucleon EM FFs, differing from each other by the different spacelike behaviour of $G_{Ep}(t)$. In the following, we will predict from them the behavior of strange nucleon FFs by the specific eight-resonance unitary and analytic model, and look for distinctive features.

The main idea [4] of the prediction of the strange-nucleon FFs behavior is based on two assumptions:

i. that ω-ϕ mixing is valid also for the coupling constants between EM (quark) current and vector-mesons:

$$\frac{1}{f_\omega} = \frac{1}{f_{\omega_0}}\cos\epsilon - \frac{1}{f_{\phi_0}}\sin\epsilon, \qquad \frac{1}{f_\phi} = \frac{1}{f_{\omega_0}}\sin\epsilon + \frac{1}{f_{\phi_0}}\cos\epsilon; \tag{7}$$

ii. that the quark current of some flavor couples with universal strength κ exclusively to the vector-meson wave function component of the same flavor:

$$\langle 0|\bar{q}_r\gamma q_r|(\bar{q}_t q_t)_V\rangle = \kappa m_V^2 \delta_{rt}\varepsilon_\mu\,, \tag{8}$$

which result in the relations

$$(f^{(i)}_{\omega NN}/f^s_\omega) = -\sqrt{6}\frac{\sin\epsilon}{\sin(\epsilon+\theta_0)}(f^{(i)}_{\omega NN}/f^e_\omega), \qquad (f^{(i)}_{\phi NN}/f^s_\phi) = -\sqrt{6}\frac{\cos\epsilon}{\cos(\epsilon+\theta_0)}(f^{(i)}_{\phi NN}/f^e_\phi), \qquad (i=1,2), \tag{9}$$

where f^s_ω, f^s_ϕ are strange-current $\leftrightarrow$ $(V=\omega,\phi)$ coupling constants, and $\epsilon = 3.7°$ is a deviation from the "ideal" mixing angle $\theta_0 = 35.3°$ [4].

If one knows from the fit of the nucleon FF data free parameters $(f^{(i)}_{\omega NN}/f^e_\omega)$, $(f^{(i)}_{\phi NN}/f^e_\phi)$ $(i=1,2)$ of the applicable eight-resonance model for the isoscalar parts of the Dirac and Pauli FFs

$$\begin{aligned}F_1^{I=0}[V(t)] &= \left(\frac{1-V^2}{1-V_N^2}\right)^4 \left\{\frac{1}{2}L(V_{\omega''})L(V_{\omega'})\right.\\ &+\left[L(V_{\omega''})L(V_\omega)\frac{(C_{\omega''}-C_\omega)}{(C_{\omega''}-C_{\omega'})} - L(V_{\omega'})L(V_\omega)\frac{(C_{\omega'}-C_\omega)}{(C_{\omega''}-C_{\omega'})}\right.\\ &\left.-L(V_{\omega''})L(V_{\omega'})\right](f^{(1)}_{\omega NN}/f^e_\omega)\\ &+\left[L(V_{\omega''})L(V_\phi)\frac{(C_{\omega''}-C_\phi)}{(C_{\omega''}-C_{\omega'})} - L(V_{\omega'})L(V_\phi)\frac{(C_{\omega'}-C_\phi)}{(C_{\omega''}-C_{\omega'})}\right.\\ &\left.\left.-L(V_{\omega''})L(V_{\omega'})\right](f^{(1)}_{\phi NN}/f^e_\phi)\right\},\end{aligned} \tag{10}$$

$$\begin{aligned}F_2^{I=0}[V(t)] &= \left(\frac{1-V^2}{1-V_N^2}\right)^6 \left\{L(V_{\omega''})L(V_{\omega'})L(V_\omega)\right.\\ &\times\left[1-\frac{C_\omega}{(C_{\omega''}-C_{\omega'})}\left(\frac{(C_{\omega''}-C_\omega)}{C_{\omega'}} - \frac{(C_{\omega'}-C_\omega)}{C_{\omega''}}\right)\right]\\ &\times(f^{(2)}_{\omega NN}/f^e_\omega) + L(V_{\omega''})L(V_{\omega'})L(V_\phi)\\ &\times\left[1-\frac{C_\phi}{(C_{\omega''}-C_{\omega'})}\left(\frac{(C_{\omega''}-C_\phi)}{C_{\omega'}} - \frac{(C_{\omega'}-C_\phi)}{C_{\omega''}}\right)\right]\\ &\left.\times(f^{(2)}_{\phi NN}/f^e_\phi)\right\},\end{aligned} \tag{11}$$

with

$$L(V_r) = \frac{(V_N-V_r)(V_N-V_r^*)(V_N-1/V_r)(V_N-1/V_r^*)}{(V-V_r)(V-V_r^*)(V-1/V_r)(V-1/V_r^*)},$$

$(r=\omega,\phi,\omega',\omega'')$,

$$C_r = \frac{(V_N-V_r)(V_N-V_r^*)(V_N-1/V_r)(V_N-1/V_r^*)}{-(V_r-1/V_r)(V_r^*-1/V_r^*)},$$

$(r=\omega,\phi,\omega',\omega'')$,

$$V(t) = i\frac{\sqrt{\mathcal{A}^{1/2}+\mathcal{B}^{1/2}} - \sqrt{\mathcal{A}^{1/2}-\mathcal{B}^{1/2}}}{\sqrt{\mathcal{A}^{1/2}+\mathcal{B}^{1/2}} + \sqrt{\mathcal{A}^{1/2}-\mathcal{B}^{1/2}}}, \tag{12}$$

where

$$\mathcal{A} = \frac{t_{N\bar{N}} - t_0^{I=0}}{t_0^{I=0}}; \qquad \mathcal{B} = \frac{t - t_0^{I=0}}{t_0^{I=0}};$$

$$V_N = V(t)_{|t=0};$$

$$V_r = V(t)_{|t=(m_r - i\Gamma_r/2)^2}; \quad (r=\omega,\phi,\omega',\omega''), \tag{13}$$

and $t_{N\bar{N}} = 4m_N^2$, then the unknown free parameters $(f^{(i)}_{\omega NN}/f^s_\omega)$, $(f^{(i)}_{\phi NN}/f^s_\phi)$ of the strange nucleon FF model

$$\begin{aligned}F_1^s[V(t)] &= \left(\frac{1-V^2}{1-V_N^2}\right)^4\\ &\times\left\{\left[L(V_{\omega''})L(V_\omega)\frac{(C_{\omega''}-C_\omega)}{(C_{\omega''}-C_{\omega'})} - L(V_{\omega'})L(V_\omega)\frac{(C_{\omega'}-C_\omega)}{(C_{\omega''}-C_{\omega'})}\right.\right.\\ &\left.-L(V_{\omega''})L(V_{\omega'})\right](f^{(1)}_{\omega NN}/f^s_\omega)\\ &+\left[L(V_{\omega''})L(V_\phi)\frac{(C_{\omega''}-C_\phi)}{(C_{\omega''}-C_{\omega'})} - L(V_{\omega'})L(V_\phi)\frac{(C_{\omega'}-C_\phi)}{(C_{\omega''}-C_{\omega'})}\right.\\ &\left.\left.-L(V_{\omega''})L(V_{\omega'})\right](f^{(1)}_{\phi NN}/f^s_\phi)\right\},\end{aligned} \tag{14}$$

$$\begin{aligned}F_2^s[V(t)] &= \left(\frac{1-V^2}{1-V_N^2}\right)^6 \left\{L(V_{\omega''})L(V_{\omega'})L(V_\omega)\right.\\ &\times\left[1-\frac{C_\omega}{(C_{\omega''}-C_{\omega'})}\left(\frac{(C_{\omega''}-C_\omega)}{C_{\omega'}} - \frac{(C_{\omega'}-C_\omega)}{C_{\omega''}}\right)\right]\\ &\times(f^{(2)}_{\omega NN}/f^s_\omega) + L(V_{\omega''})L(V_{\omega'})L(V_\phi)\\ &\times\left[1-\frac{C_\phi}{(C_{\omega''}-C_{\omega'})}\left(\frac{(C_{\omega''}-C_\phi)}{C_{\omega'}} - \frac{(C_{\omega'}-C_\phi)}{C_{\omega''}}\right)\right]\\ &\left.\times(f^{(2)}_{\phi NN}/f^s_\phi)\right\}\end{aligned} \tag{15}$$

of the same analytic structure, but with different normalization of the Dirac FF, are calculated by the relations (9).

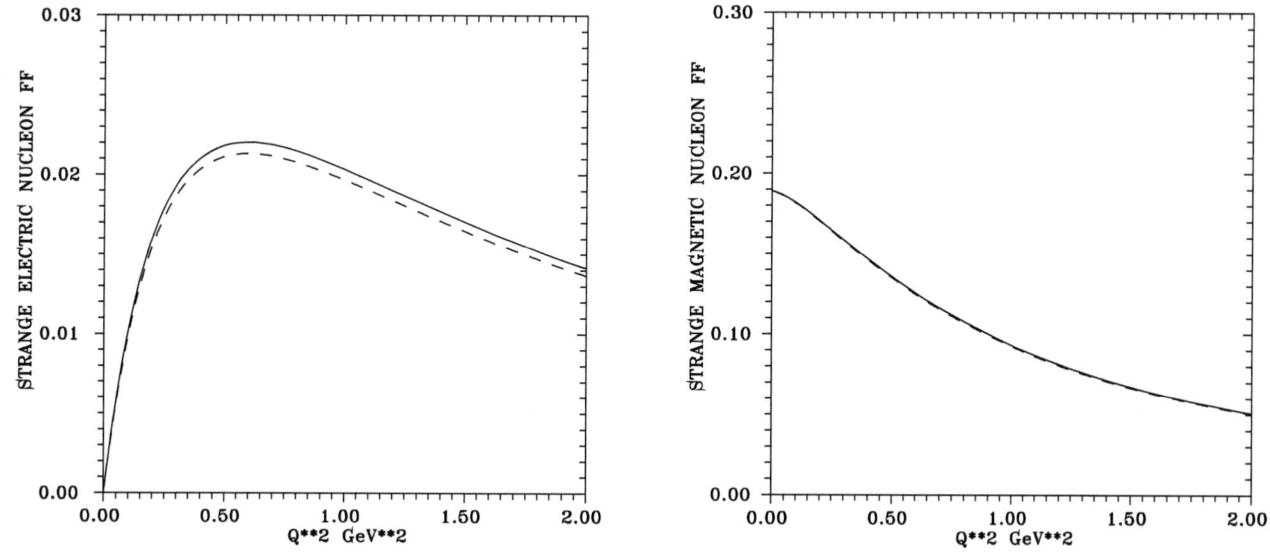

Fig. 1. The predicted behaviors of strange nucleon electric (left) and magnetic (right) form factors.

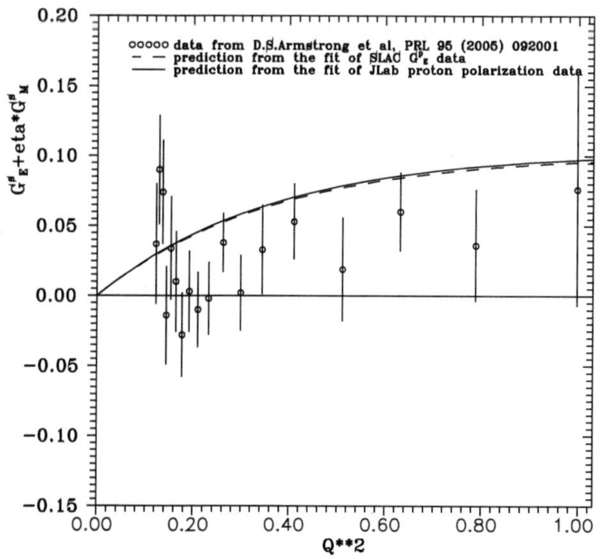

Fig. 2. Predicted combination $G_E^s(Q^2) + \eta(Q^2) G_M^s(Q^2)$ from Rosenbluth and JLab data by the 8-resonance unitary and analytic model and its comparison with the $G0$ Collaboration data.

The expressions (10) and (11) for $F_1^{I=0}$, $F_2^{I=0}$, together with similar expressions [5] for $F_1^{I=1}$, $F_2^{I=1}$ to be saturated by ρ, ρ', ρ'', ρ''' isovector vector-mesons, have been used

- first, to describe Rosenbluth $G_{Ep}(t)$ data in the $t < 0$ region together with all other existing nucleon EM FF data, with the result $\chi^2/(\text{ndf}) = 1.76$;
- then, instead of the Rosenbluth $G_{Ep}(t)$ data in the $t < 0$ region, the JLab proton polarization data on $\mu_p G_{Ep}(Q^2)/G_{Mp}(Q^2)$ for $-5.54\,\text{GeV}^2 \le t \le -0.49\,\text{GeV}^2$ together with all other existing nucleon EM FF data were analyzed with the result $\chi^2/(\text{ndf}) = 1.34$.

Finally, the strange coupling constant ratios $(f^{(i)}_{\omega NN}/f^s_\omega)$, $(f^{(i)}_{\phi NN}/f^s_\phi)$ according to the prescribed procedure given by (9) were calculated, and the behavior of $G_E^s(Q^2)$, $G_M^s(Q^2)$ by means of (14) and (15), as presented in fig. 1, are predicted. As one can see from fig. 1, a reasonable positive value of the strangeness nucleon magnetic moment is found to be $\mu_s = +0.19\,[\mu_N]$.

A reasonable description of the recent data [6] on the combination $G_E^s(Q^2) + \eta(Q^2) G_M^s(Q^2)$ for $0.12\,\text{GeV}^2 < Q^2 < 1.0\,\text{GeV}^2$ is achieved (see fig. 2) as well. Similar results were obtained also by Bijker [7] recently, exploiting a very simple parametrization [8] of the nucleon EM FF in the spacelike region.

As one can see from figs. 1 and 2, the predicted behavior of the strange-nucleon FFs by the special eight-resonance unitary and analytic model is not very sensitive to the difference in the contradictory behavior of $G_{Ep}(t)$ in the spacelike region.

The work was in part supported by Slovak Grant Agency for Sciences, Gr. No 2/4099/26.

References

1. M.K. Jones *et al.*, Phys. Rev. Lett. **84**, 1398 (2000).
2. O. Gayou *et al.*, Phys. Rev. C **64**, 038292 (2001).
3. O. Gayou *et al.*, Phys. Rev. Lett. **88**, 092301 (2002).
4. R.L. Jaffe, Phys. Lett. B **229**, 275 (1989).
5. S. Dubnička, A.Z. Dubničková, P. Weisenpacher, J. Phys. G **29**, 405 (2003).
6. D.S. Armstrong *et al.*, Phys. Rev. Lett. **95**, 092001 (2005).
7. R. Bijker, J. Phys. G **32**, L49 (2006).
8. R. Bijker, F. Iachello, Phys. Rev. C **69**, 068201 (2004).

Flavor content of nucleon form factors in a VMD approach

R. Bijker[a]

Instituto de Ciencias Nucleares, Universidad Nacional Autónoma de México, AP 70-543, 04510 México DF, Mexico

Paper reprinted from Eur. Phys. J. A with permission

Abstract. The strange form factors of the nucleon are studied in a two-component model consisting of a three-quark intrinsic structure surrounded by a meson cloud. A comparison with the available experimental world data from the SAMPLE, PVA4, HAPPEX and G0 Collaborations shows a good overall agreement. It is shown that the strangeness contribution to the electric and magnetic form factors is of the order of a few percent. In particular, the strange quark contribution to the charge radius is small $\langle r_s^2 \rangle_E = 0.005\,\mathrm{fm}^2$ and to the magnetic moment it is positive $\mu_s = 0.315\ \mu_N$.

PACS. 13.40.Gp Electromagnetic form factors – 12.40.Vv Vector-meson dominance – 14.20.Dh Protons and neutrons – 13.40.Em Electric and magnetic moments

1 Introduction

The flavor content of the electromagnetic form factors of the nucleon can be studied by combining the nucleon's response to the electromagnetic and weak neutral vector currents [1]. In recent experiments, parity-violating elastic electron-proton scattering has been used to probe the contribution of strange quarks to the structure of the nucleon [2,3]. The strange quark content of the form factors can be determined assuming charge symmetry and combining parity-violating asymmetries with measurements of the electric and magnetic form factors of the proton and neutron. The study of the strange quark content is of special interest because it is exclusively part of the quark-antiquark sea.

The experimental results from the SAMPLE, PVA4, HAPPEX and G0 Collaborations have shown evidence for a nonvanishing strange quark contribution to the structure of the nucleon. In particular, evidence was found that the strange magnetic moment of the proton is positive [4], suggesting that the strange quarks reduce the proton's magnetic moment. This is an unexpected and surprising finding, since the majority of theoretical studies favors a negative value [5].

The aim of this contribution is to study the flavor content of nucleon form factors in a VMD approach in which the two-component model of electromagnetic nucleon form factors of [6] is extended to the strange sector. The strangeness content is determined via the coupling of the strange current to the ϕ and ω mesons [7]. A comparison with the available experimental world data shows a good overall agreement for $0 < Q^2 < 1\,(\mathrm{GeV}/c)^2$.

[a] e-mail: bijker@nucleares.unam.mx

2 Nucleon form factors

Electromagnetic and weak form factors contain the information about the distribution of electric charge and magnetization inside the nucleon. These form factors arise from matrix elements of the corresponding vector current operators

$$\langle N \,|V_\mu|\, N\rangle = \bar{u}_N \left[F_1(Q^2)\,\gamma_\mu + F_2(Q^2)\,\frac{i\sigma_{\mu\nu}q^\nu}{2M_N} \right] u_N. \quad (1)$$

Here F_1 and F_2 are the Dirac and Pauli form factors which are functions of the squared momentum transfer $Q^2 = -q^2$. The electric and magnetic form factors, G_E and G_M, are obtained from F_1 and F_2 by the relations $G_E = F_1 - \tau F_2$ and $G_M = F_1 + F_2$ with $\tau = Q^2/4M_N^2$.

The Dirac and Pauli form factors are parametrized according to a two-component model of the nucleon [6] in which the external photon couples both to an intrinsic three-quark structure described by the form factor $g(Q^2)$ and to a meson cloud through the intermediate vector mesons ρ, ω and ϕ. In the original version of the two-component model [8], the Dirac form factor was attributed to both the intrinsic structure and the meson cloud, and the Pauli form factor entirely to the meson cloud. In a modified version [6], it was shown that the addition of an intrinsic part to the isovector Pauli form factor, as suggested by studies of relativistic constituent quark models in the light-front approach [9], improves the results for the electromagnetic form factors of the neutron considerably.

In order to incorporate the contribution of the isocalar (ω and ϕ) and isovector (ρ) vector mesons, it is convenient to first introduce the isoscalar and isovector current operators

$$V_\mu^{I=0} = \frac{1}{6}\left(\bar{u}\gamma_\mu u + \bar{d}\gamma_\mu d - 2\bar{s}\gamma_\mu s\right),$$
$$V_\mu^{I=I} = \frac{1}{2}\left(\bar{u}\gamma_\mu u - \bar{d}\gamma_\mu d\right). \qquad (2)$$

The corresponding isoscalar Dirac and Pauli form factors depend on the couplings to the ω and ϕ mesons:

$$F_1^{I=0}(Q^2) = \frac{1}{2}g(Q^2)\Big[1 - \beta_\omega - \beta_\phi + \beta_\omega \frac{m_\omega^2}{m_\omega^2+Q^2} + \beta_\phi \frac{m_\phi^2}{m_\phi^2+Q^2}\Big], \qquad (3)$$
$$F_2^{I=0}(Q^2) = \frac{1}{2}g(Q^2)\left[\alpha_\omega \frac{m_\omega^2}{m_\omega^2+Q^2} + \alpha_\phi \frac{m_\phi^2}{m_\phi^2+Q^2}\right],$$

and the isovector ones on the coupling to the ρ meson [6]:

$$F_1^{I=1}(Q^2) = \frac{1}{2}g(Q^2)\left[1 - \beta_\rho + \beta_\rho \frac{m_\rho^2}{m_\rho^2+Q^2}\right],$$
$$F_2^{I=1}(Q^2) = \frac{1}{2}g(Q^2)\left[\frac{\mu_p - \mu_n - 1 - \alpha_\rho}{1+\gamma Q^2} + \alpha_\rho \frac{m_\rho^2}{m_\rho^2+Q^2}\right]. \qquad (4)$$

The proton and neutron form factors correspond to the sum and difference of the isoscalar and isovector contributions, $F_i^p = F_i^{I=0} + F_i^{I=1}$ and $F_i^n = F_i^{I=0} - F_i^{I=1}$, respectively. This parametrization ensures that the three-quark contribution to the anomalous magnetic moment is purely isovector, as given by $SU(6)$. The intrinsic form factor is a dipole $g(Q^2) = (1+\gamma Q^2)^{-2}$ which coincides with the form used in an algebraic treatment of the intrinsic three-quark structure [10]. The large width of the ρ meson which is crucial for the small-Q^2 behavior of the form factors is taken into account in the same way as in [6, 8]. For small values of Q^2 the form factors are dominated by the meson dynamics, whereas for large values they satisfy the asymptotic behavior of p-QCD, $F_1 \sim 1/Q^4$ and $F_2 \sim 1/Q^6$ [11].

3 Flavor content

The strange quark content of the nucleon form factors arises through the coupling of the strange current

$$V_\mu^s = \bar{s}\gamma_\mu s, \qquad (5)$$

to the intermediate isoscalar vector mesons ω and ϕ (using the convention of Jaffe [7]). The wave functions of the ω and ϕ mesons are given by

$$|\omega\rangle = \cos\epsilon\,|\omega_0\rangle - \sin\epsilon\,|\phi_0\rangle,$$
$$|\phi\rangle = \sin\epsilon\,|\omega_0\rangle + \cos\epsilon\,|\phi_0\rangle, \qquad (6)$$

where the mixing angle ϵ represents the deviation from the ideally mixed states $|\omega_0\rangle = (u\bar{u} + d\bar{d})/\sqrt{2}$ and $|\phi_0\rangle = s\bar{s}$.

Under the assumption that the strange form factors have the same form as the isoscalar ones, the Dirac and Pauli form factors that correspond to the strange current are expressed as the product of an intrinsic part $g(Q^2)$ and a contribution from the vector mesons

$$F_1^s(Q^2) = \frac{1}{2}g(Q^2)\left[\beta_\omega^s \frac{m_\omega^2}{m_\omega^2+Q^2} + \beta_\phi^s \frac{m_\phi^2}{m_\phi^2+Q^2}\right],$$
$$F_2^s(Q^2) = \frac{1}{2}g(Q^2)\left[\alpha_\omega^s \frac{m_\omega^2}{m_\omega^2+Q^2} + \alpha_\phi^s \frac{m_\phi^2}{m_\phi^2+Q^2}\right]. \qquad (7)$$

The isocalar and strange couplings appearing in eqs. (3) and (7) are not independent of one another, but depend on the same nucleon-meson and current-meson couplings [7]. In addition, they are constrained by the electric charges and magnetic moments of the nucleon which leads to two independent isoscalar couplings

$$\alpha_\omega = \mu_p + \mu_n - 1 - \alpha_\phi,$$
$$\beta_\phi = -\beta_\omega \tan\epsilon / \tan(\theta_0 + \epsilon), \qquad (8)$$

from which the strange couplings can be obtained as [7]

$$\beta_\omega^s/\beta_\omega = \alpha_\omega^s/\alpha_\omega = -\sqrt{6}\,\sin\epsilon/\sin(\theta_0+\epsilon),$$
$$\beta_\phi^s/\beta_\phi = \alpha_\phi^s/\alpha_\phi = -\sqrt{6}\,\cos\epsilon/\cos(\theta_0+\epsilon). \qquad (9)$$

with $\tan\theta_0 = 1/\sqrt{2}$. The mixing angle ϵ can be determined either from the radiative decays of the ω and ϕ mesons [12–14] or from their strong decays [15]. The value used here is $\epsilon = 0.053$ rad (3.0°) [12].

Finally, the contributions of the up and down quarks to the electromagnetic form factors can be obtained from

$$G_{E/M}^u = 2G_{E/M}^p + G_{E/M}^n + G_{E/M}^s,$$
$$G_{E/M}^d = G_{E/M}^p + 2G_{E/M}^n + G_{E/M}^s. \qquad (10)$$

4 Results

In order to calculate the nucleon form factors in the two-component model the five coefficients, γ from the intrinsic form factor, β_ω and α_ϕ from the isoscalar couplings, and β_ρ and α_ρ from the isovector couplings, are determined in a least-square fit to the electric and magnetic form factors of the proton and the neutron using the same data set as in [6]. The electromagnetic form factor of the proton and neutron are found to be in good agreement with experimental data [16]. According to eq. (9), the strange couplings can be determined from the fitted values of the isoscalar couplings to be $\beta_\phi^s = -\beta_\omega^s = 0.202$, $\alpha_\phi^s = 0.648$ and $\alpha_\omega^s = -0.018$ [16,17].

Figure 1 shows the strange Dirac and Pauli form factors as a function of the momentum transfer Q^2. Whereas the Pauli form factor is dominated by the coupling to the ϕ meson ($\alpha_\phi^s \gg \alpha_\omega^s$), the Dirac form factor is small due to a cancelation between the contributions from the ω and ϕ mesons ($\beta_\phi^s = -\beta_\omega^s$). The qualitative features of these form

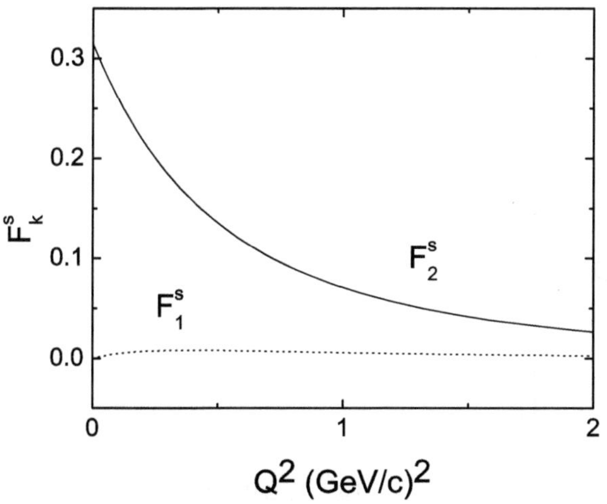

Fig. 1. Strange Dirac and Pauli form factors, F_1^s (dotted line) and F_2^s (solid line).

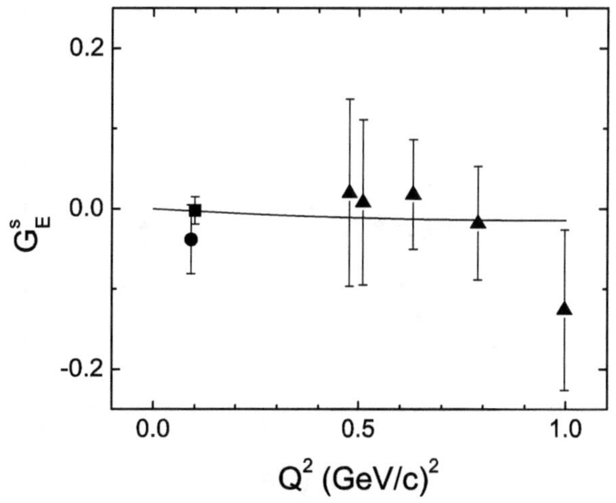

Fig. 2. Comparison between theoretical and experimental values of the strange electric form factor. The experimental values are taken from [25] (circle), [28] (triangle) and [26] (square).

factors can be understood in the limit of ideally mixed mesons, *i.e.* zero mixing angle $\epsilon = 0$ (in comparison to the value of $\epsilon = 3.0°$ used in fig. 1). Since in this case $\beta_\phi^s = \beta_\omega^s = \alpha_\omega^s = 0$, the Dirac form factor vanishes identically and the Pauli form factor depends only on the tensor coupling to the ϕ meson, α_ϕ^s.

The behavior of F_1^s and F_2^s in fig. 1 is quite different from that obtained in other theoretical approaches, especially for the strange Pauli form factor. Almost all calculations give negative values for F_2^s for the same range of Q^2 values [7,18–22], with the exception of the meson exchange model [23] and the $SU(3)$ chiral quark-soliton model [24]. In the former case, the values of F_2^s are about two orders of magnitude smaller than the present ones, whereas in the latter F_2^s is positive for small values of Q^2, but changes sign around $Q^2 = 0.1$–$0.3\,(\text{GeV}/c)^2$.

Figures 2 and 3 show the strange electric and magnetic form factors as a function of Q^2. The theoretical values for G_E^s are small and negative, in agreement with the experimental results of the HAPPEX Collaboration in which G_E^s was determined in parity-violating electron scattering from ^{4}He. The experimental values, $G_E^s = -0.038 \pm 0.042 \pm 0.010$ measured at $Q^2 = 0.091\,(\text{GeV}/c)^2$ [25] and, more recently, $G_E^s = -0.002 \pm 0.017$ at $Q^2 = 0.1\,(\text{GeV}/c)^2$ [26] are consistent with zero.

The values of G_M^s are positive, since they are dominated by the contribution from the Pauli form factor. Experimental evidence from the SAMPLE and HAPPEX Collaborations gives a positive value of the strange magnetic form factor at $Q^2 = 0.1\,(\text{GeV}/c)^2$ of $G_M^s = 0.37 \pm 0.20 \pm 0.26 \pm 0.07$ [27] and $G_M^s = 0.12 \pm 0.24$ [26], respectively. The other experimental values of G_E^s and G_M^s in figs. 2 and 3 for $0.4 < Q^2 < 1.0\,(\text{GeV}/c)^2$ were obtained [28,29] by combining the (anti)neutrino data from E734 [30] with the parity-violating asymmetries from HAPPEX [31] and G0 [32]. The theoretical values are in good overall agreement with the experimental ones for the entire range $0 < Q^2 < 1\,(\text{GeV}/c)^2$.

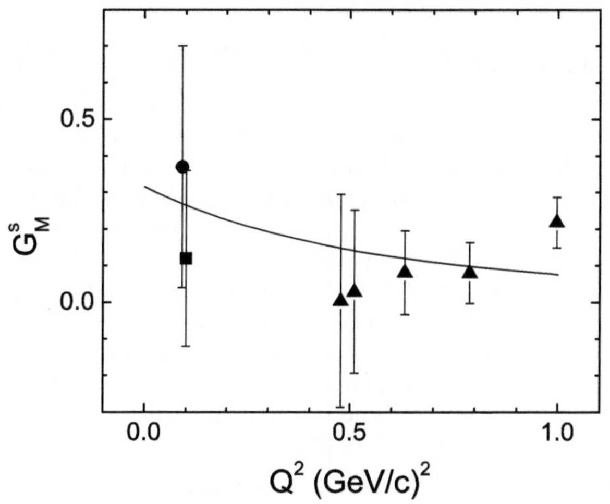

Fig. 3. Comparison between theoretical and experimental values of the strange magnetic form factor. The experimental values are taken from [27] (circle), [28] (triangle) and [26] (square).

Table 1. Comparison between theoretical and experimental values of strange form factors $G_E^s + \eta G_M^s$.

Q^2 $(\text{GeV}/c)^2$	η	$G_E^s + \eta G_M^s$ Present	Experiment	Reference
0.099	0.080	0.019	0.030 ± 0.028	[4]
0.108	0.106	0.025	0.071 ± 0.036	[33]
0.230	0.225	0.042	0.039 ± 0.034	[34]
0.477	0.392	0.047	0.014 ± 0.022	[31]

Table 1 and fig. 4 show the results obtained by the PVA4, HAPPEX and G0 Collaborations for a linear combination of the strange electric and magnetic form factors $G_E^s + \eta G_M^s$. Also in this case, there is a good agreement between the calculated values and the experimental data.

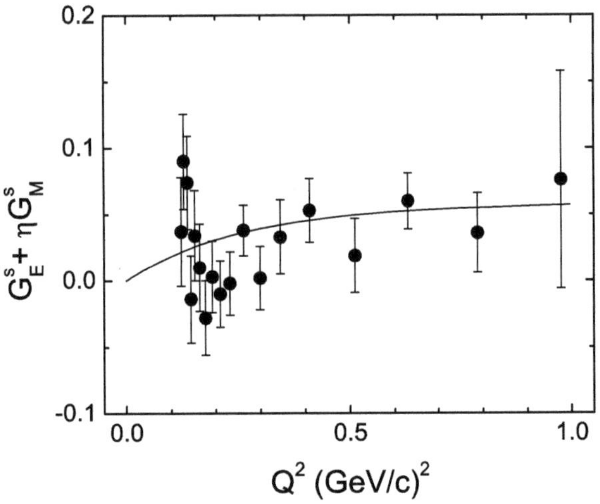

Fig. 4. Comparison between theoretical and experimental values of strange form factors $G_E^s + \eta G_M^s$. The experimental values were measured by the G0 Collaboration [32].

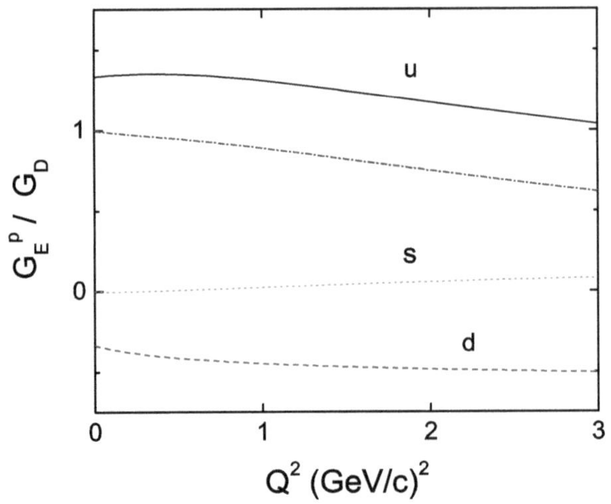

Fig. 5. Flavor decomposition of the proton electric form factor G_E^p/G_D with $G_D = 1/(1+Q^2/0.71)^2$.

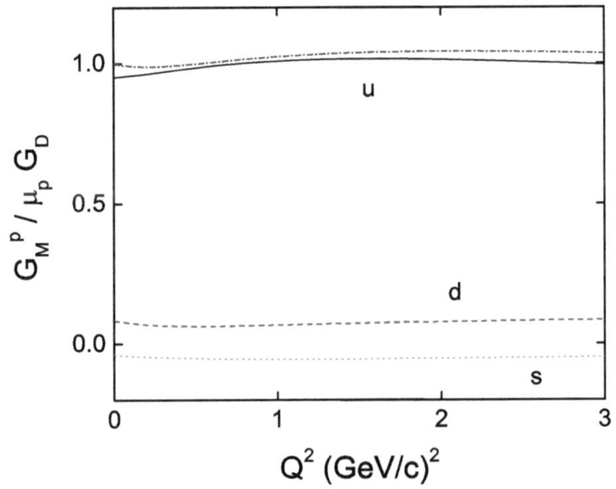

Fig. 6. Flavor decomposition of the proton magnetic form factor $G_M^p/\mu_p G_D$ with $G_D = 1/(1+Q^2/0.71)^2$.

In the majority of theoretical analyses, the strangeness contribution to the nucleon is discussed in terms of the static properties, the strange magnetic moment μ_s and the strangeness radius $\langle r_s^2 \rangle$. Most theoretical studies agree on a small negative strangeness radius and a moderate negative strange magnetic moment [5], whereas the results of a combined fit of the strange electric and magnetic form factors measured by SAMPLE, PVA4 and HAPPEX at $Q^2 \sim 0.1\,(\mathrm{GeV}/c)^2$, $G_M^s(0.1) = 0.55 \pm 0.28$ and $G_E^s(0.1) = -0.01 \pm 0.03$ [4], indicate the opposite sign for both μ_s and $\langle r_s^2 \rangle$. Recent lattice calculations give small negative values of the strange magnetic moment $\mu_s = G_M^s(0) = -0.046 \pm 0.019\,\mu_N$ [35] and the strange electric form factor $G_E^s(0.1) = -0.009 \pm 0.028$ [36].

Figures 5 and 6 show the flavor decomposition of the electric and magnetic form factors of the proton. Note, that in comparison with figs. 2 and 3 the flavor form factors have been multiplied by the quark electric charges, so that their sum gives the total form factor. The contribution of the strange quarks to the proton form factors is small for the entire range of Q^2 values and of the order of a few percent of the total.

In the present approach, the strangeness contribution to the magnetic moment and the charge and magnetic radii is given by [17]

$$\begin{aligned}
\mu_s &= \frac{1}{2}(\alpha_\omega^s + \alpha_\phi^s) = 0.315\,\mu_N, \\
\langle r_s^2 \rangle_E &= 3\beta_\phi^s \left(\frac{1}{m_\phi^2} - \frac{1}{m_\omega^2} \right) + \frac{3}{4M_N^2}(\alpha_\omega^s + \alpha_\phi^s) \\
&= 0.005\,\mathrm{fm}^2, \\
\langle r_s^2 \rangle_M &= 6 \left[2\gamma + \frac{\beta_\phi^s + \alpha_\phi^s}{\alpha_\omega^s + \alpha_\phi^s} \frac{1}{m_\phi^2} + \frac{\beta_\omega^s + \alpha_\omega^s}{\alpha_\omega^s + \alpha_\phi^s} \frac{1}{m_\omega^2} \right] \\
&= 0.410\,\mathrm{fm}^2. \qquad (11)
\end{aligned}$$

The strange magnetic moment does not depend on the mixing angle ϵ [17] and its sign is determined by the sign of the tensor coupling α_ϕ^s ($\gg \alpha_\omega^s$). The sign of the strangeness contribution to the magnetic moment and the charge radius is in agreement with the available experimental data. A positive value of the strange magnetic moment seems to preclude an interpretation in terms of a $uuds\bar{s}$ fluctuation into a ΛK configuration [37]. On the other hand, an analysis of the magnetic moment of $uuds\bar{s}$ pentaquark configurations belonging to the antidecuplet gives a positive strangeness contribution for states with angular momentum and parity $J^P = 1/2^+$, $1/2^-$, and negative for $3/2^+$ states [38].

5 Summary and conclusions

In this contribution the flavor content of nucleon form factors was studied in a VMD approach in which the two-component model of Bijker and Iachello for the electromagnetic nucleon form factors [6] is combined with the method proposed by Jaffe to determine the strangeness content via the coupling of the strange current to the ϕ and ω mesons [7]. The strange couplings are completely fixed by the electromagnetic form factors of the proton and neutron.

The good overall agreement between the theoretical and experimental values for the electromagnetic form factors of the nucleon and their strange quark content shows that the two-component model provides a simultaneous and consistent description of the electromagnetic and weak vector form factors of the nucleon. It was shown that the strangeness contribution to the charge and magnetization distributions is of the order of a few percent. In particular, the strange magnetic moment is found to be positive, in contrast with most theoretical studies, but in agreement with the presently available experimental information from parity-violating electron scattering experiments.

Future experiments on parity-violating electron scattering to backward angles and neutrino scattering will make it possible to determine the contributions of the different quark flavors to the electric, magnetic and axial form factors, and thus to provide new insight into the complex internal structure of the nucleon.

This work was supported in part by a research grant from CONACYT, Mexico.

References

1. D.B. Kaplan, A. Manohar, Nucl. Phys. B **310**, 527 (1988); R.D. McKeown, Phys. Lett. B **219**, 140 (1989); D.H. Beck, Phys. Rev. D **39**, 3248 (1989).
2. D.H. Beck, R.D. McKeown, Annu. Rev. Nucl. Part. Sci. **51**, 189 (2001).
3. E.J. Beise, M.L. Pitt, D.T. Spayde, Prog. Part. Nucl. Phys. **54**, 289 (2005).
4. K.A. Aniol *et al.*, Phys. Lett. B **635**, 275 (2006).
5. D.H. Beck, B.R. Holstein, Int. J. Mod. Phys. E **10**, 1 (2001).
6. R. Bijker, F. Iachello, Phys. Rev. C **69**, 068201 (2004).
7. R.L. Jaffe, Phys. Lett. B **229**, 275 (1989).
8. F. Iachello, A.D. Jackson, A. Lande, Phys. Lett. B **43**, 191 (1973).
9. M.R. Frank, B.K. Jennings, G.A. Miller, Phys. Rev. C **54**, 920 (1996); E. Pace, G. Salmè, F. Cardarelli, S. Simula, Nucl. Phys. A **666**, 33c (2000).
10. R. Bijker, F. Iachello, A. Leviatan, Ann. Phys. (N.Y.) **236**, 69 (1994); Phys. Rev. C **54**, 1935 (1996).
11. G.P. Lepage, S.J. Brodsky, Phys. Rev. Lett. **43**, 545 (1979); Phys. Rev. D **22**, 2157 (1980).
12. P. Jain, R. Johnson, U.-G. Meissner, N.W. Park, J. Schechter, Phys. Rev. D **37**, 3252 (1988).
13. F. Iachello, D. Kusnezov, Phys. Rev. D **45**, 4156 (1992).
14. M. Harada, J. Schechter, Phys. Rev. D **54**, 3394 (1996).
15. C. Gobbi, F. Iachello, D. Kusnezov, Phys. Rev. D **50**, 2048 (1994).
16. R. Bijker, arXiv:nucl-th/0511004.
17. R. Bijker, J. Phys. G: Nucl. Part. Phys. **32**, L49 (2006) [arXiv:nucl-th/0511060].
18. N.W. Park, H. Weigel, Nucl. Phys. A **541**, 453 (1992).
19. G.T. Garvey, W.C. Louis, D.H. White, Phys. Rev. C **48**, 761 (1993).
20. H. Forkel, M. Nielsen, X. Jin, T.D. Cohen, Phys. Rev. C **50**, 3108 (1994).
21. H.-W. Hammer, U.-G. Meissner, D. Drechsel, Phys. Lett. B **367**, 323 (1996).
22. V.E. Lyubovitskij, P. Wang, Th. Gutsche, A. Faessler, Phys. Rev. C **66**, 055204 (2002).
23. U.-G. Meissner, V. Mull, J. Speth, J.W. Van Orden, Phys. Lett. B **408**, 381 (1997).
24. A. Silva, H.-Ch. Kim, K. Goeke, Eur. Phys. J. A **22**, 481 (2004).
25. K.A. Aniol *et al.*, Phys. Rev. Lett. **96**, 022003 (2006).
26. HAPPEX 2 Collaboration (K. Paschke, D.S. Armstrong, K. De Jager *et al.*), private communication (2006).
27. D.T. Spayde *et al.*, Phys. Lett. B **583**, 79 (2004).
28. S.F. Pate, G. MacLachlan, D. McKee, V. Papavassiliou, arXiv:hep-ex/0512032.
29. S.F. Pate, Phys. Rev. Lett. **92**, 082002 (2004).
30. L.A. Ahrens *et al.*, Phys. Rev. D **35**, 785 (1987).
31. K.A. Aniol *et al.*, Phys. Rev. C **69**, 065501 (2004).
32. D.S. Armstrong *et al.*, Phys. Rev. Lett. **95**, 092001 (2005).
33. F.E. Maas *et al.*, Phys. Rev. Lett. **94**, 152001 (2005).
34. F.E. Maas *et al.*, Phys. Rev. Lett. **93**, 022002 (2004).
35. D.B. Leinweber, S. Boinepalli, I.C. Cloet, A.W. Thomas, A.G. Willams, R.D. Young, J.M. Zanotti, J.B. Zhang, Phys. Rev. Lett. **94**, 212001 (2005).
36. D.B. Leinweber, S. Boinepalli, A.W. Thomas, P. Wang, A.G. Willams, R.D. Young, J.M. Zanotti, J.B. Zhang, arXiv:hep-lat/0601025.
37. B.S. Zou, D.O. Riska, Phys. Rev. Lett. **95**, 072001 (2005).
38. R. Bijker, M.M. Giannini, E. Santopinto, Phys. Lett. B **595**, 260 (2004).

Isospin breaking in the vector current of the nucleon

R. Lewis[a]

Department of Physics, University of Regina, Regina, Saskatchewan, S4S 0A2 Canada

Abstract. Extraction of the nucleon's strange form factors from experimental data requires a quantitative understanding of the unavoidable contamination from isospin violation. A number of authors have addressed this issue during the past decade, and their work is reviewed here. The predictions from early models are largely consistent with recent results that rely as much as possible on input from QCD symmetries and related experimental data. The resulting bounds on isospin violation are sufficiently precise to be of value to on-going experimental and theoretical studies of the nucleon's strange form factors.

PACS. 13.40.Gp Electromagnetic form factors – 14.20.Dh Protons and neutrons – 12.39.Fe Chiral Lagrangians – 12.39.Ki Relativistic quark model

1 Motivation

Isospin violation is generally a small effect. For example, consider the nucleon mass splitting, $(m_n - m_p)/m_p = 0.1\%$. One similarly expects isospin violation to have a small impact on the nucleon's electromagnetic and weak form factors. However, this does not imply that isospin violation must be small relative to strangeness effects. To illustrate, recall that an explicit calculation in the electroweak theory leads to

$$G_X^{p,Z}(q^2) = (1-4\sin^2\theta_W)G_X^p(q^2) - G_X^n(q^2) - G_X^s(q^2) - G_X^{u,d}(q^2) \quad (1)$$

for electric ($X = E$) and magnetic ($X = M$) form factors. Experimental studies [1–6] show that the sum of the last two terms on the right-hand side is small. The size of isospin violation, $G_X^{u,d}(q^2)$, relative to strangeness, $G_X^s(q^2)$, is not obtained from these experiments.

In what follows, theoretical studies of $G_X^{u,d}(q^2)$ will be reviewed [7–11]. (Our entire discussion of isospin violation also fits within the more restrictive category called "charge symmetry breaking" and that language is used, for example, in ref. [8].) If the current understanding of these isospin-violating effects is sufficiently precise, then the data from refs. [1–6] allow for a determination of the authentic strange quark effects, $G_X^s(q^2)$, which are of great interest to many people at present.

Independent of any chosen theoretical approach, each isospin-violating form factor is simply the difference of isoscalar and isovector terms,

$$G_X^{u,d}(q^2) \equiv G_X^{\not{s}}(q^2) - G_X^{\not{v}}(q^2)\,, \quad (2)$$

where $G_X^{\not{s}}$ is obtained from

$$\langle p\,|\bar{u}\gamma_\mu u + \bar{d}\gamma_\mu d|\,p\rangle - \langle n\,|\bar{u}\gamma_\mu u + \bar{d}\gamma_\mu d|\,n\rangle \quad (3)$$

and $G_X^{\not{v}}$ is obtained from

$$\langle p\,|\bar{u}\gamma_\mu u - \bar{d}\gamma_\mu d|\,p\rangle + \langle n\,|\bar{u}\gamma_\mu u - \bar{d}\gamma_\mu d|\,n\rangle \quad (4)$$

in a straightforward manner (see refs. [7,11] for details). Furthermore, we know that all isospin violation is ultimately a consequence of unequal quark masses, $m_u \neq m_d$, ("strong breaking") and unequal quark electric charges, $e_u \neq e_d$ ("electromagnetic breaking"). The task for each theoretical approach is to determine the combinations of nucleon matrix elements shown in eqs. (3) and (4), with both types of breaking included.

Since the sum of strangeness and isospin violation in eq. (1) is measured to be a small fraction of the total form factors, and since isospin violation itself is expected to be a small fraction of the total form factors, it is reasonable to neglect contributions containing *both* strangeness and isospin violation as doubly (*i.e.* negligibly) small. This allows $G_X^{u,d}(q^2)$ to be calculated without dynamical strange-quark effects. Such an approach is clearly advantageous for chiral perturbation theory, where addition of a dynamical strange quark leads to severe degradation of convergence properties of the chiral expansion. All of the theoretical studies to date have computed isospin violation without dynamical strange quarks.

Our discussion will be approximately chronological. The constituent quark model studies of Dmitrašinović and Pollock [7] and Miller [8], are discussed in sect. 2, followed by Ma's use of a light-cone meson-baryon fluctuation model [9] in sect. 3. The constraints of chiral symmetry are discussed in sect. 4, based on a collaboration with

[a] e-mail: randy.lewis@uregina.ca

Mobed [10] using chiral perturbation theory. Section 5 reviews the recent results from work with Kubis [11] that combines chiral perturbation theory with resonance saturation and information from dispersion analyses. The final section, sect. 6, provides a brief summary.

2 Constituent quark model

In a constituent quark model, we might expect the scale of strong breaking to be set by *constituent* quark masses, which the authors of ref. [7] take to be $(m_D - m_U)/m_Q \sim 1.2\%$, and the scale of electromagnetic breaking to be set by $\alpha \sim 0.7\%$. To do better than this order-of-magnitude guess, an explicit calculation is required.

The first calculation was carried out by Dmitrašinović and Pollock [7] using oscillator confinement and a Coulomb potential,

$$H = H_0 + V_{EM}\,, \tag{5}$$

$$H_0 = \sum_{i=1}^{3} \frac{\mathbf{p}_i^2}{2m_i} + \frac{k}{2}\sum_{i<j}^{3}(\mathbf{r}_i - \mathbf{r}_j)^2\,, \tag{6}$$

$$V_{EM} = \sum_{i<j}^{3} \frac{e_i e_j}{4\pi\,|\mathbf{r}_i - \mathbf{r}_j|}\,. \tag{7}$$

Recall that this choice for H_0 produces Gaussian spatial wave functions. With parameters fixed to be $m_Q = 330\,\text{MeV}$, $m_D - m_U = 4\,\text{MeV}$, and k determined from the experimental mass difference between $\Delta(1232)$ and nucleon, ref. [7] finds

$$G_M^{\not{s}}(0) = G_M^{\not{c}}(0) = \frac{1}{3}\left[\frac{m_D}{m_U} - \frac{m_U}{m_D}\right] \approx 0.008\,, \tag{8}$$

$$G_M^{u,d}(0) = G_M^{\not{s}}(0) - G_M^{\not{c}}(0) = 0\,, \tag{9}$$

$$\frac{\delta\langle r_E^2\rangle}{\langle r_E^2\rangle} = 1.1\%\,, \tag{10}$$

$$\frac{\delta\langle r_M^2\rangle}{\langle r_M^2\rangle} = 0.4\%\,, \tag{11}$$

where r_X^2 denotes a squared radius as usual. These estimates are compatible with our order-of-magnitude guesses, but what confidence level should be assigned to the precise values?

Here is a list of some limitations of this model (all of which are mentioned explicitly in ref. [7]):

- The chosen parameters lead to, $m_n - m_p = 3\,\text{MeV}$, 230% above experiment.
- The nucleon charge radius, $\sqrt{\langle r_E^2\rangle} = 0.62$ fm, is 30% below experiment.
- Gaussian spatial wave functions cause both $G_E(q^2)$ and $G_M(q^2)$ to be unrealistic at large $\mathbf{q}^2$.
- The strong hyperfine interaction is omitted.
- Chiral symmetry is absent.

In sect. IV B of ref. [7], the authors conclude, "... we may expect to have calculated the correct sign and order of magnitude of the effects of interest."

Table 1. Parameter values for the three Gaussian models of ref. [8].

Model	1	2	3
$\sqrt{\beta}$ (fm)	0.7	0.6	0.5
α_s	2.3	1.2	0.35
$m_D - m_U$ (MeV)	5.2	3.8	2.3

Subsequently, Miller [8] used a more complete constituent quark model and explicitly addressed each of the limitations listed above, except chiral symmetry. (Miller says chiral symmetry is implicit in the charge symmetry conserving pion cloud of this model [12].) His Hamiltonian is

$$H = K + V_{\text{con}} + V_{\text{em}} + V_g\,, \tag{12}$$

$$K = \sum_{i=1}^{3}\left(m_i + \frac{p_i^2}{2m_i}\right), \tag{13}$$

$$V_{\text{em}} = \alpha\sum_{i<j} q_i q_j\left(\frac{1}{r_{ij}} - \frac{\pi}{2}\delta(\boldsymbol{r}_{ij})\left[\frac{2}{\bar{m}^2} + \frac{4}{3}\frac{\boldsymbol{\sigma}(i)\cdot\boldsymbol{\sigma}(j)}{\bar{m}^2}\right]\right), \tag{14}$$

$$V_g = -\alpha_s\sum_{i<j}\lambda_i\cdot\lambda_j\left[\frac{\pi}{2}\delta(\boldsymbol{r}_{ij})\left(\frac{1}{m_i^2} + \frac{1}{m_j^2} + \frac{4}{3}\frac{\boldsymbol{\sigma}(i)\cdot\boldsymbol{\sigma}(j)}{m_i m_j}\right)\right]. \tag{15}$$

Two options for V_{con} are studied; one gives Gaussian form factors,

$$\Psi(\rho,\lambda) = N\exp\left(\frac{\rho^2+\lambda^2}{-2\beta}\right), \tag{16}$$

$$\boldsymbol{\rho} = \frac{1}{\sqrt{2}}(\boldsymbol{r}_1 - \boldsymbol{r}_2)\,, \tag{17}$$

$$\boldsymbol{\lambda} = \frac{1}{\sqrt{6}}(\boldsymbol{r}_1 + \boldsymbol{r}_2 - 2\boldsymbol{r}_3)\,, \tag{18}$$

and the other gives power law form factors,

$$\Psi^2(R) = \frac{2\sqrt{6}\Lambda^7}{\pi^3 6^5} R K_1\left(\sqrt{\frac{2}{3}}\Lambda R\right)$$
$$\Rightarrow\ G_E(Q^2) = \left(\frac{\Lambda^2}{Q^2+\Lambda^2}\right)^4. \tag{19}$$

For the Gaussian case, ref. [8] uses the experimental proton magnetic moment to determine $\bar{m} = 337\,\text{MeV}$, the experimental $m_\Delta - m_N$ value to fix α_s as a function of β, and then considers three choices for β which define three models as shown in table 1. Figures 4, 5, 6 and 7 of ref. [8] display the resulting effects of isospin violation on $G_E(q^2)$ and $G_M(q^2)$, including the separate contributions from K, V_{em} and V_g. In each case, the isospin-violating contributions vanish for $q^2 = 0$, and are less than 0.2% in magnitude for a momentum transfer of $0.1\,\text{GeV}^2$. For power law form factors, ref. [8] obtains $\Lambda = 5.90/\text{fm}$ from the experimental value of $\langle r_E^2\rangle$, and all isospin-violating effects remain small.

The combined work of refs. [7,8] provides an excellent understanding of isospin violation within the constituent quark model, but we must now ask which features of the results are a true reflection of nature, and which are model dependent. For example, there are symmetries in these quark models that lead to $G_M^{u,d}(0) = 0$, recall eq. (9), but, as we will see in sect. 4, this is not a symmetry of nature. It is therefore interesting to explore other theoretical approaches as well.

3 Light-cone meson-baryon fluctuation model

To discuss the method used by Ma [9], we must think at the hadron level rather than the constituent quark level. Fluctuations of a nucleon into a virtual pion-nucleon pair, $p \to \pi^+ n$ and $n \to \pi^- p$, occur commonly. As always, isospin violation arises from strong breaking (at the hadron level, this means $m_n - m_p \neq 0$) and from electromagnetic breaking (which now means the Coulomb attraction of a charged/charged $\pi^- p$ pair is different from the charged/neutral $\pi^+ n$ pair). The expression for isospin violation in the magnetic form factor at vanishing momentum transfer is

$$G_M^{u,d}(0) = \Big(P(n \to \pi^- p) - P(p \to \pi^+ n)\Big)\Big(\mu^n_{\pi^- p} - \mu_n\Big), \tag{20}$$

where $P(\ )$ denotes a fluctuation probability, and $\mu^n_{\pi^- p}$ is the magnetic moment for the neutron's fluctuation.

To determine $\mu^n_{\pi^- p}$, begin with the fact that the total angular momentum is the orbital angular momentum plus proton spin,

$$\left|\frac{1}{2},\frac{1}{2}\right\rangle_J = \sqrt{\frac{2}{3}}\left|1,1\right\rangle_L\left|\frac{1}{2},-\frac{1}{2}\right\rangle_S - \sqrt{\frac{1}{3}}\left|1,0\right\rangle_L\left|\frac{1}{2},\frac{1}{2}\right\rangle_S . \tag{21}$$

This leads directly to

$$\begin{aligned}\mu^n_{\pi^- p} &= -\frac{\mu_p}{3} + \frac{\mu_n}{3}\left(\frac{2m_n}{m_{\pi^-}+m_p}\right)\left(\frac{m_{\pi^-}}{m_p} - \frac{m_p}{m_{\pi^-}}\right)\\ &= -4.75 \qquad \Rightarrow \qquad \mu^n_{\pi^- p} - \mu_n = -2.84 .\end{aligned} \tag{22}$$

To determine the difference of fluctuation probabilities needed for eq. (20), a light-cone Gaussian wave function is used,

$$\psi = Ae^{-(\mathcal{M}^2 - m_N^2)/(8\alpha^2)}, \tag{23}$$

where

$$\mathcal{M}^2 = \sum_{i=1}^{2}\frac{\mathbf{k}^2_{\perp i} + m_i^2}{x_i} \tag{24}$$

is the invariant mass of the meson-baryon state. (Note the implicit assumption that A is independent of which nucleon is fluctuating; ref. [9] points out that this assumption is not required, but the uncertainties associated with relaxing it are difficult to estimate.) From the experimental Gottfried sum rule, one finds

$$P(p \to \pi^+ n) \approx P(n \to \pi^- p) \approx 0.15 . \tag{25}$$

The final remaining parameter is the radius, α, and it leads to a large uncertainty. Reference [9] uses two bounds,

$$\alpha = 300\,\text{MeV} \quad \Rightarrow \quad P(n = \pi^- n) - P(p = \pi^+ n) = 0.2\% , \tag{26}$$

and

$$\left.\begin{aligned}\alpha(n \to \pi^- p) &= 205\,\text{MeV}\\ \alpha(p \to \pi^+ n) &= 200\,\text{MeV}\end{aligned}\right\} \quad \Rightarrow \quad P(n = \pi^- p) - P(p = \pi^+ n) = 3\% , \tag{27}$$

to arrive at

$$G_M^{u,d}(0) \approx 0.006 \to 0.088 . \tag{28}$$

However, Miller suggests that $(205 - 200)/200 = 2.5\%$ is too large for a Coulomb effect, and states: "A reasonable estimate of the effect could be 0.03 nuclear magnetons." [8].

Thus, the light-cone meson-baryon fluctuation model provides an explicit example of a model that does not lead to $G_M^{u,d}(0) = 0$. In particular, this model predicts $G_M^{u,d}(0) > 0$.

4 Chiral perturbation theory

Chiral perturbation theory is not a model; it merely administrates the global symmetries of QCD. This has the advantage of retaining no model dependence in predictions, and it has the disadvantage that quantities not determined by global symmetries remain completely unknown. In ref. [10], isospin violation in the nucleon's vector form factors is studied within chiral perturbation theory to leading order for the electric case, and to next-to-leading order for the magnetic case.

Because isospin violation arises from both strong breaking and electromagnetic breaking, our chiral perturbation theory must include dynamical photons as well as dynamical pions. A single baryon will flow through the process, though its identity may change from proton to neutron or even to $\Delta(1232)$ by absorbing or emitting a pion (recall the fluctuations discussed in sect. 3). Reference [10] uses the formalism of heavy-baryon chiral perturbation theory [13,14], and the perturbative expansion is defined in powers of electric charge e, momentum q/Λ, pion mass m_π/Λ, and mass difference $(m_\Delta - m_N)/\Lambda$, where Λ denotes either $4\pi F_\pi$ or m_N. Writing the Lagrangian in standard notation,

$$\mathcal{L}_{\text{ChPT}} = \mathcal{L}^{(1)} + \mathcal{L}^{(2)} + \mathcal{L}^{(3)} + \mathcal{L}^{(4)} + \mathcal{L}^{(5)} + \dots , \tag{29}$$

Reference [10] shows that the isospin violating form factors begin in $\mathcal{L}^{(4)}$ (plus the corresponding loop diagrams), and next-to-leading effects are in $\mathcal{L}^{(5)}$ (plus the corresponding loop diagrams). For a general observable at this high order, two-loop diagrams routinely appear and the most general effective Lagrangian contains literally hundreds of low-energy constants (*i.e.* parameters whose numerical values are not constrained by global symmetries, and hence unknown to chiral perturbation theory).

For the specific case of $G_M^{u,d}(q^2)$, ref. [10] shows that the situation is much simpler than for a general observable due to five key observations:

- $G_M^{u,d}(q^2)$ receives no two-loop contributions up to next-to-leading order. (In particular, "photon+photon" loop diagrams lack the required spin operators, "photon+ pion" loop diagrams sum to *anticommutators* of Pauli-Lubanski operators, and "pion+pion" loop diagrams have only higher-order isospin violation.)
- All one-loop photon effects in $G_M^{u,d}(q^2)$ can be absorbed into the physical value of $m_n - m_p$.
- Explicit $\Delta(1232)$ effects are found, numerically, to be smaller than systematic uncertainties.
- $G_M^{u,d}(q^2)$ contains just one combination of low-energy constants at leading order —it is a simple additive constant— and no additional low-energy constants at next-to-leading order. (The fact that no global symmetry forces this constant to vanish is not realized in some constituent quark models, as discussed in sect. 2.)
- All isospin breaking in loops, up to next-to-leading order, is ultimately due to $m_n - m_p$.

Based on these observations, a parameter-free prediction for the difference $G_M^{u,d}(q^2)-G_M^{u,d}(0)$ is obtained in ref. [10].

Isospin violation in the electric form factor is discussed in ref. [15], but only to leading order because in this case there *are* two-loop contributions and additional unknown constants at next-to-leading order.

Though it is useful to have these "pion cloud contributions" computed within chiral perturbation theory, the missing combination of low-energy constants indicates that chiral perturbation theory alone will not meet all of our goals, and we must turn our attention to a method for understanding the physics that resides within the low-energy constants.

5 Chiral perturbation theory with resonance saturation

The chiral perturbation theory calculations of ref. [10] were reproduced in ref. [11] using two separate formalisms: a repeat of the heavy-baryon calculation, and a newer method known as infrared regularization [16]. This newer approach is simpler to manage and needs fewer Feynman diagrams, but physical results must be identical. Reference [11] finds and corrects some errant factors of 2 in ref. [10], and confirms all of the features mentioned in sect. 4.

With the results of chiral perturbation theory now firmly in hand, ref. [11] addresses the issue of unknown low-energy constants. They are not constrained by chiral symmetry, nor by any other global symmetry of QCD. Since chiral perturbation theory is the low-energy effective theory of QCD, where pions (and in our case, photons too) are dynamical, the low-energy constants must account for all of the higher-energy QCD dynamics. Most important would be the exchange of the lightest resonances that are too heavy to appear explicitly in the effective theory: for vector form factors, these are the ρ and ω. Figure 1 sketches this notion of resonance saturation. Mathematically, it is easy to see that the exchange of a heavy particle is a low-energy constant at leading order,

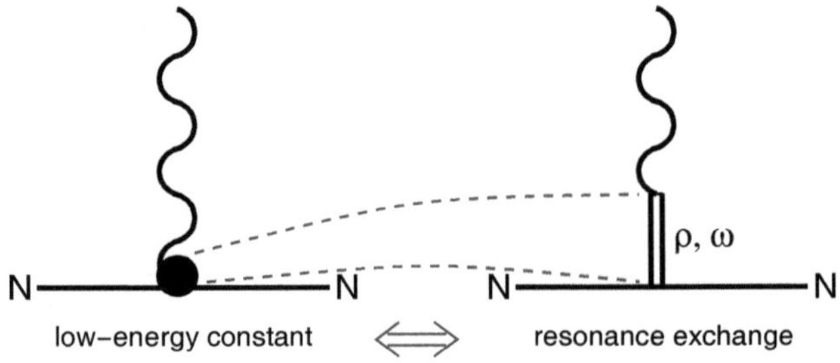

Fig. 1. The basic idea of resonance saturation for a nucleon's vector form factor.

Fig. 2. Resonance saturation contributions to isospin violation in the nucleon's vector form factors.

$$\text{propagator} \sim \frac{1}{M^2-q^2} = \frac{1}{M^2} + O\left(\frac{q^2}{M^2}\right). \tag{30}$$

Of course these resonances have to be coupled in a way consistent with chiral symmetry.

This type of resonance saturation was shown long ago to work very well in the meson chiral Lagrangian [17–19], though it has been shown that the nucleon's vector form factors require inclusion of extra resonances beyond merely the ρ and ω [20]. Nevertheless, our present goal is only to compute the isospin-violating pieces of these form factors; they come from ρ-ω mixing, and ref. [11] points out that any effects of higher resonances are more severely suppressed in this case.

The contributions of resonance saturation to the isospin-violating form factors are shown diagrammatically in fig. 2. The corresponding equations are

$$\delta G_E^{u,d}(q^2) = \frac{\Theta_{\rho\omega}}{M_V(M_V^2-q^2)^2}\left[\left(1+\frac{\kappa_\omega M_V^2}{4m_N^2}\right)g_\omega F_\rho q^2 - \left(1+\frac{\kappa_\rho M_V^2}{4m_N^2}\right)g_\rho F_\omega q^2\right], \tag{31}$$

$$\delta G_M^{u,d}(q^2) = \frac{\Theta_{\rho\omega}}{M_V(M_V^2-q^2)^2}\left[\left(q^2+\kappa_\omega M_V^2\right)g_\omega F_\rho - \left(q^2+\kappa_\rho M_V^2\right)g_\rho F_\omega\right], \tag{32}$$

where M_V, F_ρ and F_ω are the vector meson mass and decay constants, and the ρ-ω mixing parameter has been determined from experimental masses and branching ratios to be [21]

$$\Theta_{\rho\omega} = (-3.75 \pm 0.36) \times 10^{-3}\,\text{GeV}^{-2}. \tag{33}$$

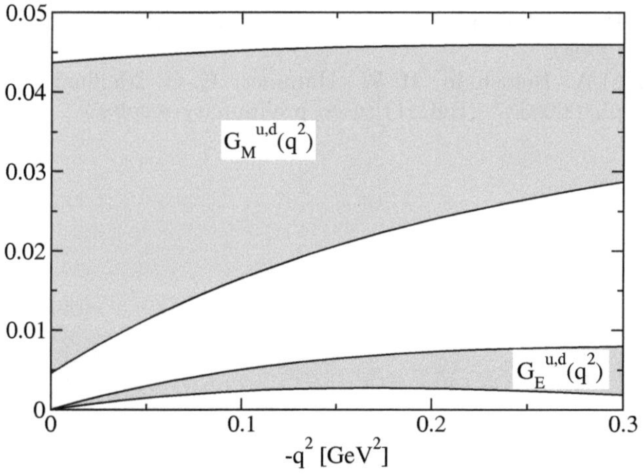

Fig. 3. Isospin violation in the electromagnetic form factors, as obtained from ref. [11].

If we had numerical values for the couplings of vector mesons to nucleons, g_ρ, κ_ρ, g_ω and κ_ω, then eq. (32) could be expanded in powers of q^2 to obtain the desired low-energy constant (at $q^2 = 0$) as well as additional contributions that are technically of higher order in the chiral expansion. We immediately see that eq. (31) vanishes at $q^2 = 0$ as required by global symmetries, but nonzero contributions from higher orders in the chiral expansion would be obtained.

In principle, there is a wide assortment of techniques for attempting to quantify g_ρ, κ_ρ, g_ω and κ_ω, but finding an approach with rigorously quantifiable uncertainties is difficult. In an attempt to rely as directly as possible on experimental data rather than models, ref. [11] uses values extracted from dispersive analyses of nucleon electromagnetic form factors. For the ρ couplings, these analyses must account for the non-resonant two-pion continuum in addition to the Breit-Wigner ρ-resonance. Data from refs. [22,23] lead to the ranges

$$4.0 < g_\rho < 6.2\,, \tag{34}$$
$$5.1 < \kappa_\rho < 6.8\,. \tag{35}$$

For the ω couplings, it is sufficient to consider only pure zero-width resonance pole residues, and refs. [22,24,25] produce the ranges

$$41.8 < g_\omega < 43.0\,, \tag{36}$$
$$-0.16 < \kappa_\omega < 0.57\,. \tag{37}$$

The final results of ref. [11], reproduced in fig. 3, include the worst-case errors bars obtained from spanning the ranges in eqs. (34)-(37) above; the poorly known κ_ω dominates the uncertainties. Notice that the range of $G_M^{u,d}(0)$ does not include zero, while $G_E^{u,d}(0) = 0$ is required. Both form factors are positive over the momentum range considered.

6 Summary

Early studies of isospin violation in the nucleon's vector form factors led to a clear understanding within specific quark models. Use of chiral perturbation theory avoids all model dependence, but leaves some parameters undetermined. Phenomenologically, those parameters are saturated by resonances, and numerical values are obtained with minimal model dependence by using dispersive analyses. The results in fig. 3 represent a conservative determination of isospin violation, as obtained from worst-case error bars.

Because it is the sum $G_X^s(q^2) + G_X^{u,d}(q^2)$ from eq. (1) that is measured in experiments, fig. 3 provides a theoretical error bar for the extraction of $G_X^s(q^2)$. As shown explicitly in table III of ref. [11], modern experiments are already approaching this level of precision.

I am grateful to the organizers of *PAVI 2006* for the opportunity to participate in such an excellent conference, and to my isospin-breaking collaborators, Bastian Kubis and Nader Mobed. The critical reading of this manuscript by Bastian Kubis and Gerald Miller is greatly appreciated. The work was supported in part by the Canada Research Chairs Program and the Natural Sciences and Engineering Research Council of Canada.

References

1. SAMPLE Collaboration (D.T. Spayde *et al.*), Phys. Lett. B **583**, 79 (2004).
2. A4 Collaboration (F.E. Maas *et al.*), Phys. Rev. Lett. **93**, 022002 (2004).
3. F.E. Maas *et al.*, Phys. Rev. Lett. **94**, 152001 (2005).
4. HAPPEX Collaboration (K.A. Aniol *et al.*), Phys. Rev. Lett. **96**, 022003 (2006).
5. HAPPEX Collaboration (K.A. Aniol *et al.*), Phys. Lett. B **635**, 275 (2006).
6. G0 Collaboration (D.S. Armstrong *et al.*), Phys. Rev. Lett. **95**, 092001 (2005).
7. V. Dmitrašinović, S.J. Pollock, Phys. Rev. C **52**, 1061 (1995).
8. G.A. Miller, Phys. Rev. C **57**, 1492 (1998).
9. B.-Q. Ma, Phys. Lett. B **408**, 387 (1997).
10. R. Lewis, N. Mobed, Phys. Rev. D **59**, 073002 (1999).
11. B. Kubis, R. Lewis, Phys. Rev. C **74**, 015204 (2006).
12. G.A. Miller, private communication.
13. E. Jenkins, A.V. Manohar, Phys. Lett. B **255**, 558 (1991).
14. V. Bernard, N. Kaiser, U.-G. Meißner, Int. J. Mod. Phys. E **4**, 193 (1995).
15. R. Lewis, N. Mobed, PiN Newslett. **15**, 144 (1999).
16. T. Becher, H. Leutwyler, Eur. Phys. J. C **9**, 643 (1999).
17. G. Ecker, J. Gasser, A. Pich, E. de Rafael, Nucl. Phys. B **321**, 311 (1989).
18. G. Ecker, J. Gasser, H. Leutwyler, A. Pich, E. de Rafael, Phys. Lett. B **223**, 425 (1989).
19. J.F. Donoghue, C. Ramirez, G. Valencia, Phys. Rev. D **39**, 1947 (1989).
20. B. Kubis, U.-G. Meißner, Nucl. Phys. A **679**, 698 (2001).
21. A. Kucurkarslan, U.-G. Meißner, Mod. Phys. Lett. A **21**, 1423 (2006).

22. P. Mergell, U.-G. Meißner, D. Drechsel, Nucl. Phys. A **596**, 367 (1996).
23. M.A. Belushkin, H.-W. Hammer, U.-G. Meißner, Phys. Lett. B **633**, 507 (2006).
24. H.-W. Hammer, U.-G. Meißner, Eur. Phys. J. A **20**, 469 (2004).
25. M.A. Belushkin, H.-W. Hammer, U.-G. Meißner, hep-ph/0608337. (Ref. [11] used preliminary results.)

Interpretation of the NuTeV experiment

J.T. Londergan[a]

Department of Physics and Nuclear Theory Center, Indiana University, Bloomington, IN 47404, USA

Abstract. The NuTeV neutrino measurement of the Weinberg angle differs by 3 standard deviations from measurements at the Z pole. We review the status of various possibilities to explain this result, including contributions from physics outside the standard model; effects of radiative corrections; nuclear corrections to parton distributions; isospin-violating effects; and effects due to strange quarks.

PACS. 12.15.Ji Applications of electroweak models to specific processes – 12.15.Lk Electroweak radiative corrections – 13.15.+g Neutrino interactions

1 The NuTeV experiment

The NuTeV experiment [1] measured total cross-sections for neutrinos and antineutrinos scattered from an iron target. Precise measurements were made for the ratios

$$R^{\nu} = \frac{\sigma^{\nu}_{NC}}{\rho_0^2 \sigma^{\nu}_{CC}}; \qquad R^{\overline{\nu}} = \frac{\sigma^{\overline{\nu}}_{NC}}{\rho_0^2 \sigma^{\overline{\nu}}_{CC}}. \tag{1}$$

In eq. (1), R^{ν} is the ratio of the neutral-current (NC) to charged-current (CC) total cross-section for neutrinos, while $R^{\overline{\nu}}$ is the analogous quantity for antineutrinos, and $\rho_0 \equiv M_W/(M_Z \cos\theta_W)$ is one in the standard model. CC and NC reactions were differentiated according to the length of the subsequent track and the amount of missing energy. Events were selected using the criterion that the visible energy in the calorimeter satisfied the relation $20 < E_{vis} < 180\,\text{GeV}$, and that the event vertex was contained within the fiducial volume. The NuTeV group measured reactions for 1.62 million ν's, and 0.351 million $\overline{\nu}$'s. Paschos and Wolfenstein [2] had shown that the NC to CC ratios on an isoscalar target N_0 could provide an independent measurement of the Weinberg angle, through the relation

$$R^{PW} \equiv \frac{\langle \sigma^{\nu N_0}_{NC}\rangle - \langle \sigma^{\overline{\nu} N_0}_{NC}\rangle}{\rho_0^2 \left(\langle \sigma^{\nu N_0}_{CC}\rangle - \langle \sigma^{\overline{\nu} N_0}_{CC}\rangle\right)} = \frac{1}{2} - \sin^2\theta_W = \frac{R^{\nu} - R^{\overline{\nu}}}{1 - r\,R^{\nu}}. \tag{2}$$

The quantity r in eq. (2) is the ratio of $\bar{\nu}$ to ν CC total cross-sections.

The NuTeV group obtained the neutral/charged ratios $R^{\nu} = 0.3916 \pm 0.0007$ and $R^{\overline{\nu}} = 0.4050 \pm 0.0016$. Differences between cuts and acceptances for ν and $\overline{\nu}$ reactions, and the necessity to take into account charm mass corrections, make it impractical to construct directly the Paschos-Wolfenstein (PW) ratio. The NuTeV group calculated the Weinberg angle using Monte Carlo simulations of the neutrino reactions. Thus, corrections to the Weinberg angle calculated using the PW ratio will not accurately predict the corresponding changes for the NuTeV experiment. The NuTeV group [3] has published functionals that provide the sensitivity of their observables to a given effect.

The NuTeV result was $\sin^2\theta_W = 0.2277 \pm 0.0013$ (stat)$\pm$ 0.0009 (syst). This value is three standard deviations above the Weinberg angle value determined from electroweak (EW) processes near the Z pole, $\sin^2\theta_W = 0.2227 \pm 0.00037$ [4]. In a given renormalization scheme, radiative and loop corrections will produce a Q^2-dependence for the effective value of the weak mixing angle [5,6]. At the time of the NuTeV measurement, this Q^2-dependence had not been experimentally established; however, measurement of Møller scattering in polarized ee reactions at SLAC [7], together with atomic parity violation experiments in cesium [8], have clearly established this Q^2-dependence. Figure 1 shows the effective Weinberg angle *vs.* Q, calculated by Erler and Ramsey-Musolf [6]. This curve includes additional experimental points from the SLAC parity-violating eD scattering [9], a forward-backward asymmetry measurement from the reaction $\overline{p} + p \to \ell^+ + \ell^-$ at the TeVatron [10], and estimated errors for the Q-Weak ep measurement [11]; the NuTeV Weinberg angle result lies 3σ above the curve.

2 "New Physics" beyond the standard model

The simple form of the PW ratio in eq. (2) results from a large number of assumptions. Determination of the error bars in the Weinberg angle in the NuTeV experiment

[a] e-mail: tlonderg@indiana.edu

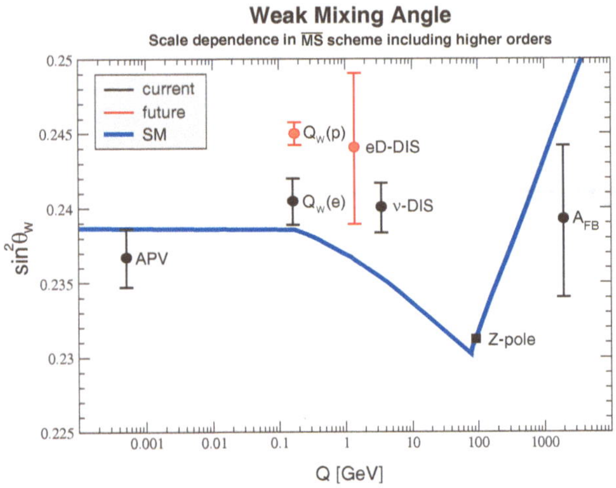

Fig. 1. Effective value for $\sin^2\theta_W$ *vs.* Q. The curve is from ref. [6], calculated in $\overline{MS}$ scheme. Experimental points represent atomic parity violation (APV) [8], Møller scattering ($Q_w(e)$) [7], measurements at the Z pole [4], NuTeV (ν-DIS) [1], e-D parity-violating scattering (eD-DIS) [9], forward-backward asymmetry from CDF (A_{FB}) [10], and an error estimate for the Q-Weak experiment ($Q_w(p)$) [11].

requires a careful assessment of the uncertainty resulting from each effect. One possibility is that the NuTeV result arises from contributions of physics beyond the standard model. The observed ratios of NC to CC cross-sections corresponds to a 1.2% decrease in the left-handed coupling g_L^2 of quarks to the EW current. It is quite a challenge for new particles to resolve the NuTeV neutrino measurement (at the 1% level), while leaving all other measured quantities within their experimental limits, since some observables measured at the Z pole are known to a precision at the level of 0.1% [4].

Davidson *et al.* undertook a comprehensive review of possible contributions to the NuTeV result from "new physics" [12]. They find that fitting the NuTeV Weinberg angle result plus existing observables would require delicate adjustment of the number, masses and couplings of these new particles. We briefly review a few of these attempts. The first candidates are the lightest supersymmetric particles; this possibility was also studied by Kurylov, Ramsey-Musolf and Su [13]. Both groups find that in general, SUSY loops in minimal supersymmetric models tend to have the wrong sign, and increase the size of the anomaly. Figure 2 plots contributions from minimal SUSY loops from light sleptinos and gauginos. Kurylov *et al.* also study possible R parity-violating interactions, and find that although they are capable of generating terms that would decrease the Weinberg angle anomaly, they tend to violate existing constraints from other EW observables.

High-mass extra Z' vector bosons, unmixed with the existing Z boson, could in principle resolve the NuTeV anomaly. Davidson *et al.* found that the most likely possibility was a Z' satisfying the gauge symmetry $B - 3L_\mu$, where B is the baryon number and L_μ the lepton number. However, the masses and couplings are tightly constrained by muon $g-2$ measurements. Leptoquarks are

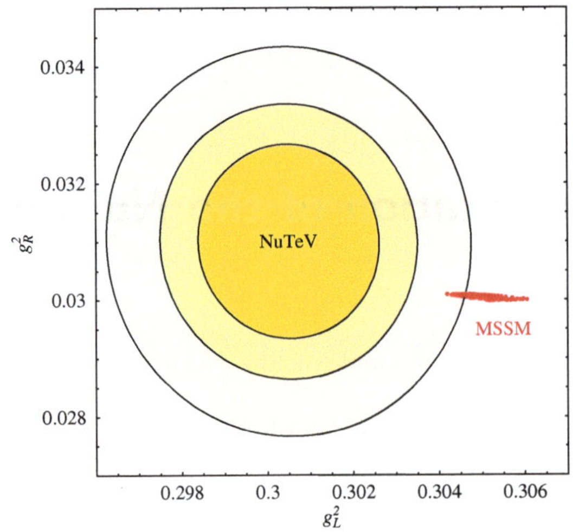

Fig. 2. (Color online) Deviations from standard model for various types of new particles, without conflicting by more than 1σ with existing EW bounds, from ref. [12], plotted *vs.* couplings of quarks to EW currents g_L^2 and g_R^2. Circles represent standard deviations from NuTeV result. Minimal supersymmetry loops from light sleptinos and gauginos.

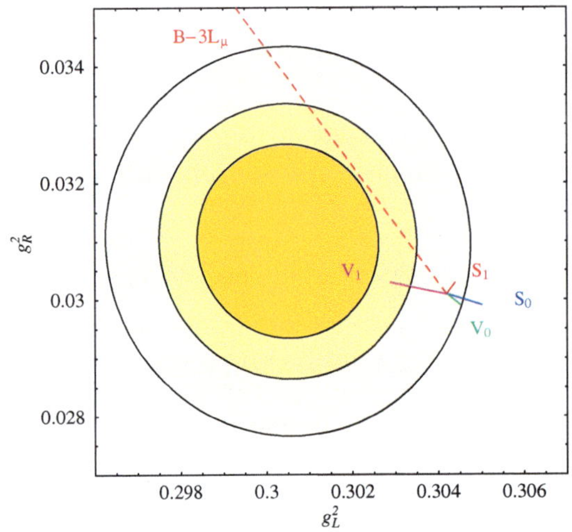

Fig. 3. (Color online) Deviations from standard model for various types of new particles, from ref. [12]; the notation is that of fig. 2. Solid lines: various leptoquark couplings; dashed line: extra $B - 3L_\mu$ gauge boson.

scalar or vector bosons that couple to both leptons and quarks; their existence could be tested at the LHC. Among couplings that conserve baryon and lepton number it was found that only triplet $SU(2)_L$ leptoquarks, with carefully chosen mass splittings, could explain the NuTeV effect. Figure 3 shows contributions from an extra Z' with gauge symmetry $B - 3L_\mu$ (dashed curve), and from minimal leptoquarks (solid curves). These are plotted *vs.* the left and right-handed couplings g_L^2 and g_R^2. Of the various scalar and vector $SU(2)_L$ symmetries for leptoquarks, Davidson *et al.* found that only the vector V_1 had the possibility of removing the NuTeV anomaly [12]. Oblique corrections (from high-mass particles that couple only to vector bosons) are tightly constrained by EW data, and were unable to resolve the NuTeV anomaly. The upcoming Q-Weak experiment with polarized e-p scattering at Jefferson Lab [11] should provide an independent test of physics beyond the standard model in determining the Weinberg angle.

3 "QCD Corrections" to the NuTeV result

Because existing EW experiments place tight constraints on physics outside the standard model, recent investigations have focused on the role of "QCD Corrections" within the standard model. We review the status of the most promising effects.

3.1 Radiative and nuclear corrections

Since the PW ratio requires an isoscalar target, one must evaluate corrections to the Weinberg angle due to the excess neutrons in the iron target. The NuTeV Collaboration have taken this correction into account. The correction is large (of order −0.008), but should be known to within a couple of percent. Radiative corrections, which involve coupling of soft photons to the final muon line, are important for CC events, and constitute a substantial correction. Recently, Diener, Dittmaier and Hollik have re-calculated the radiative corrections, including all corrections of order $\mathcal{O}(\alpha)$, and a number of additional higher-order corrections [14,15]. They find some differences from the older radiative correction program of Bardin and Dokuchaeva [16]. Diener *et al.* also include new terms that result from including EM coupling in the QCD evolution equations. Such terms have recently been included by the MRST group [17], and by Glück *et al.* [18]; we will review these terms in our discussion of isospin-violating effects. Diener *et al.* estimate that radiative correction effects would remove about one-fourth of the NuTeV anomaly [15]. However, such corrections are renormalization scheme dependent, and Diener *et al.* point out that it is unclear whether their calculations and the parton distribution functions (PDFs) that they use have both been calculated in fully consistent schemes. The NuTeV group is currently reanalyzing their data, using a radiative corrections code provided by Diener *et al.* [19].

The NuTeV measurements require nuclear corrections for the structure functions. Kumano [20] calculated a "modified PW relation" for nuclei, and Hirai *et al.* [21] estimate that nuclear effects could remove up to one-third of the NuTeV anomaly. They assume that nuclear shadowing for neutrinos is identical to that for charged leptons [22]. Miller and Thomas [23] point out that ν shadowing effects could differ significantly from shadowing of muons [24]. They also emphasize that shadowing produces different effects for CC and NC events. Brodsky *et al.* [25] made a detailed calculation of shadowing and antishadowing arising from multigluon exchange. They conclude that nuclear shadowing effects could account for roughly 20% of the NuTeV anomaly.

3.2 Isospin-violating effects

Charge symmetry is an isospin rotation that corresponds to interchange of up and down quarks (simultaneously interchanging neutron and proton labels). In nuclear physics, this symmetry is particularly well respected; it is generally obeyed to better than 1% [26,27]. There is no direct evidence for charge symmetry violation (CSV) in high-energy experiments [28], and until recently all phenomenological analyses assumed charge symmetric PDFs. Valence quark CSV PDFs are defined by $\delta d_{\rm v}(x) \equiv d^p_{\rm v}(x) - u^n_{\rm v}(x)$ and $\delta u_{\rm v}(x) \equiv u^p_{\rm v}(x) - d^n_{\rm v}(x)$. CSV and strange quark corrections to the PW ratio have the form

$$\Delta R^{PW} \approx \left[1 - \frac{7}{3}\sin^2\theta_W\right] \frac{\delta[U^-] - \delta[D^-] - [S^-]}{2([U^-] + [D^-])} . \quad (3)$$

In eq. (3), $[Q^-] = \langle x(q-\bar{q})\rangle$ represents the total momentum carried by valence PDFs of a given flavor, and the denominator contains the total momentum carried by up and down valence quarks. We have for simplicity omitted a small QCD radiative correction proportional to α_s [12]. The first two terms in eq. (3) are CSV corrections, while the last term represents a strange quark contribution. Sea quark contributions to the NuTeV result are much smaller than valence effects (only valence PDFs contribute to the PW ratio).

The MRST group [29] have now included isospin violation in phenomenological PDFs. They carried out a global fit of PDFs using a wide variety of high-energy data. In this fit, MRST assumed that $\delta u_{\rm v}$ and $\delta d_{\rm v}$ were equal and opposite, and given by the form

$$\begin{aligned} \delta u_{\rm v}(x) &= -\delta d_{\rm v}(x) = \kappa(1-x)^4 x^{-0.5}\,(x - 0.0909), \\ \delta[U^-] &\equiv \int_0^1 x\delta u_{\rm v}(x)\,{\rm d}x. \end{aligned} \quad (4)$$

The functional form was chosen to make the MRST CSV PDFs agree qualitatively with the valence quark PDFs at small and large x. Equation (4) also insures that the first moment of the CSV PDFs is zero, a necessary condition to preserve valence quark normalization. The coefficient κ was varied in a global fit to high-energy data. MRST obtained a shallow χ^2 minimum about the value $\kappa = -0.2$, with the 90% confidence level for the range $-0.8 \le \kappa \le +0.65$.

The MRST group also found evidence for sea quark CSV, although the sea quark correction to the Weinberg angle is much smaller than valence CSV. Using the best-fit CSV values obtained by MRST would remove roughly 1/3 of the NuTeV anomaly. Note that the form chosen by MRST, eq. (4), insures that $\delta u_{\rm v}$ and $\delta d_{\rm v}$ add coherently in the correction to the PW ratio, eq. (3). The value $\kappa = -0.6$, within the 90% confidence limit found by MRST, would completely remove the NuTeV anomaly, while $\kappa = +0.6$ would double the discrepancy. The MRST results show that isospin-violating PDFs could completely remove the NuTeV anomaly in the Weinberg angle, or make it twice as large, without serious disagreement with any of the high-energy data used to extract quark and gluon PDFs.

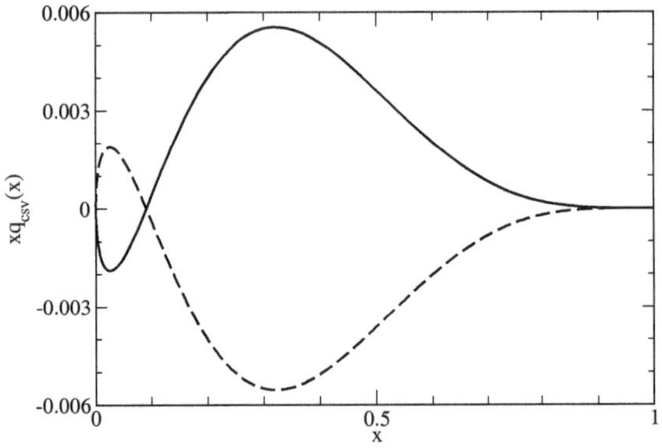

Fig. 4. MRST valence CSV PDFs from ref. [29], corresponding to best fit value $\kappa = -0.2$ in eq. (4). Solid curve: $x\delta d_v(x)$; dashed curve: $x\delta u_v(x)$.

Sather [30] and Rodionov *et al.* [31] had produced theoretical estimates of valence parton CSV. Sather derived an analytic approximation to valence quark PDFs, while Rodionov *et al.* calculated valence CSV expressions from quark models. In both cases, the resulting CSV corrections tend to decrease the size of the NuTeV anomaly, and they remove roughly 30%, or 1σ of the discrepancy in the Weinberg angle. Figure 4 shows the valence quark CSV PDFs for the MRST best fit value $\kappa = -0.2$; this is in good qualitative agreement with both Sather and Rodionov theoretical results [32], and all of these are substantially larger than earlier estimates of CSV effects by the NuTeV group [3].

A new mechanism for isospin violation in PDFs has been explored by the MRST group [17] and Glück *et al.* [18]. An important term in QCD evolution occurs when a quark radiates a gluon. An analogous term arises when a quark radiates a photon; both groups include these "QED evolution" terms in the DGLAP evolution equations. These terms produce an additional source of isospin violation in quark PDFs, and they also give rise to a "photon parton distribution." With reasonable assumptions about the starting scale for these QED effects, both groups estimate that QED evolution effects should remove roughly 1/3 of the NuTeV anomaly. This effect, which amounts to evolving the QED effects while freezing the QCD distributions, should be added to the theoretical estimates of parton CSV in QCD distributions. At present, the best estimate is that the combination of the two CSV effects would remove about 2/3 of the NuTeV anomaly.

3.3 Strange quark contribution

The contribution to the PW ratio from strange quarks depends on the momentum asymmetry between strange quarks and antiquarks. From the definitions

$$S^-(x) \equiv x\,(s(x) - \bar{s}(x))\,,$$
$$[S^-] = \int_0^1 S^-(x)\,\mathrm{d}x, \qquad (5)$$

then eq. (3) shows that the strange quark correction to the PW ratio is proportional to $[S^-]$, and that a positive value for $[S^-]$ will decrease the Weinberg angle anomaly. The most precise determinations of strange quark PDFs are obtained from opposite-sign dimuons produced from ν or $\bar{\nu}$ beams, where the former (latter) is sensitive to the s ($\bar{s}$) PDFs. The CCFR [33] and NuTeV [34] Collaborations have measured these cross-sections. The CTEQ group [35, 36] analyzed this in a global fit that allowed the possibility of $s \neq \bar{s}$. Their most likely value was $[S^-] \sim 0.0017$, which would remove roughly one-fourth of the NuTeV anomaly. The CTEQ group found the strange momentum asymmetry to be within the range $-0.001 < [S^-] < +0.004$; thus $[S^-]$ is consistent with zero.

The initial NuTeV analysis reported a negative value for $[S^-]$ [3]. Subsequently, the NuTeV and CTEQ groups collaborated to produce consistent analyses. Although a few differences still persist in the analyses, the latest results from a new analysis by NuTeV give $[S^-] = +0.00176 \pm 0.00043\,(\mathrm{stat})$ [37,38]. Thus, NuTeV and CTEQ now agree on the strange quark asymmetry. Both obtain a slightly positive best value for the strange quark asymmetry, which would remove roughly one-fourth of the NuTeV anomaly; however, both results are also consistent with zero strange quark momentum asymmetry. Catani *et al.* found an additional source of strange quark asymmetry [39]. They showed that, at NNLO, one can perturbatively generate nonzero values for $[S^-]$ even if this quantity is zero at the starting scale (this is not possible in LO or NLO). They generate a small negative value for $[S^-]$ from perturbative three-loop contributions.

4 Conclusions

We have reviewed several possible effects that might contribute to the NuTeV result for the Weinberg angle from neutrino interactions. It is difficult to find physics outside the standard model that can reconcile the NuTeV measurement of the Weinberg angle, while agreeing with the extremely precise EW data at the Z pole. In particular, SUSY loops tend to produce effects that would increase the size of the NuTeV anomaly. Recent attention has focused on QCD corrections. Of these effects, the most viable single effect is corrections from isospin-violating PDFs. Our best estimates are that CSV effects could remove 2/3 of the NuTeV anomaly. MRST global fits show that CSV effects are capable of removing 100% of the NuTeV anomaly, or they could double the size of that anomaly, without serious disagreement with any high-energy data. Partonic CSV effects large enough to explain the NuTeV anomaly would be visible at the few percent level in future experiments [32,40]. Another possible explanation would be a strange quark momentum asymmetry. Analyses by the CTEQ and NuTeV groups are now in agreement; their best estimate is that strange quark effects would remove roughly 1/4 of the NuTeV anomaly, but the results are also consistent with zero contribution from strange quarks. Another correction arises from nuclear effects in neutrino scattering; Brodsky *et al.* [25]

estimate that this could remove about 20% of the NuTeV anomaly.

The NuTeV group is currently re-analyzing their data, including a new package to calculate radiative corrections in their experiment. The results of this new analysis will be very useful. The NuTeV experiment has sparked much interest in analyzing various possible contributions to their result. We now have detailed estimates of the contributions from a series of potential effects. It appears that potential contributions from various QCD effects, particularly isospin-violating effects in parton distributions, are considerably larger than were initially estimated. In the absence of a complete re-analysis of the NuTeV experiment, it is not clear whether these effects would change the NuTeV value for the Weinberg angle; however it does suggest that the error bars on this measurement are larger than have been quoted.

Theoretical work cited here was carried out with A.W. Thomas and D.P. Murdock. The author thanks R.H. Bernstein, S. Dittmaier, F. Olness, S.A. Page, M.J. Ramsey-Musolf, W-K Tung and R.S. Thorne for useful discussions. Research sponsored in part by US NSF under grant NSF PHY0244822.

References

1. NuTeV Collaboration (G.P. Zeller *et al.*), Phys. Rev. Lett. **88**, 091802 (2002).
2. E.A. Paschos, L. Wolfenstein, Phys. Rev. D **7**, 91 (1973).
3. NuTeV Collaboration (G.P. Zeller *et al.*), Phys. Rev. D **65**, 111103 (2002).
4. D. Abbaneo *et al.*, arXiv:hep-ex/0112021.
5. A. Czarnecki, W.J. Marciano, Phys. Rev. D **53**, 1066 (1996).
6. J. Erler, M.J. Ramsey-Musolf, Phys. Rev. D **72**, 073003 (2005).
7. SLAC E158 Collaboration (P.L. Anthony *et al.*), Phys. Rev. Lett. **95**, 081601 (2005).
8. S.C. Bennett, C.E. Wieman, Phys. Rev. Lett. **82**, 2484 (1999).
9. C.Y. Prescott *et al.*, Phys. Lett. B **77**, 377 (1978).
10. CDF Collaboration (D. Acosta *et al.*), Phys. Rev. D **71**, 052002 (2005).
11. Q-Weak Collaboration, Thomas Jefferson National Laboratory; see talk by S. Page at this conference.
12. S. Davidson, S. Forte, P. Gambino, N. Rius, A. Strumia, JHEP **202**, 037 (2002).
13. A. Kurylov, M.J. Ramsey-Musolf, S. Su, Nucl. Phys. B **667**, 321 (2003).
14. K-P.O. Diener, S. Dittmaier, W. Hollik, Phys. Rev. D **69**, 073005 (2004).
15. K-P.O. Diener, S. Dittmaier, W. Hollik, Phys. Rev. D **72**, 093002 (2005).
16. D.Y. Bardin, V.A. Dokuchaeva, JINR-E2-86-260 (unpublished).
17. MRST Collaboration (A.D. Martin, R.G. Roberts, W.J. Stirling, R.S. Thorne), Eur. Phys. J. C **39**, 155 (2005).
18. M. Glück, P. Jimenez-Delgado, E. Reya, Phys. Rev. Lett. **95**, 022002 (2005).
19. R.H. Bernstein, private communication.
20. S. Kumano, Phys. Rev. D **66**, 111301 (2002).
21. M. Hirai, S. Kumano, T.-H. Nagai, Phys. Rev. D **71**, 113007 (2005).
22. M. Hirai, S. Kumano, T.-H. Nagai, Phys. Rev. C **70**, 044905 (2004).
23. G.A. Miller, A.W. Thomas, Int. J. Mod. Phys. A **20**, 95 (2005).
24. C. Boros, J.T. Londergan, A.W. Thomas, Phys. Rev. D **58**, 114030 (1998).
25. S.J. Brodsky, I. Schmidt, J.-J. Yang, Phys. Rev. D **70**, 116003 (2004).
26. G.A. Miller, B.M.K. Nefkens, I. Slaus, Phys. Rep. **194**, 1 (1990).
27. E.M. Henley, G.A. Miller, *Mesons in Nuclei*, edited by M. Rho, D.H. Wilkinson (North-Holland, Amsterdam, 1979).
28. J.T. Londergan, A.W. Thomas, *Progress in Particle and Nuclear Physics*, Vol. **41**, edited by A. Faessler (Elsevier Science, Amsterdam, 1998) p. 49.
29. MRST Collaboration (A.D. Martin, R.G. Roberts, W.J. Stirling, R.S. Thorne), Eur. Phys. J. C **35**, 325 (2004).
30. E. Sather, Phys. Lett. B **274**, 433 (1992).
31. E. Rodionov, A.W. Thomas, J.T. Londergan, Mod. Phys. Lett. A **9**, 1799 (1994).
32. J.T. Londergan, D.P. Murdock, A.W. Thomas, Phys. Rev. D **72**, 036010 (2005).
33. CCFR Collaboration (A.O. Bazarko *et al.*), Z. Phys. C **65**, 189 (1995).
34. NuTeV Collaboration (M. Goncharov *et al.*) Phys. Rev. D **64**, 112006 (2001).
35. S. Kretzer, F. Olness, J. Pumplin, D. Stump, W.-K. Tung, M.H. Reno, Phys. Rev. Lett. **93**, 041802 (2004).
36. F. Olness, J. Pumplin, D. Stump, J. Huston, P. Nadolsky, H.L. Lai, S. Kretzer, J.F. Owens, W.-K. Tung, Eur. Phys. J. C **40**, 145 (2005).
37. D.A. Mason, University of Oregon, PhD Thesis (2006, unpublished) `http://lss.fnal.gov/archive/thesis/fermilab-thesis-2006-01.shtml`.
38. D.A. Mason, in *Deep Inelastic Scattering (DIS 2006), Proceedings of the 14th International Workshop, Tsukuba, Japan, 20–24 April 2006*, edited by M. Kuze, H. Nagano, K. Tokoshuku (World Scientific, Tokyo, 2006) p. 165.
39. S. Catani, D. de Florian, G. Rodrigo, W. Vogelsang, Phys. Rev. Lett. **93**, 152003 (2005).
40. J.T. Londergan, D.P. Murdock, A.W. Thomas, Phys. Rev. D **73**, 076004 (2006).

Nucleon form factors and dispersion relations

S. Pacetti[a]

Laboratori Nazionali di Frascati dell'INFN, Frascati, Italy

Abstract. We define a dispersive technique, based on data, analyticity and dispersion relations to handle the nucleon form factors. Two applications are explicitly considered. In the first case, we use the available theoretical and experimental information on the ratio between the electric and magnetic proton form factors, as inputs in the dispersive procedure that gives a complex description of this ratio in the whole q^2 complex plane. In the second case, by inverting a dispersion relation and considering the data and the pQCD asymptotic behaviour as inputs, the magnetic nucleon form factors are reconstructed in their unphysical region. Resonances and phases in agreement with the expectations are found.

PACS. 13.40.Gp Electromagnetic form factors – 11.55.Bq Analytic properties of S matrix – 11.55.Fv Dispersion relations – 13.66.Bc Hadron production in e^-e^+ interactions

1 Introduction

Nucleon electromagnetic form factors (ffs) play a fundamental role in our understanding of the hadronic dynamics. In fact, for over 40 years, many attempts [1] have been made through dispersion relations (DRs) either to unravel these structure functions in the unphysical region or to obtain information on their asymptotic behaviour.

We consider two applications where DRs are used to handle nucleon ffs.

In the first case, the DRs are applied in "direct way" to give a complete description of the ratio $R = \mu_p G_E^p / G_M^p$. Theoretical and experimental constraints have been imposed on an analytic parameterization of $R(q^2)$. We obtained, for this ratio, a description defined in the whole q^2 complex plane.

In the second application the magnetic proton and neutron ffs G_M^p and G_M^n are considered. In this case a special DR, *i.e.* the DR for the logarithm [2], has been inverted to define an integral equation, which has, as inputs, the space- and time-like data with the asymptotic behaviour provided by the pQCD [3] and, as outcome, the modulus of the ffs in their unphysical region.

2 Analyticity and experimentally accessible regions

The physical ffs are the limit values for real q^2's (blue on-line line in fig. 1) of analytic functions, defined in the

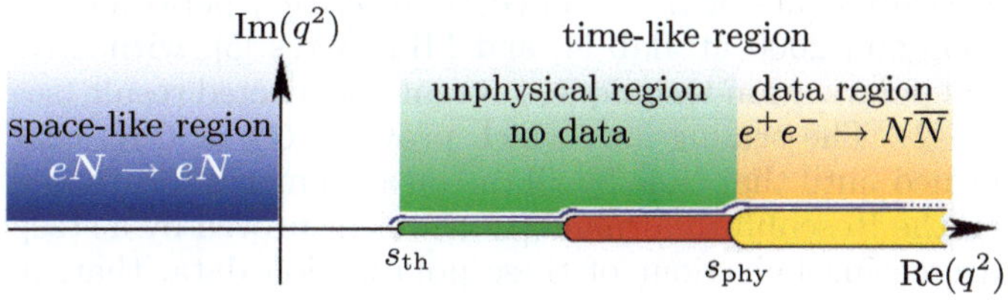

Fig. 1. The q^2 complex plane. Cuts on the real axis and, experimentally accessible and forbidden regions are shown.

q^2 complex plane with a discontinuity, on the real axis, from the theoretical threshold $s_{\rm th} = (2M_\pi)^2$ up to infinity. Such a discontinuity is the superposition of infinite cuts (see fig. 1), each of them corresponds to the opening of a channel for every allowed final state. In addition, below $s_{\rm th}$ the ffs are real and above this threshold they are complex. From the experimental point of view, data on the nucleon ffs may be extracted from the scattering $\sigma(eN \to eN)$ and the annihilation $\sigma(e^+e^- \to N\overline{N})$ cross-sections. More precisely following the schematic representation in fig. 1:

- the scattering gives information on the real values of the ffs in the space-like region ($q^2 \leq 0$);
- the annihilation allows to measure the moduli of the ffs above the physical threshold $s_{\rm phy} = (2M_N)^2$;
- it remains the energy interval, called unphysical region, which goes from $s_{\rm th}$ up to $s_{\rm phy}$, where the ffs are not experimentally accessible.

[a] e-mail: `simone.pacetti@lnf.infn.it`

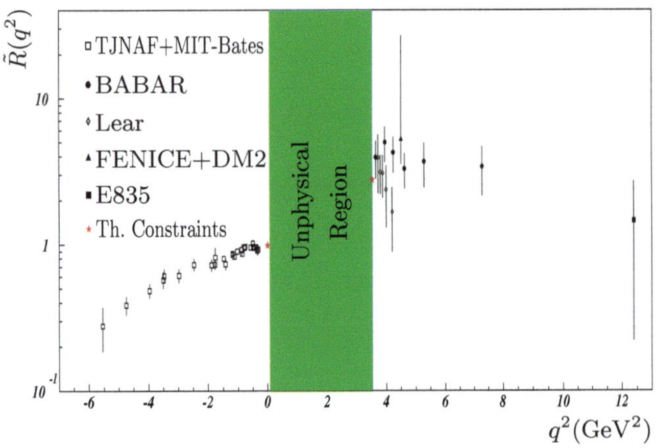

Fig. 2. Data in the space-like (TJNAF [4] and MIT-Bates [5]) and time-like (BABAR [6], Lear [7], FENICE [8]+DM2 [9] and E835 [10]) regions. The stars are the theoretical constraints and the shadowed area is the unphysical region. The label $\tilde{R}(q^2)$ of the ordinate axis is defined as: $\tilde{R}(q^2) = R(q^2)$ for $q^2 \leq s_{\rm th}$ and $\tilde{R}(q^2) = |R(q^2)|$ for $q^2 > s_{\rm th}$.

3 Dispersive description of $G_E^p(q^2)/G_M^p(q^2)$

3.1 Motivations

Measurements of the ratio $G_E^p(q^2)/G_M^p(q^2)$ performed in 1999 and 2000 at Jlab [4] and MIT-Bates [5], with a recoil polarization technique, gave an unexpected result (see fig. 2). The scaling law, which provides $G_E \simeq G_M$, confirmed until that time by all the measurements [11] (based on the Rosenbluth formula [12]), was disproved by a steep decreasing behaviour of these polarization data, that, in addition, seems to prelude to a space-like zero lying somewhere around $q^2 \sim 15\,\mathrm{GeV}^2$ [13].

Recently, new data for the modulus $|R(q^2)|$ became available (see fig. 2). The surprising space-like Jlab data have brought a great deal of interest in the nucleon ffs field. Many models [14–17] have been built in order to fit these data and to justify [18] the incompatibility with the previous measurements and then with the scaling law.

But, while in the space-like region, over the interval of the data $q^2 \sim (-6\text{–}0)\,\mathrm{GeV}^2$, all these models agree, their analytic continuations in the time-like region are far from each other (fig. 4b). The reason for this time-like incompatibility must be sought in the functional structure of these fit functions that are not rigorously analytic in q^2.

3.2 Dispersion relations

Since a generic ff $f(q^2)$ is an analytic function, which vanishes asymptotically as a positive power of $(1/q^2)$, when $|q^2|$ diverges, it verifies the DR for the imaginary part:

$$f(q^2) = \frac{1}{\pi}\int_{s_{\rm th}}^{\infty} \frac{\mathrm{Im}[f(s)]}{s-q^2}\mathrm{d}s, \tag{1}$$

with $q^2 \leq 0$. This expression connects the space-like values of $f(q^2)$ to its imaginary part over the cut.

We used the relation of eq. (1) for the ratio $R(q^2)$, which inherits all the properties of both the functions $G_E^p(q^2)$ and $G_M^p(q^2)$, except for the asymptotic behaviour and, in some cases, the analyticity itself. In fact, the presence of zeros for $G_M^p(q^2)$, which represent poles for $R(q^2)$, would reduce the analyticity domain of the ratio. However, the relation of eq. (1) can be easily generalized to the case of a function with a set of isolated poles if their positions are known. But, since we verified that $G_M^p(q^2)$ has no zeros in the q^2 physical sheet [19] (see sect. 4.4), we consider $R(q^2)$ as an analytic function.

Concerning the asymptotic behaviour, the pQCD [3] prescription is to use for both the electric and magnetic ff a power law $\sim (1/q^2)^2$, when $q^2 \to \infty$. It follows that the ratio $R(q^2)$ has a constant limit. Hence we use the DR

$$R(q^2) = R(0) + \frac{q^2}{\pi}\int_{s_{\rm th}}^{\infty} \frac{\mathrm{Im}[R(s)]}{s(s-q^2)}\mathrm{d}s, \tag{2}$$

subtracted at $q^2 = 0$ and where $R(0) = 1$. Once the imaginary part is known, the DR

$$\mathrm{Re}[R(s)] = 1 + \frac{s}{\pi}\mathcal{P}\int_{s_{\rm th}}^{\infty} \frac{\mathrm{Im}[R(s')]}{s'(s'-s)}\mathrm{d}s' \tag{3}$$

gives the real part of $R(s)$ over the cut ($s > s_{\rm th}$).

3.3 The dispersive approach

Contrary to the previous models, that are constructed starting from the data in the space-like region, we start from the imaginary part of $R(q^2)$, which is different from zero only over the cut $[s_{\rm th}, \infty[$. We parametrize this function in a general and model-independent way, *i.e.* by means of two series of orthogonal polynomials:

$$\mathrm{Im}[R(s)] \equiv \mathcal{I}(\boldsymbol{C},\boldsymbol{D},s) = \begin{cases} \sum_i^N C_i T_i(x), & x = \frac{2s - s_{\rm phy} - s_{\rm th}}{s_{\rm phy} - s_{\rm th}}, \\ & s_{\rm th} \leq s \leq s_{\rm phy}, \\ \sum_j^M D_j T_j(y), & y = \frac{s_{\rm phy}}{s} - 1, \\ & s > s_1, \end{cases} \tag{4}$$

where $s_{\rm phy} = 4M_p^2$ is the proton physical threshold. The two series, with coefficients $\boldsymbol{C} = (C_1, C_2, \ldots, C_M)$ and $\boldsymbol{D} = (D_1, D_2, \ldots, D_N)$, cover two distinguished intervals: the unphysical region (green on-line area in figs. 1, 2), where the resonance contributions to both G_E^p and G_M^p are smoothed out in the ratio, and the remaining part of the cut $[s_{\rm phy}, \infty[$ (orange on-line area in fig. 1). The advantage in using this procedure is twofold:

- some important theoretical constraints may be directly imposed on the imaginary part of $R(q^2)$;
- DRs (eqs. (2), (3)) can be used to perform the rigorous analytic continuation

$$R(q^2) = 1 + \frac{q^2}{\pi}\mathcal{P}\int_{s_{\rm th}}^{\infty} \frac{\mathcal{I}(\boldsymbol{C},\boldsymbol{D},s)}{s(s-q^2)}\mathrm{d}s + i\mathcal{I}(\boldsymbol{C},\boldsymbol{D},q^2), \tag{5}$$

which holds for all values of q^2 and where the dependence on the parameters $\boldsymbol{C}$ and $\boldsymbol{D}$ is explicitly shown

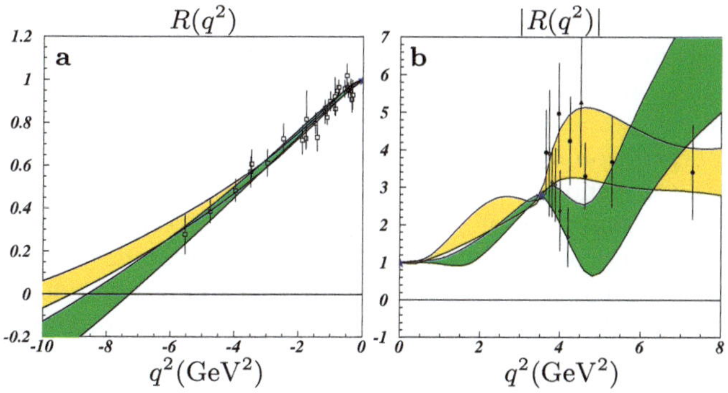

Fig. 3. (Colour on-line) Description for $R(q^2)$ in the space-like (a) and time-like (b) region. The band represents the error. Lear case in green and BABAR case in yellow.

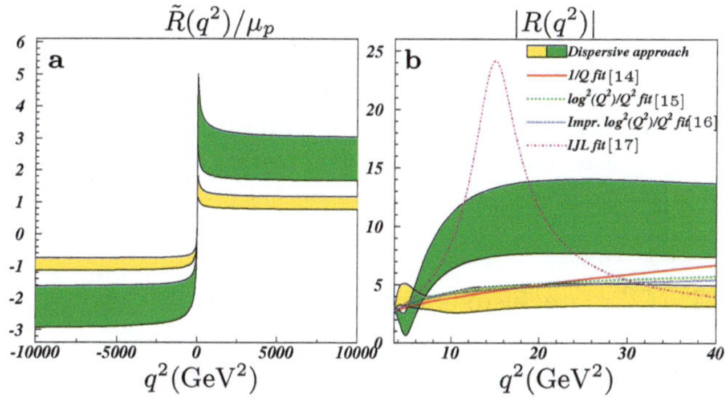

Fig. 4. a) Ratio $G_E^p(q^2)/G_M^p(q^2)$ [$\equiv R(q^2)/\mu_p$] over a wide range of q^2. b) Ratio $R(q^2)$ in the time-like region with $q^2 \geq s_{\text{phy}}$ and comparison with the time-like analytic continuations of various models ($Q^2 = -q^2$). $\tilde{R}(q^2)$ is defined as in fig. 2.

(eq. (4)). Once these parameters are fixed, the function $R(q^2)$ is completely known.

In order to select the best values for $\boldsymbol{C}$ and $\boldsymbol{D}$, a χ^2 with both experimental and theoretical contributions has been defined. While the experimental constraints shape the function $R(q^2)$ in the data regions, the theoretical conditions provide a correct analytic structure, which allows to extend the definition of $R(q^2)$ to all values of q^2.

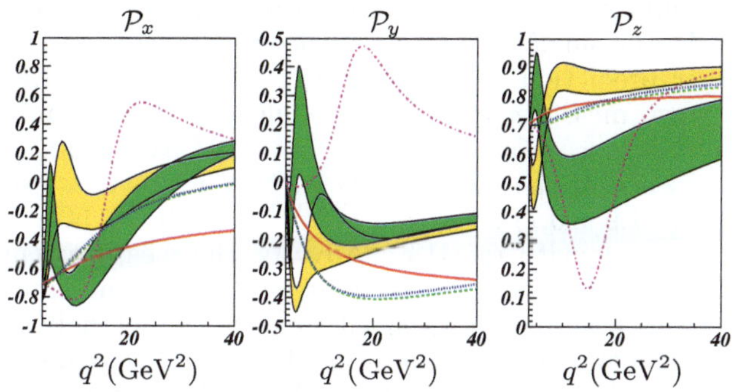

Fig. 5. Predictions ($\theta = 45°$, $P_e = 1$) for polarizations $\mathcal{P}_x$, $\mathcal{P}_y$ and $\mathcal{P}_z$ in the time-like region compared with those of the other models (same references of fig. 4b).

3.4 Results and predictions

In the pictures of fig. 3 the obtained dispersive descriptions for the ratio $R(q^2)$ is shown. Due to the incompatibility of the BABAR and Lear time-like data, the dispersive procedure has been applied by considering separately these two sets of data. More precisely, in the Lear case, in the time-like region *only* the Lear data are used, while in the BABAR case we consider in the time-like region all other measurements, *i.e.*, the BABAR data, the point of FENICE+DM2 and that of E835. Figure 3a shows the results in the space-like region.

Here, the descriptions obtained with the BABAR and Lear data, are compatible, both the cases predict a space-like zero at:

$$t_0^{\text{BABAR}} = (-10 \pm 1)\,\text{GeV}^2, \quad t_0^{\text{Lear}} = (-7.9 \pm 0.7)\,\text{GeV}^2. \quad (6)$$

This means that the unique set of space-like data constrains the two incompatible sets of time-like data to be analytically consistent with the hypothesis of a space-like zero.

Figure 3b shows the behaviour of $|R(q^2)|$ in the time-like region. It is just in this region that the difference between the two cases is more evident and it will be more and more in the asymptotic regime.

According to the Phragmèn-Lindelöff theorem [20], that has been implemented by imposing a vanishing asymptotic value for $\text{Im}[R(q^2)]$, we obtain a real time-like limit for the ratio $R(q^2)$, as $q^2 \to +\infty$. As it is shown in fig. 4a, the space- and time-like limits have in modulus the same value, but they have opposite sign, *i.e.*:

$$\lim_{q^2 \to \pm\infty} \left| \frac{G_E^p(q^2)}{G_M^p(q^2)} \right| = \left[\frac{G_E^p}{G_M^p} \right]_\infty = \begin{cases} 0.95 \pm 0.20 & \text{BABAR} \\ 2.3 \pm 0.7 & \text{Lear} \end{cases}, \quad (7)$$

where $[G_E^p/G_M^p]_\infty$ is the real and positive time-like limit for $G_E^p(q^2)/G_M^p(q^2)$. By considering the result of eq. (7), the scaling law, providing $G_E^p \simeq G_M^p$, could be restored at least in particular conditions. In fact, in the BABAR case and in the time-like asymptotic regime we get: $G_E^p \simeq G_M^p$, while, as $q^2 \to -\infty$: $G_E^p \simeq -G_M^p$. This opposite sign is due to the space-like zero.

We can also give a prediction for the asymptotic behaviour of the ratio between the Pauli and Dirac ffs as

$$\lim_{q^2 \to \infty} \frac{q^2}{4M_p^2} \frac{F_2^p(q^2)}{F_1^p(q^2)} = \begin{cases} -0.05 \pm 0.20 & \text{BABAR} \\ 1.3 \pm 0.7 & \text{Lear} \end{cases}. \quad (8)$$

This means that, as q^2 diverges, $F_2^p(q^2)/F_1^p(q^2)$ scales like $(1/q^2)$ or faster in the BABAR case. The knowledge of the complex structure of $R(q^2)$ allows to give predictions concerning also quantities, such as the polarization observables, which depend explicitly on the real and imaginary parts. In fact, in the time-like region, when we consider the annihilation $e^+e^- \to p\bar{p}$, the complex phase of the ffs gives rise to special polarization effects (see fig. 5). The outgoing proton may be polarized even if there are no polarized initial leptons. This polarization depends on the relative phase between $G_E^p(q^2)$ and $G_M^p(q^2)$, *i.e.* the phase $\Phi(q^2)$ of $R(q^2)$. The polarization vector $\boldsymbol{\mathcal{P}}$ has the components [21]:

$$\boldsymbol{\mathcal{P}} = \left(-P_e \frac{2\sin(\theta)|R|\cos(\Phi)}{D\sqrt{\tau}}, -\frac{\sin(2\theta)|R|\sin(\Phi)}{D\sqrt{\tau}}, P_e \frac{2\cos(\theta)}{D} \right), \quad (9)$$

where $\tau = q^2/4M_p^2$ and $D = [1+\cos^2(\theta)+\frac{1}{\tau}|\tau|^2 \sin^2(\theta)]/\mu_p$. P_e is the longitudinal polarization of the initial lepton, θ is the scattering angle, z is the direction of the outgoing proton in CoM and y is that orthogonal to the scattering plane. $\mathcal{P}_y$ does not depend on P_e, while the longitudinal polarization $\mathcal{P}_z$ does not depend on the phase Φ.

3.5 G_E^p/G_M^p: conclusions and perspectives

We have used a dispersive approach to construct an expression for the ratio $R(q^2)$, defined in the whole q^2 complex plane. This parametrization verifies all the constraints imposed by the theory and available data. The knowledge of the complex structure of $R(q^2)$ allows to formulate a wide range of predictions as the low energy and asymptotic behaviour of the ratio between Dirac and Pauli ffs, the single and double nucleon polarizations, etc. Another prediction that should be confirmed or refuted soon (new measurements are scheduled at Jlab to push down the space-like limit at $q^2 \simeq -10\,\text{GeV}^2$ [22]), is the presence of the space-like zero, that we estimate at $t_0^{\text{Lear}} = (-7.9 \pm 0.7)\,\text{GeV}^2$ and $t_0^{\text{BABAR}} = (-10 \pm 1)\,\text{GeV}^2$ (in agreement with ref. [13]). Anyway, the measurement of $|R(q^2)|$ and of the polarizations (eq. (9)) in the time-like region would have a crucial significance not only to disentangle among the models (fig. 4b), but also to gain a rather complete experimental knowledge of the proton ffs.

4 The "inverse" problem: extracting unphysical nucleon magnetic ffs from time-like and space-like data

4.1 Motivations

As we have already stated in sect. 2 and sketched in fig. 1, there exists an energy interval, called unphysical region, where the ffs are not experimentally accessible.

Nevertheless, by means of the DRs (sect. 3.2) we can relate values of the ffs in different regions of q^2. Then, in principle, it is possible to extract the values of a ff on a certain interval of q^2, by knowing its values in other regions.

To use directly the space- and time-like data on $G_M^{p,n}(q^2)$, we consider the DR for the logarithm [1,2,19]:

$$\ln[G(q^2)] = \frac{\sqrt{s_{\text{th}} - q^2}}{\pi} \int_{s_{\text{th}}}^{\infty} \frac{\ln|G(s)|\mathrm{d}s}{(s - q^2)\sqrt{s - s_{\text{th}}}} \qquad (10)$$

with $q^2 \leq s_{\text{th}}$. This integral contains just measurable quantities in the time-like region: the modulus of the ffs. By means of space- and time-like data, analyticity and DRs, we define a practical strategy to determine ffs in the unphysical region in a model-independent way, without bias towards expected resonances.

Uncovering the nucleon unphysical region is an interesting task by many points of view:

- the time-like data on $|G_M^p(q^2)|$ and $|G_M^n(q^2)|$, above the physical threshold s_{phy}, show a steep behaviour, that, in agreement with the multi-hadronic cross-section, could be interpreted as a resonance (or a superposition of more structures) lying near, but below, this threshold;
- all the main vector meson contributions to the nucleon ffs should be in the nucleon unphysical region;
- the strange ff, *i.e.* the $s\bar{s}$ content in the nucleons, could be revealed by the presence of the $\phi(1020)$ resonance.

4.2 The integral equation

Consider the DR of eq. (10) for a normalized ff $G(q^2) \equiv G^N(q^2)/\mu^N$ ($N = p$ or n) with a subtraction at $q^2 = 0$:

$$\ln[G(q^2)] = \frac{q^2\sqrt{s_{\text{th}} - q^2}}{\pi} \int_{s_{\text{th}}}^{\infty} \frac{\ln|G(s)|\mathrm{d}s}{s(s - q^2)\sqrt{s - s_{\text{th}}}}. \qquad (11)$$

The subtraction, that makes the integral less dependent on the ff asymptotic behaviour, does not involve any additional term, being $\ln[G(0)] = 0$. For $q^2 \geq s_{\text{th}}$, over the cut, the phase of the ff, defined through $G(q^2) = |G(q^2)|e^{\delta(q^2)}$, is given by the principal value integral

$$\delta(q^2) = -\frac{q^2\sqrt{q^2 - s_{\text{th}}}}{\pi}\mathcal{P} \int_{s_{\text{th}}}^{\infty} \frac{\ln|G(s)|\mathrm{d}s}{s(s - q^2)\sqrt{s - s_{\text{th}}}}. \qquad (12)$$

The method consists in using the DR of eq. (11) to define an integral equation, where the modulus of the ffs in the unphysical region is the unknown and, data and asymptotic behaviour, as predicted by the pQCD, represent the inputs. More precisely, the integration interval of eq. (11) is split in the two sub-intervals: $[s_{\text{th}}, s_{\text{phy}}]$, where the ffs are to be determined, and $]s_{\text{phy}}, \infty[$, where data and the pQCD asymptotic behaviour are used. We obtain the first kind integral equation, linear in the unknown $\ln|G(q^2)|$:

$$\ln[G(q^2)] - I(q^2) = \int_{s_{\text{th}}}^{s_{\text{phy}}} K(q^2, s) \ln|G(s)|\mathrm{d}s, \qquad (13)$$

where $K(q^2, s) = \frac{q^2\sqrt{s_{\text{th}}-q^2}}{\pi s(s-q^2)\sqrt{s-s_{\text{th}}}}$ is the kernel and

$$I(q^2) = \int_{s_{\text{phy}}}^{\infty} K(q^2, s) \ln|G(s)|\mathrm{d}s \qquad (14)$$

can be computed by means of time-like data and pQCD high-energy behaviour. To avoid instability, usually met in solving integral equations of the first kind, we use some prescriptions devoted to improving smoothness in the solutions and also a regularization scheme. To summarize:

- $I(t)$ is evaluated using time-like data and pQCD asymptotic behaviour. The subtraction at $q^2 = 0$ makes the results less dependent on the assumed asymptotic behaviour.
- To avoid instabilities across the threshold s_{phy}, the upper limit of the unphysical region is raised to $s'_{\text{phy}} = s_{\text{phy}} + \Delta$, where $\Delta \approx 0.5\,\text{GeV}^2$.

– The regularization consists in limiting the local curvature of the ff in the unphysical region

$$R_2 = \int_{s_{\rm th}}^{s'_{\rm phy}} \left[\frac{{\rm d}^2|G(q^2)|}{({\rm d}q^2)^2}\right]^2 {\rm d}q^2. \qquad (15)$$

– Equation (13) is linearized by transforming the integrals into sums over $M = 50$ suitable sub-intervals in q^2, with amplitudes increasing with the energy. This introduces further smoothness by effectively integrating over any structure with a narrower half-width.
– The quantity to be minimized, without the "smoothness improving terms" is:

$$R_0 = \sum_{i=1}^{L}\left[\sum_{j=1}^{M} F_j \int_{s_j}^{s_{j+1}} K(t_i, s){\rm d}s + I(t_i) - \ln G(t_i)\right]^2, \qquad (16)$$

where $F_j = \ln |G[(s_{j+1} + s_j)/2]|$, with $j = 1, 2, \ldots, M$, is evaluated in the middle of the j-th sub-interval and the values t_i, with $i = 1, 2, \ldots, L$, correspond to the experimental points available in the space-like region.
– Finally, we minimize

$$R_{\rm total} = R_0 + \tau^6 R_2 + C, \qquad (17)$$

where R_2 is the regularization term and C contains all the continuity constraints. The "dumping parameter" τ, in unit of energy, has to be chosen by trial (see sect. 4.3).
– The uncertainties in the solution of eqs. (13) and (12), due to experimental errors, were estimated by simulating new space- and time-like data according to the quoted errors and then solving the integral equation for each simulated set.

4.3 Test of the regularization method

To test the entire procedure and also to get a suitable range for the parameter τ, we use the well-known pion ff.

The test consists in re-obtaining the pion ff in the nucleon unphysical region, where it is assumed as unknown, by means of the previously described procedure.

By using the data on the modulus $|F_\pi(q^2)|$, only above the shifted nucleon threshold $s'_{\rm phy}$, the pQCD asymptotic behaviour and the space-like real values, as computed via DR (yellow on-line bands in fig. 6), we obtain a result for $|F_\pi(q^2)|$ in the nucleon unphysical region, which can be compared with the data, not used for the computation. The good agreement with data (dark band in fig. 6), not only represents a positive check for the procedure but, in addition, it allows to choose a working value for the dumping parameter, which turns out: $\tau \approx M_\pi$. The phase of the pion ff, computed via eq. (12), approaches, just above $q^2 = 4\,{\rm GeV}^2$, its expected asymptotic value of 180 degrees (dark band in fig. 7).

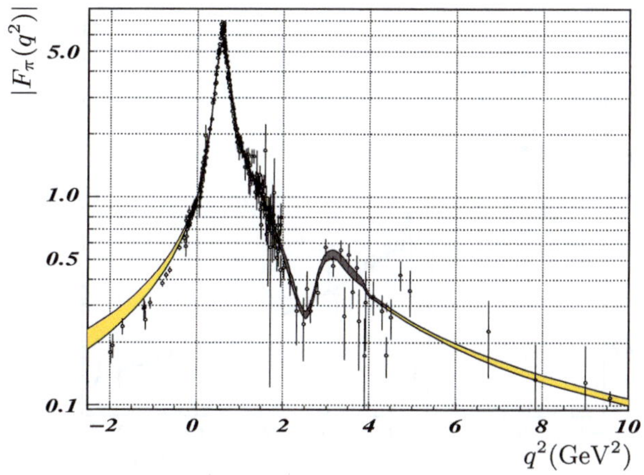

Fig. 6. (Colour on-line) Pion ff, the dark band is the solution of the integral equation. The yellow bands are the inputs.

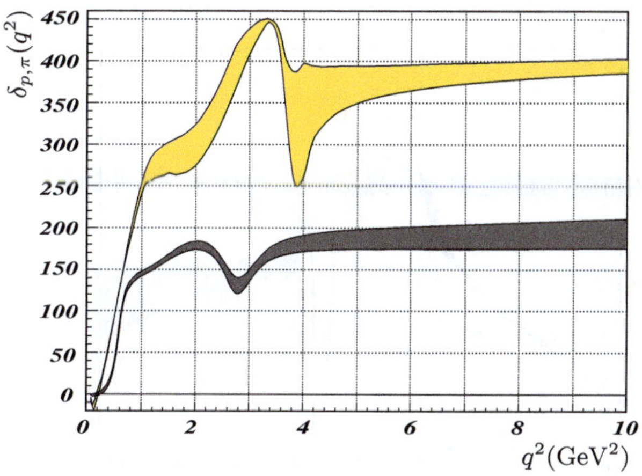

Fig. 7. (Colour on-line) Phase of proton (yellow band) and pion (dark band) ff computed via DR of eq. (12).

4.4 Results for the unphysical nucleon ffs

Below we summarize some salient features of our findings.

– For the first time, resonant structures have been generated from "smooth" inputs (see figs. 8, 9). The method is stable and reliable.
– The combined (ρ + ω) peaks and the $\rho'(1600)$ are generated at the right mass. However, the ρ peak is much broader. Earlier analyses [23] had also found a similar discrepancy.
– No $\phi(1020)$ signal is visible thus signaling a very small $s\bar{s}$ content in the nucleon.
– Phases for the nucleon are consistent with expectations: $\delta_N \rightarrow 360°$ within the error bands (see fig. 7).
– There is an interference pattern near threshold ($M \approx 1.88\,{\rm GeV}$) which may be related to baryonium, figs. 8, 9.

4.5 The "inverse problem": conclusions

Nucleon time-like magnetic ffs have been obtained in an model-independent way by means of DR for $\ln G(q^2)$ and, space- and time-like data.

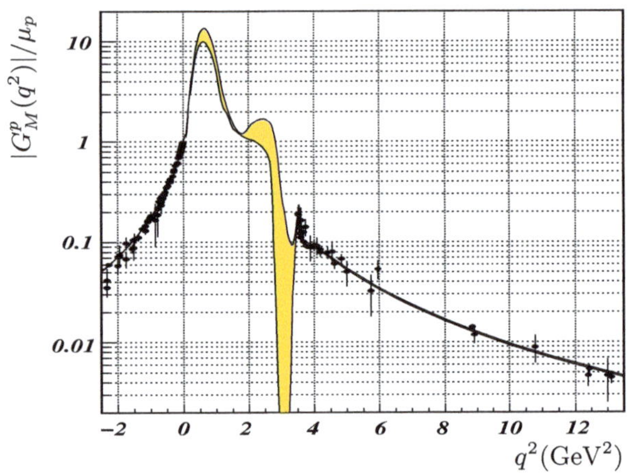

Fig. 8. (Colour on-line) Proton magnetic ff computed via DR (yellow band). The dark bands are the inputs.

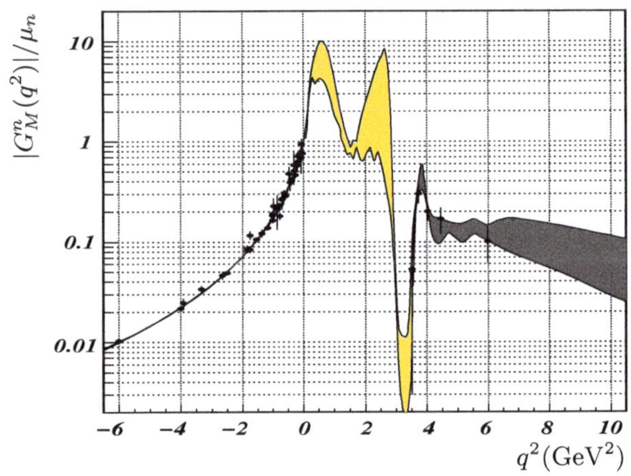

Fig. 9. (Colour on-line) Neutron magnetic ff computed via DR (yellow band). The dark bands are the inputs.

Resonances have been found consistent with the $\rho(770)$ and $\rho'(1600)$ masses. However, a very large ρ width is obtained as in previous DR analyses. Further work is in progress to understand the sources of discrepancies as well as the relationship of our results with other DR analyses [24]. New data on proton magnetic ffs are now available [6], these data are in agreement with the old ones, however we are upgrading the analysis by including them.

5 Future measurements of the nucleon ffs at Frascati

The scientific program of the DAΦNE accelerator at Frascati is decided up to 2008. During the next two years, the detectors FINUDA [25] and SIDDHARTA [26] will alternate in taking data to collect an integrated luminosity: $\int \mathcal{L}dt = 1\,\mathrm{fb}^{-1}$ each.

An important upgrade is planned for DAΦNE, aiming at a machine able to deliver $\int \mathcal{L}dt = 50\,\mathrm{fb}^{-1}$ at the ϕ resonance, in 4–5 years of data taking, starting in 2011, and to operate with 1 GeV–2.5 GeV center-of-mass energy.

This new accelerator, called DANAE, could give the definitive answers to the still open questions about the nucleon ffs. It could provide the first accurate measurement of the time-like $|G_E^{p,n}|$ and $|G_M^{p,n}|$ and also the relative phase between G_E^p and G_M^p, by observing the polarization of the outgoing proton [27].

We thank C. Bini, P. Gauzzi, M. Mirazita, M. Negrini, G. Pancheri and Y. Srivastava for discussions on the subject of this paper.

References

1. M. Gourdin, Phys. Rep. **11**, 29 (1974) and references therein.
2. B.V. Geshkenbein, Yad. Fiz. **9**, 1232 (1969).
3. V.A. Matveev, R.M. Muradyan, A.N. Tavkhelidze, Lett. Nuovo Cimento **7**, 719 (1973); S.J. Brodsky, G.R. Farrar, Phys. Rev. Lett. **31**, 1153 (1973); Phys. Rev. D **11**, 1309 (1975); S.J. Brodsky, P.G. Lepage, Phys. Rev. D **22**, 2157 (1980); T. Gousset, B. Pire, Phys. Rev. D **51**, 15 (1995).
4. M.K. Jones *et al.*, Phys. Rev. Lett. **84**, 1398 (2000); O. Gayou *et al.*, Phys. Rev. C **64**, 038202 (2001); O. Gayou *et al.*, Phys. Rev. Lett. **88**, 092301 (2001).
5. B.D. Milbrath *et al.*, Phys. Rev. Lett. **80**, 452 (1998); **82**, 2221 (1999)(E).
6. B. Aubert *et al.*, Phys. Rev. D **73**, 012005 (2006).
7. G. Bardin *et al.*, Nucl. Phys. B **411**, 3 (1994).
8. A. Antonelli *et al.*, Nucl. Phys. B **517**, 3 (1998).
9. D. Bisello *et al.*, Nucl. Phys. B **224**, 379 (1983).
10. G. Stancari, PhD Thesis; M. Andreotti, Laurea Thesis, Ferrara University (2001).
11. R.C. Walker *et al.*, Phys. Rev. D **49**, 5671 (1994); L. Andivahis *et al.*, Phys. Rev. D **50**, 5491 (1994); J. Litt *et al.*, Phys. Lett. B **31**, 40 (1970); C. Berger *et al.*, Phys. Lett. B **35**, 87 (1971); L.E. Price *et al.*, Phys. Rev. D **4**, 45 (1971); W. Bartel *et al.*, Nucl. Phys. B **58**, 429 (1973).
12. M.N. Rosenbluth, Phys. Rev. **79**, 615 (1950).
13. S. Dubnicka, A.Z. Dubnickova, Fizika B **13**, 287 (2004).
14. J.P. Ralston, P. Jain, Phys. Rev. D **69**, 053008 (2004); J.P. Ralston, P. Jain, R.V. Buniy, AIP Conf. Proc. **549**, 302 (2000); G.A. Miller, M.R. Frank, Phys. Rev. C **65**, 065205 (2002); M.R. Frank, B.K. Jennings, G.A. Miller, Phys. Rev. C **54**, 920 (1996); G. Holzwarth, Z. Phys. A **356**, 339 (1996); F. Cardarelli, S. Simula, Phys. Rev. C **62**, 065201 (2000); R.F. Wagenbrunn *et al.*, Eur. Phys. J. A **14**, 17 (2002).
15. A.V. Belitsky, X. Ji, F. Yuan, Phys. Rev. Lett. **91**, 092003 (2003).
16. S.J. Brodsky *et al.*, Phys. Rev. D **69**, 054022 (2004).
17. F. Iachello, A.D. Jackson, A. Lande, Phys. Lett. B **43**, 191 (1973); M. Gari, W. Kruempelmann, Z. Phys. A **322**, 689 (1985); E.L. Lomon, Phys. Rev. C **66**, 045501 (2002); **64**, 035204 (2001).
18. P.A. Guichon, M. Vanderhaeghen, Phys. Rev. Lett. **91**, 142303 (2003); P.G. Blunden, W. Melnitchouk, J.A. Tjon, Phys. Rev. Lett. **91**, 142304 (2003); J. Arrington, Phys. Rev. C **69**, 032201(R) (2004); M.P. Rekalo, E. Tomasi-Gustafsson, Eur. Phys. J. A **22**, 331 (2004); A.V. Afanasev *et al.*, Phys. Rev. D **72**, 013008 (2005).
19. R. Baldini *et al.*, Eur. Phys. J. C **11**, 709 (1999).

20. E.C. Tichmarsch, *The Theory of Functions* (Oxford University Press, London, 1939).
21. A.Z. Dubnickova, S. Dubnicka, P.M. Rekalo, Nuovo Cimento A **109**, 241 (1966).
22. C.F. Perdrisat *et al.*, Jefferson Lab. experiment E01-109.
23. See, for example, B. Orman, Phys. Rev. **145**, 1140 (1966) and other references therein.
24. P. Mergell, U.-G. Meissner, D. Drechsel, Nucl. Phys. A **596**, 367 (1996); H. Forkel, Prog. Part. Nucl. Phys. **36**, 229 (1996); U.-G. Meissner, Nucl. Phys. A **623**, 340 (1997); H.W. Hammer, U.G. Meissner, D. Drechsel, Phys. Lett. B **385**, 343 (1996); H.W. Hammer, Eur. Phys. J. A **28**, 49 (2006).
25. `http://www.lnf.infn.it/esperimenti/finuda/finuda.html`.
26. `http://www.lnf.infn.it/esperimenti/siddharta/`.
27. F. Ambrosino *et al.*, hep-ex/0603056.

Nucleon form factors of the isovector axial-vector current

Situation of experiments and theory

M.R. Schindler[a] and S. Scherer

Institut für Kernphysik, Johannes Gutenberg-Universität, D-55099 Mainz, Germany

Paper reprinted from Eur. Phys. J. A with permission

Abstract. The theoretical and experimental status of the isovector axial-vector current form factors $G_A(q^2)$ and $G_P(q^2)$ of the nucleon is reviewed. We also describe a new calculation of these form factors in manifestly Lorentz-invariant chiral perturbation theory (ChPT) with the inclusion of axial-vector mesons as explicit degrees of freedom.

PACS. 11.40.-q Currents and their properties – 12.39.Fe Chiral Lagrangians

1 Introduction

The structure of the nucleon is encoded in several form factors. For example, the electromagnetic Dirac and Pauli form factors $F_1(q^2)$ and $F_2(q^2)$, or equivalently the electric and magnetic Sachs form factors $G_E(q^2)$ and $G_M(q^2)$, parameterize the matrix elements of the electromagnetic current operator and are well known over a wide region of momentum transfer squared q^2. For recent reviews on the experimental status of the electromagnetic form factors see, *e.g.*, [1–3]. In contrast to the electromagnetic case, the three form factors of the isovector axial-vector current, $G_A(q^2)$, $G_P(q^2)$, and $G_T(q^2)$, are not as well known. They encode the structure of the matrix elements of the isovector axial-vector current $A^{\mu,a}(x)$ which, in the $SU(2)$ case, is given by

$$A^{\mu,a}(x) \equiv \bar{q}(x)\gamma^\mu\gamma_5\frac{\tau^a}{2}q(x), \quad q=\begin{pmatrix}u\\d\end{pmatrix}, \quad a=1,2,3. \quad (1)$$

$A^{\mu,a}(x)$ is a Hermitian operator that transforms as an isovector under isospin transformations and as an axial vector under Lorentz transformations. The corresponding matrix element between initial and final nucleon states is parameterized as

$$\langle N(p')|A^{\mu,a}(0)|N(p)\rangle = \bar{u}(p')\left[\gamma^\mu\gamma_5\,G_A(q^2)+\frac{q^\mu}{2m_N}\gamma_5\,G_P(q^2) +i\frac{\sigma^{\mu\nu}q_\nu}{2m_N}\gamma_5\,G_T(q^2)\right]\frac{\tau^a}{2}u(p)\,, \quad (2)$$

where $q=p'-p$ and m_N denotes the nucleon mass. $G_A(t)$ is called the axial form factor, $G_P(t)$ is the induced pseudoscalar form factor, and $G_T(t)$ denotes the induced pseudotensorial form factor. For the isospin-symmetric case of equal up- and down-quark masses, $m_u=m_d$, the strong interactions are invariant under $\mathcal{G}$ conjugation, which is a combination of charge conjugation $\mathcal{C}$ and a rotation by π about the 2 axis in isospin space (charge symmetry operation),

$$\mathcal{G}=\mathcal{C}\exp(i\pi I_2). \quad (3)$$

As a consequence the pseudotensorial form factor $G_T(q^2)$ vanishes in the limit of perfect isospin symmetry. So far only upper limits have been established on the size of G_T [4]. We will neglect G_T in the following discussion. From the hermiticity of the current operator one infers that G_A and G_P are real functions for space-like momentum transfer squared ($q^2<0$).

In the Breit system, where $\boldsymbol{p}=-\frac{1}{2}\boldsymbol{q}=-\boldsymbol{p}'$ and $q_0=0$, the zeroth component of the matrix element of eq. (2) vanishes,

$$\left\langle N\left(\frac{\boldsymbol{q}}{2}\right)|A^{0,a}(0)|N\left(-\frac{\boldsymbol{q}}{2}\right)\right\rangle=0, \quad (4)$$

while the spatial components can be written as[1]

$$\left\langle N\left(\frac{\boldsymbol{q}}{2}\right)|\boldsymbol{A}^{a}(0)|N\left(-\frac{\boldsymbol{q}}{2}\right)\right\rangle = \left[2E\boldsymbol{\sigma}_\perp G_A(-\boldsymbol{q}^2)+2m_N\boldsymbol{\sigma}_\parallel D(-\boldsymbol{q}^2)\right]\frac{\tau^a}{2},$$

where $E=\sqrt{m_N^2+\boldsymbol{q}^2/4}$, $\boldsymbol{\sigma}_\parallel=\boldsymbol{\sigma}\cdot\hat{q}\hat{q}$, $\boldsymbol{\sigma}_\perp=\boldsymbol{\sigma}-\boldsymbol{\sigma}_\parallel$, and

$$D(q^2)=G_A(q^2)+\frac{q^2}{4m_N^2}G_P(q^2).$$

[a] e-mail: schindle@kph.uni-mainz.de

[1] We adopt the normalization $\bar{u}u=2m_N$.

We have thus separated the spatial components of the matrix element into its transverse and longitudinal parts. As we will see below, the longitudinal part vanishes in the chiral limit.

For recent reviews on the form factors of the isovector axial-vector current, see [5,6].

2 Axial form factor $G_A(q^2)$

$G_A(q^2)$ is the best known of the isovector axial-vector current form factors. Its value at zero momentum transfer squared is defined as the axial-vector coupling constant g_A,

$$G_A(0) = g_A, \tag{5}$$

which has been determined from neutron beta decay to be [7]

$$G_A(q^2 \approx 0) = g_A = 1.2695 \pm 0.0029. \tag{6}$$

There are two distinct experimental ways of determining the q^2-dependence of G_A. The first is the (quasi)elastic (anti)neutrino scattering (referred to as simply neutrino scattering in the following). For the analysis of experimental data, $G_A(q^2)$ is conventionally parameterized using a dipole form as

$$G_A(q^2) = \frac{g_A}{\left(1 - \frac{q^2}{M_A^2}\right)^2}, \tag{7}$$

where M_A is the so-called axial mass. The global average for the axial mass extracted from neutrino scattering experiments given in [5] is

$$M_A = (1.026 \pm 0.021)\,\text{GeV}, \tag{8}$$

whereas a recent analysis [8] finds a slightly smaller value

$$M_A = (1.001 \pm 0.020)\,\text{GeV}. \tag{9}$$

The second method makes use of the so-called Adler-Gilman relation [9] which provides a chiral Ward identity establishing a connection between charged pion electroproduction at threshold and the isovector axial-vector current evaluated between single-nucleon states (see, *e.g.*, [10, 11] for more details). At threshold (the spatial components of) the center-of-mass transition current for pion electroproduction can be written in terms of two s-wave amplitudes E_{0+} and L_{0+}:

$$e\boldsymbol{M}\big|_{\text{thr}} = \frac{4\pi W}{m_N}\left[i\boldsymbol{\sigma}_\perp E_{0+}(k^2) + i\boldsymbol{\sigma}_\parallel L_{0+}(k^2)\right],$$

where W is the total center-of-mass energy, k^2 is the four-momentum transfer squared of the virtual photon, and $\boldsymbol{\sigma}_\parallel = \boldsymbol{\sigma}\cdot\hat{k}\hat{k}$ and $\boldsymbol{\sigma}_\perp = \boldsymbol{\sigma} - \boldsymbol{\sigma}_\parallel$. The reaction $p(e, e'\pi^+)n$ has been measured at MAMI at an invariant mass of $W = 1125\,\text{MeV}$ (corresponding to a pion center-of-mass momentum of $|\boldsymbol{q}^*| = 112\,\text{MeV}$) and photon four-momentum transfers of $-k^2 = 0.117, 0.195$ and $0.273\,\text{GeV}^2$ [12]. Using an effective-Lagrangian model an axial mass of

$$\bar{M}_A = (1.077 \pm 0.039)\,\text{GeV}$$

was extracted, where the bar is used to distinguish the result from the neutrino scattering value. In the meantime, the experiment has been repeated including an additional value of $-k^2 = 0.058\,\text{GeV}^2$ [13] and is currently being analyzed. The global average from several pion electroproduction experiments is given by [5]

$$\bar{M}_A = (1.068 \pm 0.017)\,\text{GeV}. \tag{10}$$

It can be seen that the values of eqs. (8) and (9) for the neutrino scattering experiments are smaller than eq. (10) for the pion electroproduction experiments. The discrepancy was explained in heavy-baryon chiral perturbation theory [14]. It was shown that at order $\mathcal{O}(q^3)$ pion loop contributions modify the k^2-dependence of the electric dipole amplitude E_{0+} from which $\bar{M}_A$ is extracted. These contributions result in a change of

$$\Delta M_A = 0.056\,\text{GeV}, \tag{11}$$

bringing the neutrino scattering and pion electroproduction results for the axial mass into agreement.

3 Induced pseudoscalar form factor $G_P(q^2)$

The induced pseudoscalar form factor $G_P(q^2)$ is less known than the axial form factor $G_A(q^2)$. A complete overview over the theoretical and experimental situation can be found in [6].

The axial-vector current is divergence free in the chiral limit, $\partial_\mu A^{\mu,a} = 0$. In combination with the Dirac equation one obtains from eq. (2) in the chiral limit that the equation

$$4\,\overset{\circ}{m}{}_N^2 \overset{\circ}{G}_A(q^2) + q^2\,\overset{\circ}{G}_P(q^2) = 0 \tag{12}$$

must hold, where $\circ$ denotes quantities in the chiral limit (see, *e.g.*, [15] for a detailed discussion). Equation (12) has two possible solutions. The first solution is that $\overset{\circ}{m}_N = 0$ and $\overset{\circ}{G}_P(q^2) = 0$. However, the nucleon mass in the chiral limit does not vanish. The second solution is that $\overset{\circ}{G}_A(q^2)$ and $\overset{\circ}{G}_P(q^2)$ are related via

$$\overset{\circ}{G}_P(q^2) = -\frac{4\,\overset{\circ}{m}{}_N^2 \overset{\circ}{G}_A(q^2)}{q^2}. \tag{13}$$

Since $G_A(0) = g_A \approx 1.27$, it follows that $\overset{\circ}{G}_P(q^2)$ has a pole for $q^2 \to 0$. The behavior of $\overset{\circ}{G}_P(q^2)$ for $q^2 \to 0$ is interpreted as stemming from a pion pole contribution (see fig. 1). The most general expression for the diagram of fig. 1 in the chiral limit is given by

$$-\frac{4\,\overset{\circ}{m}_N\, F(q^2)\, \overset{\circ}{g}_{\pi N}(q^2)}{q^2 - \overset{\circ}{\Sigma}(q^2)}\,\bar{u}(p')\frac{q^\mu}{2\,\overset{\circ}{m}_N}\gamma_5\frac{\tau^a}{2}u(p), \tag{14}$$

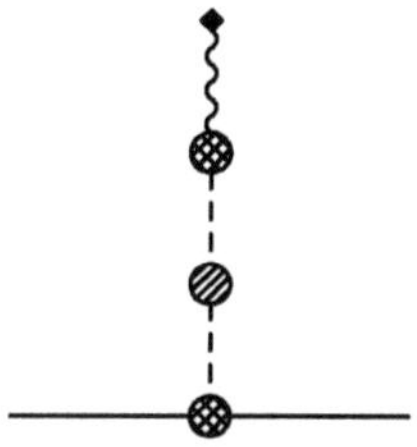

Fig. 1. Pion pole contribution to $\overset{\circ}{G}_P(q^2)$.

where $F(q^2)$ denotes the coupling of the axial source to the pion, $\overset{\circ}{g}_{\pi N}(q^2)$ is the pion nucleon coupling in the chiral limit and $\overset{\circ}{\tilde{\Sigma}}(q^2)$ stands for the pion self-energy with $\overset{\circ}{\tilde{\Sigma}}(0) = \overset{\circ}{\tilde{\Sigma}}'(0) = 0$. Comparing with eq. (2) one finds the pion pole diagram contribution

$$-\frac{4\,\overset{\circ}{m}_N\,F(q^2)\,\overset{\circ}{g}_{\pi N}(q^2)}{q^2 - \overset{\circ}{\Sigma}'(q^2)} \tag{15}$$

to $\overset{\circ}{G}_P(q^2)$. The limit $q^2 \to 0$ is given by

$$\lim_{q^2\to 0} q^2\,\overset{\circ}{G}_{P,\pi\text{-}pole}(q^2) = -4\,\overset{\circ}{m}_N\,F\,\overset{\circ}{g}_{\pi N}, \tag{16}$$

which, by comparison with eq. (12), leads to the famous Goldberger-Treiman relation [16,17]

$$\frac{\overset{\circ}{g}_A}{F} = \frac{\overset{\circ}{g}_{\pi N}}{\overset{\circ}{m}_N}. \tag{17}$$

While the Goldberger-Treiman relation contains quantities in the chiral limit, it is interesting to note that it is satisfied to about 2% in the real world.

Information on $G_P(q^2)$ is mainly extracted from muon capture experiments. The induced pseudoscalar coupling g_P is defined as

$$g_P = \frac{m_\mu}{2m_N} G_P(q^2 = -0.88m_\mu^2). \tag{18}$$

It is mostly this quantity that has been determined by experiments, although one pion production experiment measured the q^2-dependence of $G_P(q^2)$ [18]. The weighted world average from ordinary muon capture (OMC),

$$\mu^- + p \to \nu_\mu + n, \tag{19}$$

is [5]

$$g_P = 8.79 \pm 1.92\,. \tag{20}$$

Most OMC experiments have used a liquid-hydrogen target, which results in uncertainties due to the formation of $p\mu p$ molecules. For an exact determination of g_P the ortho-para transition rate has to be known. The current MuCap experiment at PSI [19] is using a hydrogen gas target and thereby avoiding these complications.

The pseudoscalar coupling g_P can also be determined from radiative muon capture (RMC),

$$\mu^- + p \to \nu_\mu + n + \gamma. \tag{21}$$

The value of g_P extracted from these experiments is [6]

$$g_P = 12.3 \pm 0.9 \tag{22}$$

and does not agree with the OMC result. However, a recent measurement at TRIUMF [20] found a new value for the ortho-para transition rate in the $p\mu p$ molecule of

$$\Lambda_{op}^{new} = (11.1 \pm 1.7 \pm^{0.9}_{0.6}) \times 10^4\,\mathrm{s}^{-1}, \tag{23}$$

which is significantly larger than the previous result [21] of

$$\Lambda_{op}^{old} = (4.1 \pm 1.4) \times 10^4\,\mathrm{s}^{-1}. \tag{24}$$

With the new value the RMC experiment gives [20]

$$g_P = 10.6 \pm 1.1, \tag{25}$$

while the average of the liquid-hydrogen results is modified to be

$$g_P = 5.6 \pm 4.1. \tag{26}$$

Theoretically, $G_P(q^2)$ has been determined using heavy-baryon chiral perturbation theory [22,23], which at order $\mathcal{O}(q^3)$ gives

$$g_P = 8.23. \tag{27}$$

4 $G_A(q^2)$ and $G_P(q^2)$ in Lorentz-invariant ChPT

With the introduction of renormalization schemes such as infrared regularization [24] or the extended-on-mass-shell scheme [25] the calculation of the isovector axial-vector current form factors in a manifestly Lorentz-invariant formulation of baryon chiral perturbation theory has been made possible [26].

The results of such a calculation up to and including order $\mathcal{O}(q^4)$ for $G_A(q^2)$ and $G_P(q^2)$ [27] are shown in fig. 2 and fig. 3, respectively. The analytic expression for $G_A(q^2)$ reads

$$G_A(q^2) = g_A + \frac{1}{6} g_A \langle r^2 \rangle_A q^2 + \frac{g_A^3}{4F^2} H(q^2), \tag{28}$$

where $\langle r^2 \rangle_A$ is the mean square axial radius and $H(q^2)$ contains contributions from loop diagrams with $H(0) = H'(0) = 0$. One can see that the contributions from $H(q^2)$ are small and the data is only described for very low values of momentum transfer. The calculation at order $\mathcal{O}(q^4)$ gives

$$g_P = 8.09 \tag{29}$$

and one can clearly see the pion pole contribution in fig. 3.

The situation for the axial form factor $G_A(q^2)$ can be compared to the electromagnetic form factors, which in ChPT at order $\mathcal{O}(q^4)$ only describe the data for very low momentum transfers as well. In [28] a reformulation of infrared renormalization was used to include vector mesons as explicit degrees of freedom. This resulted in a better description of the data up to about $-q^2 \approx 0.4\,\mathrm{GeV}^2$. Similar to the case of the electromagnetic form factors one can

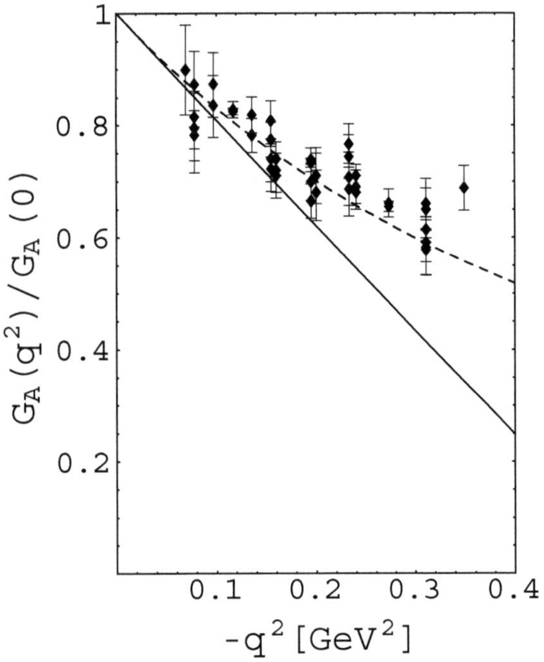

Fig. 2. $G_A(q^2)$ in chiral perturbation theory at $\mathcal{O}(q^4)$ (solid line). The dashed line is a dipole fit. The data compilation is taken from [5].

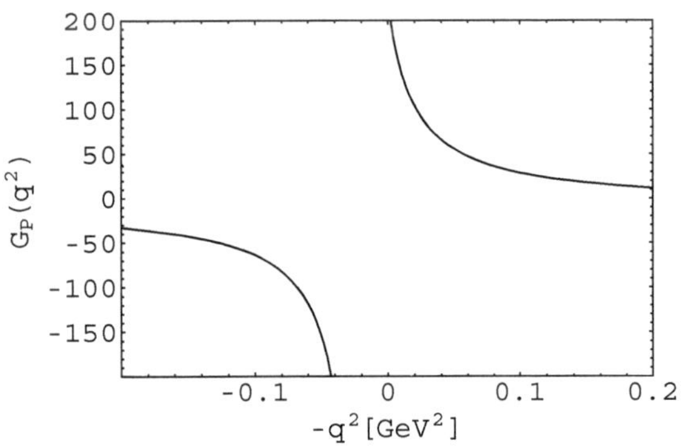

Fig. 3. $G_P(q^2)$ in chiral perturbation theory at $\mathcal{O}(q^4)$.

include axial-vector mesons in the formalism of baryon ChPT [27]. The Lagrangian for the coupling of the axial-vector meson to pions reads [29]

$$\mathcal{L}_{\pi A}^{(3)} = f_A \,\mathrm{Tr}[A_{\mu\nu} F_-^{\mu\nu}], \tag{30}$$

while the coupling to the nucleon is given by

$$\mathcal{L}_{AN}^{(0)} = G_{AN} \bar{\Psi} A^\mu \gamma_\mu \gamma_5 \Psi. \tag{31}$$

The contributions of the diagrams containing the axial vector meson are

$$G_A^{AVM}(q^2) = -8 f_A G_{AN} \frac{q^2}{q^2 - M_{a_1}^2} \tag{32}$$

to the axial form factor and

$$G_P^{AVM}(q^2) = 32 m_N^2 f_A G_{AN} \frac{1}{q^2 - M_{a_1}^2} \tag{33}$$

to the induced pseudoscalar form factor, respectively. One sees that effectively only one new coupling constant, namely $f_A G_{AN}$, appears. This coupling constant can be fitted to the data [27].

5 Summary

A short overview over the experimental and theoretical situation of the nucleon form factors of the isovector axial-vector current was given. The axial form factor $G_A(q^2)$ has been determined by two types of experiments, neutrino scattering and pion electroproduction. The results from these two methods agree once pion loop corrections to the electroproduction amplitude have been taken into account. The situation for the induced pseudoscalar form factor $G_P(q^2)$ is less clear. The results obtained for the induced pseudoscalar coupling g_P from ordinary and radiative muon capture do not agree. However, a recent measurement of the ortho-para transition rate in $p\mu p$ molecules results in a significant change of the previous results for g_P. A reformulation of the infrared renormalization in baryon ChPT allows for the inclusion of axial-vector mesons in the calculation of the form factors, which could result in a better description of the experimental data. Only one new low-energy constant effectively appears, which can be fitted to the data.

This work was supported by the Deutsche Forschungsgemeinschaft (SFB 443).

References

1. H.y. Gao, Int. J. Mod. Phys. E **12**, 1 (2003); 567 (2003)(E).
2. J. Friedrich, Th. Walcher, Eur. Phys. J. A **12**, 607 (2003).
3. C.E. Hyde-Wright, K. de Jager, Annu. Rev. Nucl. Part. Sci. **54**, 217 (2004).
4. D.H. Wilkinson, Eur. Phys. J. A **7**, 307 (2000).
5. V. Bernard, L. Elouadrhiri, U.-G. Meißner, J. Phys. G **28**, R1 (2002).
6. T. Gorringe, H.W. Fearing, Rev. Mod. Phys. **76**, 31 (2004).
7. W.-M. Yao *et al.*, J. Phys. G **33**, 1 (2006).
8. H. Budd, A. Bodek, J. Arrington, arXiv:hep-ex/0308005.
9. S.L. Adler, F.J. Gilman, Phys. Rev. **152**, 1460 (1966).
10. S. Scherer, J.H. Koch, Nucl. Phys. A **534**, 461 (1991).
11. T. Fuchs, S. Scherer, Phys. Rev. C **68**, 055501 (2003).
12. A1 Collaboration (A. Liesenfeld *et al.*), Phys. Lett. B **468**, 20 (1999).
13. D. Baumann, PhD Thesis, Johannes Gutenberg-Universität, Mainz (2004).
14. V. Bernard, N. Kaiser, U.-G. Meißner, Phys. Rev. Lett. **69**, 1877 (1992).
15. S. Scherer, in *Advances in Nuclear Physics*, Vol. **27**, edited by J.W. Negele, E.W. Vogt (Kluwer Academic/Plenum, New York, 2003) pp. 277-538.
16. M.L. Goldberger, S.B. Treiman, Phys. Rev. **110**, 1178 (1958).
17. M.L. Goldberger, S.B. Treiman, Phys. Rev. **111**, 354 (1958).
18. S. Choi *et al.*, Phys. Rev. Lett. **71**, 3927 (1993).

19. MuCap Collaboration (Spokespersons: P. Kammel, C. Petitjean), `http://www.npl.uiuc.edu/exp/mucapture/`.
20. J.H.D. Clark *et al.*, Phys. Rev. Lett. **96**, 073401 (2006).
21. G. Bardin *et al.*, Phys. Lett. B **104**, 320 (1981).
22. V. Bernard, N. Kaiser, U.-G. Meißner, Phys. Rev. D **50**, 6899 (1994).
23. H.W. Fearing, R. Lewis, N. Mobed, S. Scherer, Phys. Rev. D **56**, 1783 (1997).
24. T. Becher, H. Leutwyler, Eur. Phys. J. C **9**, 643 (1999).
25. T. Fuchs, J. Gegelia, G. Japaridze, S. Scherer, Phys. Rev. D **68**, 056005 (2003).
26. S.i. Ando, H.W. Fearing, Phys. Rev. D **75**, 014025 (2007).
27. M.R. Schindler, T. Fuchs, J. Gegelia, S. Scherer, Phys. Rev. C **75**, 025202 (2007).
28. M.R. Schindler, J. Gegelia, S. Scherer, Eur. Phys. J. A **26**, 1 (2005).
29. G. Ecker, J. Gasser, H. Leutwyler, A. Pich, E. de Rafael, Phys. Lett. B **223**, 425 (1989).

The controversial role of strangeness in the spin structure of the nucleon*

E. Leader[a]

Imperial College London, Prince Consort Road, London SW7 2BW, UK

Abstract. Although the strangeness content of the nucleon is small, it has played a major role in provoking puzzles and controversies in our understanding of the internal structure of the nucleon, particularly as concerns the spin structure. We recall the role of the strange polarization in precipitating the "spin crisis in the parton model" and discuss our present knowledge of the shape and sign of $\Delta s(x)$.

PACS. 13.60.-r Photon and charged-lepton interactions with hadrons – 12.38.-t Quantum chromodynamics

1 Introduction: the "spin crisis in the parton model"

A misjudgement of the significance of strangeness was behind the "spin crisis in the parton model" [1] which arose from the famous EMC experiment on polarized deep inelastic scattering of leptons on protons [2] in 1988.

We shall use the following notation:

$$\Delta q = \int_0^1 \mathrm{d}x \Delta q(x), \tag{1}$$

where $\Delta q(x)$ is the polarized parton density *i.e.* the difference in the number densities of quarks polarized along and opposite to the longitudinal polarization of a proton taken to be moving along the OZ axis and 100% polarized in that direction. Note that Δq is referred to as the first moment of the quark density.

There are three particularly useful flavour combinations of the first moments:

$$a_3 = \Delta u + \Delta \bar{u} - \Delta d - \Delta \bar{d} \tag{2}$$

which transforms like the third component of an isospin triplet,

$$a_8 = \Delta u + \Delta \bar{u} + \Delta d + \Delta \bar{d} - 2(\Delta s + \Delta \bar{s}) \tag{3}$$

which transforms like the eighth component of a flavour octet, and

$$\Delta \Sigma = \sum_f (\Delta q_f + \Delta \bar{q}_f) \tag{4}$$

which is a flavour singlet.

Because a_3 and a_8 are directly connected to the hadronic matrix elements of two of the currents belonging to the octet of axial-vector Cabibbo currents that control weak interactions, there is a very beautiful connection between the high-energy physics of deep inelastic scattering (DIS) and the low-energy physics of β-decay. Thus, according to the Bjorken sum rule, a_3 is known from neutron β-decay: $a_3 = 1.2670 \pm 0.0035$, and assuming $SU(3)$ flavour symmetry holds for hyperon β -decays, $a_8 = 0.585 \pm 0.025$.

Now the EMC measured the first moment of the spin-dependent structure function g_1 of the proton

$$\Gamma_1^p = \int \mathrm{d}x g_1^p(x) \tag{5}$$

which is given by

$$\Gamma_1^p = \frac{1}{12}\left[a_3 + \frac{1}{3}(a_8 + 4a_0)\right], \tag{6}$$

where a_0 is the hadronic matrix element of a flavour singlet operator.

Knowing the values of a_3 and a_8, the EMC measurement implied

$$a_0^{EMC} \simeq 0. \tag{7}$$

But in the naive parton model

$$a_0 = \Delta \Sigma, \tag{8}$$

where $\Delta\Sigma$ is given by (4).

In 1974 Ellis-Jaffe [3] suggested that one could ignore $\Delta s + \Delta \bar{s}$ implying that

$$a_0 \simeq a_8 \simeq 0.59. \tag{9}$$

* Work done in collaboration with A.V. Sidorov, Bogoliubov Theoretical Laboratory, Joint Institute for Nuclear Research, Dubna and D.B. Stamenov, Institute for Nuclear Research and Nuclear Energy, Bulgarian Academy of Sciences, Sofia.

[a] e-mail: `e.leader@imperial.ac.uk`

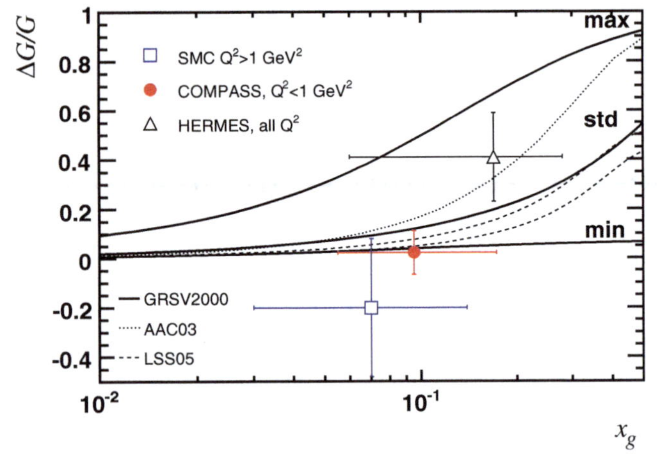

Fig. 1. Comparison of $\Delta G/G$ measurements from COMPASS, SMC and HERMES. Curves show various parametrizations from NLO fits in the $\overline{MS}$ scheme.

Thus, the EMC result (7) is in gross contradiction with Ellis-Jaffe.

Moreover, given the physical meaning of the $\Delta q(x)$ it is clear that

$$\Delta\Sigma = 2S_z^{quarks}, \tag{10}$$

where S_z^{quarks} is the longitudinal component of the total spin of the quarks. The EMC measurement therefore seems to imply $S_z^{quarks} = 0$ and there appears to be a crisis, since, naively, from one's experience of building quark models of the nucleon and baryon resonances, one would expect almost all of the proton's spin to be carried by the spin of its quarks. This inspired Anselmino and me to write a paper entitled *A crisis in the parton model: where, oh where, is the proton's spin?* [1].

2 The effect of QCD

The QCD world is not the same as the naive parton model and two new features have to be taken into account: Q^2-dependence and renormalization scheme dependence.

Thus:

1) In most schemes $\Delta\Sigma$ varies with Q^2 so maybe it does not make sense to compare it to the spin carried by the quarks.

2) The expression for a_0 depends on the scheme. In *physical schemes*, where $\Delta\Sigma$ does *not* vary with Q^2

$$a_0 = \Delta\Sigma - \frac{3\alpha_s(Q^2)}{2\pi}\Delta G(Q^2), \tag{11}$$

where $\Delta G(Q^2)$ is the first moment of the polarized gluon density.

It was thus hoped that one could have a reasonable $\Delta\Sigma \simeq 0.6$ and still obtain a very small a_0, say $\simeq 0.2$ at $Q^2 = 1\,\mathrm{GeV}^2$. But this requires $\Delta G \simeq 1.7$ at $Q^2 = 1\,\mathrm{GeV}^2$. Alas the latest information (still pretty rough) suggests a much smaller value of ΔG than needed. Figure 1 (from Bradamante [4]) shows the current situation.

Moreover, these new measurements are consistent with the results from several analyses of polarized DIS, which typically give $\Delta G(Q^2 = 1\,\mathrm{GeV}^2) \simeq 0.29 \pm 0.32$, whereas we hoped for something like 1.7.

Conclusion: the crisis sparked by a misjudgement of $\Delta s + \Delta\bar{s}$ is still with us!

3 Attempts to measure Δs(x)

There are two possibilities: via polarized DIS or polarized semi-inclusive DIS (SIDIS).

Recall that DIS only depends on $\Delta q(x) + \Delta\bar{q}(x)$. So we can obtain information on $\Delta s(x) + \Delta\bar{s}(x)$.

In SIDIS we could, in principle obtain $\Delta s(x)$ and $\Delta\bar{s}(x)$ separately, but that is for the future!

3.1 Results from polarized DIS

Aside from one small issue there is general agreement among several analyses: see fig. 2.

What causes the disagreement at moderate to large x? Surprisingly —*positivity i.e.* the requirement that

$$|\Delta s(x)| \leq s(x). \tag{12}$$

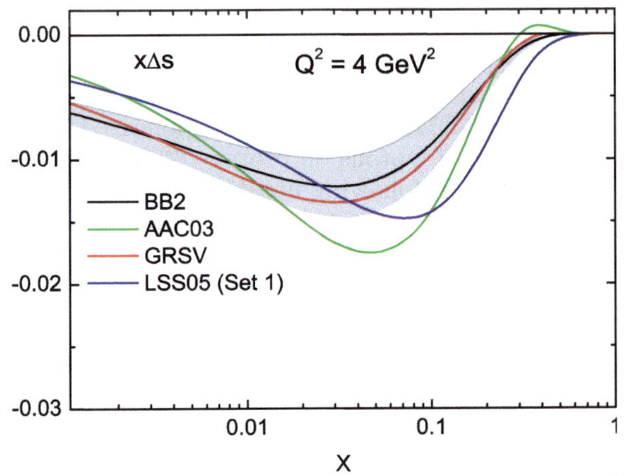

Fig. 2. $x\Delta s(x)$ *vs.* x from various NLO fits to polarized DIS in the $\overline{MS}$ scheme.

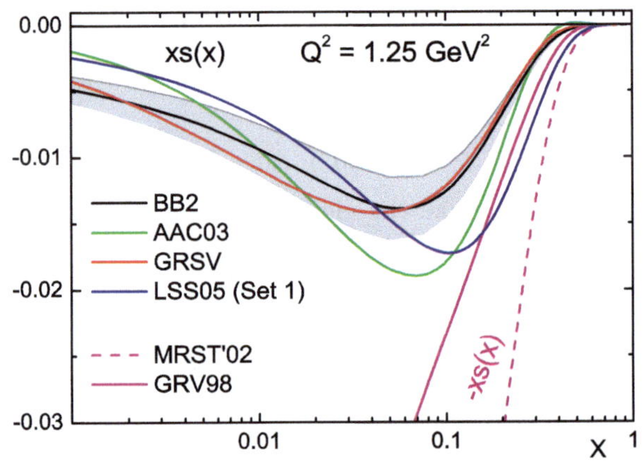

Fig. 3. Comparison of $x\Delta s(x)$ and two versions of $xs(x)$.

As shown in fig. 3 the data seem to want a large negative $\Delta s(x)$ at moderate values of x. So there is a clash with positivity and the result is that the shape of $\Delta s(x)$ is sensitive to the input *unpolarized* density. In the figure the polarized analyses BB2, AAC03 and GRSV utilized the unpolarized strangeness density of GRV98, whereas LSS05 (Set 1) used the unpolarized strangeness density of MRST'02. It is seen that LSS05 is incompatible with the unpolarized GRV density.

There is an important lesson to be learned here. Analyzers of the polarized data should be aware of this sensitivity and should use the latest available unpolarized densities when imposing positivity.

3.2 Results from polarized SIDIS

Before looking at results consider the following constraint [5] on the first moment:

$$\delta_s \equiv [\Delta s + \Delta \bar{s}]. \tag{13}$$

We can rewrite the expression for Γ_1^p as

$$\Gamma_1^p(Q^2) = \frac{1}{6}\left[\frac{1}{2}a_3 + \frac{5}{6}a_8 + 2\delta_s(Q^2)\right] \tag{14}$$

or

$$a_8 = \frac{6}{5}\left[6\Gamma_1^p(Q^2) - \frac{1}{2}a_3 - 2\delta_s(Q^2)\right]. \tag{15}$$

We know a_3 very accurately. Using the measured values of $\Gamma_1^p(Q^2)$ we show that $\delta_s(Q^2) \geq 0$ implies an unacceptable value for a_8.

We have to decide what value to use for $\Gamma_1^p(Q^2)$, since the result depends on the extrapolation to $x = 0$. We take two extremes:

A) Assume perturbative QCD holds at small x (E155 etc.). This yields

$$\Gamma_1^p(Q^2 = 5) = 0.118 \pm 0.004 \pm 0.007. \tag{16}$$

B) Assume Regge behaviour at small x (E143 etc.). This gives

$$\Gamma_1^p(Q^2 = 3) = 0.133 \pm 0.003 \pm 0.009. \tag{17}$$

Results: If δ_s is *positive* we find

A) $\quad a_8 \leq 0.089 \pm 0.058,$

B) $\quad a_8 \leq 0.197 \pm 0.068.$

Recall that hyperon β-decay is adequately described by $SU(3)_F$ and this leads to $a_8 = 0.585 \pm 0.025$.

Thus $\delta_s(Q^2) \geq 0$ implies a dramatic breaking of $SU(3)_F$, and we onclude that it is *almost* impossible to have $\delta_s(Q^2) \geq 0$.

Now, HERMES has extracted $\Delta s(x) + \Delta \bar{s}(x)$ from a study of SIDIS [6]. The results are shown in fig. 4.

Within errors results are consistent with zero, and HERMES quote

$$\delta_s(Q^2 = 2.5) = 0.028 \pm 0.033 \pm 0.009. \tag{18}$$

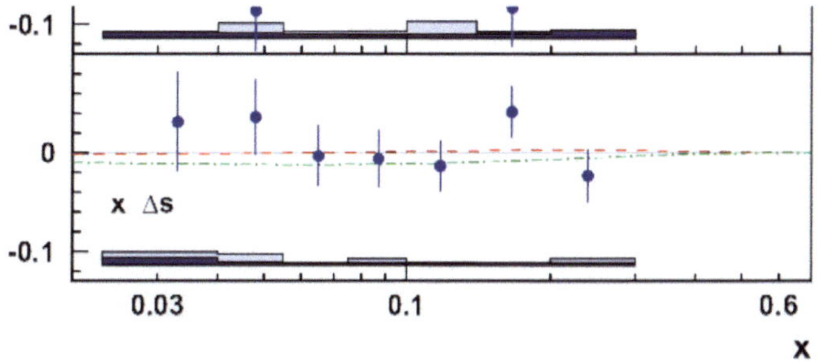

Fig. 4. HERMES results for $x[\Delta s(x) + \Delta \bar{s}(x)]$.

The previous discussion suggests that the central value *cannot* be the true value unless we have totally failed to understand the connection between DIS and SIDIS. If the latter is not the case, how can we understand the HERMES results?

I think it is important to remember that HERMES uses a LO method based on so-called *purities*. I suspect that such an approach is unreliable at the values of Q^2 involved, and that the errors on the purities are somewhat underestimated in the analysis. So I strongly believe that this new "strange quark crisis" will prove to be illusory.

4 Conclusions

Though the polarized strange quark density is undoubtedly small, thinking about it or misunderstanding it has had a mighty effect:

1) It inspired the Ellis-Jaffe sum rule, without which the appreciation of the significance of the EMC experiment might have been delayed for ages.

2) The failure to obtain consistent results for $\Delta s + \Delta \bar{s}$ from DIS and SIDIS *might* be signalling a failure in our understanding of the connection between these kinds of processes, but, as explained above, I suspect that a more mundane explanation will emerge.

Clearly, measuring $\Delta s(x)$ more accurately remains a very important and challenging task for experimentalists.

This research was supported by a UK Royal Society International Joint Project Grant.

References

1. M. Anselmino, E. Leader, Z. Phys. C **41**, 239 (1988).
2. J. Ashman *et al.*, Phys. Lett. B **206**, 364 (1988); Nucl. Phys. B **328**, 1 (1989).
3. J. Ellis, R.L. Jaffe, Phys. Rev. D **9**, 1444 (1974).
4. F. Bradamante, talk given at the *Workshop on Contribution of the Gluon Spin to the Proton Spin, RIKEN, Wako, Saitama, Japan, December 2005.*
5. E. Leader, D.B. Stamenov, Phys. Rev. D **67**, 037503 (2003).
6. HERMES Collaboration (A. Airapetian *et al.*), Phys. Rev. D **71**, 012003 (2005).

Strangeness contributions to nucleon form factors

R.D. Young[a]

Jefferson Lab, 12000 Jefferson Ave., Newport News, VA 23606, USA

Abstract. We review a recent theoretical determination of the strange quark content of the electromagnetic form factors of the nucleon. These are compared with a global analysis of current experimental measurements in parity-violating electron scattering.

PACS. 11.30.Er Charge conjugation, parity, time reversal, and other discrete symmetries – 14.20.Dh Protons and neutrons

1 Introduction

The determination of the strange quark content of the nucleon offers a unique probe to measure the nonperturbative structure of the nucleon. As the nucleon carries zero net strangeness, the influence of strange quarks arises entirely through interaction with the vacuum. Technically speaking, strange quarks directly probe the role of the fermion determinant in QCD. While strangeness measurements in nucleon structure have been difficult to isolate, the contribution of the neutral weak current in elastic scattering offers perhaps the most direct measurement of the strange quark content of the nucleon [1].

Here we review recent progress in the study of the strange quark contributions to the nucleon form factors. In sect. 2 we discuss the theoretical developments in the chiral extrapolation of lattice simulation results that have enabled a precise determination of the strangeness form factors. An outline of this determination is provided in sect. 3. In sect. 4 this theoretical prediction is compared with a global analysis of the experimental measurements searching for strangeness in the nucleon.

2 Chiral applications in lattice QCD

The computational expense of incorporating the effects of the fermion determinant has restricted modern lattice QCD simulations to the use of pion masses that are typically $m_\pi \gtrsim 500\,\mathrm{MeV}$. Recent progress has seen nucleon 3-point functions simulated with pion masses pushing down to the 350 MeV range [2–4], yet a reliable extrapolation in the pion mass is still required to compare with reality —until the physical point is readily accessible.

[a] e-mail: `young@jlab.org`

Ultimately, chiral perturbation theory offers the potential to deliver model-independent quark mass extrapolations of lattice results. As disappointing as it may be, there is mounting evidence that applications of low-order chiral expansions should be taken with serious caution beyond pion masses of the order 300 MeV [5,6]. Further, the situation could be significantly worse for observables which are particularly singular near the chiral limit, such as magnetic moments [7], charge radii or polarisabilities.

In the future, chiral extrapolations will be constrained *model independently* by precision, large-volume lattice calculations in the chiral regime. Until then, one requires methods which can reliably extrapolate from the moderately heavy quark mass regime, while maintaining all the constraints of the effective field theory. The best available solution is to reformulate the effective field theory using finite-range regularisation (FRR) [5].

In extrapolating lattice simulation results from beyond the chiral regime, one cannot guarantee that results will be independent of regularisation scheme. By choosing a particular scheme, one has necessarily introduced a model —whether it be FRR, or a more traditional regularisation. The advantages of FRR have been quantitatively demonstrated for the nucleon mass. Using lattice results over the range $0.25 \gtrsim m_\pi^2 \gtrsim 1.0\,\mathrm{GeV}^2$, the FRR extrapolated nucleon mass at the physical point displays less than 1% variation associated with the truncation between successive orders in the chiral expansion. Further, the sensitivity to the choice of the functional form of FRR is also less than 1% [8]. Until sufficient lattice results are available in the chiral regime, when the choice of regularisation becomes superfluous, FRR offers *independent-of-model* chiral extrapolations.

Because of the cost of simulating the fermion determinant, historically it has been common in lattice QCD to ignore this contribution to the path integral. This is the quenched "approximation", where the influence of quark-antiquark pair creation in the vacuum is neglected. Fortunately, the study of the chiral extrapolation of baryon masses in quenched and dynamical simulations has revealed a remarkable phenomenological relation between these simulations. The differences between quenched and dynamical baryon masses are well described by the differences in the Goldstone boson loop corrections of the low-energy effective field, when evaluated with an appropriate finite-range regulator [9]. Although this is not a field-theoretic connection, the numerical success does mean that one has substantial confidence in obtaining physical estimates from quenched lattice results.

Beyond the baryon masses, the technique of chiral unquenching has been extended to the nucleon magnetic moments [7]. Here it was predicted that there should be very little difference in the quenched and dynamical nucleon magnetic moments over a large range of quark masses, with significant differences only anticipated near the chiral limit. These findings have been recently supported by first calculations with 2-flavour dynamical lattices [3,4].

With the success of chiral extrapolations and the estimation of the effects of the quark determinant, we look to the extraction of the strangeness contributions to the nucleon electromagnetic form factors.

3 Strangeness calculation

Direct lattice QCD calculations of the strangeness content of the nucleon have been unable to produce a conclusive determination [10]. It is hoped that the next generation of calculations could shed light on this elusive signal. This may require further development of emerging lattice techniques. One potential gain could be seen by utilising background field methods [11], where a weak signal could be enhanced by coupling a strong electromagnetic field to the vacuum strange quarks. The method to evaluate the all-to-all propagator developed by the Dublin group offers significantly improved precision over traditional stochastic estimators [12], and it would be interesting to see this applied to a strangeness form factor calculation.

While awaiting the development of these techniques, one must rely on more indirect methods for an accurate extraction of the strangeness form factors. By manipulating the QCD path integral, one can isolate the various quark contributions to baryon 3-point functions [13]. Using this decomposition, combined with charge symmetry [14] and the experimentally measured hyperon radii, the sensitivity to lattice systematics can be significantly reduced [15]. The strangeness magnetic moment can be written as

$$G_M^s = \frac{{}^lR_d^s}{1-{}^lR_d^s}\left[2p+n-\frac{u^p}{u^\Sigma}(\Sigma^+-\Sigma^-)\right], \quad (1)$$

$$G_M^s = \frac{{}^lR_d^s}{1-{}^lR_d^s}\left[p+2n-\frac{u^n}{u^\Xi}(\Xi^0-\Xi^-)\right], \quad (2)$$

where p, n, $\Sigma^\pm$ and $\Xi^{0/-}$ denote the experimentally measured magnetic moments of the respective baryon. The formulae each rely on two inputs from lattice simulations.

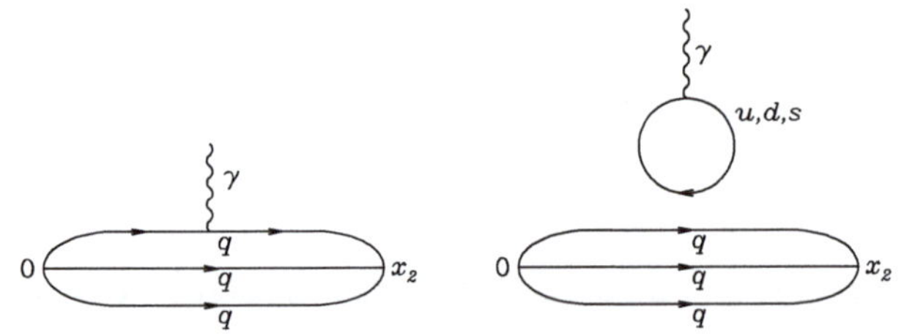

Fig. 1. Connected (left) and disconnected (right) contributions to baryon 3-point functions.

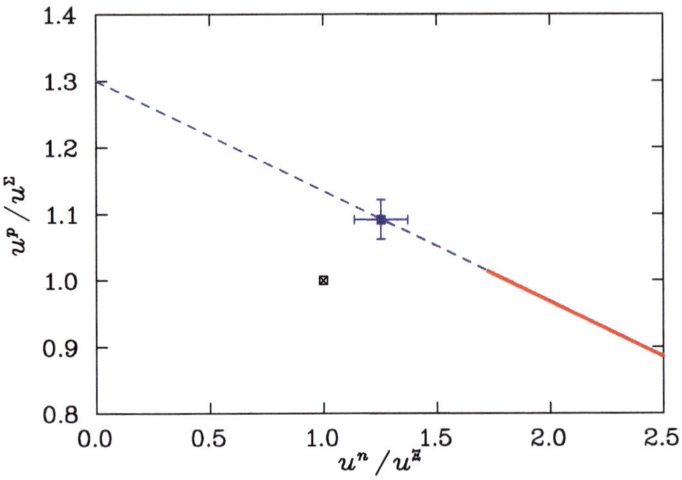

Fig. 2. The line (dashed $G_M^s(0) < 0$, solid $G_M^s(0) > 0$) indicates the charge symmetry constraint on the ratios u^p/u^Σ and u^n/u^Ξ. The crossed square indicates the point corresponding to environment-independent quark moments. Our determination is illustrated by the filled square [16].

The first is the ratio u^p/u^Σ, which measures the relative strength of the valence (fig. 1, left) u-quark contribution in the proton relative to the Σ^+ —or similarly u^n/u^Ξ in eq. (2). The second is ${}^lR_d^s$, which describes the ratio of the strange-to-light disconnected (fig. 1, right) contributions. Equating eqs. (1) and (2) and using the experimental magnetic moments, produces a linear relationship between the two unknown valence ratios. This constraint, a result of charge symmetry alone, is displayed in fig. 2. The line is divided by two segments, where the sign of G_M^s can be determined under the quite general assumption that $0 > {}^lR_d^s > 1$. Recently, it has been suggested that there could be a sign change in this ratio between the heavy-quark limit and naive expectations in the Goldstone boson sector [17]. Given that the properties of the kaon are much more Goldstone-like than a heavy-light meson, and that the heavy-quark limit of μ_p/μ_n is approached very slowly [18], it should not be expected that the strange quark could be reliably described by heavy-quark effective theory.

The techniques discussed in sect. 2 were applied to determine the ratios u^p/u^Σ and u^n/u^Ξ, appearing in eqs. (1) and (2). The analysis has utilised a high-precision numerical study of the baryon electromagnetic form factors in quenched lattice QCD [19]. Upon performing finite-volume corrections, adjustments for the quenched approximation and a controlled chiral extrapolation, the resulting

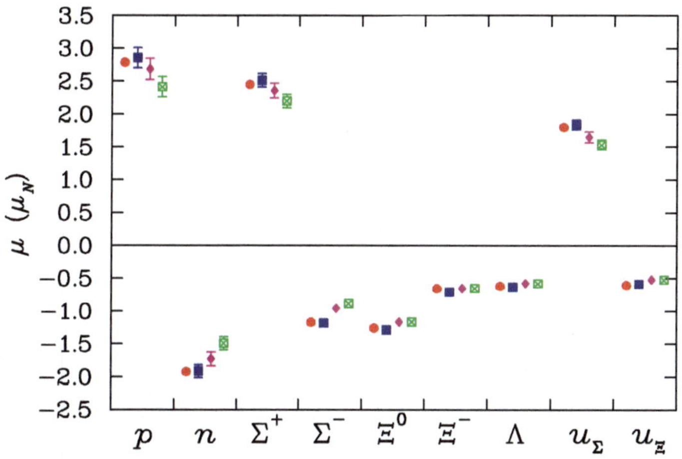

Fig. 3. The extracted full-QCD magnetic moments (■) [16] display excellent agreement with the experimental moments (•). To indicate the size of corrections, the quenched (♦) and finite-volume quenched (⊠) magnetic moments are also displayed.

ratios are compared with the experimental constraint in fig. 2 [16]. The excellent agreement with the constraint from experiment is a first check on the consistency of our calculation.

Using the same procedure which enabled the correction from quenched to dynamical within the valence sector, we have also estimated the disconnected contributions to reconstruct the full magnetic moments in QCD. In fig. 3, the full magnetic moments are shown with the experimentally measured values for the entire baryon octet. The agreement with experiment is remarkable, offering further support for the validity of this analysis. Further, fig. 3 also displays our excellent reproduction of the two experimental valence moments, u^Σ and u^Ξ.

With the valence ratios determined, the final input required from eqs. (1) and (2) is the ratio ${}^lR_d^s$. As discussed above, there have been technical difficulties in extracting a signal for the disconnected insertion directly. Therefore we have used the relative magnitude of the strange-to-light disconnected insertions, estimated through the same method that applied the unquenching corrections and constructed the full moments displayed in fig. 3. Given that the magnitude of the valence sector is so well reproduced, particularly by u^Σ and u^Ξ, and that the overall scale of the disconnected insertions is observed to correctly adjust the full moments from the valence-only sector, we have substantial confidence that our estimate is accurate.

The final result for the strange magnetic moment of the proton is $G_M^s = -0.046 \pm 0.022\mu_N$ [16], an error of just two hundredths of a nuclear magneton. The error quoted includes the best possible estimates of the statistical and systematic errors in the calculation [20]. The announcement by that G0 Collaboration that G_M^s was positive, at the 95% confidence level [21], posed a significant challenge to this theoretical result.

Recently, the same techniques have been applied to theoretically extract the strange electric form factor of the proton [22]. The absence of accurate data for octet charge radii meant that it was necessary to use absolute values of the unquenched contributions of the u and d quarks to the charge radius of the proton. Again the result was particularly accurate, with $G_E^s(Q^2 = 0.1\,\mathrm{GeV}^2) = +0.001 \pm 0.004 \pm 0.004$. This is in quite good agreement with the published measurement by the HAPPEX Collaboration at Jefferson Lab, $G_E^s(Q^2 = 0.1\,\mathrm{GeV}^2) = -0.01 \pm 0.03$ [23].

With the increasing collection of strangeness measurements, and the promise of even more accurate data from HAPPEx in 2006, it is especially timely to see a consolidated treatment of the strange form factor extraction. Recently, a systematic analysis of the published world data has been performed in ref. [24]. The next section briefly summarizes the input to that analysis and its main conclusions.

4 Global analysis of experimental data

Since the first results reported by the SAMPLE Collaboration in 1997 [25], many measurements of the parity-violating contribution to the elastic form factors of the nucleons have now been completed. These experiments have been performed on several targets, the proton, deuteron and helium-4, and at various kinematic configurations. The measured parity-violating electron scattering (PVES) asymmetries are sensitive to the strange electric and magnetic form factors in different linear combinations. By combining the results of these measurements, one can separate the electric and magnetic contributions.

In addition to probing the strange vector current, the experimental asymmetries are also sensitive to the weak axial current in the nucleon, which become increasingly more significant at backward scattering angles. The axial-Z coupling to the nucleon can be controlled through semileptonic decays, deep-inelastic scattering and careful treatment of radiative corrections. In addition, there is a parity-violating photon coupling, which is sensitive to an odd-parity component of the nucleon wavefunction —the anapole form factor [26]. The latter being nonperturbative in origin, one is forced to introduce another unknown form factor which must be determined by data. Because a single stand-alone experiment is sensitive to several unknown form factors, the strangeness content has previously been reported by imposing theoretical estimates for the anapole contributions [27]. With the extensive experimental programs now completed, a global fit of all the data permits the extraction of all unknown form factors [24], without any need for theoretical input.

The parity-violating asymmetries have been constructed with a consistent set of inputs to avoid introducing any systematic distortion in the extracted form factors. In particular, the asymmetries have all been computed using the same set of electromagnetic form factors [28] and the latest radiative corrections [29].

The cleanest separation of the unknown form factors is available at $Q^2 \sim 0.1\,\mathrm{GeV}^2$, where the most substantial coverage of measurements have been performed [23,30]. With further data collected in the near-Q^2 vicinity, particularly by the G0 Collaboration [21], it is useful to use as much of the available data as possible. Using measurements over a range of Q^2 values necessitates introducing a parameterisation of the Q^2 evolution of the form factors. A Taylor expansion of the strange electric and magnetic form factors has been utilised, defining

$$G_E^s = \rho_s Q^2 + \rho_s' Q^4 + \dots, \qquad G_M^s = \mu_s + \mu_s' Q^2 + \dots. \quad (3)$$

Provided that the range of Q^2 values does not extend too high, this approach gives a systematic technique to combine a large set of data.

Because the anapole contribution contributes together with the asymmetry arising from the axial charges, for simplicity the same dipole form is chosen for the axial and anapole contributions, with

$$\tilde{G}_A^N = \tilde{g}_A^N (1 + Q^2/M_A^2)^{-2}, \quad (4)$$

and

$$\tilde{g}_A^N = (\xi_A^{T=1} g_A \tau_3 + \xi_A^{T=0} a_8 + \xi_A^0 a_s) + (A_{\mathrm{ana}}^{T=1} \tau_3 + A_{\mathrm{ana}}^{T=0}). \quad (5)$$

The factors, ξ, denote the tree level plus radiative corrections multiplying the various axial charges. These charges are relatively well known, with $g_A = 1.2695$, $a_8 = 0.58 \pm 0.03 \pm 0.12$ and $a_s = -0.07 \pm 0.04 \mp 0.05$. The second error in a_8 and a_s denotes a correlated uncertainty allowing for potential violations of $SU(3)$-flavour symmetry in semileptonic hyperon decay. The second bracketed term describes the anapole form factor, for which there is only limited phenomenological information. Zhu *et al.* [27] have estimated its magnitude, and Maekawa *et al.* [31] have estimated the leading momentum dependence. The approach taken here is to determine this contribution from the data, free from theory input. Given the size of the uncertainties of the charges, $\tilde{g}_A^p$ and $\tilde{g}_A^n$, the error is dominated by the poor knowledge of the anapole contributions.

The fit to the complete set of world PVES data at $Q^2 < 0.3\,\mathrm{GeV}^2$ yields best-fit parameters

$$\tilde{g}_A^p = 0.05 \pm 1.38 \mp 0.29, \quad (6)$$

$$\tilde{g}_A^n = 2.61 \pm 2.27 \mp 0.37, \quad (7)$$

$$\rho_s = -0.06 \pm 0.41 \mp 0.00\,\mathrm{GeV}^{-2}, \quad (8)$$

$$\mu_s = 0.12 \pm 0.55 \pm 0.07, \quad (9)$$

where the first error denotes the uncorrelated experimental uncertainty and the second the correlated uncertainty in the G0 experiment. The joint determination of the strangeness electric and magnetic form factors at $Q^2 = 0.1\,\mathrm{GeV}^2$ is shown in fig. 4, where we also compare with the theoretical prediction described above.

The 68% and 95% confidence intervals for the determination of G_M^s against $\tilde{G}_A^p$ is shown in fig. 5 and $\tilde{G}_A^n$–$\tilde{G}_A^p$ in fig. 6. This helps provide some picture as to the four-dimensional parameter space that is being constrained by the data. Within the experimental uncertainty on the determination of the axial form factors, fig. 4 shows the strangeness form factors mapping out a long, yet narrow, region of parameter space —including the point of vanishing strangeness. Going to the space of the axial form

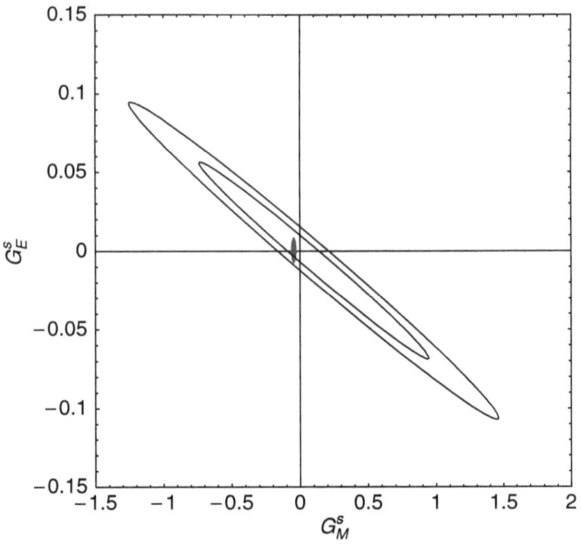

Fig. 4. Determination of the strangeness magnetic and electric form factors at $Q^2 = 0.1\,\mathrm{GeV}^2$: 68% and 95% confidence intervals are shown by the contours. The solid ellipse depicts the theory result described in sect. 3 [16,22].

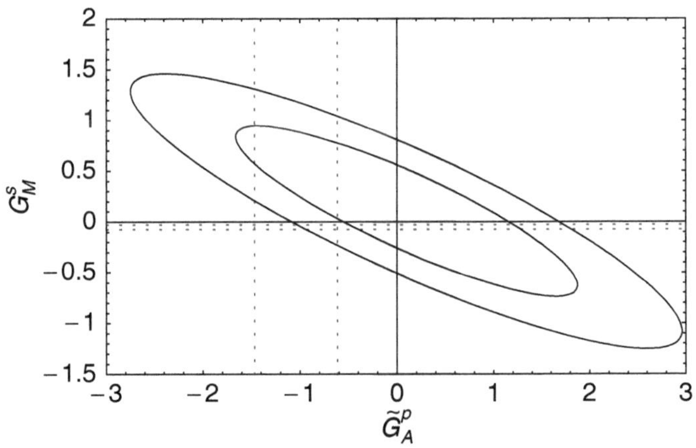

Fig. 5. The contours display the 68% and 95% confidence intervals for the joint determination of $\tilde{G}_A^p$ and G_M^s at $Q^2 = 0.1\,\mathrm{GeV}^2$. The horizontal and vertical bands display the theory results of Leinweber *et al.* [16] and Zhu *et al.* [27], respectively.

factors, within the strangeness determination, these are quite poorly constrained in comparison to the theoretical result of Zhu *et al.* [27]. Nevertheless, they are totally consistent with this calculation, which includes the point of vanishing anapole form factor.

Independently of each other, the strangeness and anapole contributions appear to be consistent with zero, and hence in agreement with both the theoretical results in question. The left panel in fig. 6 indicates that the two theory results appear to be outside the 68% confidence level to be simultaneously supported by the data. In the complete four-dimensional space, it is found that there is a 92% support for a nonzero value in at least one of the strange or anapole form factors [24].

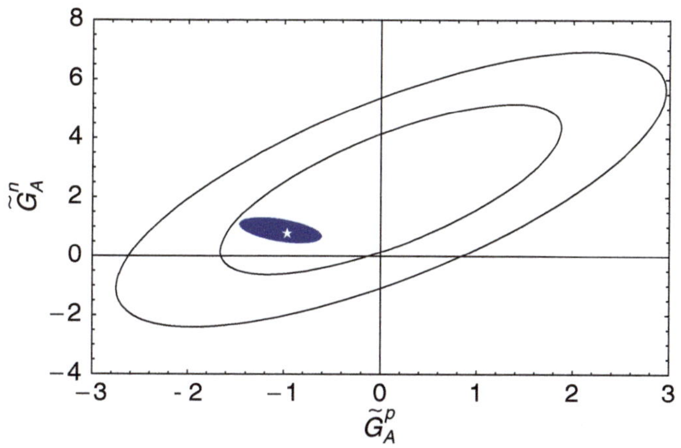

Fig. 6. The contours display the 68% and 95% confidence intervals for the joint determination of $\tilde{G}_A^p$ and $\tilde{G}_A^n$ at $Q^2 = 0.1\,\mathrm{GeV}^2$. The disc depicts the result of Zhu *et al.* [27], with the white star indicating a null anapole form factor.

5 Summary

A series of developments in the study of the chiral extrapolation problem in lattice QCD, combined with a high-precision numerical calculation of baryon electromagnetic form factors, has enabled a detailed study of the strangeness content of the nucleon. A precise, small negative value has been predicted for the strangeness magnetic moment. The strangeness electric contribution is found to be bounded within half a percent of the total mean-square charge radius of the proton.

The determined strangeness form factors are found to be in good agreement with a global analysis of the world strangeness measurements. The anapole contributions in parity-violating elastic scattering are also consistent with modern theoretical estimates —although there is a small hint that both theory results cannot both be supported by the experimental data simultaneously. We look forward to future measurements which will further expand our view of the flavour structure of the nucleon.

Thanks are extended to the many collaborators who have been involved in the research presented here, particularly D. Leinweber and A. Thomas. This work was supported by U.S. DOE Contract No. DE-AC05-06OR23177, under which Jefferson Science Associates, LLC operate Jefferson Lab.

References

1. R.D. McKeown, Phys. Lett. B **219**, 140 (1989); D.H. Beck, Phys. Rev. D **39**, 3248 (1989).
2. LHPC Collaboration (R.G. Edwards *et al.*), Phys. Rev. Lett. **96**, 052001 (2006) [arXiv:hep-lat/0510062].
3. C. Alexandrou *et al.*, Phys. Rev. D **74**, 034508 (2006) [arXiv:hep-lat/0605017].
4. M. Gockeler *et al.*, arXiv:hep-lat/0609001.
5. R.D. Young *et al.*, Prog. Part. Nucl. Phys. **50**, 399 (2003) [arXiv:hep-lat/0212031].
6. S.R. Beane, Nucl. Phys. B **695**, 192 (2004) [arXiv:hep-lat/0403030]; D.B. Leinweber *et al.*, Nucl. Phys. A **755**, 59 (2005) [arXiv:hep-lat/0501028]; V. Bernard, U.G. Meissner, Phys. Lett. B **639**, 278 (2006) [arXiv:hep-lat/0605010]; J.A. McGovern, M.C. Birse, arXiv:hep-lat/0608002.
7. R.D. Young *et al.*, Phys. Rev. D **71**, 014001 (2005) [arXiv:hep-lat/0406001].
8. D.B. Leinweber *et al.*, Phys. Rev. Lett. **92**, 242002 (2004) [arXiv:hep-lat/0302020].
9. R.D. Young *et al.*, Phys. Rev. D **66**, 094507 (2002) [arXiv:hep-lat/0205017].
10. S.J. Dong *et al.*, Phys. Rev. D **58**, 074504 (1998) [arXiv:hep-ph/9712483]; N. Mathur, S.J. Dong, Nucl. Phys. Proc. Suppl. **94**, 311 (2001) [arXiv:hep-lat/0011015]; R. Lewis *et al.*, Phys. Rev. D **67**, 013003 (2003) [arXiv:hep-ph/0210064].
11. W. Detmold, Phys. Rev. D **71**, 054506 (2005) [arXiv:hep-lat/0410011]; F.X. Lee *et al.*, Phys. Lett. B **627**, 71 (2005) [arXiv:hep-lat/0509067].
12. J. Foley *et al.*, Comput. Phys. Commun. **172**, 145 (2005) [arXiv:hep-lat/0505023].
13. D.B. Leinweber, Phys. Rev. D **53**, 5115 (1996) [arXiv:hep-ph/9512319].
14. G.A. Miller *et al.*, Phys. Rep. **194**, 1 (1990); B. Kubis, R. Lewis, Phys. Rev. C **74**, 015204 (2006) [arXiv:nucl-th/0605006].
15. D.B. Leinweber, A.W. Thomas, Phys. Rev. D **62**, 074505 (2000) [arXiv:hep-lat/9912052].
16. D.B. Leinweber *et al.*, Phys. Rev. Lett. **94**, 212001 (2005) [arXiv:hep-lat/0406002].
17. X. Ji, D. Toublan, arXiv:hep-ph/0605055.
18. D.B. Leinweber, A.W. Thomas, R.D. Young, Phys. Rev. Lett. **86**, 5011 (2001) [arXiv:hep-ph/0101211].
19. S. Boinepalli *et al.*, arXiv:hep-lat/0604022.
20. D.B. Leinweber *et al.*, Eur. Phys. J. A **24**, s2, 79 (2005) [arXiv:hep-lat/0502004].
21. G0 Collaboration (D.S. Armstrong *et al.*), Phys. Rev. Lett. **95**, 092001 (2005) [arXiv:nucl-ex/0506021].
22. D.B. Leinweber *et al.*, Phys. Rev. Lett. **97**, 022001 (2006) [arXiv:hep-lat/0601025].
23. HAPPEX Collaboration (K.A. Aniol *et al.*), Phys. Lett. B **635**, 275 (2006) [arXiv:nucl-ex/0506011]; Phys. Rev. Lett. **96**, 022003 (2006) [arXiv:nucl-ex/0506010].
24. R.D. Young, J. Roche, R.D. Carlini, A.W. Thomas, Phys. Rev. Lett. **97**, 102002 (2006) [arXiv:nucl-ex/0604010].
25. SAMPLE Collaboration (B. Mueller *et al.*), Phys. Rev. Lett. **78**, 3824 (1997) [arXiv:nucl-ex/9702004].
26. W.C. Haxton *et al.*, Phys. Rev. Lett. **63**, 949 (1989).
27. S.L. Zhu *et al.*, Phys. Rev. D **62**, 033008 (2000) [arXiv:hep-ph/0002252].
28. J.J. Kelly, Phys. Rev. C **70**, 068202 (2004).
29. Particle Data Group (S. Eidelman *et al.*), Phys. Lett. B **592**, 1 (2004).
30. SAMPLE Collaboration (D.T. Spayde *et al.*), Phys. Lett. B **583**, 79 (2004) [arXiv:nucl-ex/0312016]; F.E. Maas *et al.*, Phys. Rev. Lett. **94**, 152001 (2005) [arXiv:nucl-ex/0412030].
31. C.M. Maekawa *et al.*, Phys. Lett. B **488**, 167 (2000) [arXiv:hep-ph/0006181].

Probing nucleon structure on the lattice

The QCDSF/UKQCD Collaboration

M. Göckeler[1], Ph. Hägler[2], R. Horsley[3], Y. Nakamura[4], D. Pleiter[4], P.E.L. Rakow[5], A. Schäfer[1], G. Schierholz[4,6], W. Schroers[4], H. Stüben[7], and J.M. Zanotti[3,a]

[1] Institut für Theoretische Physik, Universität Regensburg, 93040 Regensburg, Germany
[2] Institut für Theoretische Physik T39, Physik-Department der TU München, 85747 Garching, Germany
[3] School of Physics, University of Edinburgh, Edinburgh EH9 3JZ, UK
[4] John von Neumann-Institut für Computing NIC/DESY, 15738 Zeuthen, Germany
[5] Theoretical Physics Division, Department of Mathematical Sciences, University of Liverpool, Liverpool L69 3BX, UK
[6] Deutsches Elektronen-Synchrotron DESY, 22603 Hamburg, Germany
[7] Konrad-Zuse-Zentrum für Informationstechnik Berlin, 14195 Berlin, Germany

Paper reprinted from Eur. Phys. J. A with permission

Abstract. The QCDSF/UKQCD Collaboration has an ongoing program to calculate nucleon matrix elements with two flavours of dynamical $\mathcal{O}(a)$ improved Wilson fermions. Here we present recent results on the electromagnetic form factors, the quark momentum fraction $\langle x \rangle$ and the first three moments of the nucleon's spin-averaged and spin-dependent generalised parton distributions, including preliminary results with pion masses as low as 320 MeV.

PACS. 12.38.Gc Lattice QCD calculations – 13.40.Gp Electromagnetic form factors

1 Introduction

The ability of generalised parton distributions (GPDs) [1] to describe both exclusive and inclusive processes has led to an enormous amount of interest in these functions both experimentally and theoretically. Not only do GPDs encompass the ordinary electromagnetic form factors and parton distribution functions, but they also allow for the computation of the total quark contribution to the nucleon spin [2] as well as revealing important information on the transverse structure of the nucleon [3,4]. A full mapping of the parameter space spanned by GPDs is an extremely extensive task which needs support from non-perturbative techniques like lattice simulations.

Substantial progress has already been made in computing the first three moments of unpolarised, polarised [5–7] and tensor [8] GPDs on the lattice.

In this paper we present recent results from the QCDSF/UKQCD Collaboration. In sect. 2 we investigate the q^2-dependence of the Dirac and Pauli electromagnetic form factors, while sect. 3 contains preliminary results for the average fraction of the nucleon's momentum carried by the quarks, $\langle x \rangle$. Finally, in sect. 4 we present results for the first three moments of the GPDs H and $\tilde{H}$.

[a] e-mail: `jzanotti@ph.ed.ac.uk`; Spokesperson.

2 Electromagnetic form factors

The study of the electromagnetic properties of hadrons provides important insights into the non-perturbative structure of QCD. The EM form factors reveal important information on the internal structure of hadrons including their size, charge distribution and magnetisation. Phenomenological interest in these form factors has been revived by recent Jefferson Lab polarisation experiments [9] measuring the ratio of the proton electric to magnetic form factors, $\mu^{(p)} G_e^{(p)}(q^2)/G_m^{(p)}(q^2)$. These experiments show that this ratio unexpectedly decreases almost linearly with increasing q^2, indicating that the proton's electric form factor falls off faster than the magnetic form factor.

A lattice calculation of the q^2-dependence of the proton's electromagnetic form factors can not only allow for a comparison with experiment, but also help in the understanding of the asymptotic behaviour of these form factors. Such a lattice calculation would also allow for the extraction of other phenomenologically interesting quantities such as magnetic and electric charge radii and magnetic moments.

2.1 Lattice techniques

On the lattice, we determine the form factors $F_1(q^2)$ and $F_2(q^2)$ by calculating the following matrix element of the electromagnetic current:

$$\langle p',\, s'|j_\mu|p,\, s\rangle = \bar{u}(p',\, s')\Big[\gamma_\mu F_1(q^2) + i\sigma_{\mu\nu}\frac{q_\nu}{2m_N}F_2(q^2)\Big]u(p,\, s)\,, \qquad (1)$$

where $u(p,\,s)$ is a Dirac spinor with momentum p and spin polarisation s, $q = p' - p$ is the momentum transfer with $Q^2 = -q^2$, m_N is the nucleon mass and j_μ is the electromagnetic current.

The form factors of the proton are obtained by using

$$j_\mu^{(p)} = \frac{2}{3}\bar{u}\gamma_\mu u - \frac{1}{3}\bar{d}\gamma_\mu d\,, \qquad (2)$$

while for iso-vector (*i.e.* proton-neutron) form factors

$$j_\mu^v = \bar{u}\gamma_\mu u - \bar{d}\gamma_\mu d\,. \qquad (3)$$

It is common to rewrite the form factors F_1 and F_2 as

$$\mathcal{G}_e(q^2) = F_1(q^2) + \frac{q^2}{(2m_N)^2}F_2(q^2), \qquad (4)$$

$$\mathcal{G}_m(q^2) = F_1(q^2) + F_2(q^2), \qquad (5)$$

which are known as the electric and magnetic Sachs form factors, respectively.

At zero momentum transfer, $F_1(0) = \mathcal{G}_e(0)$ gives the electric charge (*e.g.*, 1 for the proton), while

$$\mathcal{G}_m^{(p)}(0) = \mu^{(p)} = 1 + \kappa^{(p)} \qquad (6)$$

gives the magnetic moment, where $F_2^{(p)}(0) = \kappa^{(p)}$ is the anomalous magnetic moment.

In order to extract the non-forward matrix elements from our lattice simulations, we compute ratios of three- and two-point functions

$$\mathcal{R}(t,\tau;\boldsymbol{p}',\boldsymbol{p};\mathcal{O}) = \frac{C_\Gamma(t,\tau;\boldsymbol{p}',\boldsymbol{p},\mathcal{O})}{C_2(t,\boldsymbol{p}')} \times \left[\frac{C_2(\tau,\boldsymbol{p}')C_2(t,\boldsymbol{p}')C_2(t-\tau,\boldsymbol{p})}{C_2(\tau,\boldsymbol{p})C_2(t,\boldsymbol{p})C_2(t-\tau,\boldsymbol{p}')}\right]^{\frac{1}{2}}, \qquad (7)$$

which for large time separations, $0 \ll \tau \ll t \lesssim \frac{1}{2}L_T$, where L_T is the temporal extent of our lattice, is proportional to the matrix element we are interested in, $\langle p'|\mathcal{O}_q|p\rangle$. The nucleon two- and three-point functions are given, respectively, by

$$C_2(\tau,\boldsymbol{p}) = \mathrm{Tr}\left[\Gamma_{\mathrm{unpol}}\langle B(\tau,\boldsymbol{p})\overline{B}(0,\boldsymbol{p})\rangle\right],$$
$$C_\Gamma(t,\tau;\boldsymbol{p}',\boldsymbol{p},\mathcal{O}) = \mathrm{Tr}\left[\Gamma\langle B(t,\boldsymbol{p}')\mathcal{O}(\tau)\overline{B}(0,\boldsymbol{p})\rangle\right]. \qquad (8)$$

Here t and τ are the Euclidean times of the nucleon sink and operator insertion, respectively, $\boldsymbol{p}'$ ($\boldsymbol{p}$) is the nucleon momentum at the sink (source), and $\mathcal{O}$ is the local vector current

$$\mathcal{O}(\tau) = \psi(\tau)\gamma_\mu\psi(\tau)\,, \qquad (9)$$

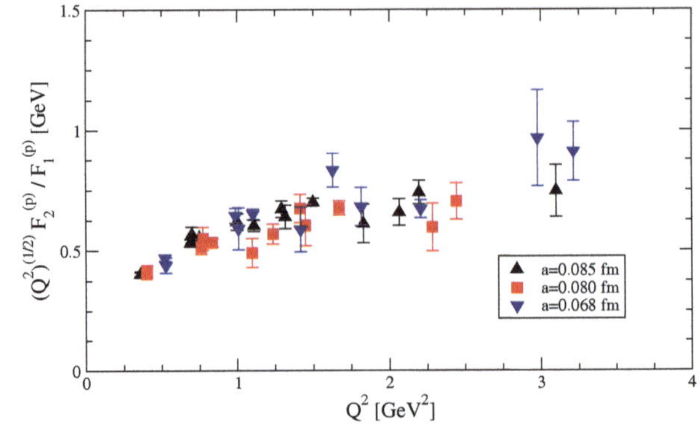

Fig. 1. $\sqrt{Q^2}\,F_2/F_1$ form factor ratio on three datasets with the same pion mass ($\approx$ 550 MeV), but with different lattice spacings, $a = 0.085$, 0.080, 0.068 fm.

which we renormalise non-perturbatively [10]. The trace in eq. (8) is over spinor indices and the Γ matrix determines the polarisation of the nucleon with $\Gamma_{\mathrm{unpol}} = \frac{1}{2}(1+\gamma_4)$. We note here that in the calculation of nucleon matrix elements, we neglect contributions coming from disconnected quark diagrams as these are extremely computationally demanding. Hence, in the following we mainly restrict ourselves to the calculation of iso-vector matrix elements where the disconnected quark contributions cancel.

Finally, we use the Sommer parameter, r_0, to set the scale with $r_0 = 0.5$ fm.

2.2 Results

Of particular interest is the need to understand the behaviour of the form factor $F_2(Q^2)$. The question arises which is the best way to fit the form factor since such a fitting function also allows an extrapolation of the form factor to $Q^2 = 0$. This is a necessary ingredient to find the anomalous magnetic moment of the nucleon, κ.

Based on perturbative QCD, F_1 should scale asymptotically as $1/Q^4$, while $F_2 \sim 1/Q^6$ [11,12]. It is difficult to obtain lattice data with high enough precision over a large enough range of Q^2 values to distinguish between a dipole or tripole behaviour. It may, however, be instructive to consider the form factor ratio $F_2(Q^2)/F_1(Q^2)$ since asymptotically this ratio should scale as $1/Q^2$. Spin polarisation experiments have instead found that the data is compatible with

$$\frac{F_2(Q^2)}{F_1(Q^2)} \sim \frac{1}{\sqrt{Q^2}}\,. \qquad (10)$$

To investigate the asymptotic behaviour of the form factor ratio $F_2(Q^2)/F_1(Q^2)$, we plot in fig. 1 the results for $\sqrt{Q^2}F_2/F_1$ obtained at three working points with approximately the same pion mass, but with different values of the lattice spacing. Here we observe the lattice data to be consistent with a constant for $Q^2 > 1.5\,\mathrm{GeV}^2$, similar to the experimental data. Multiplying these results by an extra factor of $\sqrt{Q^2}$, as suggested by perturbative QCD, would clearly destroy the plateau. Quantitatively,

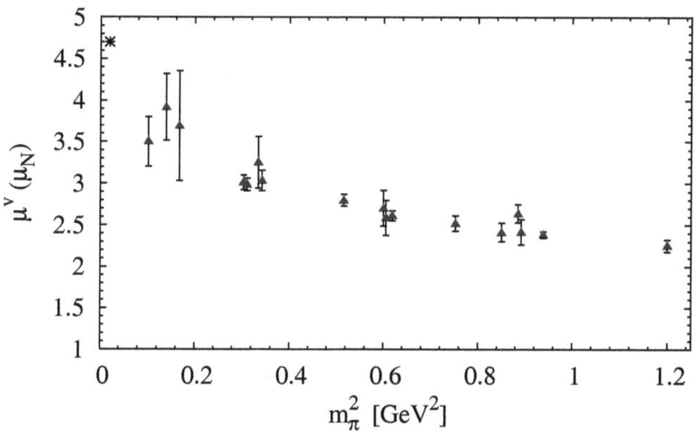

Fig. 2. Results for the isovector magnetic moment as a function of m_π^2. The experimental value is denoted by the asterisk.

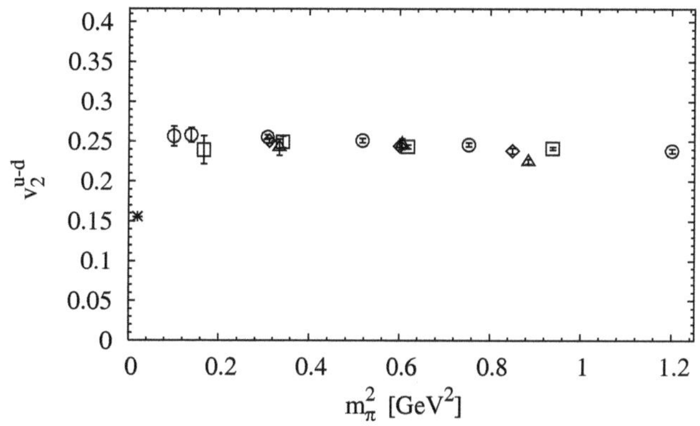

Fig. 3. Isovector $\langle x \rangle$ as a function of m_π^2 in the $\overline{\text{MS}}$ scheme at $\mu^2 = 4\,\text{GeV}^2$. These preliminary results are obtained at four different lattice spacings (in fm): 0.092 (triangles), 0.085 (diamonds), 0.080 (circles) and 0.068 (squares). The asterisk indicates the phenomenological result of the MRST analysis [20] as given in [18]. This is in agreement with a recent higher-order analysis [21].

though, the lattice data is higher than the corresponding experimental ratios, cf. [13]. This shows that the lattice simulations are able to reproduce the qualitative features of the experimental data, but for a quantitative reproduction the pion mass is still unrealistically large.

In the following we fit F_1 and F_2 with a dipole ansatz

$$F_i^{(v)}(q^2) = \frac{F_i(0)}{(1 - q^2/M_i^2)^2}\,, \tag{11}$$

where $F_1^{(v)}(0) = 1$, $F_2^{(v)}(0) = \kappa^{(v)}$ and M_i is the fitted dipole mass for the form factor, i.

We display our results for the isovector magnetic moment in fig. 2 as a function of m_π^2. Our results are in good agreement with recent quenched [14–16] and $N_f = 2$ [16] results, which indicates that there appears to be little effect due to quenching on the magnetic moments, as predicted in [17]. The experimental value is indicated by a asterisk at the physical pion mass. We clearly see that a linear extrapolation would miss the experimental point. This, however, is not completely unexpected as results from chiral perturbation theory suggest that we should observe a dramatic increase in the results at lighter pion masses [14,17]. The new points at lighter pion masses, $m_\pi^2 < 0.2\,\text{GeV}^2$, are beginning to show a hint of such curvature, although more work needs to be done to reduce the error bars.

3 Quark momentum fraction, ⟨x⟩

Forward matrix elements (no momentum transfer) provide moments of quark distributions in some scheme, $\mathcal{S}$, at some scale, M:

$$\langle N(\boldsymbol{p})|\mathcal{O}_q^{\{\mu_1\cdots\mu_n\}}|N(\boldsymbol{p})\rangle^{\mathcal{S}} = 2v_n^{(q)\mathcal{S}}(g^{\mathcal{S}}(M))\, p^{\mu_1}\cdots p^{\mu_n}, \tag{12}$$

where

$$\mathcal{O}_q^{\{\mu_1\cdots\mu_n\}} = \overline{q}\, i^{n-1}\gamma^{\{\mu_1} \overset{\leftrightarrow}{D}{}^{\mu_2} \cdots \overset{\leftrightarrow}{D}{}^{\mu_n\}} q\,, \tag{13}$$

$\overset{\leftrightarrow}{D} = \frac{1}{2}(\vec{D} - \overset{\leftarrow}{D})$ and $\{\cdots\}$ indicates symmetrisation of indices and removal of traces.

Matrix elements with no momentum transfer are determined from a simplified version of the ratio of three-point to two-point correlation functions given in eq. (7). See [18] for additional details.

We use non-perturbative renormalisation as outlined in sect. 5.2.3 of [18] to convert our lattice results to the $\overline{\text{MS}}$ scheme at $\mu^2 = 4\,\text{GeV}^2$.

In the language of the parton model, v_n^q is often denoted by $\langle x^{n-1}\rangle^q$

$$\langle x^{n-1}\rangle^q = \int_0^1 \mathrm{d}x\, x^{n-1}\,[q(x) + (-1)^n \bar{q}(x)] = v_n^q\,. \tag{14}$$

Of particular interest is the first ($n = 2$) moment, $v_2^q = \langle x\rangle^q$, which determines the fraction of the nucleon's momentum carried by the quark, q. This quantity is notorious on the lattice for producing values much larger than phenomenologically accepted results. These discrepancies can possibly be explained by the fact that all lattice calculations to date have been performed at quark masses that are much larger than the physical masses [19]. Hence, it is a challenge for current lattice simulations to calculate $\langle x\rangle$ at small enough quark masses in order to search for the severe curvature predicted in ref. [19].

Figure 3 displays preliminary results for $\langle x\rangle^{(u-d)}$ with pion masses as low as $\sim$ 320 MeV. Before we can draw any conclusions on the behaviour at small quark masses, we need to study scaling violations and finite-size effects more carefully. Indeed, it has been suggested [22,23] that a volume of at least $(4\,\text{fm})^3$ is required to confirm the predicted chiral curvature.

4 Generalised parton distributions

4.1 Matrix elements and moments of GPDs

For a lattice calculation of GPDs, we work in Mellin space to relate matrix elements of local operators to Mellin moments of the GPDs. The non-forward matrix elements of the twist-2 operator in eq. (13) specifies the $(n-1)^{th}$ moments of the spin-averaged generalised parton distributions. Replacing γ with $\gamma_5\gamma$ leads to moments of the spin-dependent GPDs. In particular, for the unpolarised GPDs, we have

$$\int_{-1}^{1} \mathrm{d}x x^{n-1} H^q(x,\xi,t) = H_n^q(\xi,t), \quad \int_{-1}^{1} \mathrm{d}x x^{n-1} E^q = E_n^q, \tag{15}$$

where [2]

$$H_n^q(\xi,t) = \sum_{i=0}^{\lfloor\frac{n-1}{2}\rfloor} A_{n,2i}^q(t)(-2\xi)^{2i} + C_n^q(t)(-2\xi)^n|_{n\ \mathrm{even}},$$
$$E_n^q(\xi,t) = \sum_{i=0}^{\lfloor\frac{n-1}{2}\rfloor} B_{n,2i}^q(t)(-2\xi)^{2i} - C_n^q(t)(-2\xi)^n|_{n\ \mathrm{even}}. \tag{16}$$

Here we denote the invariant of the momentum transfer by $t=\Delta^2=(p'-p)^2$. The generalised form factors $A_{n,2i}^q(t)$, $B_{n,2i}^q(t)$ and $C_n^q(t)$ for the lowest three moments are extracted from non-forward nucleon matrix elements of the operators in eq. (13) [6].

For the lowest moment, A_{10} and B_{10} are just the Dirac and Pauli form factors F_1 and F_2, respectively

$$\int_{-1}^{1} \mathrm{d}x\, H^q(x,\xi,t) = A_{10}^q(t) = F_1(t), \tag{17}$$
$$\int_{-1}^{1} \mathrm{d}x\, E^q(x,\xi,t) = B_{10}^q(t) = F_2(t), \tag{18}$$

while $\widetilde{A}_{10}$ and $\widetilde{B}_{10}$ are the usual axial-vector and pseudoscalar form factors, respectively

$$\int_{-1}^{1} \mathrm{d}x\, \widetilde{H}^q(x,\xi,t) = \widetilde{A}_{10}^q(t) = g_A(t), \tag{19}$$
$$\int_{-1}^{1} \mathrm{d}x\, \widetilde{E}^q(x,\xi,t) = \widetilde{B}_{10}^q(t) = g_P(t). \tag{20}$$

We also observe that in the forward limit ($t=\xi=0$), the moments of H_q reduce to the moments of the unpolarised parton distribution $A_{n0}(0)=\langle x^{n-1}\rangle$.

4.2 Results for generalised form factors

Burkardt [4] has shown that the spin-independent and spin-dependent generalised parton distributions $H(x,0,t)$ and $\widetilde{H}(x,0,t)$ gain a probability interpretation when Fourier transformed to impact parameter space at longitudinal momentum transfer $\xi=0$

$$q(x,\boldsymbol{b}_\perp) = \int \frac{\mathrm{d}^2\Delta_\perp}{(2\pi)^2}\, \mathrm{e}^{-i\boldsymbol{b}_\perp\cdot\boldsymbol{\Delta}_\perp} H(x,0,-\Delta_\perp^2), \tag{21}$$

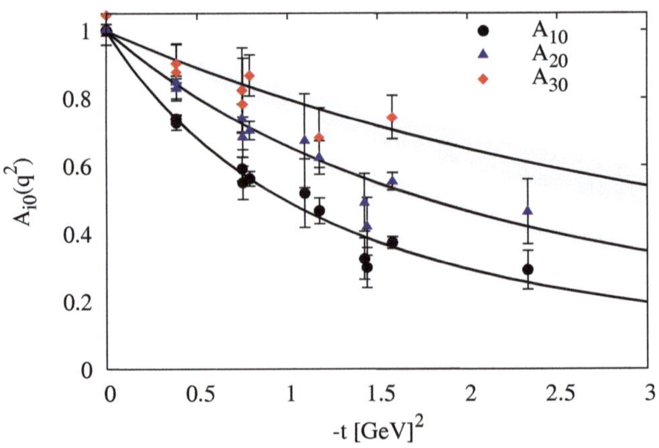

Fig. 4. Generalised form factors A_{10}^{u-d}, A_{20}^{u-d}, A_{30}^{u-d} together with a dipole fit. All form factors have been normalised to unity.

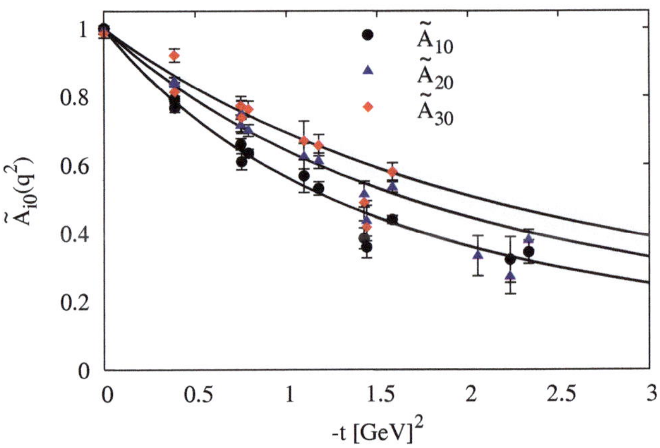

Fig. 5. Generalised form factors $\widetilde{A}_{10}^{u-d}$, $\widetilde{A}_{20}^{u-d}$, $\widetilde{A}_{30}^{u-d}$ together with a dipole fit. All form factors have been normalised to unity.

(and similar for the polarised $\Delta q(x,\boldsymbol{b}_\perp)$) where $q(x,\boldsymbol{b}_\perp)$ is the probability density for a quark with longitudinal momentum fraction x and at transverse position (or impact parameter) $\boldsymbol{b}_\perp$.

Burkardt [4] also argued that $H(x,0,-\Delta_\perp^2)$ becomes $\Delta_\perp^2$-independent as $x\to 1$ since, physically, we expect the transverse size of the nucleon to decrease as x increases, *i.e.* $\lim_{x\to 1} q(x,\boldsymbol{b}_\perp) \propto \delta^2(\boldsymbol{b}_\perp)$. As a result, we expect the slopes of the moments of $H(x,0,-\Delta_\perp^2)$ in $\Delta_\perp^2$ to decrease as we proceed to higher moments. This is also true for the polarised moments of $\widetilde{H}(x,0,-\Delta_\perp^2)$, so from eq. (16) with $\xi=0$, we expect that the slopes of the generalised form factors $A_{n0}(t)$ and $\widetilde{A}_{n0}(t)$ should decrease with increasing n.

In figs. 4 and 5, we show the t-dependence of $A_{n0}(t)$ and $\widetilde{A}_{n0}(t)$, respectively, $n=1,2,3$, for $\beta=5.40$, $\kappa_{\mathrm{sea}}=\kappa_{\mathrm{val}}=0.13500$. The form factors have been normalised to unity to make a comparison of the slopes easier and we fit the form factors with a dipole form as in eq. (11). We observe here that the form factors for the unpolarised mo-

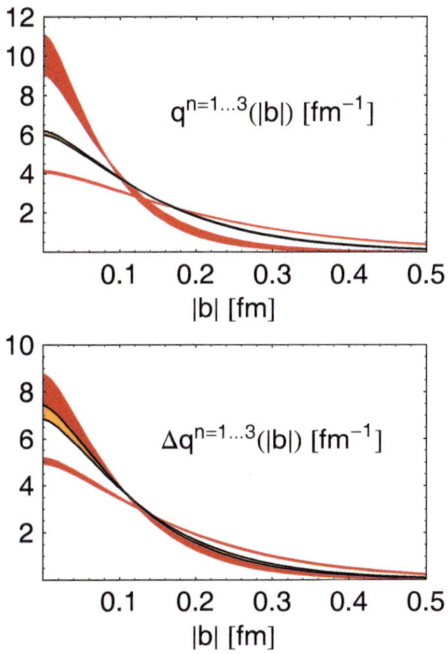

Fig. 6. The lowest three moments of the GPD $H(x, \xi = 0, t)$ (top) and $\widetilde{H}(x, \xi = 0, t)$ (bottom) in impact parameter space as a function of impact parameter, b.

ments are well separated and that their slopes do indeed decrease with increasing n as predicted. For the polarised moments, we observe a similar scenario, however here the change in slope between the form factors is not as large. The flattening of the GFFs $A_{n0}(t)$ has first been observed in ref. [7], where at the same time practically no change in slope was seen going from $\widetilde{A}_{20}(t)$ to $\widetilde{A}_{30}(t)$.

Although fitting the form factors with a dipole is purely phenomenological (see ref. [24] for an alternative ansatz), it does provide us with a useful means to measure the change in slope of the form factors by monitoring the extracted dipole masses as we proceed to higher moments. We have calculated these generalised form factors on a subset of our full complement of (β, κ) combinations and have extracted the corresponding dipole masses. Recall that A_{10} is the Dirac form factor F_1, while $\widetilde{A}_{10}$ is the axial form factor g_A. Hence the dipole fits can be compared with experiment. A linear extrapolation produces a result larger than experiment for both the polarised and unpolarised case, although the findings of ref. [25] suggest that the chiral extrapolation of the dipole masses of the electromagnetic form factors may be non-linear.

In fig. 6 we show the lowest three moments of the GPD $H(x, \xi = 0, t)$ (top) and $\widetilde{H}(x, \xi = 0, t)$ (bottom) in impact parameter space. The curves correspond to the Fourier-transformation of our dipole ansatz eq. (11), with the dipole masses extrapolated linearly to the chiral limit, to $b_\perp$-space, and the shaded error band is a result of the errors in the extrapolated dipole masses at the physical pion mass. The curves have been normalised so that they represent line densities with $\int \mathrm{d}b\, q^n(b) = 1$. The top figure of fig. 6 clearly shows how the $u-d$ quark distribution narrows as we proceed to higher moments n and thereby larger values of the average momentum fraction, while for the polarised case in the bottom figure, the narrowing of the distribution is not so severe.

The numerical calculations have been performed on the Hitachi SR8000 at LRZ (Munich), the Cray T3E at EPCC (Edinburgh) [26] the APE*1000* and apeNEXT at NIC/DE-SY (Zeuthen), the BlueGeneL at NIC/Jülich and the BlueGeneL at EPCC (Edinburgh). Some of the configurations at the small pion mass have been generated on the Blue GeneL at KEK by the Kanazawa group as part of the DIK research programme. This work was supported in part by the DFG, by the EU Integrated Infrastructure Initiative Hadron Physics (I3HP) under contract number RII3-CT-2004-506078. Ph.H. acknowledges support by the DFG Emmy-Noether program.

References

1. D. Müller *et al.*, Fortschr. Phys. **42**, 101 (1994) [hep-ph/9812448]; A.V. Radyushkin, Phys. Rev. D **56**, 5524 (1997) [hep-ph/9704207]; M. Diehl *et al.*, Phys. Lett. B **411**, 193 (1997) [hep-ph/9706344].
2. X.D. Ji, Phys. Rev. Lett. **78**, 610 (1997) [hep-ph/9603249]; X.D. Ji, J. Phys. G **24**, 1181 (1998) [hep-ph/9807358].
3. M. Diehl, Eur. Phys. J. C **25**, 223 (2002); **31**, 277 (2003) [hep-ph/0205208].
4. M. Burkardt, Int. J. Mod. Phys. A **18**, 173 (2003) [hep-ph/0207047].
5. M. Göckeler *et al.*, Phys. Rev. Lett. **92**, 042002 (2004) [hep-ph/0304249]; Nucl. Phys. Proc. Suppl. **128**, 203 (2004), [hep-ph/0312104]; **140**, 399 (2005), [hep-lat/0409162]; Few-Body Syst. **36**, 111 (2005) [hep-lat/0410023].
6. Ph. Hägler *et al.*, Phys. Rev. D **68**, 034505 (2003) [hep-lat/0304018].
7. Ph. Hägler *et al.*, Phys. Rev. Lett. **93**, 112001 (2004) [hep-lat/0312014]; J.W. Negele *et al.*, Nucl. Phys. Proc. Suppl. **128**, 170 (2004) [hep-lat/0404005]; Ph. Hägler *et al.*, Eur. Phys. J. A **24**, s1, 29 (2005) [hep-ph/0410017]; R.G. Edwards *et al.*, PoS (LAT2005), 056 (2006) [hep-lat/0509185].
8. M. Göckeler *et al.*, Nucl. Phys. A **755**, 537 (2005) [hep-lat/0501029]; M. Göckeler *et al.*, Phys. Lett. B **627**, 113 (2005) [hep-lat/0507001]; M. Göckeler *et al.*, Nucl. Phys. Proc. Suppl. **153**, 146 (2006) [hep-lat/0512011]; M. Diehl *et al.*, hep-ph/0511032.
9. M.K. Jones *et al.*, Phys. Rev. Lett. **84**, 1398 (2000) [nucl-ex/9910005]; O. Gayou *et al.*, Phys. Rev. C **64**, 038202 (2001); O. Gayou *et al.*, Phys. Rev. Lett. **88**, 092301 (2002) [nucl-ex/0111010].
10. T. Bakeyev *et al.*, Phys. Lett. B **580**, 197 (2004) [hep-lat/0305014].
11. S.J. Brodsky, G.R. Farrar, Phys. Rev. D **11**, 1309 (1975).
12. G.P. Lepage, S.J. Brodsky, Phys. Rev. D **22**, 2157 (1980).
13. A.V. Belitsky *et al.*, Phys. Rev. Lett. **91**, 092003 (2003) [hep-ph/0212351].
14. M. Göckeler *et al.*, Phys. Rev. D **71**, 034508 (2005) [hep-lat/0303019].
15. S. Boinepalli *et al.*, hep-lat/0604022.
16. C. Alexandrou *et al.*, Phys. Rev. D **74**, 034508 (2006) [hep-lat/0605017].

17. R.D. Young *et al.*, Phys. Rev. D **71**, 014001 (2005) [hep-lat/0406001].
18. M. Göckeler *et al.*, Phys. Rev. D **71**, 114511 (2005) [hep-ph/0410187].
19. W. Detmold *et al.*, Phys. Rev. Lett. **87**, 172001 (2001) [hep-lat/0103006].
20. A.D. Martin *et al.*, Eur. Phys. J. C **23**, 73 (2002) [hep-ph/0110215].
21. J. Blumlein *et al.*, hep-ph/0607200.
22. W. Detmold, C.J. Lin, Phys. Rev. D **71**, 054510 (2005) [hep-lat/0501007].
23. W. Detmold *et al.*, Mod. Phys. Lett. A **18**, 2681 (2003) [hep-lat/0310003].
24. M. Diehl *et al.*, Eur. Phys. J. C **39**, 1 (2005) [hep-ph/0408173].
25. J.D. Ashley *et al.*, Eur. Phys. J. A **19**, 9 (2004) [hep-lat/0308024].
26. C.R. Allton *et al.*, Phys. Rev. D **65**, 054502 (2002) [hep-lat/0107021].

New results from the HAPPEX Experiments at $Q^2 = 0.1\,\mathrm{GeV}/c^2$

D.S. Armstrong[a]

For the HAPPEX Collaboration
Department of Physics, College of William & Mary, Williamsburg, VA 23187-8795, USA

Abstract. The nucleon's strange electric and magnetic form factors G_E^s and G_M^s can be probed via parity-violating electron scattering. The HAPPEX Collaboration has made new measurements of the parity-violating asymmetry A_{PV} in elastic scattering of 3 GeV electrons off hydrogen and ^{4}He targets with $\langle\theta_{lab}\rangle \approx 6.0°$. For ^{4}He the preliminary result is $A_{\mathrm{PV}} = (+6.43 \pm 0.23\ (\mathrm{stat}) \pm 0.22\ (\mathrm{syst})) \times 10^{-6}$. For hydrogen the preliminary result is $A_{\mathrm{PV}} = (-1.60 \pm 0.12\ (\mathrm{stat}) \pm 0.05\ (\mathrm{syst})) \times 10^{-6}$. From these values we extract $G_E^s = 0.004 \pm 0.014 \pm 0.013$ at $\langle Q^2\rangle = 0.077\,\mathrm{GeV}/c^2$, and $G_E^s + 0.09\,G_M^s = 0.004 \pm 0.011 \pm 0.005$ at $\langle Q^2\rangle = 0.109\,\mathrm{GeV}/c^2$, both consistent with zero, providing stringent new limits on the role of strange quarks in the vector structure of the nucleon.

PACS. 25.30.Bf Elastic electron scattering – 12.15.-y Electroweak interactions – 11.30.Er Charge conjugation, parity, time reversal, and other discrete symmetries – 14.20.Dh Protons and neutrons

1 Introduction

Although QCD is widely accepted as the correct theory of the strong interaction, it is not yet possible to describe precisely all the properties of the nucleon in terms of quarks and gluons, the fundamental degrees of freedom appearing in the Lagrangian. Measurements of deep-inelastic lepton scattering give information on quark distribution functions and the quark spin structure. These data indicate that the sea quarks, in particular strange quarks in the form of $s\bar{s}$ pairs, may contribute significantly to certain observables in the perturbative regime. However, the contribution of the sea quarks in the non-perturbative realm is far from being understood. Strange quarks effects provide a direct probe of the sea in the nucleon. Although the up and down quarks and anti-quarks also contribute to the sea, at low momentum transfer they are also intimately connected with the special role of the pion as the Goldstone boson of chiral symmetry breaking. In particular, there is relatively little known about the contribution of strange quarks to vector matrix elements (*i.e.*, to the electric and magnetic form factors of the nucleon).

Much of the original impetus for examining strange quark form factors of the nucleon arose when initial attempts to understand the "spin-crisis" appeared to indicate that Δs was large, *i.e.*, that a large fraction of the proton's spin might be due to strange quarks (for an update, see [1]). This suggested that other (*e.g.* vector) strange quark matrix elements could also be sizable. Indeed, many calculations, using a variety of nucleon models, were found to allow (or even prefer) large vector strange form factors (for reviews, see [2]).

Parity-violating electron scattering opens a unique experimental window into the strange quark vector form factors. The differing couplings of the quark flavors to the γ and the Z^0 allow one to separate the contributions from the various flavors, using the electroweak $\gamma - Z^0$ interference term. This interference leads to a parity-violating asymmetry in the scattering cross-section, proportional to the neutral-current amplitude. Precise measurements of this asymmetry can be interpreted in terms of the strange electric (G_E^s) and magnetic (G_M^s) form factors of the nucleon. For the elastic scattering from the proton, the parity-violating asymmetry between cross-sections for positive and negative helicity electrons has contributions from weak electric, magnetic and axial vector currents, and is given in the standard model by [3]

$$A_{\mathrm{PV}} = \frac{\sigma^+ - \sigma^-}{\sigma^+ + \sigma^-} = -\frac{G_F Q^2}{4\pi\alpha\sqrt{2}} \times \Bigg\{ (1 - 4\sin^2\theta_W) - \frac{\epsilon G_E^{\gamma p} G_E^{\gamma n} + \tau G_M^{\gamma p} G_M^{\gamma n}}{\epsilon (G_E^{\gamma p})^2 + \tau (G_M^{\gamma p})^2} - \frac{\epsilon G_E^{\gamma p} G_E^s + \tau G_M^{\gamma p} G_M^s}{\epsilon (G_E^{\gamma p})^2 + \tau (G_M^{\gamma p})^2} - \frac{(1 - 4\sin^2\theta_W)\,\epsilon'\, G_M^{\gamma p}}{\epsilon (G_E^{\gamma p})^2 + \tau (G_M^{\gamma p})^2} \times \left[-2(1+R_A^{T=1})G_A^{T=1} + (\sqrt{3}R_A^{T=0})G_A^{T=0}\right] \Bigg\}, \quad (1)$$

where $G_{E(M)}^{\gamma p(n)}$ are the proton (neutron) electric (magnetic)

[a] e-mail: armd@jlab.org

form factors, $G_A^{T=1(0)}$ is the isovector (isoscalar) proton axial form factor, G_F is the Fermi constant, α is the fine-structure constant, and θ_W is the electroweak mixing angle. All form factors are functions of Q^2, the square of the four-momentum transfer; for the hydrogen experiment the kinematical quantities are $\epsilon = 0.994$, $\tau = 0.028$, $\epsilon' = 0.018$. The vector electroweak radiative corrections are well known, and are not shown, for clarity; the R_A factors parametrize the axial electroweak radiative corrections [3]. In deriving eq. (1), the usual assumption of charge symmetry has been adopted [4]; recent work has reexamined this assumption, and has shown that the present experimental precision is approaching the level at which charge symmetry violating effects become important (see ref. [5] and R. Lewis, this conference).

Forward-angle scattering is used in the present experiments, which suppresses the final term involving the axial form factors. Thus, given adequate knowledge of the ordinary electromagnetic form factors, a measurement of A_{PV} at a given Q^2 will yield a linear combination of the strange form factors G_E^s and G_M^s. In order to disentangle G_E^s from G_M^s at a given Q^2, we have chosen to make a separate measurement on a ^{4}He target, which is solely sensitive to G_E^s.

Elastic electron scattering from ^{4}He is an isoscalar $0^+ \to 0^+$ transition and therefore has no contributions from magnetic or axial-vector currents. The same one-body transition densities appear in the matrix elements of the weak and electromagnetic operators. When the ratio comprising an asymmetry is formed, these transition densities cancel out, as long as two-body (meson-exchange) currents are negligible. Thus the nuclear many-body physics divides out and only the single nucleon form factors G_E^s and $G_E^{\gamma T=0}$ remain; the parity-violating asymmetry at tree level is therefore given by [3]

$$A_{\mathrm{PV}}^{\mathrm{He}} = \frac{G_F Q^2}{4\pi\alpha\sqrt{2}} \left(4\sin^2\theta_W + \frac{G_E^s}{G_E^{\gamma T=0}} \right), \qquad (2)$$

where $G_E^{\gamma T=0} = (G_E^{\gamma p} + G_E^{\gamma n})/2$ is the isospin-zero electric form factor (adequately known from other experiments). Nuclear model dependence in $A_{\mathrm{PV}}^{\mathrm{He}}$ due to isospin-mixing [6] and D-state admixtures [7] in the ^{4}He ground state is negligible, as are meson exchange current contributions at the low Q^2 chosen here [8].

2 Experiment overview

The present measurements represent another step in the HAPPEX program in Hall A at the Thomas Jefferson National Accelerator Facility. Results from the first round of measurements at the present kinematics, taken in 2004, have been published [9,10]. Here we report results from a significantly higher statistics data set ($\times 10$ for ^{4}He and $\times 3$ for hydrogen) obtained in 2005. The experimental approach has been documented in detail elsewhere [11]. In brief, the twin high-resolution spectrometers (HRS) in Hall A are used for independent measurements of the asymmetry for forward-angle elastic scattering of longitudinally polarized electrons. The focal-plane detectors, situated in heavily shielded detector huts in each HRS, are total absorption calorimeters consisting of alternating layers of brass and synthetic quartz, viewed by photomultiplier tubes (PMTs). For the hydrogen experiment a two-segment detector was used to cover the full flux of elastically scattered electrons (see fig. 1); for helium, the elastic locus is smaller, and a single detector segment was used in each arm. The spectrometers are arranged to create an approximately left-right symmetric acceptance. New septum magnets were installed to allow detection of electrons with small ($\langle\theta_{lab}\rangle \sim 6°$) scattering angles, in order to reach down to $Q^2 \sim 0.1\,\mathrm{GeV}/c^2$.

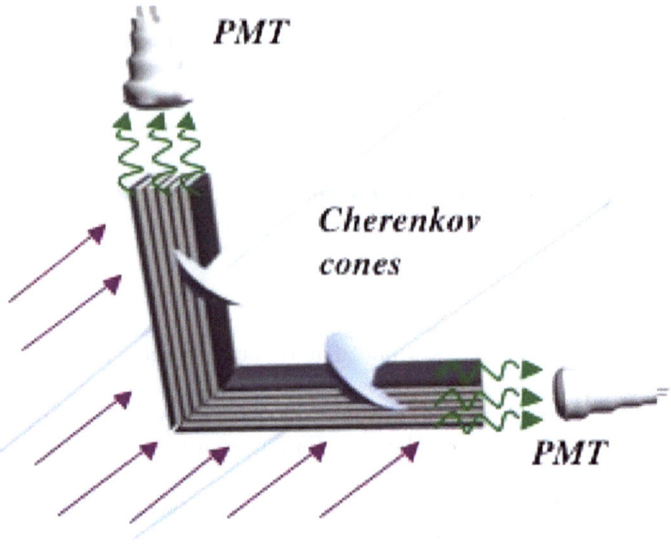

Fig. 1. Sketch of a HAPPEX focal plane detector, consisting of a layered brass-quartz calorimeter viewed by photomultiplier tubes. The "L"-shaped configuration shown was used for the hydrogen experiment; for helium only a single segment and PMT was used.

The beam helicity was flipped in a pseudorandom pattern at 30 Hz. The PMT response was integrated separately for each arm, along with the response of beam current and beam position monitors, during the 33.3 ms periods of constant beam helicity. To help to control systematic effects, the standard technique was adopted of periodically inserting a half-wave ($\lambda/2$) plate in the laser optical path at the electron source, which thereby passively reverses the sign of the beam polarization. Each such subset of data is referred to as a "slug"; there were 41 slugs of data for hydrogen and 121 for ^{4}He.

The cryogenic targets used 20 cm long cells with transverse flow of the cryogenic fluid (liquid hydrogen, and cold ^{4}He gas at 6.7 K and 200 psi, respectively) and was capable of operating with 400 W of deposited beam power. Average beam currents of 55 μA (hydrogen) and 35 μA (^{4}He) 3 GeV electrons were used; the beam current was limited by beam-induced heating of the septum magnets. Target density fluctuations did not contribute signficantly to the statistical width of the measured asymmetry in either case. The electron beam polarization was continu-

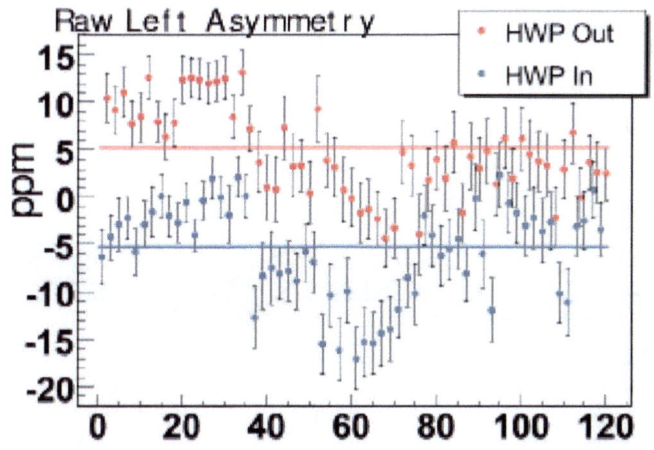

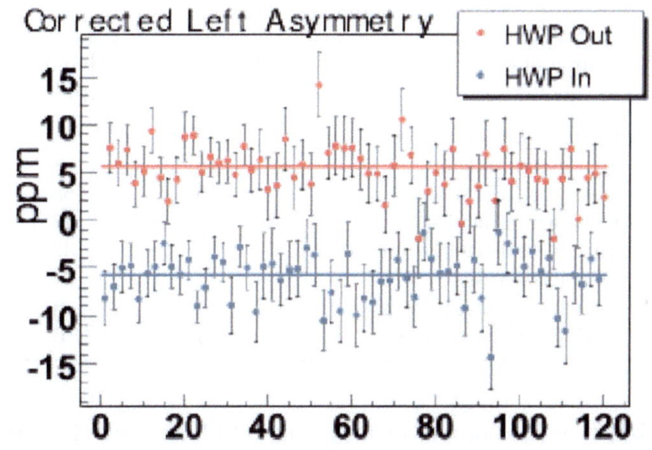

Fig. 2. Left: measured raw asymmetry in the left HRS for ^{4}He, for both settings of the insertable half-wave plate (HWP), as a function of "slug" of data. The helicity driver signal was reversed at slug 37, and the electronics cross-talk fixed at slug 77, whereupon the data stabilize; Right: same data after correction for helicity-correlated beam parameters. Results for the right HRS are similar.

Table 1. Averaged helicity-correlated beam parameters.

Parameter of error	Hydrogen	Helium
Intensity asymmetry	0.406 ppm	−0.377 ppm
Energy asymmetry	0.2 ppb	3 ppb
x-position difference	0.5 nm	−0.2 nm
x-angle difference	−0.2 nrad	4.4 nrad
y-position difference	1.7 nm	−26 nm
y-angle difference	0.2 nrad	−4.4 nrad

ously monitored using the Hall-A Compton polarimeter; the run-averaged polarization was $(86.7 \pm 2)\%$ for the hydrogen data and $(84.0 \pm 2.5)\%$ for ^{4}He. Periodic measurements using the Hall-A Møller polarimeter and the 5 MeV Mott polarimeter were consistent with these results.

Loose cuts were placed on the data to remove periods of poor beam quality, or when a particular spectrometer was not functioning; no cuts were placed based on any helicity-dependent quantities. In the setup of the electron source, careful attention was paid to the laser optics that brings the light on the "superlattice" GaAs photocathode, so as to minimize helicity-correlated changes in beam parameters such as intensity, energy, and horizontal (x) and vertical (y) positions and angles at the target. These procedures are extensively discussed in K. Paschke's contribution to this conference.

For the hydrogen data, the run-averaged values for the helicity-correlated beam parameters sets a new standard in the field (see table 1). For the helium data the results were not as impressive, while still quite acceptable (see table 1). A problem was uncovered mid-way through the helium data taking, in which an in-time helicity signal was found to be "picked up" by devices at an early stage in the accelerator. This electronics cross-talk caused significant helicity-correlated steering of the beam, and thus large (of order 400 nm) position differences at the target. While the effect largely cancelled out with half-wave plate insertion, and also largely cancelled between the right and left spectrometers, it was still large enough to be of concern. The problem was diagnosed about 1/3 of the way through the data taking by reversing the sign of the helicity driver signal (whereupon it also reversed, adding another layer of cancellation), and the offending signal cable was removed about 2/3 of the way through the data taking and the problem disappeared. The effect is visible in the "raw" asymmetry data for one arm as shown in fig. 2. However, the effect on the asymmetry is essentially invisible if one averages data from the right and left spectrometers. Note that dedicated tests showed that any helicity-correlated electronics noise in the DAQ was below the 1 part per billion (ppb) level and thus negligible.

Correction of the raw asymmetry for the effects of the five correlated beam parameter differences listed in table 1 requires determination of the sensitivity of the detected signal rates to these beam parameters. Two independent methods were used to calculate these sensitivities. The first relied on linear regression of the observed response of the detector PMTs to natural beam fluctuations. The other used calibration data in which the beam was deliberately modulated (by large amounts compared to the intrinsic beam fluctuations) using steering magnets and an accelerating cavity. The differences between the sensitivities calculated using these two methods were much smaller than corresponding statistical errors on the corrections. The net helicity-correlated correction to the helium data set was 120 ppb (compared to a statistical error of 191 ppb) and for hydrogen was 11 ppb (compared to a statistical error of 105 ppb).

The data were found to be perfectly behaved statistically under a number of tests. For example, the results agreed within errors for both settings of the half-wave plate, the results were consistent for right- and left-arm spectrometers, the asymmetry distributions of individual helicity pairs, after the correction for helicity-correlated beam parameters discussed above, were perfectly Gaus-

Table 2. Systematic uncertainties.

Source of error	Hydrogen (ppb)	Helium (ppb)
Beam polarization	37	192
Q^2 determination	16	58
Detector linearity	15	58
False asymmetries	17	48
Target wall background (Al)	15	32
Quasielastic scattering		24
Transverse asymmetry	4	8
Radiative corrections	3	6

sian over four orders of magnitude. Subdividing the data into slugs or into one-hour long runs gave consistent distributions with a reduced χ^2 close to one in each case.

An important feature of the HAPPEX apparatus is the fact that the spectrometers focus the elastically scattered electrons onto a compact region in the focal plane, where the detectors are located. Backgrounds from inelastic scattering are essentially not intercepted by the detectors and thus contributed less than 1% to the detected signal rate. The thin aluminum walls of the target cells lead to only a modest background from alumimun (0.76% for hydrogen and 1.8% for helium). A detailed discussion of the determination of the small background contributions and of the central kinematics (Q^2) is available in B. Moffit's contribution to this conference; their effects on the systematic errors are given in table 2.

Dedicated measurements of the transverse beam spin asymmetry were made for both targets (see L.J. Kaufman, this conference). Combined with limits placed on the degree of transverse polarization, obtained using the Møller polarimeter, the results indicate only a small background to A_{PV} in each case (see table 2). The smallness of the effect is due both to the fact that the production beam was almost perfectly longitudinal, and to the large cancellation in the effect of the transverse asymmetry between the left- and right-arm spectrometers.

3 Results

For ^{4}He the preliminary result is

$$A_{\mathrm{PV}} = +6.43 \pm 0.23\ (\mathrm{stat}) \pm 0.22\ (\mathrm{syst})\ \ \mathrm{ppm}$$

at $\langle Q^2\rangle = 0.077\,\mathrm{GeV}/c^2$. The expected asymmetry in the absence of strange form factors is 6.37 ppm. For hydrogen the preliminary result is

$$A_{\mathrm{PV}} = -1.60 \pm 0.12\ (\mathrm{stat}) \pm 0.05\ (\mathrm{syst})\ \ \mathrm{ppm}$$

at $\langle Q^2\rangle = 0.109\,\mathrm{GeV}/c^2$. The expected asymmetry in the absence of strange form factors is -1.64 ppm. This latter includes the small contribution from the axial form factor (see eq. (1)) of -0.037 ± 0.018 ppm, determined using the the axial radiative corrections calculated by Zhu *et al.* [12]. In interpreting our asymmetries, we adopt the Friedrich

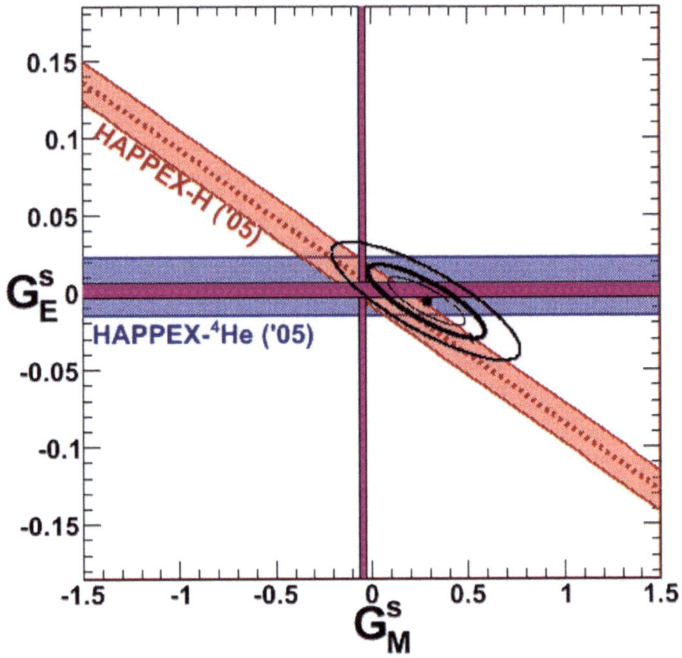

Fig. 3. The results, with 1 σ error bands, of the 2005 HAPPEX data taking on hydrogen and helium targets, expressed in the G_M^s *vs.* G_E^s plane. Also shown are the 68% and 95% confidence level allowed regions from a combined fit to the world data at this Q^2 (see text). Shown as vertical and horizontal bands are the theoretical predictions of Leinweber *et al.* [19] using charge symmetry, hyperon data, and lattice QCD input.

and Walcher parameterization of the electromagnetic form factors [13], supplemented by the recent BLAST data for G_E^n [14].

The contributions to the systematic error in each case are listed in table 2. The dominant systematic for both targets is the knowledge of the beam polarization, which at present has an error of 2% for hydrogen and 2.5% for helium; continued refinement of the data analysis techniques for the Compton polarimeter promises to reduce these errors significantly for the final results.

From our 2005 data alone, we extract $G_E^s = 0.004 \pm 0.014 \pm 0.013$ and $G_E^s + 0.09\,G_M^s = 0.004 \pm 0.011 \pm 0.005$. These results are shown in fig. 3. The results are in excellent agreement with (but much more precise than) our previously published data from the runs in 2004 [9,10]. To extrapolate to a common $Q^2 = 0.1\,\mathrm{GeV}/c^2$, we assumed that $G_E^s \propto Q^2$ and that G_M^s was constant over this small range of Q^2. A fit to the present data combined with our 2004 results yields

$$G_E^s = -0.002 \pm 0.017$$

and

$$G_M^s = 0.12 \pm 0.24.$$

Also shown in fig. 3 is a simple-minded fit to the world data at this Q^2, including the SAMPLE [15], PV-A4 [16], and G0 [17] results, as well as our 2004 results, which gives $G_E^s = -0.006 \pm 0.016$ and $G_M^s = 0.28 \pm 0.20$. Notice that the HAPPEX results dominate the precision of the world fit. Correlated errors and correlated assumptions have not been taken into account in this global fit. A more sophisticated combined fit, which makes different assumptions about the axial form factors and incorporates data up to $Q^2 = 0.3\,\mathrm{GeV}/c^2$ was presented by R. Young at this conference (see also [18]); the results of that fit agree within errors with ours, however the central value for G_M^s is much closer to zero in that fit.

4 Interpretation and conclusion

The present high-precision results place significant constraints on the strange vector form factors of the nucleon. In particular, the HAPPEX-^{4}He experiment is the *first* measurement made to date that is sensitive to only a single, isolated strange form factor (G_E^s). The results obtained at low Q^2 are gratifyingly consistent with the data of the other Collaborations (SAMPLE, PV-A4, G0) that have contributed measurements at or near this kinematic point, using a variety of different experimental approaches. While a number of model calculations are ruled out by the present data, a number are still consistent with the result. Going beyond models, the HAPPEX data are in spectacular agreement with the predictions of Leinweber *et al.* (ref. [19], and R. Young, this conference), which adopt charge symmetry, and combine input from lattice gauge calculations with hyperon data to yield the results that are also shown in fig. 3. The excellent mutual consistency of all the data, and the high precision, are good news for the Q_{weak} experiment (S. Page, this conference), as they indicate that the form factor uncertainty in interpreting that measurement is under adequate control.

The present results for G_E^s indicate that, to a remarkable degree, the spatial distribution of strange quarks in the nucleon mirrors that of the anti-strange quarks; there appears to be little tendency (if any) for one or the other to preferentially appear at larger radii, unlike what one expect in a naive model in which the nucleon has a significant Fock space component as a hyperon-kaon state. The result for G_M^s is consistent with zero, although the central value is positive. This slight favoring of a positive value is in agreement with the (less precise) indications from earlier experiments (*e.g.*, ref. [15]). While the new data significantly improve constraints on G_M^s, there is still room for non-trivial values; at the 1σ level the HAPPEX results allow as much as 4.3% of the nucleon's magnetic moment to arise from strangeness, or a decidely significant 27% of the isoscalar part of magnetic moment.

The HAPPEX Collaboration has an approved experiment planned to measure forward-angle parity-violating electron scattering at $Q^2 = 0.63\,\mathrm{GeV}^2$, which, when combined with the G0 backward-angle measurement at the same Q^2, will allow a similarly precise separation of the strange form factors at this larger Q^2. The measurements underway now (PV-A4, G0) at backward angles and intermediate Q^2 should provide a third kinematic point with separated form factors, as well as valuable information on the axial form factor; together these results should provide a comprehensive picture of the strange quark contributions to the vector structure of the nucleon. The present results show that these contributions remain stubbornly elusive.

The author thanks the organizers for the invitation to contribute to this stimulating and superbly organized workshop in a very pleasant setting. This work was supported by DOE contract DE-AC05-84ER40150, Modification No. M175 under which the Southeastern Universities Research Association (SURA) operates the Thomas Jefferson National Accelerator Facility. The author gratefully acknowledges financial support from the U.S. National Science Foundation under grant PHY-0400583, and the talents and dedication of the members of the HAPPEX Collaboration.

References

1. E. Leader, this conference.
2. K.S. Kumar, P.A. Souder, Prog. Part. Nucl. Phys. **45**, S333 (2000); D.H. Beck, B.R. Holstein, Int. J. Mod. Phys. E **10**, 1 (2000); D.H. Beck, R.D. McKeown, Annu. Rev. Nucl. Part. Sci. **51**, 189 (2001).
3. M.J. Musolf *et al.*, Phys. Rep. **239**, 1 (1994).
4. Gerald A. Miller, Phys. Rev. C **57**, 1492 (1998).
5. B. Kubis, R. Lewis, Phys. Rev. C **74**, 015204 (2006).
6. S. Ramavataram, E. Hadjimichael, T.W. Donnelly, Phys. Rev. C **50**, 1175 (1994).
7. M.J. Musolf, T.W. Donnelly, Phys. Lett. B **318**, 263 (1993).
8. M.J. Musolf, R. Schiavilla, T.W. Donnelly, Phys. Rev. C **50**, 2173 (1994).
9. HAPPEX Collaboration (K.A. Aniol *et al.*), Phys. Rev. Lett. **96**, 022003 (2006).
10. HAPPEX Collaboration (K.A. Aniol *et al.*), Phys. Lett. B **635**, 275 (2006).
11. HAPPEX Collaboration (K.A. Aniol *et al.*), Phys. Rev. C. **69**, 065501 (2004).
12. S.-L. Zhu, S.J. Puglia, B.R. Holstein, M.J. Ramsey-Musolf, Phys. Rev. D **62**, 033008 (2000).
13. J. Friedrich, Th. Walcher, Eur. Phys. J. A **17**, 607 (2003).
14. V. Ziskin, PhD Thesis, MIT, 2005 (unpublished); R. Milner, private communication.
15. SAMPLE Collaboration (D.T. Spayde *et al.*), Phys. Lett. B **583**, 79 (2004).
16. PV-A4 Collaboration (F.E. Maas *et al.*), Phys. Rev. Lett. **94**, 152001 (2005).
17. G0 Collaboration (D.S. Armstrong *et al.*), Phys. Rev. Lett. **95**, 092001 (2005).
18. R.D. Young, J. Roche, R.D. Carlini, A.W. Thomas, Phys. Rev. Lett. **97**, 102002 (2006).
19. D.B. Leinweber, S. Boinepalli, I.C. Cloet, A.W. Thomas, A.G. Williams, R.D. Young, J.M. Zanotti, J.B. Zhang, Phys. Rev. Lett. **94**, 212001 (2005); D.B. Leinweber, S. Boinepalli, A.W. Thomas, P. Wang, A.G. Williams, R.D. Young, J.M. Zanotti, J.B. Zhang, Phys. Rev. Lett. **97**, 022001 (2006).

The parity violation A4 experiment at forward and backward angles

S. Baunack[a]

For the A4 Collaboration
Institut für Kernphysik, Universität Mainz, Germany

Abstract. The A4 Collaboration at the MAMI accelerator in Mainz measures the parity-violating asymmetry in the cross-section of elastic scattering of longitudinally polarized electrons off unpolarized protons from which the contribution of strange sea quarks to the vector form factors of the proton can be derived. Measurements at forward angles $30° \leq \theta \leq 40°$ and two different momentum transfers Q^2 of 0.23 $(\mathrm{GeV}/c)^2$ and 0.11 $(\mathrm{GeV}/c)^2$ have been performed in the past. Measurements at backward angles $140° \leq \theta \leq 150°$ are underway for a separate determination of the strange electric and the strange magnetic form factor at $Q^2 = 0.23$ $(\mathrm{GeV}/c)^2$.

PACS. 13.40.Gp Electromagnetic form factors – 11.30.Er Charge conjugation, parity, time reversal, and other discrete symmetries – 13.40.-f Electromagnetic processes and properties – 14.20.Dh Protons and neutrons

1 Introduction

In constituent quark models the nucleon is made up of three quarks with the flavours up, u, and down, d. Within these models the baryon spectrum and properties of the nucleon like spin and magnetic moment can be explained. In 1988 Kaplan and Manohar [1] showed that strange sea quarks could also contribute to the vector matrix elements of the nucleon. Parity-violating electron scattering can be used as a tool to determine the contribution of strange quarks to the vector form factors of the nucleon. Such contributions can be expressed in terms of the strange electric and magnetic form factors G_E^s and G_M^s [2]. The asymmetry A_{PV} in the cross-section $\sigma_{+,-}$ of elastic scattering of, respectively, right- and left-handed polarized electrons off unpolarized protons can be written, omitting radiative corrections, as a sum of three terms:

$$A_{PV} = \frac{\sigma^+ - \sigma^-}{\sigma^+ + \sigma^-} \tag{1}$$

$$= \frac{G_\mu Q^2}{4\pi\alpha\sqrt{2}} \frac{A_V + A_S + A_A}{\epsilon(G_E^p)^2 + \tau(G_M^p)^2} \tag{2}$$

with G_μ the Fermi coupling constant from muon decay, α the fine-structure constant, $\tau = \frac{Q^2}{4M_p^2}$, Q^2 the negative squared momentum transfer, $\epsilon = [1 + 2(1+\tau)\tan^2\frac{\theta_e}{2}]^{-1}$ and θ_e the electron scattering angle in the laboratory frame. A_V represents the asymmetry due to the vector coupling at the proton vertex without strangeness contribution, A_S is the asymmetry arising only from the strangeness contribution and A_A represents the asymmetry due to the axial coupling at the proton vertex. The three terms are

$$A_V = (1 - 4\hat{s}_Z^2) - \epsilon G_E^p G_E^n + \tau G_M^p G_M^n, \tag{3}$$

$$A_A = (1 - 4\hat{s}_Z^2)\sqrt{1-\epsilon^2}\sqrt{\tau(1+\tau)} G_M^p \tilde{G}_A^p, \tag{4}$$

$$A_S = \epsilon G_E^p G_E^s + \tau G_M^p G_M^s, \tag{5}$$

with $\hat{s}_Z^2 = 0.23113(15)$ the weak mixing angle and $\tilde{G}_A^p$ the weak axial form factor. One can see from eqs. (1)–(5) that by measuring the asymmetry A_{PV} and by using the known electromagnetic form factors of the nucleon it is possible to determine a linear combination of the strange electric and strange magnetic form factors. For a separate determination of G_E^s and G_M^s it is necessary to perform at least two measurements at the same momentum transfer Q^2, but with different scattering angles. In the A4 kinematics in forward-scattering configuration, the experiment is sensitive to $G_E^s + \eta \cdot G_M^s$, where η is given by the momentum transfer Q^2 measured in $(\mathrm{GeV}/c)^2$. In the backward-scattering configuration, the experiment is mainly sensitive to G_M^s. The A4 Collaboration performed two forward-angle measurements at momentum transfer values of 0.23 $(\mathrm{GeV}/c)^2$ and 0.11 $(\mathrm{GeV}/c)^2$. After the rearrangement of the setup, a backward-angle measurement at a momentum transfer of 0.23 $(\mathrm{GeV}/c)^2$ is currently under way.

[a] e-mail: baunack@kph.uni-mainz.de

2 Experimental setup

The measurements take place at the MAMI accelerator facility. A strained layered GaAs electron source produces a polarized electron beam with an average polarization P_e of 80%. The electron beam intensity is 20 μA. The degree of polarization is measured once a week with a Møller polarimeter situated in another experimental hall. In the mean time the polarization is monitored with a transmission Compton polarimeter situated in the experimental hall A4 and by frequent Mott measurements directly at the source. The spin of the electrons is reversed every 20 ms following a randomly selected pattern, $(+--+)$ or $(-++-)$. The electrons are scattered off a liquid-hydrogen target with a 10 cm length in the forward-angle and 23 cm length in the backward-angle configuration. The luminosity is monitored by eight water-Cherenkov monitors at small scattering angles $4.4° \le \theta_e \le 10°$, covering the full azimuthal range. Due to the smallness of the physical asymmetries ($\approx 10^{-6}$) special care has to be taken to minimize false asymmetries coming from helicity-correlated beam differences in position, angle, energy or intensity. Several monitoring and stabilization systems have been installed along the accelerator to control and minimize such beam fluctuations.

The scattered electrons are detected in a total absorbing calorimeter that consists of 1022 individual lead fluoride (PbF_2) crystals placed in 7 rings and 146 rows. It covers a solid angle of $\Delta\Omega = 0.6$ sr. A CAD drawing of the calorimeter is shown in fig. 1. PbF_2 is a pure Cherenkov radiator and hence intrisically fast. The photomultiplier pulse of an event is only 20 ns long. For each accepted event the signals of 3×3 neighbouring crystals are summed by an analogue circuit, then digitized and stored in histograms.

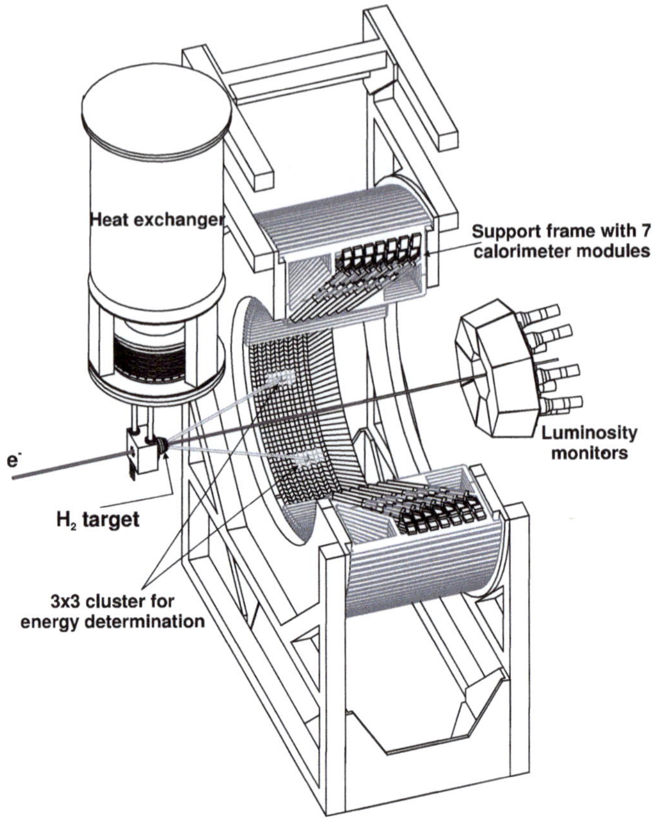

Fig. 1. Drawing of the total absorbing A4 calorimeter. The 1022 PbF_2 crystals are arranged in 146 frames and 7 rings covering the full azimuthal angle and a polar scattering angle from 30° to 40° in the forward or 140° to 150° in the backward configuration. Also shown is the target construction and the luminosity monitor system.

3 Forward-angle measurements

In the forward-angle configuration, the PbF_2 calorimeter detects the electrons at scattering angles θ_e between 30° and 40°. Two measurements have been taken at beam energies of 854.3 MeV and 570.4 MeV. This corresponds to two mean momentum transfer values of 0.23 $(\mathrm{GeV}/c)^2$ and 0.11 $(\mathrm{GeV}/c)^2$, respectively. Figure 2 shows an energy spectrum of a single module taken within five minutes at the beam energy of 854.3 MeV. One can identify the peak of the elastic-scattering events at the right of the spectrum. The positions of the π^0-threshold and of the Δ-resonance are also indicated. One can see that a clean separation of the elastic events from the inelastic events is possible by applying cuts in the energy spectrum. The elastic events that we extracted for the analysis are marked by the grey area under the peak. The measured linear combinations of the strange electric and the strange magnetic form factors for the forward-angle measurements are [3,4]:

$$G_E^s + 0.225 G_M^s|_{Q^2=0.23(\mathrm{GeV}/c)^2} = 0.039 \pm 0.034, \quad (6)$$

$$G_E^s + 0.107 G_M^s|_{Q^2=0.11(\mathrm{GeV}/c)^2} = 0.069 \pm 0.036. \quad (7)$$

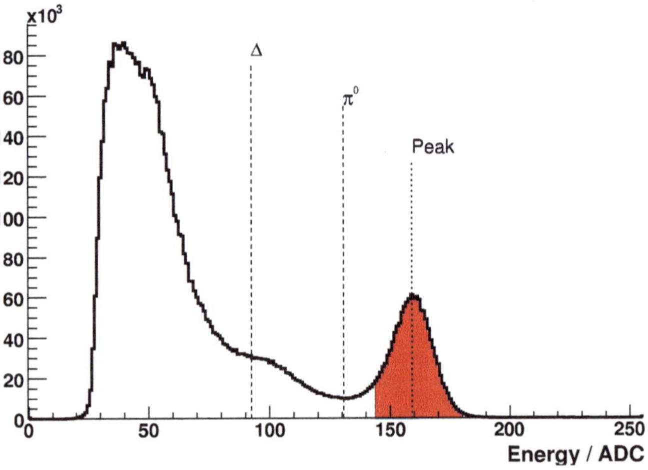

Fig. 2. Energy spectrum of the PbF_2 calorimeter at a forward scattering angle $\theta_e = 34°$ and a beam energy of $E = 854.3$ MeV. One can clearly see the peak of the elastic-scattering events. The π^0-production threshold and the position of the Δ-resonance are also indicated. The elastic events can be extracted by applying a cut to the energy spectrum (light-grey area under the peak).

Both linear combinations are positive and differ from zero less than two standard deviations. To extract the physical asymmetries from the data, a multi-dimensional regression analysis has been performed and the correlation matrix has been used to determine the uncertainties due to the corrections of the helicity-correlated beam parameters. All other systematic effects have been summed up quadratically. The theoretical uncertainties arise mainly from the uncertainties in the electromagnetic and axial form factors and have also been summed up quadratically.

4 Backward-angle measurements

Between August 2004 and May 2005 we mounted the lead fluoride calorimeter on a rotatable platform to allow an easy change from forward- to backward-angle measurements. This rearrangement is described in detail in the contribution of B. Gläser at this conference. In this article, the modifications are described briefly and then the first backward-angle measurement results are presented.

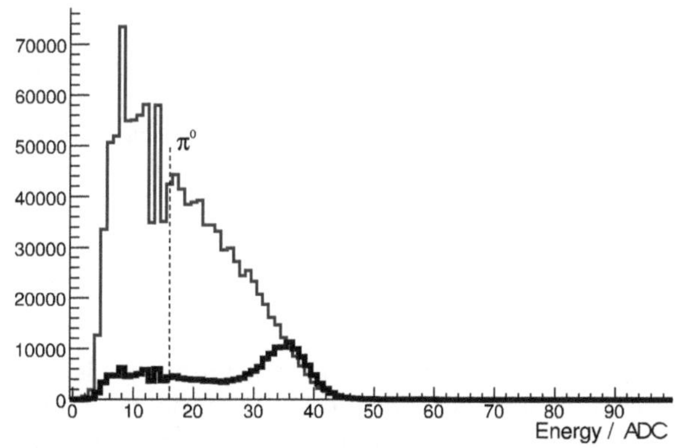

Fig. 3. Energy spectrum of the PbF_2 calorimeter at a backward scattering angle $\theta_e = 146°$ and a beam energy of $E = 315.1\,\mathrm{MeV}$. There are two histograms shown: the thick-black line shows the spectrum of events in coincidence with the electron tagging detector, the thin-grey line indicates the spectrum of the events without coincidence that is generated by neutral particles. The position of the threshold for the π^0-production is also indicated.

4.1 Modification of the setup

In the backward-scattering configuration, the calorimeter covers scattering angles θ_e from 140° to 150°. To obtain a momentum transfer of $0.23\ (\mathrm{GeV}/c)^2$ a beam energy of 315.1 MeV was chosen. There were two major changes in the experimental setup for these measurements. First, the target length was more than doubled from 10 cm to 23 cm to partially compensate for the smaller cross-section for elastic scattering at backward angles. Second, the installation of an additional trigger detector system was necessary. First measurements as well as Monte Carlo simulations showed that in the region of the elastic peak there are non-negligible contributions from inelastic events. These events are mainly due to photons from π^0-decay which are not distinguishable from electrons in our calorimeter because they produce a very similar electromagnetic shower in the PbF_2 crystals.

To discriminate the electrons (charged particles) from the photons (neutral particles), a plastic scintillator system has been installed between the scattering chamber and the calorimeter. Events are stored in different histograms depending on whether a signal from the scintillator is detected in coincidence with a signal in the PbF_2 calorimeter or not. Figure 3 shows an example of the resulting histograms. One can see, on the one hand, the energy spectrum of charged particles, the coincidence spectrum, and on the other hand, the energy spectrum of neutral particles, the non-coincidence spectrum. In the coincidence spectrum the elastic peak is clearly visible around ADC channel #36. The contribution of inelastic electrons to this spectrum is small; for example the π^0-production threshold is located at around the ADC channel #16.

4.2 Background subtraction

Currently there are Monte Carlo simulations under way in order to understand all contributions to the coincidence as well as to the non-coincidence spectra. The main source of background contribution to the coincidence spectrum is identified as photons from π^0-decay that underlie e^+e^- pair production in the material between the hydrogen target and the scintillator, mainly in the wall of the scattering chamber which consists of aluminium with an effective thickness of 1 cm in the path of the photons. In the coincidence histogram in fig. 3 one can observe these pair conversion contributions explicitly in the range between ADC channels #16 and #25, where only small contributions from electrons are expected. From the data we can estimate that a fraction of about 10% of the photons convert and are counted in the coincidence spectra. Another background contribution to the coincidence spectra is from accidental coincidences. With an out-of-gate measurement we found that the fraction of accidental coincidences is less than 1%.

For the determination of the asymmetry in the elastic scattering, as in the forward-angle measurements, cut limits in the region of the elastic peak in the coincidence spectra are defined and the number of events N^+_{coinc} and N^-_{coinc} for the two helicities are extracted. The asymmetry in these coincidence events can be written as

$$A_{\mathrm{coinc}} = \frac{N^+_{\mathrm{coinc}} - N^-_{\mathrm{coinc}}}{N^+_{\mathrm{coinc}} + N^-_{\mathrm{coinc}}}. \tag{8}$$

This asymmetry has two contributions: one from the asymmetry in the elastic *ep*-scattering A_{PV} and one from the asymmetry in the background events A_{back}. If one denotes the ratio of background events to all coincidence events with $\epsilon = N_{\mathrm{back}}/N_{\mathrm{coinc}}$, one can write for the asymmetry A_{coinc}:

$$A_{\mathrm{coinc}} = (1-\epsilon)\cdot A_{PV} + \epsilon \cdot A_{\mathrm{back}}. \tag{9}$$

Hence, for a determination of A_{PV} one needs to know the fraction ϵ of background events in the elastic cut region and the asymmetry in the background events. The latter can be easily found under the assumption of a 100% efficiency of the scintillator by applying the same cuts in the non-coincidence spectrum as in the coincidence spectrum and then extracting the number of non-coincident events N^+_{non} and N^-_{non}:

$$A_{\text{back}} = \frac{N^+_{\text{non}} - N^-_{\text{non}}}{N^+_{\text{non}} + N^-_{\text{non}}}. \qquad (10)$$

The statistical accuracy for this asymmetry is of the same order as for the coincidence asymmetry. The ratio ϵ can be determined by subtracting a fraction r of the non-coincidence spectrum from the coincidence spectrum. First, the ratio r of converted to non-converted photons is calculated from the two spectra in an energy region where only few electrons are expected, for example for the spectra of fig. 3 in the energy range from 100 MeV to 130 MeV, corresponding to an ADC channel range from #14 to #22:

$$r = \frac{\sum_{\text{ADC=low}}^{\text{high}} N_{\text{coinc}}(\text{ADC})}{\sum_{\text{ADC=low}}^{\text{high}} N_{\text{non}}(\text{ADC})}. \qquad (11)$$

Then, a subtraction spectrum of charged events only (N_{sub}) and a spectrum of converted photons only (N_{back}) can be calculated:

$$N_{\text{sub}}(\text{ADC}) = N_{\text{coinc}}(\text{ADC}) - r \cdot N_{\text{non}}(\text{ADC}), \qquad (12)$$

$$N_{\text{back}}(\text{ADC}) = r \cdot N_{\text{non}}(\text{ADC}). \qquad (13)$$

Figure 4 demonstrates this procedure. One can see the coincidence spectrum of a single module as a thin-grey line. The ratio r in this example has been determined to be $r = 10.2\%$. The resulting subtraction spectrum following eq. (12) is indicated by the thick-black line together with a Gaussian fit to the elastic peak. Also shown by a grey dashed line are the background events coming from the converted photons. Evidently, the fraction of background

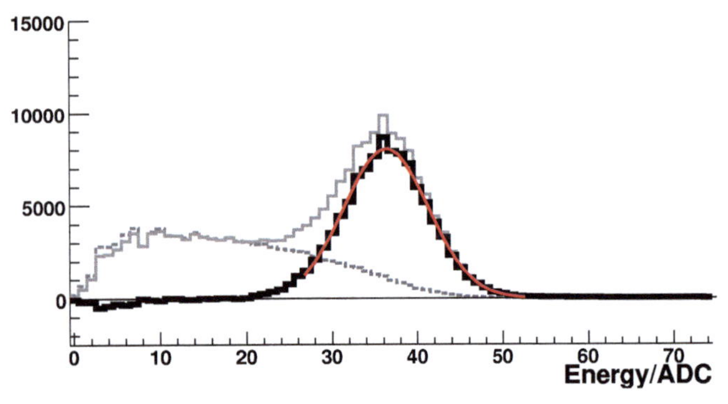

Fig. 4. Illustration of the background subtraction method. The coincidence spectrum is shown in light grey and the resulting spectrum after applying the substraction in black, together with a Gaussian fit to the elastic peak. The contribution from background events to the coincidence spectrum is also shown by the grey-dashed line.

Table 1. Achieved helicity-correlated asymmetries and differences in the beam parameters for the 315 MeV backward-angle measurements with about 550 h of data taking together with an estimate of the false asymmetries linked to them.

Beam parameter	Value	Est. false asymmetry
Current asymmetry	−0.31 ppm	0.31 ppm
Horiz. pos. difference	−0.10 μm	0.15 ppm
Vert. pos. difference	−0.04 μm	0.06 ppm
Horiz. angle difference	−7.41 nrad	0.03 ppm
Vert. angle difference	−3.00 nrad	0.01 ppm
Energy difference	−0.21 eV	0.00 ppm

events under the elastic peak depends on the choice of the lower-energy cut. For the forward-angle measurements, the energy resolution was determined by fitting the elastic peak and then the lower cut was set $k \cdot \sigma$ away from the π^0-production threshold with $k = 1.8$. When applying this method to the backward measurements, we get depending on k typically a fraction of $0.1 \leq \epsilon \leq 0.2$. Since in the backward measurements the contribution from inelastic electrons is smaller, other definitions of the cuts are being investigated.

4.3 First results

In the first half of 2006 we took about 550 h of asymmetry data for the $Q^2 = 0.23$ (GeV/c)^2 backward-angle measurement. The MAMI accelerator provided a polarized electron beam with an intensity of 20 μA at an energy of 315.1 MeV that was stabilized in intensity, position and energy. Table 1 lists the relevant helicity-correlated beam differences that we achieved together with an upper estimate of the false asymmetries linked to them. We extract the physical asymmetry out of the measured asymmetry through the linear ansatz

$$A_{\text{meas}} = P_e \cdot A_{\text{phys}} + \sum_{i=1}^{6} a_i \cdot X_i \qquad (14)$$

with P_e the electron beam polarization, X_i the helicity-correlated beam differences and a_i the correlation coefficients which are determined via a multiple linear regression analysis. For the data presented here this analysis is not yet complete. We divide our data in subsets of about 80 h of data taking, the so-called data samples. There is a half-wave plate located at the electron source which can be inserted in the laser beam. This inverts the meaning of "+" and "−" helicity. Inserting or taking out this half-wave plate is done between each data sample. The resulting asymmetries should change sign when the plate is inserted.

We did the analysis as well for the coincidence as for the non-coincidence data. After extracting the raw asymmetries, corrections for beam current asymmetry and beam polarization have been applied. Figure 5 shows the resulting asymmetries for the coincidence data. One can

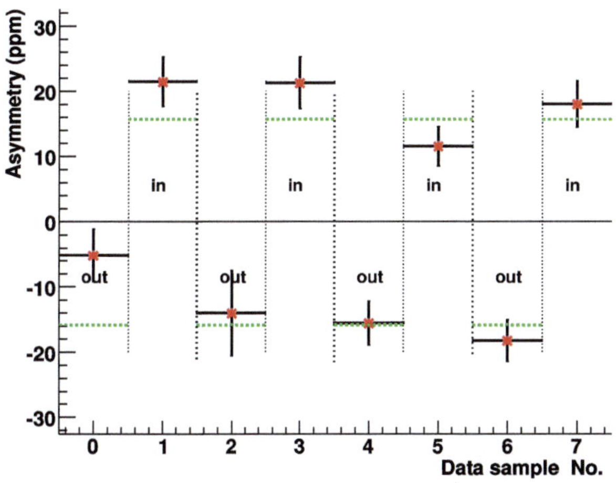

Fig. 5. Experimental asymmetries of the coincidence data arranged in so-called data samples, which represent about 80 hours of data taking each. The position of the insertable half-wave plate at the source which reverses the "+" and "−" helicity is indicated with "in" or "out". One can clearly observe the sign flip of the asymmetries when the half-wave plate is inserted. The dotted line represents the average asymmetry over all data samples where the half-wave plate sign flip has been taken into account.

observe clearly the sign flip when the half-wave plate is inserted. With the preliminary cut we applied we got $N_{\text{coinc}} = 1.6 \cdot 10^{12}$ events for the coincidence data and $N_{\text{non}} = 3.1 \cdot 10^{12}$ events for the non-coincidence data. The resulting asymmetries —where the systematic uncertainty has not yet been fully investigated— are

$$A_{\text{coinc}} = (-15.8 \pm 1.3) \cdot 10^{-6}, \quad (15)$$
$$A_{\text{non}} = (-3.9 \pm 0.9) \cdot 10^{-6}. \quad (16)$$

Clearly, the asymmetry in the non-coincidence data is about a factor of 4 smaller than the asymmetry in the coincidence data. To extract the asymmetry A_{PV} in the elastic *ep*-scattering according to eq. (9), the fraction ϵ of converted photons in the coincidence data within the cuts was estimated roughly at $\epsilon = 0.1$. Doing so, one gets the parity-violating asymmetry

$$A_{PV} = (-17.1 \pm 1.6) \cdot 10^{-6}. \quad (17)$$

This asymmetry has to be compared with the standard model prediction $A_0 = A_V + A_A$ following eqs. (1)–(5). To evaluate A_0 we used for the electromagnetic form factors the Friedrich/Walcher parametrization [5] and for the weak axial form factor the calculation of Zhu *et al.* [6]. For our kinematics, we get

$$A_0 = (-16.3 \pm 1.2) \cdot 10^{-6}. \quad (18)$$

Neglecting the strange electric form factor here, one gets for the strange magnetic form factor

$$G^s_M(Q^2 = 0.23) = -0.07 \pm 0.17. \quad (19)$$

It has to be stressed that this result is based on a preliminary analysis without full investigation of systematic effects and based on rough estimates. Nevertheless, the result suggests that with a full analysis it will be possible to put contraints on the strange contribution to the magnetic form factor.

5 Summary and outlook

The A4 Collaboration has measured the parity-violating asymmetry in elastic electron-proton scattering at two momentum transfer values at forward kinematics. In both cases the extracted linear combinations of the strange form factors turned out to be positive and compatible with zero within two standard deviations. For a separate determination of G^s_E and G^s_M the apparatus has been rearranged for measurements under backward angles and an additional electron trigger system has been installed.

For a backward-angle measurement at $Q^2 = 0.23\ (\text{GeV}/c)^2$ 1000 hours of data taking is planned. About half of the data have been taken for the results presented in this article. With a very preliminary analysis, the extracted strange magnetic moment seems to be small. For a detailed analysis Monte Carlo simulations are under way which should reproduce the energy spectra for the coincidence and for the non-coincidence case. When these simulations agree well with the experimentally observed spectra, this will allow us a precise determination of the background contribution to the events in the elastic peak. Additional studies on systematic effects are also being done now. In the months from September until November 2006 additional measurements are being performed which will give at least 1000 h of asymmetry data for the final analysis.

We also performed a first test with a liquid-deuterium target. The energy spectra in coincidence with the electron trigger showed a well-separated quasi-elastic peak. In 2007 we plan to take about 800 h of data with a beam energy of 315 MeV at backward angles for a separate determination of the strange magnetic form factor G^s_M and the weak axial form factor $\tilde{G}^p_A$.

References

1. D.B. Kaplan *et al.*, Nucl. Phys. B **310**, 527 (1988).
2. M.J. Musolf *et al.*, Phys. Rep. **239**, 1 (1994).
3. F. Maas *et al.*, Phys. Rev. Lett. **93**, 022002 (2004).
4. F. Maas *et al.*, Phys. Rev. Lett. **94**, 152001 (2005).
5. J. Friedrich *et al.*, Eur. Phys. J. A. **17**, 706 (2003).
6. S.-L. Zhu *et al.*, Phys. Rev. D **69**, 033008 (2000).

Results from the forward G0 experiment

Strange quark contribution to the nucleon structure

J.-S. Real[a]

For the G0 Collaboration[b]
LPSC, IN2P3-UJF, 53 Avenue des Martyrs, 38026 Grenoble Cedex, France

Paper reprinted from Eur. Phys. J. A with permission

Abstract. The G0 experiment is dedicated to the determination of the strange quark contribution to the electric and magnetic nucleon form factors for a large range of momentum transfers between 0.1 to $1\,(\mathrm{GeV}/c)^2$. This information is provided by the asymmetries of cross-sections measured with longitudinally polarized electrons in elastic electron-proton scattering and quasi-elastic electron-deuteron scattering. A set of measurements at two different Q^2 will allow the complete separation of the electric and magnetic weak, as well as axial nucleon form factors. This report summarizes the physics case, gives details about the dedicated set-up used, and shows the results of the combination of the strange quark contribution in the electric and magnetic form factors of the protons. The experiment was performed at the *Jefferson Laboratory*, during years 2003 and 2004, and will be completed after backward-angle measurements in 2006, 2007.

PACS. 13.60.Fz Elastic and Compton scattering – 13.40.Gp Electromagnetic Form Factors – 14.20.Dh Protons and neutrons

1 Introduction

In quantum chromodynamics, the nucleon is viewed as composed of three-valence u and d quarks and a sea of gluons and quark-antiquark pairs ($u\bar{u}, d\bar{d}, s\bar{s}, \ldots$). Because the strange quarks contribute only to the sea, their contribution to the nucleon properties is one of the open questions, which are investigated in a joint experimental and theoretical study.

Several indications already exist on the strange quark contribution to nucleon properties, as to the momentum with deep inelastic neutrino scattering, to the spin with deep inelastic electron scattering and to the mass from π-nucleon scattering. All these experiments provide contributions ranging from 0 to 30% with large theoretical and experimental uncertainties.

The parity-violating (PV) electro-weak interaction in electron-nucleon elastic scattering offer a new opportunity to study the contribution of strange quarks to the electroweak structure of the nucleon through the nucleon form factors, and in particular possible difference in the s and $\bar{s}$ spatial densities.

Several dedicated PV asymmetry experiments and apparatus have been developed during the last decade for these studies [1]. In this contribution we focus on the G0 experiment which provides a Q^2-dependence of the linear combination of the strange magnetic and electric form factors over a large range of momentum transfers (Q^2 between 0.1 to $1\,(\mathrm{GeV}/c)^2$), and will provide a complete determination of the contribution of the strange quarks to the electric, magnetic and axial form factors.

In the following, we will present the physics motivation before describing the G0 experimental set-up. Then, we will show the results of the G0 experiment in the forward-angle configuration which provide the first determination of a linear combination of the strange contribution to the electric and magnetic nucleon form factors over a large Q^2 range.

2 Elastic scattering and parity-violating experiment

In electron-nucleon scattering, the electroweak interaction takes place at first order through two diagrams, one corresponding to the exchange of a virtual photon ($\mathcal{M}_\gamma$) and one to the exchange of a Z^0 ($\mathcal{M}_Z$) with a 4-momentum transferred $-Q^2 = \nu^2 - \boldsymbol{q}^2$.

The scattering amplitudes are expressed in terms of the leptonic and hadronic currents. The hadronic currents

[a] e-mail: `real@in2p3.fr`

[b] Jefferson Lab Experiments E-05-108/06-108 and E-04-101. A list of collaborators and information about the experiment can be found at `http://www.npl.uiuc.edu/exp/G0/`.

associated to the virtual photon and the Z^0 can be decomposed through a multipole expansion in terms of the electromagnetic proton form factors $G_E^{(\gamma,p)}$ and $G_M^{(\gamma,p)}$ and the electric $G_E^{(Z,p)}$, magnetic $G_M^{(Z,p)}$ and axial G_A^{ep} weak form factors.

In order to provide information on the spatial distribution of the quark inside the nucleon, the form factors can be decomposed in terms of their quark contribution separately for the electromagnetic, vector and axial weak parts as

$$\begin{aligned} G_{E,M}^{(\gamma,N)} &= \sum_{q=u,d,s} Q_q G_{E,M}^{(q,N)}, \\ G_{E,M}^{(Z,N)} &= \sum_{q=u,d,s} C_V^q G_{E,M}^{(q,N)}, \\ G_A^{eN} &= \sum_{q=u,d,s} C_A^q G_A^{(q,N)}. \end{aligned} \tag{1}$$

The contribution of the three heavy flavors has been neglected due to their larger mass. Using the charge symmetry approximation under the exchange of u and d quarks ($G_{E,M}^{(u,p)} = G_{E,M}^{(d,n)}$) and s quark ($G_{E,M}^{(s,p)} = G_{E,M}^{(s,n)}$), it is possible to express the electric and magnetic quark form factors as a function of electromagnetic and weak form factors of the proton and neutron.

Because the weak interaction does not conserve parity, $\mathcal{M}_Z$ can be accessible through the asymmetry measurement of the cross-sections associated to elastic scattering of longitudinally polarized electrons with two opposite helicities:

$$\begin{aligned} A_{PV} &= \frac{|\mathcal{M}^+|^2 - |\mathcal{M}^-|^2}{|\mathcal{M}^+|^2 + |\mathcal{M}^-|^2} \\ &\simeq \frac{\mathcal{R}e(\mathcal{M}_\gamma \mathcal{M}_Z^*)^+ - \mathcal{R}e(\mathcal{M}_\gamma \mathcal{M}_Z^*)^-}{2|\mathcal{M}_\gamma|^2}. \end{aligned} \tag{2}$$

That can be rewritten

$$A_{PV} = a\frac{\epsilon G_E^{(\gamma,p)} G_E^{(Z,p)} + \tau\, G_M^{(\gamma,p)} G_M^{(Z,p)} - b\epsilon' G_M^{(\gamma,p)} G_A^{ep}}{\epsilon \left(G_E^{(\gamma,p)}\right)^2 + \tau \left(G_M^{(\gamma,p)}\right)^2} \tag{3}$$

with $\tau = \frac{Q^2}{4M^2}$ and

$$\epsilon = 1/(1 + 2(1+\tau)\tan^2\theta_e/2), \quad a = G_F Q^2/(4\sqrt{2}\pi\alpha),$$
$$\epsilon' = \sqrt{\tau(1+\tau)(1-\epsilon^2)}, \quad b = 1 - 4\sin^2\theta_W.$$

Using eq. (1) to express G_{EM}^Z in terms of G_{EM}^p and G_{EM}^s:

$$A_{PV} = A_0 + \alpha G_E^s + \beta G_M^s + \delta G_A^e(T=1), \tag{4}$$

where α, β and δ contain kinematics factor and known electromagnetic form factors.

Asymmetry as small as few 10^{-6} can be extracted with few % of precision thanks to the cancellation of normalization systematic errors in the ratio of the cross-section measurements. This expression shows that the asymmetry is already non-zero even if the strange quarks do not contribute to the nucleon properties.

A complete separation of the three form factors G_E^s, G_M^s and G_A^e require three independent measurements. This could be performed by varying the kinematical factors (ϵ and ϵ') at a fixed Q^2. Forward and backward (scattered electron) angles measurements correspond, respectively, to small ϵ (large ϵ') and large ϵ (small ϵ'). As a third measurement, it appears more effective [2] to measure the asymmetry in quasi-elastic scattering on deuterium, which involves the proton and the neutron.

The HAPPEX experiment [3], performed in Hall A of *Jefferson Laboratory*, has measured the PV asymmetry on the proton at forward angles for $Q^2 = 0.1$ and $0.48\,(\text{GeV}/c)^2$, which allowed the extraction of a linear combination $G_E^s + \eta G_M^s$, and on ^{4}He at $Q^2 = 0.1$ which allowed the extraction of G_E^s. The PVA4 experiment [4], performed at the MAMI facility, has measured asymmetry on the proton at forward angles for $Q^2 = 0.1$ and $0.23\,(\text{GeV}/c)^2$, and is sensitive to the combination $G_E^s + \eta G_M^s$. The SAMPLE experiment at *Bates*, has measured the asymmetries on proton and deuteron for backward angles of $Q^2 = 0.04$ and $0.1\,(\text{GeV}/c)^2$ [5], which allowed the separation of the magnetic G_M^s and axial G_A^e form factors.

The G0 Collaboration proposes to perform the complete separation of the electric G_E^s, magnetic G_M^s and axial G_A^{ep} form factors for two different momentum transfers 0.23 and 0.6 $(\text{GeV}/c)^2$. First an asymmetry measurement has been performed for *forward electron scattering* angles between 7 and 15° for a large Q^2 range between 0.1 and 1 $(\text{GeV}/c)^2$. This measurement was done by detecting the recoil protons at a fixed angle between 68° and 78° from the elastic electron-proton scattering. A second measurement is underway at *backward electron scattering angle* of about 110° for the hydrogen and deuterium targets for different incident beam energies of 362 and 687 MeV.

The axial form factor G_A^{eN} has been considered for a long time as a quantity to be predicted for the extraction of G_E^s and G_M^s (see relation (4)). But, despite its unknown Q^2 behavior which implies a needed measurement for strange contribution at high Q^2, it contains information with theoretical and experimental implication. It can be expressed through the relation

$$G_A^{eN}(Q^2) = G_A^{Z,N}(Q^2) + \eta F_A(Q^2) + R_e\,. \tag{5}$$

The first term contains the charged current contribution (accessible in neutrino scattering) and the strange contribution to the axial for factor which was measured at $Q^2 = 0$ in deep inelastic scattering. The second term is the anapole form factor which corresponds to multiquark radiative correction and the last term is a one-quark radiative correction. Except for the charged current contribution, the Q^2-dependence of all these terms is unknown. The last two terms have been calculated theoretically at $Q^2 = 0$ but with large uncertainty.

The first experimental information on G_A^{eN} has been obtained from the SAMPLE experiment at $Q^2 = 0.1\,\text{GeV}^2$, where the extracted isovector nucleon axial form factor was found to be consistent with the theoretical estimate of Zhu [6].

3 Experimental set-up in the forward-angle configuration

The G0 experiment [7] has been performed in Hall C at *Jefferson Laboratory*. It requires high statistics (about 10^{13} elastic events) in order to measure very small asymmetries between 3 to 50 ppm with a relative error close to 5%. This requires beams with high intensity and high polarization. Also small helicity-correlated effects from the beam (intensity, position, ...) must be suppressed and if possible be uncorrelated to the detector and electronics response. The design of the experimental set-up has to achieve the highest possible data taking rates with a long cryogenic target and large acceptance of the detectors.

In the following we describe in more details all the items associated to the G0 set-up from the beam properties to the magnet, target, detectors and electronics.

3.1 Electron beam

The polarized electrons are provided by the interaction of a circularly polarized laser on the strained GaAs photocathode, which allows to select a particular transition of the crystal. This method allows to achieve high polarization ($\geq$ 70%), which will be reversed at 30 Hz frequency (defining a macro-pulse period (MPS) of 33 ms) in less than 200 μs, by changing the polarity on a Pockels cell. In order to avoid time drift effects and to minimize the correlation of the helicity signal with the counting recording, the helicity will be randomly sequenced by quartet and will also be sent to the acquisition after a few MPS.

As previously mentioned, G0 requires high intensity (40 μA) at a different mode than CEBAF is usually operating (31 MHz instead of 500 MHz). This mode uses a charge per bunch 16 times larger than the normal operating, which implies new requirements on the beam optics, due to space charge effects. The beam polarization, will be measured with an accuracy of about 2% using the standard Møller polarimeter available in Hall C.

An important issue in this kind of experiment is related to the systematic errors associated with the beam properties. The resulting false asymmetries, induced by possible correlations between the counting rates in the detectors and the beam parameters, as the beam charge, positions, angles and energy, have to be corrected with a good precision.

The way to minimize these systematic errors to an acceptable level is to keep these differences as low as possible so the run-averaged correction on the asymmetry remains lower than the statistical error ($\simeq 10^{-7}$). Secondly, the beam parameters differences have to be measured accurately in such a way that the error on the correction (during the overall experiment) is kept as low as 5% ΔA_{stat} ($\simeq 5\cdot 10^{-9}$). These beam requirements, which are summarized from each beam parameter on table 1, were all met during the G0 experiment.

Table 1. Challenging beam requirements and realizations for the G0 functioning. These are the differences between the two helicity states summed over the entire experiment.

Beam parameters	Differences specifications	Differences measured
CW current	≤ 1 ppm	-0.14 ± 0.32 ppm
Position	≤ 20 nm	3 ± 4 nm
Angle	≤ 2 nrd	1 ± 1 nrd
Energy	≤ 75 eV	24 ± 4

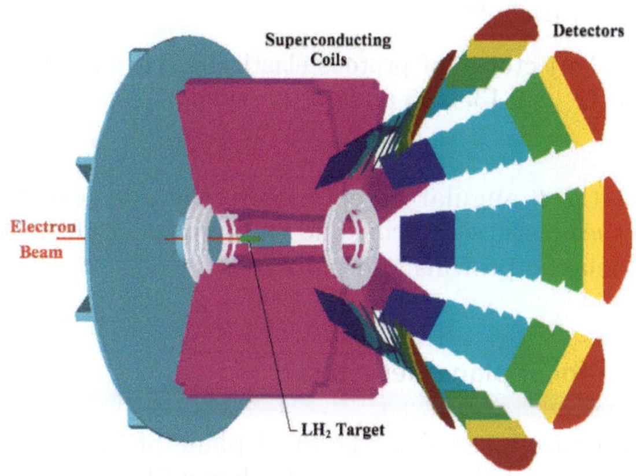

Fig. 1. G0 spectrometer.

3.2 Target and spectrometer

The G0 experiment will use a 20 cm target. Its design, which is based on the one used in the SAMPLE experiment, is optimized to reduce the energy loss on the scattered particles, and to minimize density variations up to 250 W of beam power deposited in the target. Also the target can be removed from the beam trajectory for diagnostic purposes and can be warmed independently of the spectrometer. It is fronted by a helium cell which is used to extend the entrance of the hydrogen cell and to eliminate variations in the effective LH_2 thickness with beam position.

The spectrometer is the central piece of the G0 apparatus. It consists of a superconducting toroidal magnet with eight sectors placed around the beam axis (see fig. 1). The charged particles coming from the elastic scattering, produced in the target are selected in momentum by the magnetic field so that their location on the focal plane is independent of the interaction point along the beam axis. It has been designed to achieve large acceptance in solid angle which corresponds to a large-Q^2 acceptance from 0.1 to 1 $(\mathrm{GeV}/c)^2$.

The high magnetic field produces a bend angle of 35° which is associated to a collimator geometry allowing to remove the direct view from the target and thus reduce the background coming from neutral particles in the detectors. Horizontal collimators are installed in order to define accurately the azimuthal acceptance, which is about 22° for each octant, and prevent particles to be deviated in the neighboring octants. Also vertical collimators are used to

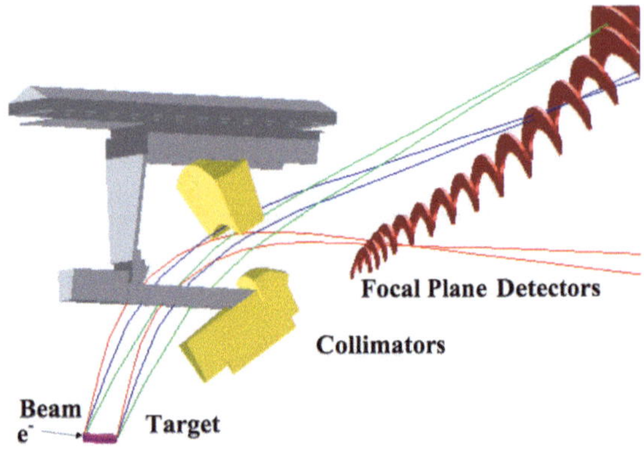

Fig. 2. Trajectories of protons elastically scattered for three different $Q^2 = 0.13$, 0.25 and 0.7 $(\mathrm{GeV}/c)^2$.

restrict the θ angular acceptance to the one corresponding to the elastic scattering and thus reduce counting rates from inelastic protons and pions.

3.3 The focal-plane detectors

Detectors are located at the focal plane of the spectrometer for each of the eight octants defined by the coils. An octant is composed of 16 pairs of plastic scintillators coupled to light guides. The arc shape of each scintillator has been calculated from simulated trajectories of elastic protons created along the target at fixed transfers momentum Q^2. As shown in fig. 2, protons related to different Q^2 are scattered at a different location in the focal plane of the spectrometer and, for a fixed Q^2, all protons coming from different part of the target will be focused in the same detector.

The detector size have been optimized to limit the counting rates to 1–2 MHz. Also each detector is composed of a pair of scintillators separated by an aluminium or plastic plate. The coincidence between these two active layers allows to reduce significantly the background related to neutral particles like neutrons and γ's produced in the target and other materials. Because PMTs (Photomultiplier tubes) must be located far away from the spectrometer in relatively low magnetic-field regions, scintillators have been connected at each extremity to long light guides whose lengths range from 40 cm up to 2 m. Their design has been carefully studied in order to optimize the light collection at the PMT.

The elastic protons are discriminated from other particles (protons, pions, etc.) using time of flight measurement (see fig. 4). In order to optimize the experimental time resolution, the detectors are placed transverse to the particle trajectory in order to reduce spread on the time of flight. Also the choice of BC408 Bicron as scintillating material has been made to insure a good timing resolution (less than 0.5 ns) required for time measurement.

The time-of-flight spectrum associated with detector 15 is different compared to the other detectors because it is located at the maximum kinematic angle allowed for elastic protons. Thus the range in Q^2 is large, from 0.5 to 1 $(\mathrm{GeV}/c)^2$ and this is reflected in a wider time distribution associated to the elastic peak. Three different binnings in the time spectra were used to define different Q^2 ranges.

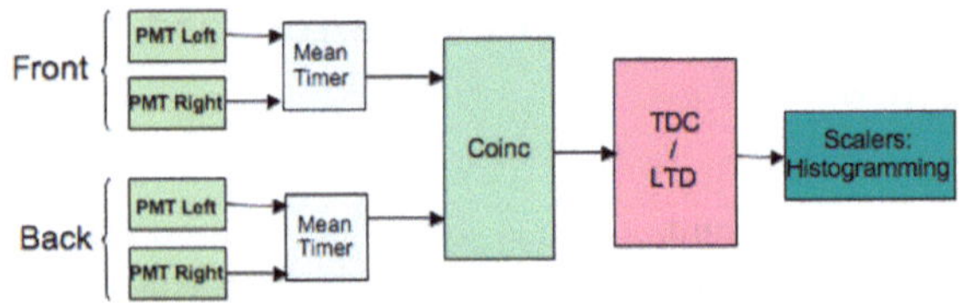

Fig. 3. Schematic design of the forward-angle electronics.

One has to notice that the French and North-American parts of the collaboration have provided each four octants which differ only in details concerning light guides and mechanical support.

3.4 Electronics

For the forward-angle measurement, the electronics needed for the asymmetry measurement has to provide time-of-flight information associated to each detector with rates ranging between 1 to 2 MHz. Also the electronics response has to be very stable whatever the experimental conditions are, to avoid any helicity-correlated systematic effects. For example the electronics deadtime has to be minimized and known with a good accuracy.

Two different electronics systems have been built by the North-American and French parts of the collaboration and each was used for half of the detectors. They allowed to study, in detail, different responses and related systematic effects. These two systems are based on the same concept as shown in fig. 3. The four analog signals, corresponding to the four PMTs of one detector, are sent into Constant Fraction Discriminators (CFD). The time information associated to left and right PMTs are meantime to cancel the time dependence *versus* the location of the interaction along the scintillator. Then the coincidence between the front and the back detectors is required to record the time-of-flight information. While the time range of 32 ns is fixed by the interval between two beam pulses, the time binning differs for the two electronics with 1 ns for the North-American and 250 ps for the French one. Then, the time information is histogrammed during one MPS and read out during the helicity reversal.

Finally, the charge and time distributions associated to each discriminator and meantimer will also be prescaled from a standard event-by-event Fastbus acquisition and will be used to monitor possible drifts of gains and time resolution of PMTs. Fastbus acquisition is also used for recording information related to the beam.

4 G0 forward-angle results

4.1 From raw to physics asymmetries

The data collected in 701 hours correspond to 101 coulombs of high-quality beams suited for parity violation

Table 2. Asymmetries correction applied to raw asymmetries.

Source	Correction and uncertainty
Residual deadtime	$\Delta A = 0.05 \pm 0.05\,\text{ppm}$
Residual helicity-correlated beam properties	$\Delta A = 0.02 \pm 0.01\,\text{ppm}$
Beam leakage	$\Delta A = 0.71 \pm 0.14\,\text{ppm}$
Beam polarization	$Pe = 73.7 \pm 1\%$
Background	$\Delta A = -1$ to 42 ± 0.2 to $9\,\text{ppm}$ (depending on Q^2)

experiments. Systematic studies show that the response of the spectrometer to beam changes is well understood. The distribution of the measured asymmetries calculated for each four helicity sequence is compatible with a Gaussian over five order of magnitude, and the measured width agrees at a level of few percent with the expected ones. Other searches for experimental asymmetries are through the polarization reversal using mechanical insertion of a $\lambda/2$ wave plate. The asymmetries integrated over detectors change sign with no offset observed at the 0.15 ppm level. Correction to beam changes, as well as the deadtime, are corrected within the analysis for each helicity state. Several sources of remaining non-physical asymmetries has been identified and corrected with very small absolute errors (see table 2). The beam leakage comes from the beam of other experimental Halls (A and B) into the Hall C. The beam leakage current is very small ($40\,\text{n}A/40\,\mu A$) but the asymmetry is large (around 340 ppm). This induce a false asymmetry depending on the time of flight (TOF) and the beam intensity. The correction to be applied is measured using a region in TOF which is signal free. From the raw asymmetries measured, the various corrections are applied to give the physics asymmetries. The Q^2 of the measurement is then determined by simulation with 1% error and allow to interpolate the electromagnetic form factors to calculate the non-strange vector asymmetry (A_{NVS}). The difference between A_{phys} and A_{NVS} gives the linear combination $G_E^s + \eta G_M^s$.

4.2 Background overview

The background in the experiment has been intensively studied using several targets to understand and measure the electro and photo production coming form the windows of the target and from the LH_2 itself. The main contribution to the asymmetries is due to protons coming from hyperon weak-decay scattering inside the spectrometer. The simulation provides a very good understanding of these contributions but cannot be used for correction. These events are rare with very large positive asymmetries.

The background asymmetries are determined in two steps. First the background's yield is fitted over a large time spectra range, using a polynomial function (see fig. 4, top panel). This gives the dilution factor $f(t) = \frac{Y_{back}(t)}{Y_{meas}(t)}$. Then, the background and elastic asymmetries are fitted

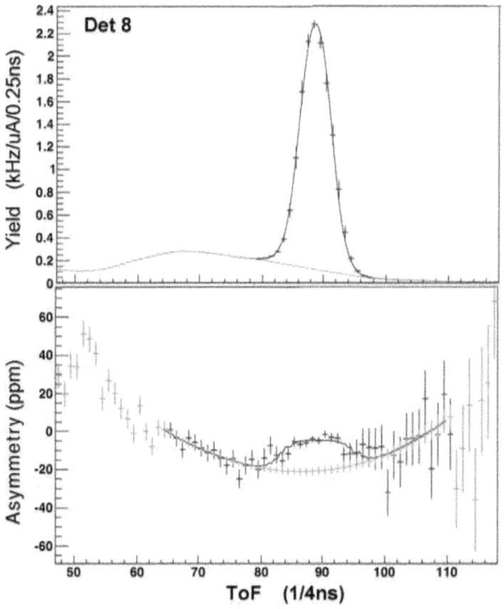

Fig. 4. Fit of the yield (top) and asymmetry (bottom) for detector 8.

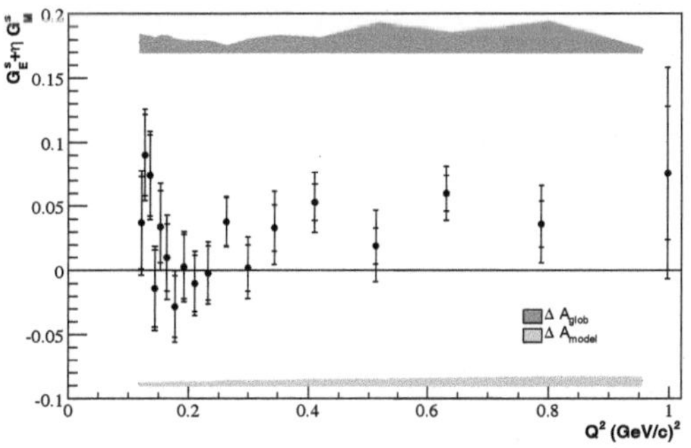

Fig. 5. Combination of electric and magnetic form factors obtained from the G0 forward-angle experiment.

(see fig. 4, bottom panel) according to

$$A_{meas}(t) = (1 - f(t))\, A_{el} + f(t)\, A_{back}(t). \quad (6)$$

The elastic asymmetry is taken as constant, while the background one is taken as a polynomial. Conservative uncertainties are obtained by varying these quantities over a reasonable range. Due to the wide Q^2 range of the detector 15 (see sect. 3.3), three differents asymmetries are fitted for three differents Q^2 values (0.51, 0.63, 0.79 $(\text{GeV}/c)^2$).

4.3 Results

The results of the experiment are shown in fig. 5. The linear combination $G_s^E + \eta G_s^M$ is the difference of the experimental asymmetry and the "no-vector-strange" asymmetry calculated from eq. (4) using $G_s^E = G_s^M = 0$ for all values of Q^2. The error bars include the statistical uncertainty (inner) and the statistical plus point-to-point systematic uncertainties added in quadrature. The error bands correspond to the global systematic uncertainties from the experiment (upper), and from the quantities entering in A_{NVS} (lower) (see [8] for details). This results show a non-trivial Q^2-dependence which may indicate two contributions of opposite sign. The G0 data are in agreement with other experiments for Q^2 values where data exist (0.1, 0.23 and 0.48 $(\mathrm{GeV}/c)^2$). Actually, independent extraction of G_s^E and G_s^M is only possible at $Q^2 = 0.1$. The last HAPPEX results [9] confirms a null G_s^E with a better precision and a positive G_s^M. New measurements are underway with the G0 apparatus in the backward angle mode [10]. New data will soon be available for $Q^2 = 0.23$ and $0.6\,(\mathrm{GeV}/c)^2$, for hydrogen and deuterium target. This will allow the separation of the two terms (electric and magnetic) over a large Q^2 domain in order to get the Q^2-dependence of the strange quark contribution.

I would like to thank the organisers of *PAVI 2006* for their great hospitality. I wish a successful running G0 data taking. The G0 experiment is supported by DOE (U.S.), CNRS/IN2P3 (France), NSERC (Canada) and NSF (U.S.).

References

1. D.H. Beck, B. Holstein, Int. J. Mod. Phys. E **10**, 1 (2001).
2. R. Tieulent, PhD Thesis, Grenoble University, ISN 02-27 (2002).
3. K.A. Aniol *et al.*, Phys. Rev. Lett. **96**, 022003 (2006); K.A. Aniol *et al.*, Phys. Lett. B **635**, 275 (2006).
4. PV A4 Collaboration (F.E. Maas *et al.*), Phys. Rev. Lett. **93**, 022002 (2004); **94**, 1520001 (2005).
5. SAMPLE Collaboration (D.T. Spayde *et al.*), Phys. Lett. B **583**, 79 (2004).
6. S. Zhu *et al.*, Phys. Rev. D **62**, 033008 (2000).
7. R. Carr *et al.*, in preparation.
8. D. Armstrong *et al.*, Phys. Rev. Lett. **95**, 092001 (2005).
9. D. Armstrong, this conference.
10. S. Kox, this conference.

The G0 backangle measurements

S. Kox[a]

For the G0 Collaboration[b]
LPSC, IN2P3-UJF, 53 Avenue des Martyrs, 38026 Grenoble Cedex, France

Abstract. This contribution presents the status, at the time of this conference, of the second phase of the G0 experiment. This experiment is being performed in Hall C at Jefferson Laboratory. It measures PV asymmetries in elastic electron-proton and quasi-elastic electron-deuteron scattering. Details of the experimental setup are given, together with the first data collected in 2006 at backward angles.

PACS. 13.60.Fz Elastic and Compton scattering – 13.40.Gp Electromagnetic form factors – 14.20.Dh Protons and neutrons

1 Introduction

A good understanding of the role of its sea of gluons and pairs of quarks and antiquarks is crucial to our knowledge of the structure of the nucleon. This is, in particular, the case for the search for possible contributions of strange quark-antiquarks pairs to nucleon properties. To this end, parity violation (PV) experiments in electron-nucleon scattering have become, over the years, a reliable and powerful technique for extracting information on the nucleon internal structure [1]. These experiments have already published a set of results [2–5], but a new round of measurements is underway, or planned, to allow for the separation of the neutral weak form factors and the effective axial current of the nucleon. Combined with the known electromagnetic form factors, these new data permit in particular the isolation of the contribution of the s quarks to the nucleon charge and magnetization densities.

Part of a worldwide program, the G0 experiment is carried out in Hall C at Jefferson Laboratory by a large collaboration of scientists from institutions in the United States, Canada, France, Armenia and Croatia. The ultimate goal of G0 is to provide data to allow a comprehensive and precise map of the neutral weak form factors of the proton over the range of momentum transfers 0.1–1.0 $(\mathrm{GeV}/c)^2$. This is done by measuring PV asymmetries in the scattering of polarized electrons on hydrogen and deuterium targets. To allow a separation of the different contributions (electric, magnetic, axial), the G0 experimental program is now performing its backward-angle measurements. In the coming months, it will provide results for two Q^2 values, 0.23 and 0.62 $(\mathrm{GeV}/c)^2$. During these measurements, with a single experimental configuration, data are also recorded for inelastic electron scattering.

Fig. 1. Photograph of the G0 set-up installed in Hall C in its "backangle" mode.

2 Experimental set-up

A photograph of the overall set-up, after the 180° rotation of the spectrometer necessary for the backangle running, is shown in fig. 1. The heart of the G0 apparatus [6] is a superconducting toroidal magnet (SMS) that focuses scattered particles through collimators onto a focal-plane array of scintillators. The magnet contains eight superconducting coils splitting the magnet and detector system into

[a] e-mail: kox@in2p3.fr

[b] Jefferson Lab Experiments E-05-108/06-108 and E-04-101. A list of collaborators and information about the experiment can be found at http://www.npl.uiuc.edu/exp/G0/.

octants. Each octant is defined by collimators placed near the target with an acceptance of 20° in azimuthal angle. Combined with the polar angle acceptance, the device has a total solid angle of about 0.9 sr. The polarized electrons are scattered from a high-power cryogenic target (liquid hydrogen or deuterium) of 20 cm length [7]. Currents up to 60 μA are used for the backangle measurements with negligible observed target density fluctuations.

The G0 octant detection systems are composed of detectors that count either recoil protons from small-angle (7–13°) *e*-*p* scattering (initial orientation of the magnetic spectrometer), or scattered electrons from large-angle (110°) *e*-*N* scattering (the backangle phase of G0 presented here). A set of 16 detectors in each octant, used in both measurements, lies near the focal surface of the SMS magnet. These detectors (referred to as the Focal-Plane Detectors, or FPDs), consist of pairs of plastic scintillators with phototubes at both ends. In the case of the forward-angle measurements, particle identification was determined by a time-of-flight technique using a special 32 ns beam time structure. This technique is not suitable for backward-angle measurements where one has to discriminate between elastic electrons and electrons or pions produced in inelastic processes with similar velocities. The standard CEBAF 2 ns beam time structure was then restored and a second array formed by 9 scintillators was added near the exit window of the magnet cryostat. For each of the eight octants, these detectors (CEDs, for Cryostat Exit Detectors) are used in coincidence with the focal-plane detectors. This CED/FPD combination correlates the momentum and scattering angle of the detected electrons after its path in the SMS magnetic field, and thus allows for the separation of elastic and inelastic events. With front-end electronics composed of gate arrays, it is possible to record events for given CED/FPD combinations, and thereby measure asymmetries for both elastic and inelastic events. However, simulations and analysis of test data [8] suggested that additional discrimination of electrons and pions was necessary. The π^- background from $n(e,\pi^-)e'p$ reactions can indeed be reduced by introducing cuts in CED-FPD space, but not to a level sufficient to isolate quasi-elastic electrons.

Therefore an aerogel Cherenkov counter has been designed to provide pion rejection across the momentum range of the full G0 program, up to at least $\sim 400\,\mathrm{MeV}/c$ for $Q^2 = 0.8\,\mathrm{GeV}^2$. This, of necessity, must be an eight-sectored array of individual Cherenkov detectors mounted in conjunction with the CED-FPD sectors. The Cherenkov counters have been located between the CEDs and FPDs. Negatively charged particles entering an octant of the G0 spectrometer pass through 5.5 cm of aerogel (SP30 from Matsushita Electronics). With the chosen index of refraction of the aerogel ($n = 1.03$), pions up to a momentum of $570\,\mathrm{MeV}/c$ do not produce any light whereas all primary electrons do. The photons emitted are collected by four phototubes (XP4572 from Photonis) by means of a diffusion box lined with an internal white reflector. From measurements performed with cosmic muons and test beams at TRIUMF, the electrons were found to generate a signal of 4 to 7 photoelectrons. The detection efficiency obtained in the present data taking is more than 80% for electrons (when requesting 2 out of the 4 PMTs being fired), whereas a rejection factor of more than 100 to 1 is reached for a $400\,\mathrm{MeV}/c$ pion. The Cherenkov detector operates in *coincidence* mode for the electron detection, and in *veto* mode for background and pion studies.

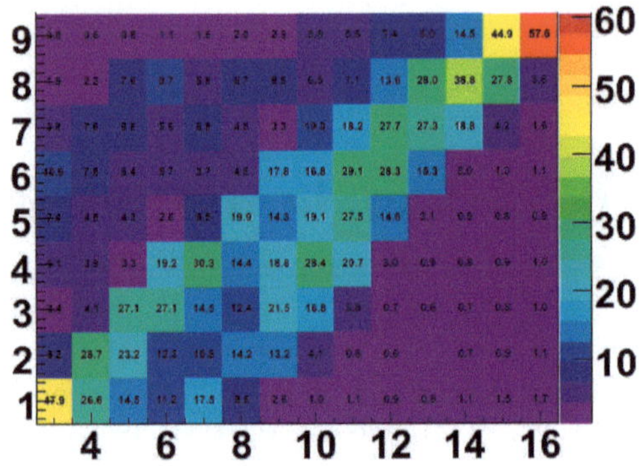

Fig. 2. Counting rates (Hz per μA of beam and per octant) in the CED-FPD plane, for particles identified as electrons by the Cherenkov counter, at $Q^2 = 0.62\,\mathrm{GeV}^2$.

The G0 experiment is also instrumented to measure beam properties (position, charge) in the Hall C line; a feedback system at the source level ensures that they are largely independent from the beam helicity state. A new more reliable fiber laser has been used for the present running. It allowed also to switch within a few hours to a 32 ns beam time structure, used for studies of the Cherenkov performances.

3 Data taking and first results

In the backangle phase of the G0 experiment, PV asymmetries are measured for several physics cases. The elastic scattering events are detected to access the weak form factors of the nucleon, the inelastic electrons produced in a kinematic range near the N-Δ transition [9] to determine the axial mass of the nucleon and finally the pions also produced in this region to study EW corrections [10]. This requires the identification of the three corresponding particles. The experimental set-up for backangle measurements has thus been upgraded with the new detectors discussed above, with corresponding changes in the electronics scheme. The particle identification is then performed by combining information provided by the CEDs, FPDs and Cherenkov detectors. Figure 2 shows an example of the discrimination of the elastic against inelastic electrons in the CED-FPD plane when the Cherenkov is fired (a similar plot is built for pions with the Cherenkov detector being used as a veto).

There have been several experimental difficulties, some only apparent with the first beam in the Hall. First of all, the experiment is running with non-standard, low beam energies (686 and 362 MeV). This has presented challenges

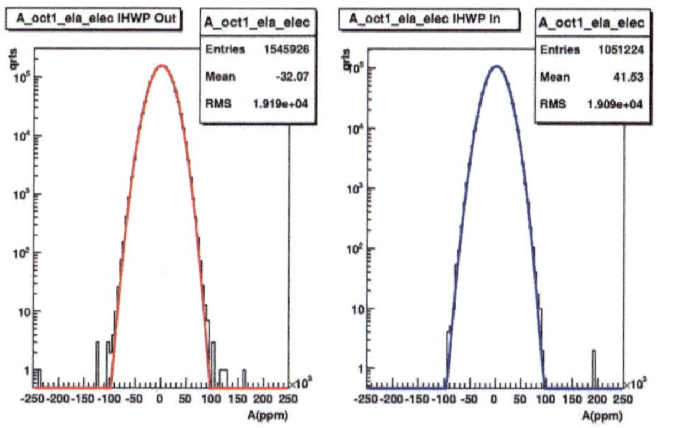

Fig. 3. Distribution of measured detector asymmetries for the two states of the IHWP (see text). A multiplicative blinding factor of 1 ± 0.25 has been applied to the data.

for the accelerator to reach the halo rates and parity quality of the beam required by the measurement. The experiment is using the highly polarized electron beams now produced routinely at Jefferson Laboratory from a superlattice GaAs cathode. The measurement of the beam polarization for these energies required a change in the geometry of the Møller polarimeter of Hall C and new operational settings. After Møller commissioning, the beam polarization (typically 85%) was measured for the 686 MeV energy with a precision of 2–3%, checked during runs with polarimeters at the injector and in other halls.

A second experimental challenge was related to large single rates observed in the counters. They are generated mainly by low-energy neutrons produced in the target and by moderating processes in the SMS collimators and in the Hall. This has been dealt with for the CED and FPD counters by using additional shielding, achieving better beam conditions and by a change in the FPD phototube arrangement (by placing one PMT from the front scintillator in coincidence with one on the opposite side of the back scintillator). The first data taken have also allowed the tuning of the experimental conditions (magnetic field, thresholds and timing). After these changes, a beam intensity of 60 μA is being used for the LH_2 target at both energies. The rates obtained are such that losses due to dead time and random coincidences are kept at a reasonable level. It was, however, not possible to use such a beam intensity with a LD_2 target for which the production of neutrons is highly enhanced. The limitation comes from the Cherenkov counters for which the very large random coincidence rate observed reduced the effectiveness of the pion rejection. It has been determined that these high rates are due to the capture of thermal neutrons in the glass (Borosilicate) of the Cherenkov phototubes. This has been checked with measurements performed at the NIST facility in Maryland and at the PEREN facility in Grenoble. The solution proposed is to replace the Cherenkov phototubes by new ones with different (quartz) glass windows. From tests with beam in the hall, rates indicate that running with the LD_2 target should be possible at the nominal 60 μA beam intensity.

There are, at this stage, no physics results to be presented. However the quality of the data being recorded can be estimated from standard PV control spectra like the distribution of the measured detector asymmetries. This is shown in fig. 3 for the events selected in the elastic band of the CED-FPD electron plane. As expected, a Gaussian distribution is observed over several orders of magnitude. Also, the sign of the asymmetry is reversed with the insertion of a mechanical device (Insertable Half-Wave Plate, IHWP) into the laser path at the source, while, as expected, its amplitude remains unchanged within the error bars.

4 Conclusions

After the completion of the forward angle measurements of the G0 program, the experimental set-up was upgraded and installed in its backward-angle configuration. The data taking started early 2006 and is presently on schedule until mid-February 2007. Several serious issues have been successfully handled for this measurement. Some of them were related to the use of a non-standard low-energy beam (optics and polarimetry) and others to large rates and anode currents due to low-energy neutron background. The accelerator has provided beams with the requested parity quality specifications. The running with the liquid-hydrogen target is underway with 60 μA beam intensity. The deuterium target running will follow, once the phototubes in the Cherenkov counters have been replaced by new ones having a quartz window.

With the recording of new backward-angle data at two different Q^2 values, the physics program of the G0 experiment is nearing completion. The new data will be combined with the data already measured at forward angles, and also with the new measurements foreseen by the HAPPEX and A4 Collaborations. The set of data then available should allow one to place strong constraints on the elusive role of the strange sea quarks in the nucleon electromagnetic structure.

This work has been made possible by many years of effort from the G0 collaborators and with essential support from the accelerator, target and Hall C groups at Jefferson Laboratory.

References

1. *Proceedings of the PAVI04 Conference*, Eur. Phys. J. A **24** (2005).
2. T. Ito *et al.*, Phys. Rev. Lett. **92**, 102003 (2004); D.T. Spayde *et al.*, Phys. Lett. B **583**, 79 (2004).
3. K. Aniol *et al.*, Phys. Rev. C **69**, 065501 (2004); Phys. Lett. B **635**, 275 (2006); Phys. Rev. Lett. **96**, 022003 (2006).
4. F. Maas *et al.*, Phys. Rev. Lett. **93**, 022002 (2004); Phys. Lett. B **583**, 79 (2004); Phys. Rev. Lett. **94**, 152001 (2005).
5. D. Armstrong *et al.*, Phys. Rev. Lett. **95**, 092001 (2005) and the contribution of J.S. Réal, this conference.

6. For details, see the PhD Thesis of J. Liu, University of Maryland, April 2006 and B. Guillon, University of Grenoble, December 2005.
7. S. Covrig *et al.*, Nucl. Instrum. Methods A **551**, 218 (2005).
8. R. Tieulent *et al.*, G0-Reports G0-01-012 and G0-01-022.
9. S. Wells, N. Simicevic (spokespersons), JLab E-04-101 Experiment.
10. J. Martin (spokesperson), JLab E-04-115 Experiment.

Kinematics and backgrounds for HAPPEX measurements

B. Moffit[a]

The College of William & Mary, Department of Physics, Williamsburg, VA 23187-8795, USA

Paper reprinted from Eur. Phys. J. A with permission

Abstract. Measurements and systematic errors of the 4-momentum transfer squared (Q^2) and backgrounds for the 2005 HAPPEX runs are described.

PACS. 12.15.-y Electroweak interactions – 25.30.Bf Elastic electron scattering – 07.05.Kf Data analysis: algorithms and implementation; data management

1 Introduction

Recent HAPPEX measurements have the goal of determining parity-violating asymmetries with very high absolute and relative precision in order to provide tight constraints on the strange electric and magnetic form factors of the proton. Accurate knowledge of the systematic errors that arise from experimental kinematics and contribution from backgrounds are needed in order to ensure that these measurements are statistically limited. The High Resolution Spectrometers (HRS) of Hall A at Jefferson Lab are the key feature in the achievement of this goal. In this paper we present the technique used for the measurement of the 4-momentum transfer squared (Q^2) between the longitudinally polarized electron and unpolarized target nucleus, with particular focus on optics optimization and the target-recoil technique used to determine the central scattering angle. A detailed view of the determination and magnitude of backgrounds that contribute to the experimental asymmetry will also be presented.

2 Kinematics determination

The 4-momentum transfer squared (Q^2) sets the kinematic scale of the asymmetry measurement, and is defined as

$$Q^2 = 2EE'(1 - \cos\theta), \qquad (1)$$

where E is the beam energy, and E' and θ are the energy and angle of the scattered electron, respectively.

The beam energy is measured using dispersion in the 34.3° bend which brings the beam into the experimental hall, detailed in ref. [1]. The parameters of the scattered electron are determined by reconstructing its path from the target to the spectrometer focal plane.

[a] Presenting on behalf of the HAPPEX Collaboration; e-mail: `moffit@jlab.org`

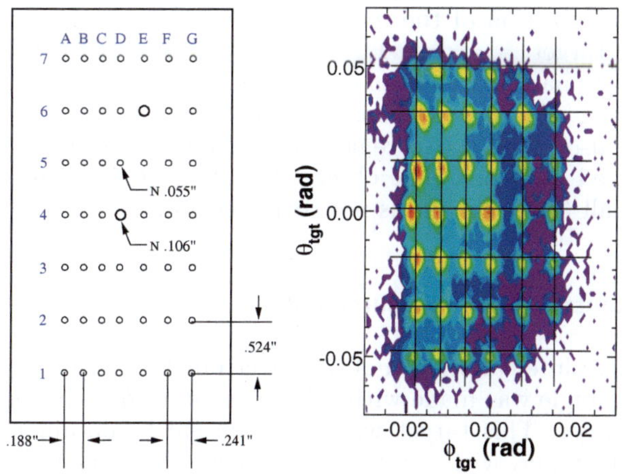

Fig. 1. Insertable sieve slit (left) and reconstructed sieve hole pattern (right).

2.1 Optics optimization

Scattered electrons are detected in the HRS using two planes of vertical-drift chambers to measure their position and angle in the focal plane. Combined with incident beam information, the track parameters at the target are reconstructed using transfer matrices optimized for the spectrometer optics. Optics optimization is aided by the use of a sieve slit: a 5 mm thick collimator placed in front of the acceptance of the first magnetic element of the spectrometer. The sieve slit (shown in fig. 1) contains a pattern of 49 pin holes (7 × 7) with two larger holes to allow for unambiguous identification of the orientation of the image at the focal plane. The known momentum, position, and angle of tracks from elastic scattering from a thin target through individual sieve-holes is used to optimize a polynomial expansion relating the trajectory in the focal plane

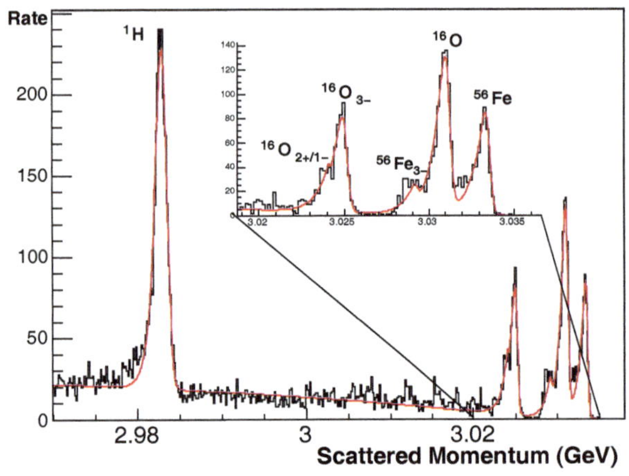

Fig. 2. Scattered momentum spectrum observed from the water cell through a single sieve hole. The solid-line fit to the data includes information about the mass of each state as well as radiative energy corrections.

to parameters at the target, as outlined in ref. [2]. Optics targets, located at $Z_{\rm lab} = -12\,{\rm cm}$, $0\,{\rm cm}$, $+12\,{\rm cm}$, allow for optimization of the reconstruction along the full axis of the target.

Systematic errors from reconstruction are evaluated by applying *ad hoc* corrections to the observed sieve-slit pattern (also in fig. 1) to remove the small deviations from expected locations. The observed shift in Q^2 is found to be small ($\leq 0.3\%$) and is not corrected for.

2.2 Central angle measurement

The absolute calibration of the spectrometer central scattering angle was made with calibration data from a water cell target. This target was a 0.5 mm thick container of flowing water, with 1 mil steel windows. From relativistic kinematics, the scattered electron energy E' can be found in terms of the scattering angle θ:

$$E' = \frac{E - \frac{1}{2m}(m^{*2} - m^2)}{1 + \frac{E}{m}(1 - \cos\theta)}, \tag{2}$$

where E is the beam energy, m is the mass of the target nucleus, and m^* is the mass of the recoil state. Figure 2 shows a fit to the resulting measured energy spectrum from this target. The momentum separation between the hydrogen and heavier recoil peaks was used to determine the central angle with an uncertainty of 0.2%.

2.3 Corrections

In production mode, the signals from the HAPPEX detector are integrated over a 33 ms gate. This means that the asymmetry measurement is implicitly weighted by the energy deposited by the incoming electron. The Q^2 measurement is performed in tracking mode, therefore each track hitting the detector must be explicitly weighted by its ADC response. An upper limit to the systematic of this weighting was taken to be the full size of the corrections: 0.1% for the helium run, and 0.5% for the hydrogen run.

A drift over time was observed for these measurements in the helium run period, with the full scale of the drift taken to be an upper limit on the uncertainty of 0.6%. Although no such drift was apparent in the hydrogen run period, this error estimate was included in both data sets.

A series of measurements was taken at various beam currents to check for possible systematic errors due to rate-dependent effects in reconstruction, resulting in an additional estimated uncertainty of 0.3% for helium and 0.1% for the hydrogen runs.

Table 1 presents a summary of the systematic errors in Q^2.

Table 1. Summary of the systematic errors in Q^2.

Source	Percent Error in Q^2
Beam Energy	0.1%
HRS Momentum Scale	0.2%
Optics Optimization	
At $Z = 0$	0.3%
Z dependence	0.1%
Scattering Angle	0.4%
ADC Weighting	He: 0.1% H: 0.5%
Drifts in time	0.6%
Rate Dependence	He: 0.3% H: 0.1%
Total Error	He: 0.9% H: 1.0%

3 Backgrounds

There are three main contributors to the backgrounds in HAPPEX: scattering from the target aluminum walls, quasielastic scattering from ^{4}He, and rescattering in the spectrometer.

3.1 Target walls

Background from scattering from the target aluminum walls was analyzed using a pair of aluminum foils positioned at $Z_{\rm lab} = \pm 10\,{\rm cm}$ (the location of the target cell's endcaps). The thickness of the foils (1.0 mm) was chosen to match the radiation length of the full target cell. The background fraction ($f_{\rm Al}$) is then determined from

$$f_{\rm Al} = x_t \frac{R_{\rm Al\ Foils}}{R_{\rm Target\ Cell}}, \tag{3}$$

where x_t is the ratio of the target cell wall thickness to the dummy cell wall thickness, and $R_{\rm Al\ Foils(Target\ Cell)}$ is the normalized scattering rate from the aluminum foils (target cell). With this method, the aluminum fraction for helium was $(1.8\pm0.2)\%$ and for hydrogen was $(0.7\pm0.3)\%$. The uncertainty is dominated by the uncertainty in the

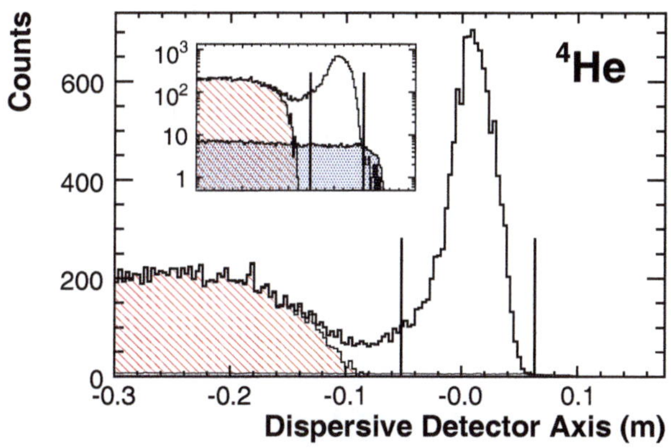

Fig. 3. Spectra of scattered electrons from ^{4}He, as observed along the dispersive detector axis. Insets show the same spectra on a logarithmic scale. Vertical lines indicate the acceptance of the detectors. The hatched region indicates the estimated quasielastic spectrum. The shaded region shows the contribution from the target aluminum endcaps.

thickness of the cryogenic cell walls. The asymmetry of this fraction, predominantly quasielastic scattering, was determined from [3]

$$A_{\mathrm{QE}}^{\mathrm{PV}} = \frac{N_p \sigma_p A_p^{\mathrm{PV}} + N_n \sigma_n A_n^{\mathrm{PV}}}{N_p \sigma_p + N_n \sigma_n}, \qquad (4)$$

which is ~ -2 ppm at these kinematics. A conservative uncertainty of 100% was used for this quantity.

3.2 ^{4}He quasielastic

The optical properties of the spectrometer act to transform momentum differences at the target to spatial differences at the focal plane. For scattering from ^{4}He, there are no excited states between the elastic peak and the nuclear-breakup threshold, which lies 19 MeV/c away. Careful positioning of the elastic peak over the HAPPEX detector, in the focal plane, is crucial to minimizing the background from the quasielastic processes.

Special runs were taken to determine the detector location and orientation in the spectrometer focal plane. This is accomplished by setting the spectrometer to fully illuminate the focal plane with the fairly isotropic distribution of inelastic scattered electrons from solid targets. Measurements made from the ^{4}He target were then made at varying spectrometer settings to optimize the elastic peak over the HAPPEX detector.

The rise of the quasielastic spectrum is then observed, in relation to the detector edge in the focal plane (shown in fig. 3). This rise is estimated with a cut on the difference in the reconstructed momentum and the calculated scattered momentum that would be expected from elastic scattering. All inelastic events lie beyond a 19 MeV threshold in this cut; events below this threshold are shown, hatched, in fig. 3.

To account for reconstruction resolution, an upper limit of quasielastic contamination was established cutting on this momentum difference at 15 MeV, resulting in a fraction from quasielastic ^{4}He to be (0.15 ± 0.15)% of the total rate observed in the detector. The asymmetry from this background is calculated to be (-1.5 ± 10.0) ppm from eq. (4).

3.3 Rescatter

Some fraction of the detector signal arises from electrons that are rescattered inside of the spectrometer. This fraction is based on the following integral:

$$f_{\mathrm{rs}} = \int_0^{E_{\mathrm{elastic}}} \mathrm{d}E' P_{\mathrm{rs}}(E') R(E'), \qquad (5)$$

where E' is the energy of the scattered electron, P_{rs} is the ADC weighted probability of an electron with this energy to rescatter in the spectrometer and produce a signal in the detector, and $R(E')$ is the ratio of the inelastic cross-section to the elastic cross-section:

$$R(E') = \frac{\left(\frac{\mathrm{d}\sigma}{\mathrm{d}\Omega \mathrm{d}E'}\right)_{\mathrm{inelastic}}}{\left(\frac{\mathrm{d}\sigma}{\mathrm{d}\Omega}\right)_{\mathrm{elastic}}}. \qquad (6)$$

Measurement of $P_{\mathrm{rs}}(E')$ is performed by increasing the spectrometer dipole field to force the elastic trajectories to trace those taken by the lower-energy inelastics at the production momentum setting. $R(E')$ is then measured by decreasing the dipole field. The resulting product between these two functions is integrated from the $E' = 0$ to the elastic energy. From this model, the background fraction from rescattering within the spectrometer is calculated to be $f_{\mathrm{rs}} = (0.25 \pm 0.15)$% for the helium run, and less than 0.10% for the hydrogen run. The asymmetry for these small fractions accounts for the estimated mix of inelastic and elastic radiative tail, at (-1.5 ± 3.0) ppm for helium and (-4 ± 4) ppm for hydrogen.

Studies of the source of rescattered detector signals were also used to estimate an upper limit on the fraction from exposed iron of $\sim 10^{-5}$. With a maximum asymmetry of ~ 200 ppm, this source of background contributed to a negligible uncertainty in each measurement.

4 Conclusion

An analysis of the Q^2 and backgrounds for the recent HAPPEX measurements has yielded precise results that keep the total systematic error in the parity-violating asymmetry less than the statistical precision (4% relative error for helium and 0.1 ppm for hydrogen). The techniques used provide a useful standard for future parity violation experiments endeavoring to obtain smaller relative measurements.

References

1. J. Alcorn *et al.*, Nucl. Instrum. Methods Phys. Res. A **522**, 294 (2004).
2. N. Liyanage, *Optics calibration of the Hall A high resolution spectrometers using the new C++ optimizer*, JLab-TN-02-012, 2002.
3. M.J. Musolf, T.W. Donnelly, Nucl. Phys. A **549**, 509 (1992).

Nucleon form factors and the BLAST experiment

R. Alarcon[a] and the BLAST Collaboration

Department of Physics, Arizona State University, Tempe, AZ 85287, USA

Paper reprinted from Eur. Phys. J. A with permission

Abstract. Measurements of the charge and magnetic form factors of the nucleon present a sensitive test of nucleon models and QCD-inspired theories. A precise knowledge of the neutron form factors at low Q^2 is also essential to reduce the systematic errors of parity-violation experiments. At the MIT-Bates Linear Accelerator Center, the nucleon form factors have been measured by means of scattering of polarized electrons from vector-polarized hydrogen and deuterium. The experiment used the longitudinally polarized stored electron beam of the MIT-Bates South Hall Ring along with an isotopically pure, highly vector-polarized internal atomic hydrogen and deuterium target provided by an atomic beam source. The measurements have been carried out with the symmetric Bates Large-Acceptance Spectrometer Toroid (BLAST) with enhanced neutron detection capability.

PACS. 25.30.-c Lepton-induced reactions – 25.30.Rw Electroproduction reactions – 29.25.-t Particle sources and targets – 29.30.-h Spectrometers and spectroscopic techniques

1 Introduction

The electromagnetic structure of the nucleon is traditionally described in terms of two form factors and it has been extensively studied by the scattering of electrons on nucleons. Experiments on electron-proton scattering have yielded abundant information on the magnetic G_M^p and charge G_E^p form factors of the proton in a range of momentum transfer squared, Q^2, up to about $30\,\mathrm{GeV}^2$. The information on the neutron form factors comes almost entirely from the scattering of electrons on deuterons or ^{3}He nuclei, and it is much less definite than that on the proton. The difficulty is due mainly to final-state interactions, mesonic currents, and the fact that the form factors of nucleons inside the deuteron or ^{3}He may differ from those of free particles.

Measurements of the nucleon form factors present a sensitive test of nucleon models and QCD-inspired theories. At low momentum transfer, $Q^2 \leq 1\,\mathrm{GeV}^2$, the pion cloud of the nucleon may play a significant role in the quantitative description of the form factors [1,2], in particular for the charge form factor of the neutron G_E^n in the absence of a net charge. Accurate measurements of the form factors at low Q^2 are required to reduce the systematic uncertainty in the extraction of the strange quark contribution to the nucleon electromagnetic structure as studied in parity-violating electron-scattering experiments [3].

In recent years the advent of polarized beams, targets, and polarimetry have made possible new classes of experiments aimed at extracting the nucleon form factors utilizing spin degrees of freedom. The general form for the differential cross-section in the exclusive scattering of longitudinally polarized electrons from a polarized target is given by [4]

$$\frac{\mathrm{d}\sigma}{\mathrm{d}e'\mathrm{d}\Omega_e'\mathrm{d}\Omega_N} = \Sigma + h\Delta, \tag{1}$$

where h is the helicity of the incident electron, and Σ and Δ are the helicity sum and helicity difference cross-sections, respectively. The Σ and Δ cross-sections for the case of elastic scattering from a polarized nucleon can be written as

$$\Sigma = c\left(\rho_L G_E^2 + \rho_T \frac{q^2}{2M^2} G_M^2\right) \tag{2}$$

and

$$\Delta = -c\left(\rho'_{LT} \frac{q}{2^{3/2}M} G_E G_M P_x + \rho'_T \frac{q^2}{2M^2} G_M^2 P_z\right), \tag{3}$$

where c is a kinematical factor, the ρ's are the virtual photon densities, and P_j indicates the polarization of the nucleon along each of the three coordinate axes, of which the z-axis has been chosen parallel to the momentum transfer $\boldsymbol{q}$. The terms containing $G_E G_M$ and G_M^2 can be completely isolated by tuning the target spin polarization. Experimentally one measures spin asymmetries defined as

$$A_{exp} = p_e p_T \frac{\Delta}{\Sigma}, \tag{4}$$

[a] e-mail: ricardo.alarcon@asu.edu

where p_e and p_T are the electron beam and target polarization, respectively. To separate both terms of the Δ cross-section the target spin is oriented perpendicular (parallel) to the direction of $\boldsymbol{q}$ (*i.e.*, selecting P_x (P_z)).

This work reports preliminary results from new measurements of the nucleon form factors over a range of four-momentum transfer Q^2 between 0.12 and 0.70 GeV^2 with the BLAST experiment at the MIT-Bates Linear Accelerator Center. The technique makes use of elastic *ep*-scattering of polarized electrons from polarized hydrogen to access the proton form factors, quasi-elastic $(\boldsymbol{e}, e'n)$ scattering from vector-polarized deuterium to get at the charge form factor of the neutron, and inclusive $(\boldsymbol{e}, e')$ scattering from vector-polarized deuterium for the magnetic form factor of the neutron.

2 The BLAST experiment

The BLAST experiment has been designed to measure spin-dependent electron scattering at intermediate energies from polarized targets in the elastic, quasi-elastic and resonance region. Based on the internal-target technique BLAST optimizes the use of a longitudinally polarized electron beam stored in the South Hall Ring of the MIT-Bates Linear Accelerator Center, in combination with an isotopically pure, highly polarized internal target for both hydrogen or deuterium. In case of deuterium the target was both vector and tensor polarized. The polarized target is provided by an Atomic Beam Source (ABS) [5]. The ejected gas molecules are first dissociated into atoms before they pass through sections of sextupole magnets and RF transition units to populate the desired single spin states through Stern-Gerlach beam splitting and induced transitions between hyperfine states (see fig. 1).

This selection process is highly efficient and thus provides nuclear polarizations of more than 70%. The spin-state selection was altered every five minutes in a random sequence to minimize systematics. The polarized atoms are injected into a 60 cm long cylindrical target cell with open ends through which the stored electron beam passes. As there are no target windows, the experiment is very clean with negligibly small background rates of only a few percent in the prominent channels.

The direction of the target spin can be freely chosen within the horizontal plane using magnetic holding fields. During BLAST data taking, the spin direction pointed at 32° and 47° to the left side of the beam axis in the 2004 and 2005 runs, respectively. At Bates beam currents of up to 225 mA were stored in the ring at 65% polarization and beam lifetimes of 20–30 minutes. The electron beam energy was 850 MeV throughout the BLAST program. The relatively thin target in combination with the high beam intensity yields a luminosity of about $5 \times 10^{31}/(\mathrm{cm}^2\,\mathrm{s})$ at an average current of 175 mA. The Bates storage ring contains a Compton polarimeter to monitor the longitudinal beam polarization in real time and without affecting the beam. The electron spin precession is compensated with a spin rotator (Siberian snake) in the ring section opposite BLAST. The helicity of the beam was flipped once before every ring fill.

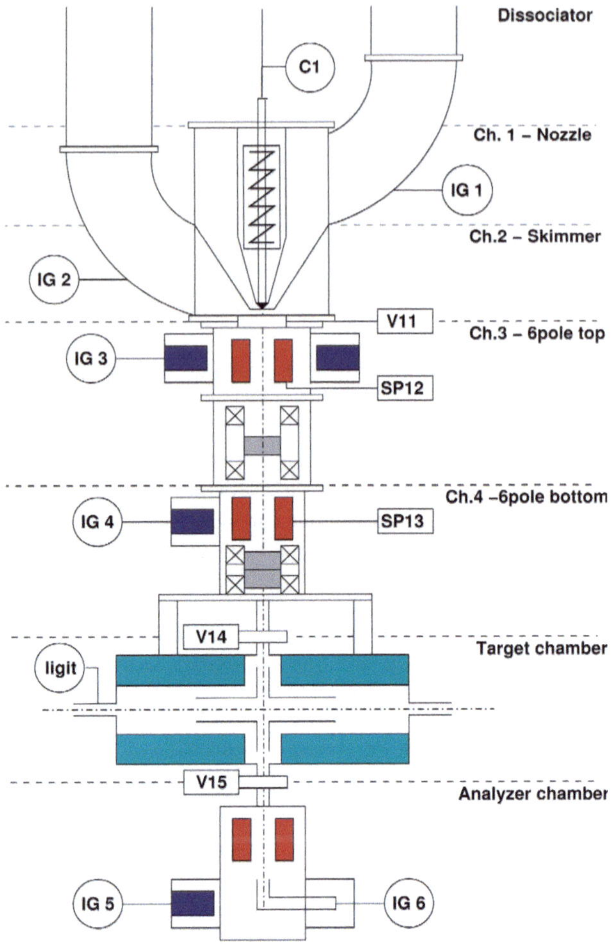

Fig. 1. Schematics of the atomic beam source and the target region. The ABS was embedded in the strong, spatially varying magnetic field of the BLAST toroid [5].

The BLAST detector is schematically shown in fig. 2. It was built as a toroidal spectrometer consisting of eight normal-conducting copper coils producing a maximum field of 0.38 T. The two in-plane sectors opposing each other are symmetrically equipped with drift chambers for the reconstruction of charged tracks, aerogel-Čerenkov detectors for e/π discrimination and 1″ thick plastic scintillators for timing, triggering and particle identification. The angular acceptance covers scattering angles between 20° and 80° as well as ±15° out of plane. The symmetric detector core is surrounded by thick large-area walls of plastic scintillators for the detection of neutrons using the time-of-flight method. The thin scintillators in combination with the voluminous wire chambers in front of the neutron detectors were used as a highly efficient veto for charged tracks, making the selection of $(e, e'n)$ events extremely clean. The neutron detectors are enhanced in the right sector with ∼ 30% neutron detection efficiency (∼ 10% in the left sector). The reason for this is the choice

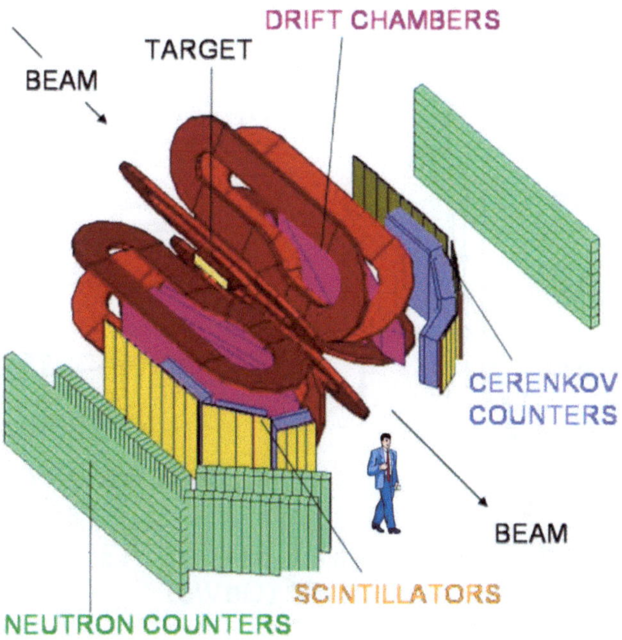

Fig. 2. Schematic, isometric view of the BLAST detector showing the main detector elements.

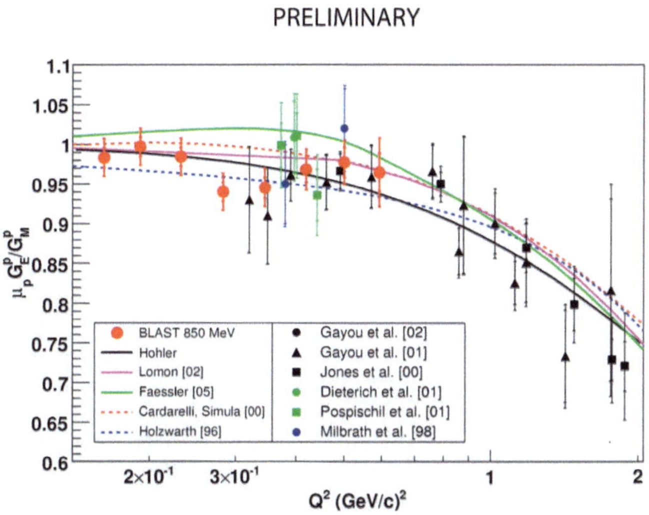

Fig. 3. Preliminary results of $\mu_p G_E^p/G_M^p$ shown with the world polarized data and several models described in the text.

of the target spin orientation. The setup allows to simultaneously measure the inclusive and exclusive channels (e,e'), $(e,e'p)$, $(e,e'n)$, $(e,e'd)$ elastic or quasi-elastic, respectively, as well as $(e,e'\pi)$ in the excitation region of the Δ-resonance. By measuring many reaction channels at the same time over a broad range of momentum transfer, the systematic errors are minimal.

3 Results

3.1 Proton charge to magnetic form factor ratio

The polarized hydrogen data were divided into eight Q^2 bins and the yield distributions were in good agreement with results from a Monte Carlo simulation, including all detector efficiencies measured from the data. The BLAST detector configuration is symmetric about the incident electron beam and the target polarization angle is oriented $\sim 45°$ to the left of the beam. Two independent asymmetries of electrons scattered into the beam-left and beam-right sectors, respectively, can be measured simultaneously. The ratio G_E^p/G_M^p can then be determined, independent of p_e and p_T, from the ratio of these experimental asymmetries measured at the same Q^2 value but corresponding to different spin orientations [6].

Preliminary results for G_E^p/G_M^p are shown in fig. 3 with the error bars due to statistical and systematic contributions added in quadrature. Also shown in fig. 3 are published recoil polarization data [7–13], together with a few selected models: a soliton model [14], an extended vector meson dominance model [15], an updated dispersion model [16], a relativistic constituent-quark model (CQM) with $SU(6)$ symmetry breaking and a constituent-quark form factor [17], and a Lorentz covariant chiral quark model [2]. Using these results and the world differential cross-section data on e-p elastic scattering at the same Q^2 values, the proton charge and magnetic form factors can be extracted and work to this extent is in progress.

3.2 Neutron magnetic form factor

The first experiments to measure G_M^n used electron scattering off unpolarized deuterium in elastic or quasi-elastic kinematics [18–21]. These methods involved significant uncertainties due to the required subtraction of the proton contribution. Alternatively, the ratio of the cross-sections from the reactions $^2\mathrm{H}(e,e'n)p$ and $^2\mathrm{H}(e,e'p)n$ in quasi-elastic kinematics was used by [22–26]. By measuring the neutron in coincidence, the theoretical uncertainties related to the proton subtraction are eliminated. Furthermore, using the cross-section ratio reduces the dependence of the measurement on nuclear structure. However, these experiments require accurate knowledge of the neutron detection efficiency, which is difficult to establish.

Exploiting the recent advances in polarization techniques [27–29] have used the inclusive quasi-elastic reaction $^3\vec{He}(\vec{e},e')$ for their measurement of G_M^n. Since in the ground state of ^3He the proton spins anti-align and cancel, the spin of the nucleus is carried by the unpaired neutron [30]. By the simultaneous use of beam and target polarization the experimental asymmetry is related to the ratio of the form factors of the neutron. The systematic uncertainties for these measurements stem mainly from Final-State Interaction (FSI) and Reaction Mechanism (Meson Exchange Currents (MEC) and Isobar Configurations (IC)) corrections.

With BLAST an alternative measurement of G_M^n has been carried out by using inclusive electron scattering with a polarized beam and a vector-polarized deuterium target, $^2\vec{H}(\vec{e},e')$, in quasi-elastic kinematics. The two-body

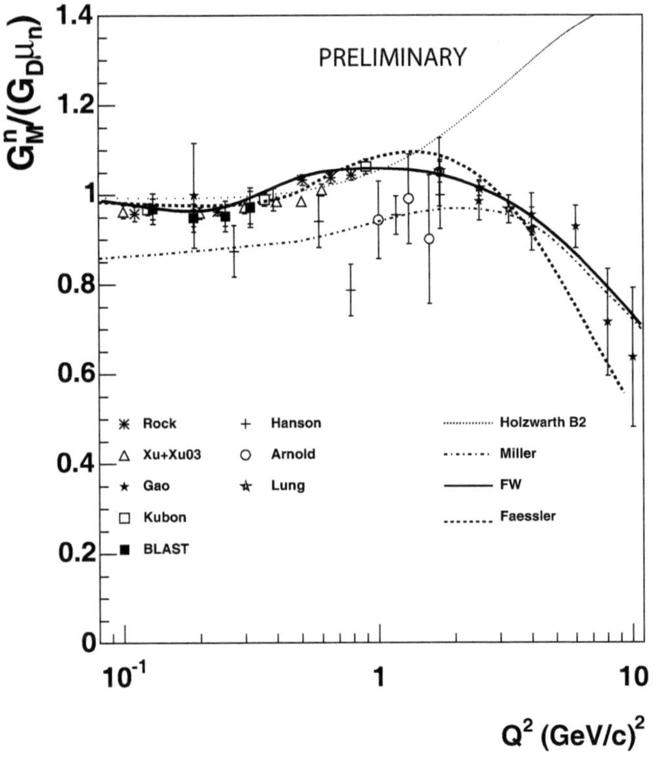

Fig. 4. Sample of the world's data on G_M^n along with the results of the BLAST experiment. The systematic and statistical errors for the BLAST data are added in quadrature. Holzwarth B1 is a Soliton model [31], Miller [32] is a cloudy bag model, Faessler [2] is a chiral perturbation theory calculation and FW is the Friedrich and Walcher parametrization [1]. The data are taken from [18–21, 25, 27–29] and the present work.

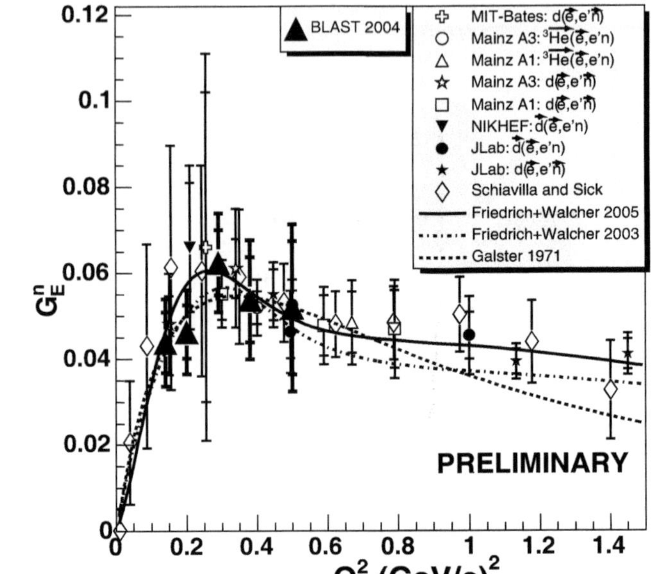

Fig. 5. Charge form factor of the neutron from polarization experiments [34] along with preliminary results from BLAST. The curves are the original parameterization by Galster *et al.* [35] and the recent parameterization by Friedrich and Walcher [1].

nature of the deuteron allows analytic solutions to its nuclear structure. Additionally, in the quasi-elastic regime, the FSI, MEC and IC are fairly small. Two data sets were collected, one with a target polarization angle of 32° in 2004 (320 kC) and one with 47° in 2005 (550 kC). The asymmetries, the ratio and the sensitivity to G_M^n vary with the target polarization angle. Therefore, the form factor was extracted separately and the results were combined, weighted by their respective error bars. The G_M^n results from this work are shown in fig. 4 along with a selection of the world's data and theoretical calculations.

The systematic uncertainties are dominated by the uncertainty in the target polarization angle which was known to 0.5°. This uncertainty amounted to 5%/degree for the 32° and 3%/degree for the 47° data sets. The second most important uncertainty was due to reconstruction and resolution which amounted to about 1.5%. Radiative effects on the asymmetries in both perpendicular and parallel kinematics were calculated using the MASCARAD software package by approximating the inclusive electrodisintegration channel as an incoherent sum of a proton and a neutron. The uncertainty due to radiative effects associated with this measurement was less than 1%.

The BLAST data agree fairly well with the recent parametrization of [1] and the calculations of [2]. Both of these show a dip in the form factor centered at $Q^2 \approx 0.2\,\mathrm{GeV}^2$ with a width of $\sim 0.2\,\mathrm{GeV}^2$.

3.3 Neutron charge form factor

The low-Q^2 region of G_E^n is an ideal testing ground for QCD- and pion-cloud inspired and other effective nucleon models. Among the four nucleon electromagnetic form factors, G_E^n is experimentally the least known one with uncertainties of typically 15–20%. Significant improvement of the experimental uncertainty is highly desirable and would set strong constraints for nucleon models. A precise knowledge of G_E^n at low Q^2 is also essential to reduce the systematic errors of parity violation experiments.

With BLAST measurements of G_E^n have been carried out by means of $(e, e'n)$ quasi-elastic scattering using polarized electrons and a vector-polarized deuterium target. The experimental double spin asymmetry is formed from the measured $(e, e'n)$ yields in each beam-target spin-state combination, properly normalized to the collected deadtime corrected beam charge. For five bins in Q^2, the experimental asymmetry as a function of missing momentum is compared with the full BLAST Monte Carlo simulation based on the deuteron electrodisintegration cross-section calculations by H. Arenhövel [33] with a consistent inclusion of reaction mechanism and deuteron structure effects. The charge form factor of the neutron is varied as an input parameter to the simulation and its measured value is extracted by a χ^2 minimization for each Q^2 bin.

Figure 5 shows the preliminary results for G_E^n from the 2004 run of BLAST along with the world data from polarization experiments [34]. Also shown is the parameterization by Galster *et al.* [35] (G) $G_E^n = 1.91\tau/(1 + 5.6\tau)G_{Dipole}$ where $\tau = Q^2/(4M_n^2)$. The excess of the data over the Galster curve at high and at low Q^2 is better accounted for by the more recent parameterization by Friedrich and Walcher [34,1] (FW), who describe all four nucleon form factors as sums of a smooth and a bump part, where the latter is attributed to the role of the pion cloud around the nucleon. The new preliminary BLAST data are quite consistent with both the bulk of existing data as well as with the parameterizations shown in fig. 5, of which the FW parameterization appears slightly favored. Note that the BLAST data are preliminary and based on only about half of the statistics that were acquired in the total run in 2004 and 2005. The preliminary results shown here comprise parts of a PhD Thesis [36] based on the BLAST data taken in 2004. This data set was used in the analysis of the new HAPPEX results presented at this workshop. Analysis of the full 2004-2005 data set is in progress [37].

4 Summary

Preliminary results for the electromagnetic form factors of the nucleon have been obtained with the BLAST experiment. The technique utilizes a combination of polarized electron beam, polarized internal gas targets, and a large-acceptance detector. These new measurements cover a range of four-momentum transfer Q^2 between 0.12 and 0.70 GeV^2. The proton charge to magnetic form factor ratio was measured by elastic *ep*-scattering of polarized electrons from polarized hydrogen. The neutron charge and magnetic form factors were extracted from simultaneous measurements of quasi-elastic $(\boldsymbol{e}, e'n)$ scattering and inclusive $(\boldsymbol{e}, e')$ scattering from vector-polarized deuterium, respectively. The systematic uncertainties are small due to the use of spin degrees of freedom.

References

1. J. Friedrich, T. Walcher, Eur. Phys. J. A **17**, 607 (2003).
2. A. Faessler, Th. Gutsche, V.E. Lyubovitskij, K. Pumsaard, Phys. Rev. D **73**, 114021 (2006).
3. D.S. Armstrong *et al.*, Phys. Rev. Lett. **95**, 092001 (2005).
4. T.W. Donnelly, A.S. Raskin, Ann. Phys. **169**, 247 (1986).
5. D. Cheever *et al.*, Nucl. Instrum. Methods A **556**, 410 (2006); L.D. van Buuren *et al.*, Nucl. Instrum. Methods A **474**, 209 (2001).
6. C.B. Crawford, PhD Thesis, Massachusetts Institute of Technology (2005).
7. M. Jones *et al.*, Phys. Rev. Lett. **84**, 1398 (2000).
8. O. Gayou *et al.*, Phys. Rev. Lett. **88**, 092301 (2002).
9. V. Punjabi *et al.*, Phys. Rev. C **71**, 055202 (2005).
10. O. Gayou *et al.*, Phys. Rev. C **64**, 038202 (2001).
11. S. Dieterich *et al.*, Phys. Lett. B **500**, 47 (2001).
12. T. Pospischil *et al.*, Eur. Phys. J. A **12**, 125 (2001).
13. B. Milbrath *et al.*, Phys. Rev. Lett. **80**, 452 (1998); **82**, 2221 (1999)(E).
14. G. Holzwarth, Z. Phys. A **356**, 339 (1996).
15. E.L. Lomon, Phys. Rev. C **66**, 045501 (2002).
16. H.-W. Hammer, Ulf-G. Meissner, Eur. Phys. J. A **20**, 469 (2004).
17. F. Cardarelli, S. Simula, Phys. Rev. C **62**, 065201 (2000); S. Simula, e-print nucl-th/0105024.
18. S. Rock *et al.*, Phys. Rev. Lett. **49**, 1139 (1982).
19. R.G. Arnold *et al.*, Phys. Rev. Lett. **61**, 806 (1988).
20. A. Lung *et al.*, Phys. Rev. Lett. **70**, 718 (1993).
21. K.M. Hanson *et al.*, Phys. Rev. D **8**, 753 (1973).
22. P. Markowitz *et al.*, Phys. Rev. C **48**, R5 (1993).
23. H. Anklin *et al.*, Phys. Lett. B **336**, 313 (1994).
24. E.E.W. Bruins *et al.*, Phys. Rev. Lett. **75**, 21 (1995).
25. G. Kubon *et al.*, Phys. Lett. B **524**, 26 (2002).
26. H. Anklin *et al.*, Phys. Lett. B **428**, 248 (1998).
27. H. Gao *et al.*, Phys. Rev. C **50**, R546 (1994).
28. W. Xu *et al.*, Phys. Rev. Lett. **85**, 2900 (2000).
29. W. Xu *et al.*, Phys. Rev. C **67**, R012201 (2003).
30. J.L. Friar *et al.*, Phys. Rev. C **42**, 2310 (1990).
31. G. Holzwarth, Z. Phys. A **356**, 339 (1996).
32. G.A. Miller, Phys. Rev. C **66**, 032201 (2002).
33. H. Arenhövel, W. Leidemann, E.L. Tomusiak, Eur. Phys. J. A **23**, 147 (2005).
34. D.I. Glazier *et al.*, Eur. Phys. J. A **24**, 101 (2005) and references therein.
35. S. Galster *et al.*, Nucl. Phys. B **32**, 221 (1971).
36. V. Ziskin, PhD Thesis, Massachusetts Institute of Technology (2005).
37. E. Geis, PhD Thesis, Arizona State University (2007).

Determination of the gluon polarization at RHIC and COMPASS

S. Procureur[a]

DSM/DAPNIA/SPhN, CEA-Saclay, Orme des merisiers, 91191 Gif-sur-Yvette, France

Abstract. The most recent determinations of the gluon polarization in the nucleon, $\frac{\Delta G}{G}$, obtained at RHIC and COMPASS experiments, are reviewed. The former accesses the gluon polarization mainly through the production of neutral pions (PHENIX) or jets (STAR) in polarized proton collisions. The latter uses the photon-gluon fusion in polarized lepton-nucleon scattering, tagged either by open charm or high-p_T hadrons production. All the results are in good agreement, and favour values of ΔG roughly between 0 and 0.5 at a few $(\mathrm{GeV}/c)^2$, thus in contradiction with what could be derived from the Ellis-Jaffe sum rule and the axial anomaly. Much stronger constraints will be obtained in a near future by both programs, helping us to clarify the role of gluons in the nucleon spin.

PACS. 13.60.Hb Total and inclusive cross sections (including deep-inelastic processes) – 13.88.+e Polarization in interactions and scattering – 14.20.Dh Protons and neutrons – 21.10.Hw Spin, parity, and isobaric spin

1 Introduction

For the last 20 years, the decomposition of the nucleon spin in terms of the contributions of its constituents has been a central topic in polarized deep-inelastic scattering (DIS), and more recently, in polarized proton-proton collision. This decomposition can be written as

$$\frac{1}{2} = \frac{1}{2}\Delta\Sigma + \Delta G + L_q + L_g, \qquad (1)$$

where $\frac{1}{2}\Delta\Sigma$ is the contribution from the spin of quarks, ΔG the contribution from the spin of the gluons, and $L_q + L_g$ is the orbital angular momentum of quarks and gluons.

The first exciting result was obtained by the EMC [1] experiment at CERN, which showed that the spin of the quarks only contributes to a small fraction of the proton spin. Other experiments confirmed this result, establishing $a_0 \sim \Delta\Sigma$ around 25%, *i.e.* in contradiction with the 60% expected by the Ellis-Jaffe sum rule within the quark-parton model. However, due to axial anomaly, $a_0 = \Delta\Sigma - \frac{3\alpha_s}{2\pi}\Delta G$ in some schemes, so $\Delta\Sigma$ can still be large if the contribution from the spin of gluons is large enough. Combining the experimental measurement of a_0 with the predicted value of $\Delta\Sigma$, $\Delta G \equiv \int \Delta G(x_G)\mathrm{d}x_G$, where x_G is the nucleon momentum fraction carried by the gluon, is expected to be around 1.5 (respectively, 2.5) at $Q^2 = 1$ (respectively, 3) $(\mathrm{GeV}/c)^2$. Up to recently, it could only be determined in inclusive DIS, using the Q^2-dependence of the polarized structure function g_1. Unfortunately, the precision of such a determination is strongly limited by the small Q^2 range of existing data, as can be seen from the most recent analysis of these data [2], $\Delta G = 0.47 \pm 1.08$. Several experiments therefore concentrated on a direct determination of the gluon polarization $\frac{\Delta G}{G}$: STAR and PHENIX at the RHIC p-p collider, and the fixed-target experiments HERMES and COMPASS. We present in this review the most recent results obtained at RHIC and COMPASS.

2 $\frac{\Delta G}{G}$ at RHIC

In 2001 the Relativistic Heavy Ion Collider of Brookhaven has started a program dedicated to the exploration of the nucleon spin, and first longitudinally polarized protons were collided in 2003. To do this, a polarized H^- source is used to create bunches of around 1×10^{12} highly polarized protons, that are then sent by the Alternating Gradient Synchrotron (AGS) to the RHIC rings. In these rings, protons are accelerated from 22 GeV to 100 GeV (250 GeV in coming years), to be collided at one of the 6 intersection points. To keep the high polarization of protons in spite of the many encountered spin resonances, a string of dipole magnets (called a *Siberian snake*) rotates the proton spin by 180° in the horizontal plane at each pass.

At a center-of-mass energy of $\sqrt{s} = 200\,\mathrm{GeV}$, proton collisions can be described in terms of parton (quark and gluon) interactions, whose cross-section can be factorized

[a] e-mail: sprocure@mail.cern.ch

Table 1. Integrated luminosity L, beam polarization P and figure of merit for the three longitudinal runs already taken at RHIC, as of the date of this presentation.

Run number	L	P	Figure of merit (P^4L)
2003	$350\,\mathrm{nb}^{-1}$	27%	$1.86\,\mathrm{nb}^{-1}$
2004	$120\,\mathrm{nb}^{-1}$	40%	$3.1\,\mathrm{nb}^{-1}$
2005	$3.8\,\mathrm{pb}^{-1}$	47%	$185\,\mathrm{nb}^{-1}$

as a convolution of parton densities and fragmentation functions with a corresponding partonic subprocess cross-section. There are essentially three possible subprocesses, qq', qg and gg, the last two being sensitive to the gluon distribution in the nucleon. More precisely, one can show that the cross-section helicity asymmetry, defined as

$$A_{||} = \frac{\sigma_{++} - \sigma_{+-}}{\sigma_{++} + \sigma_{+-}}, \quad (2)$$

exhibits a quadratic dependence on the gluon polarization (due to gg interactions), as well as a linear one (due to qg interactions). Here σ_{++} (respectively, σ_{+-}) is the cross-section of proton collisions with parallel (respectively, anti-parallel) spins. Experimentally, it is calculated using number of events in each beam spin configuration, N_{++} and N_{+-}:

$$A_{||} = \frac{1}{P_1 P_2} \frac{N_{++} - RN_{+-}}{N_{++} + RN_{+-}}. \quad (3)$$

The beam bunches (up to 110 bunches in each RHIC ring) are loaded with a selected pattern of spin directions to collect data for all helicity combinations, for each fill. R is the ratio of relative luminosities for the two beam helicity configurations; P_1 and P_2 are the polarizations of each beam, measured by independent polarimeters. Luminosities and beam polarizations of previous runs are summarized in table 1.

Different production channels can be used to select qg and gg interactions, like high-p_T photon, photon plus jet, heavy flavor, high-p_T neutral pions or inclusive jet production. As we will see, precise results were recently obtained with the latter two, by the main experiments of the RHIC spin program, PHENIX and STAR.

2.1 PHENIX experiment

PHENIX is a multipurpose experiment used both for heavy ions and spin physics programs. The setup includes two central arms at mid-rapidity ($|\eta| < 0.35$), two muon arms at large rapidities ($1.2 < |\eta| < 2.4$), and additional global detectors, as Beam-Beam Counters (BBC) and Zero-Degree Calorimeters (ZDC), essentially used for local polarimetry and for the determination of the relative luminosities of bunches. The precision of such determination is $\Delta R = 1.0 \times 10^{-4}$, and is essential for asymmetry calculation (see eq. (3)).

Neutral pions with high transverse momentum are detected in the central arms using electromagnetic calorimeters and a high-p_T photon trigger. The π^0 cross-section is

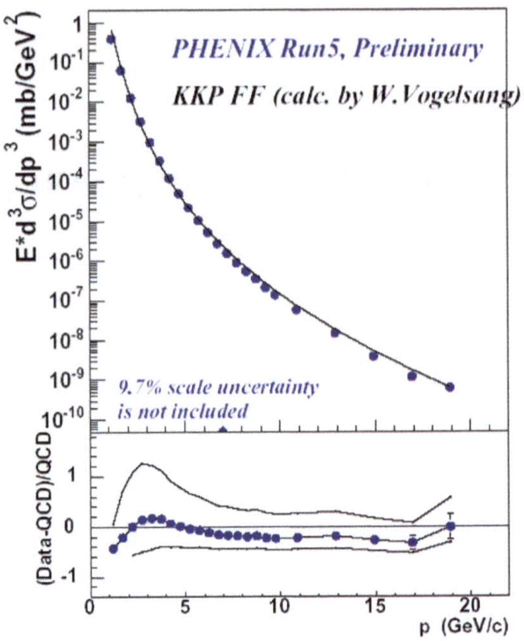

Fig. 1. Comparison of π^0 production cross-section measured by PHENIX and NLO calculations, using the KKP model for fragmentation functions; a very good agreement is obtained over 9 orders of magnitude.

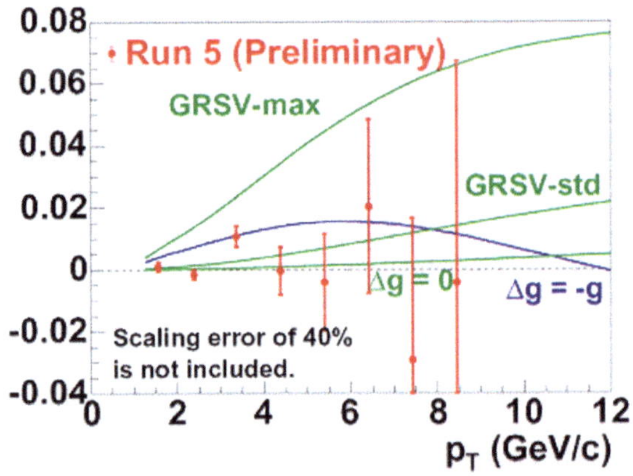

Fig. 2. Double spin asymmetries for π^0 production at PHENIX, as a function of transverse momentum. Theoretical curves correspond to NLO calculations with different input values for ΔG. Data are compatible with curves with a confidence level of around 20% for both GRSV-std and $\Delta G = 0$ scenarios, whereas GRSV-max ($\Delta G = g$) and $\Delta G = -g$ scenarios are incompatible with the data.

found to be in very good agreement with NLO pQCD calculations [3], as shown in fig. 1, allowing us to be confident in the comparison of results with theoretical expectations for ΔG. Double spin asymmetry is calculated using eq. (3) for a di-photon invariant-mass range of $\pm 25\,\mathrm{MeV}/c$ around the π^0 peak, and background asymmetry is then subtracted using mass regions on both sides of the peak. Figure 2 shows the resulting asymmetry as a function of the π^0 transverse momentum using the 2005 run ($L = 2.7\,\mathrm{pb}^{-1}$) [3]; comparison with theo-

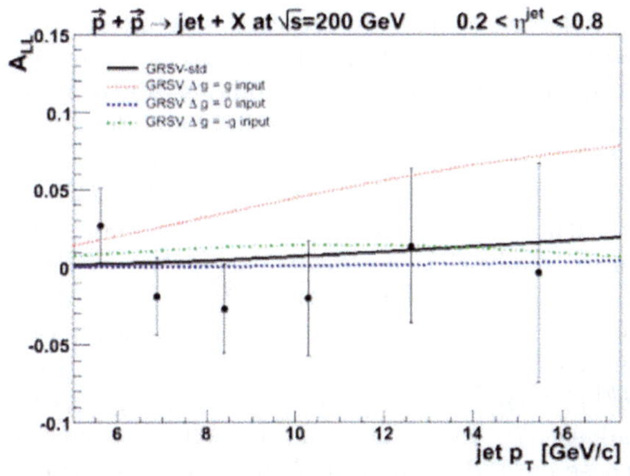

Fig. 3. Double spin asymmetries for jet production at STAR, as a function of jet transverse momentum, using 2003 and 2004 data. Comparison with theoretical curves disfavours large values of ΔG.

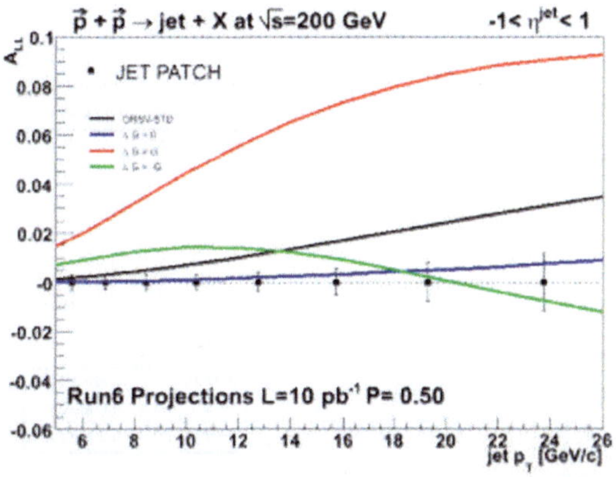

Fig. 4. Prediction for double spin asymmetry error bars in jet production at STAR, after 2006 run. The four theoretical curves will be distinguished.

retical curves using different values for ΔG allows to exclude large gluon polarizations ($\Delta G = \pm G$ at input scale $Q^2 = 0.4\,(\mathrm{GeV}/c)^2$), in favour of standard or null polarizations. The precision of these results should increase dramatically in the coming years, in particular with use of the 2006 run, for which a beam polarization around 60% is expected, and also with the second part of the spin program, with collisions at center-of-mass energy of 500 GeV.

2.2 STAR experiment

STAR is designed to be a large-acceptance detector, with a Time Projection Chamber ($|\eta| < 1.4$) and an Electromagnetic Calorimeter ($-1.0 < |\eta| < 2.0$), optimized for mid-rapidity jet reconstruction. Jets are reconstructed using a midpoint cone algorithm [4], with a cone size of 0.4 and a minimum transverse momentum of $5\,\mathrm{GeV}/c$. Additional cuts are then applied on vertex to ensure a uniform efficiency of TPC, on the jet axis to limit edge effects near the border of the calorimeter, and on the contribution of TPC to the reconstructed jet energy, to reduce contamination from beam background. A total of around 300k jets are thus selected, using 2003 and 2004 runs, during which STAR collected an integrated luminosity of $0.4\,\mathrm{pb}^{-1}$.

As for PHENIX, the inclusive jet cross-section was compared to NLO pQCD calculations, and a reasonnable agreement was found over 7 orders of magnitude. Double spin asymmetries, extracted according to eq. (3), are shown in fig. 3 as a function of the jet transverse momentum [5]. They are compared with theoretical curves obtained using the same inputs as in sect. 2.1, but in the jet production case. These results already disfavour large gluon polarization, but are not yet able to distinguish between the three other scenarios.

Much more precise results should be soon available, using the $3\,\mathrm{pb}^{-1}$ integrated luminosity recorded by STAR during the 2005 run. Figure 4 shows expected precision for asymmetry after the 2006 run, with an average polarization of 50%: the three remaining scenarios should be distinguished, thus providing a useful complement to PHENIX's results. When statistics will be high enough, STAR also expects to compute asymmetry in the production of photon plus jet, that will allow to map the x_G-dependence of the gluon polarization.

3 $\frac{\Delta G}{G}$ at COMPASS

COMPASS is a polarized DIS experiment installed at the M2 beam line of the CERN SPS which delivers a 160 GeV polarized muon beam with an intensity of $2 \cdot 10^8$ muons per SPS cycle (*i.e.* every 16.2 s). The polarized ^{6}LiD (deuteron) target is split into an upstream and a downstream cell with opposite polarizations. The position and the momentum of the incoming muons are measured via a set of trackers and a beam momentum station located upstream of the target. The particles produced by the μN interactions are detected downstream of the target in a double-stage spectrometer with high rate capability, containing around 200 tracker planes.

In polarized semi-inclusive DIS, a direct measurement of the polarization $\Delta G/G(x_G)$ of the gluons can be obtained from the cross-section helicity asymmetry $A_{||}$ of an event sample enriched with photon-gluon fusion (PGF) processes, $\gamma^* g \to q\bar{q}$. Indeed, assuming factorization, and at LO:

$$A_{||} = R_{PGF}\hat{a}_{LL}^{PGF}\frac{\Delta G}{G} + A_{backgd}\,, \qquad (4)$$

where $\hat{a}_{LL}^{PGF}$ is the helicity asymmetry of the hard lepton-gluon scattering, R_{PGF} the fraction of PGF events in the selected sample, and A_{backgd} the asymmetry of background processes.

Two different methods are used at COMPASS to tag the PGF events: one consists in selecting charmed particles, providing a pure but small sample of PGF events; the other selects events containing 2 hadrons with high transverse momentum p_T with respect to the virtual photon direction. We now present results obtained by these 2 methods, using 2002 to 2004 data.

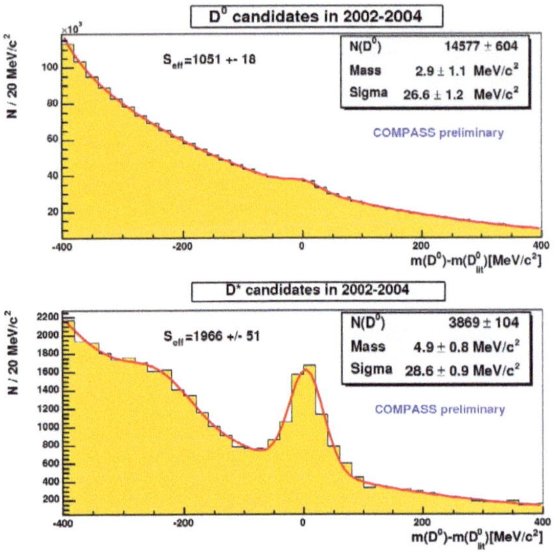

Fig. 5. Invariant-mass distribution for D^0 (top) and D^* (bottom) candidates using 2002 to 2004 data. $S_{eff} = S^2/(S+B)$, where S (respectively, B) is the number of signal (respectively, background) events.

3.1 Open charm channel

Due to its large mass ($\sim 1.5\,\mathrm{GeV}/c^2$), it is very unlikely to find a c-quark in the nucleon, or to create it during the fragmentation. Moreover, at tree level, c-quarks can only be produced in the photon-gluon fusion process. That is why the detection of a charmed hadron in the final state is a very clear signature of PGF events. Also it was checked on the data that the remaining (combinatorial) background does not contribute to the asymmetry, $A_{backgd} \approx 0$. The selection is done by reconstructing D^0 or D^* from the decays: $D^0 \to K\pi$ or $D^* \to D^0\pi_s \to K\pi\pi_s$, combinatorial background being largely suppressed in the second case (see fig. 5) due to the small available phase space for the soft pion. A crucial point of this analysis is the identification of kaons and pions in a RICH detector, using a likelihood method.

A preliminary analysis of 2002 and 2004 data gave

$$\frac{\Delta G}{G}(x_G \sim 0.15, \mu^2 \approx 13\,(\mathrm{GeV}/c)^2) = -0.57 \pm 0.041(stat), \quad (5)$$

where the analysing power of PGF events was determined with the AROMA Monte Carlo generator. Several tests were also performed to check that the systematic errors are negligible with respect to statistics.

3.2 High-p_T channel

In this case, the statistics is much larger, but a non-negligible fraction of background processes is still present in the selected sample, $A_{backgd} \neq 0$, and the fraction of PGF events is unknown. These quantities can only be estimated using a Monte Carlo simulation, thus introducing a model dependence in our determination of $\Delta G/G$.

The selected events are required to contain at least 2 charged hadrons associated to the primary vertex, in addition to the incident and scattered muons. Since the generator we use, PYTHIA, provides a reliable model for interactions of virtual photons with nucleons in the low-virtuality region [6], we select events with $Q^2 < 1\,(\mathrm{GeV}/c)^2$. For events with $Q^2 > 1\,(\mathrm{GeV}/c)^2$ (10% of the statistics), a separate analysis is performed using the LEPTO generator. To enhance the fraction of PGF events in our sample, we require the transverse momentum of the 2 hadrons to be large: $p_T^{h1} > 0.7\,\mathrm{GeV}/c$, $p_T^{h2} > 0.7\,\mathrm{GeV}/c$ and $(p_T^{h1})^2 + (p_T^{h2})^2 > 2.5\,(\mathrm{GeV}/c)^2$, as in the SMC high-$p_T$ analysis [7]. In total, around 500,000 events remain after these cuts, defining our high-p_T sample, from which we extract the following asymmetry (defined in [8]):

$$\left\langle \frac{A_{\parallel}}{D} \right\rangle = +0.004 \pm 0.013(stat) \pm 0.003(syst), \quad (6)$$

where D is the depolarization factor. The systematic error accounts for the false asymmetries related to the apparatus. Other sources of systematics errors, including the error on the beam and target polarizations, are proportional to the (small) measured asymmetry, and have been therefore neglected.

As already stated, our determination of the gluon polarization involves a Monte Carlo simulation. The events generated by PYTHIA are propagated through a GEANT model of our apparatus, and reconstructed with the same program as for real data. The cuts described above are then applied, and the selected events define our Monte Carlo high-p_T sample.

In our kinematics, PYTHIA generates 2 kinds of processes. In the *direct processes*, the virtual photon directly takes part in the hard partonic interaction, whereas in the *resolved processes*, it first fluctuates into a hadronic state, from which a parton is extracted and interacts with a parton from the nucleon. These resolved processes account for nearly 60% of our Monte Carlo high-p_T sample.

To reach a good agreement between our Monte Carlo and the data, we varied many parameters of the generator, and we finally found that only one should be changed, *i.e.* the width of the intrinsic transverse momentum distribution of partons in the resolved virtual photon was decreased from $1\,\mathrm{GeV}/c$ to $0.5\,\mathrm{GeV}/c$. Figure 6 presents a comparison between the Monte Carlo and the real data high-p_T samples.

Taking into account all the generated subprocesses, the high-p_T asymmetry can be approximately expressed as

$$\left\langle \tfrac{A_{\parallel}}{D} \right\rangle = R_{PGF} \left\langle \hat{a}_{LL}^{PGF}/D \right\rangle \tfrac{\Delta G}{G} + R_{QCDC} \left\langle \hat{a}_{LL}^{QCDC}/D \right\rangle A_1 + \textstyle\sum_{f,f'=u,d,s,\bar{u},\bar{d},\bar{s},g} R_{ff'} \left\langle \hat{a}_{LL}^{ff'} \left(\tfrac{\Delta f}{f}\right)^d \left(\tfrac{\Delta f'}{f'}\right)^\gamma \right\rangle. \quad (7)$$

The contribution of the PGF events to the high-p_T asymmetry was found to be $-0.292 \times \Delta G/G$ and the contribution of QCD Compton 0.0063 [9], using a fit on the world data for the virtual photon deuteron asymmetry A_1. For resolved photon processes, some of them involve

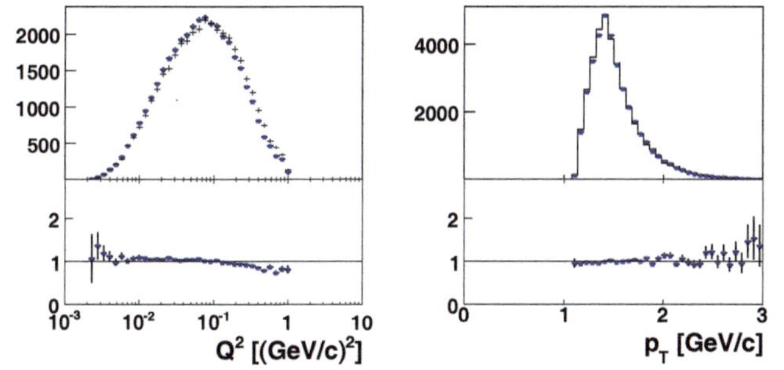

Fig. 6. The upper parts of these plots show a comparison between the simulated (histograms) and real data (points) samples of high-p_T events, normalized to the number of events. The lower part shows the corresponding data/simulation ratio. p_T is the total transverse momentum of the leading hadron. A similar agreement is obtained for the next-to-leading hadron.

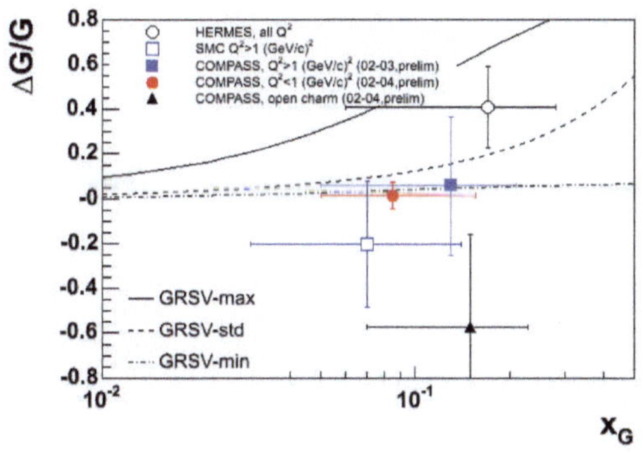

Fig. 7. Comparison between previous determinations of $\Delta G/G(x_G)$ and recent COMPASS results. Curves correspond to parametrizations using different input values for ΔG, as in sect. 2. These parametrizations were evolved at a scale of $3\,(\mathrm{GeV}/c)^2$, *i.e.* the scale of COMPASS high-p_T points. COMPASS results completely exclude large gluon polarization.

a gluon in the nucleon, and are thus part of the signal. Their analysing powers $\hat{a}_{LL}^{ff'}$ can be calculated in perturbative QCD [10], and the polarizations $(\Delta f/f)^d$ of the u, d and s quarks in the deuteron were estimated using the GRV98 [11] and GRSV2000 [12] parametrizations. On the other hand, the polarized PDFs of the virtual photon are presently unknown. However, theoretical considerations [13] provide a minimum and a maximum value for each $(\Delta f')^\gamma$. We were thus able to estimate the contribution of all the resolved photon processes to be in some ($\Delta G/G$ dependent) range.

Finally, the systematic errors related to the use of PYTHIA were estimated by varying the relevant parameters of this generator in a range where the agreement between our simulation and the data remains reasonable. It appeared that the result for $\Delta G/G$ depends essentially on the width of the distribution of the intrinsic transverse momentum for partons in the resolved photon: the variation of this parameter results indeed in a variation of 30% for R_{PGF}.

Gathering all the pieces, we obtained

$$\frac{\Delta G}{G}(x_G = 0.085, \mu^2 \approx 3\,(\mathrm{GeV}/c)^2) = 0.016 \pm 0.058(stat) \pm 0.055(syst). \quad (8)$$

The nucleon momentum fraction x_G carried by the gluon, as well as the hard scale μ^2 on which $\Delta G/G$ depends weakly, were estimated using the simulation.

Values between 0 and 0.6 for ΔG are clearly favoured, as it can be seen in fig. 7, unless $\Delta G(x_G)$ has a node around $x_G \sim 0.1$. Analysis of 2004 data at $Q^2 > 1\,(\mathrm{GeV}/c)^2$ is in progress, and will soon provide an independent constraint on ΔG. The use of 2006 data, for which a new magnet with larger acceptance will be used, will allow us to probe region of higher momentum fraction x_G, in which the different scenarios can be better distinguished. Moreover, the use of a neural network for event selection and an extraction of $\frac{\Delta G}{G}(x_G)$ for different x_G should allow us to better constrain ΔG in a near future.

4 Conclusion

Almost 20 years after the EMC result and the spin crisis, the role played by the gluons in the nucleon spin starts to be clarified. COMPASS and RHIC have recently provided the first really significant constraints on ΔG, both indicating that the contribution of gluons is less than what was expected. However, these results do not (yet!) announce a new spin crisis, as they are still compatible with a ΔG around 0.4, *i.e.* the totality of the currently missing spin of the nucleon. On the other hand, ΔG could also be zero, in which case orbital momenta would play an important role. In a couple of years, COMPASS and RHIC will both be able to distinguish between these two extreme scenarios.

References

1. European Muon Collaboration (J. Ashman *et al.*), Phys. Lett. B **206**, 364 (1988).
2. Asymmetry Analysis Collaboration (M. Hirai, S. Kumano, N. Saito), hep-ph/0603213.
3. K. Boyle, *Proceedings of PANIC05, Santa Fe, 2005*, nucl-ex/0606008.
4. G. Blazey *et al.*, *Proceedings of the Run II QCD and Weak Boson Physics Workshop*, hep-ex/0005012.
5. J. Kiryluk, *Proceedings of DIS2006, Tsukuba, 2006*, hep-ex/0608022.
6. C. Friberg, T. Sjostrand, JHEP **09**, 010 (2000).
7. Spin Muon Collaboration (B. Adeva *et al.*), Phys. Rev. D **70**, 012002 (2004).
8. COMPASS Collaboration (E.S. Ageev *et al.*), Phys. Lett. B **612**, 154 (2005).
9. COMPASS Collaboration (E.S. Ageev *et al.*), Phys. Lett. B **633**, 25 (2006).
10. C. Bourrely, J. Soffer, F.M. Renard, P. Taxil, Phys. Rep. **177**, 319 (1989).
11. M. Gluck, E. Reya, A. Vogt, Eur. Phys. J. C **5**, 461 (1998).
12. M. Gluck, E. Reya, M. Stratmann, W. Vogelsang, Phys. Rev. D **63**, 094005 (2001).
13. M. Gluck, E. Reya, C. Sieg, Eur. Phys. J. C **20**, 271 (2001).

A Monte-Carlo method to analyze the electromagnetic form factors of the nucleon

M.A. El-Yakoubi[1], J. Arvieux[1,a], R. Frascaria[1], R. Kunne[1], F. Maas[1], M. Morlet[1], S. Ong[1], J. Van de Wiele[1,b], and S. Baunack[2]

[1] Institut de Physique Nucléaire, CNRS/IN2P3 and Université Paris-Sud, 91406 Orsay-Cedex, France
[2] Institut für Kernphysik, Universität Mainz, D-55099 Mainz, Germany

Abstract. Parity violating elastic electron-nucleon scattering allows to determine the vector stangeness content of the nucleon. The final uncertainty on the strange form factors is limited, among other parameters, by the uncertainty on the electromagnetic form factors. These are usually fitted with a functional form constrained by boundary conditions at $Q^2=0$ and at large Q^2. These conditions induce huge correlations between parameters which are not taken into account to full extent by purely statistical methods. We describe here a Monte-Carlo method which accounts for correlations between parameters to all orders. We also propose a method for taking into account some systematical errors induced by the dispersion of the data.

PACS. 25.30.Bf Elastic electron scattering – 21.10.Ft Charge distribution

1 Introduction

In order to extract the strange form factors from a parity violating asymmetry, one needs to know the electromagnetic nucleon form factors G_E^p , G_M^p , G_E^n , G_M^n and their uncertainties. The Q^2-dependence of the proton and neutron form-factors is known experimentally up to a few $(GeV/c)^2$. The precision of the neutron data is still limited and the uncertainties are mainly experimental. The precision of the proton data is much higher (as shown below) and the overall accuracy depends on the statistical error and on theoretical considerations.

Electromagnetic form factors can be measured separately using a Rosenbluth method based on two experiments done at the same Q^2but in different kinematical conditions (e.g. forward/backward angle separation). This method yields large error bars at higher $Q^2(Q^2 > 1(GeV/c)^2)$, especially for the electric form-factors as the weight of the magnetic term is proportional to Q^2. The Rosenbluth method indicates that $\mu_p G_E/G_M \simeq 1$ over the range 0-6 $(GeV/c)^2$.

Another method measures the polarization of the recoil proton in ep scattering with incident polarized electrons [1]. This methods yields directly the ratio $\mu_p G_E/G_M$ with a higher precision than the Rosenbluth method. The problem is that the polarization data are in disagreement with the Rosenbluth method above $1(GeV/c)^2$, the difference increasing rapidly by more than a factor of 2 at $Q^2=5$ $(GeV/c)^2$.

For $Q^2 \leq 1$ $(GeV/c)^2$, there are essentially four parametri- zations which reproduce the experimental results: Galster [2], Friedrich-Walcher [3], Arrington [4] and Kelly [5]. Three of them [2,3,5] aim at fitting the proton and neutron form-factors. Arrington's parametrization [4] is only related to the proton data and it aims at resolving the ambiguities of the Rosenbluth/polarization methods.

The representation of the form factors is usually obtained through a χ^2 analysis of the experimental data. There are several problems arising from this procedure: 1) how to take quantitatively into account the large correlations induced when theoretical constraints are taken into account, 2) what should we do with data which are not compatible with each other and 3) how to quantify the corresponding uncertainties. The usual procedure is to suppress the data points which are away from the mainstream. This method of selecting the "good" data has been used in every recent analysis [3–7]. Its difficulty is that one has to introduce a qualitative systematic uncertainty which has no quantitative justification.

Concerning the problem of theoretical guidelines, one can use theoretical constraints such as the values at Q^2=0 or the slope at Q^2=0 which is related to the mean square radius $< r^2 >$, as shown in table 1 in which $G' = \frac{dG_E(Q^2)}{dQ^2}|_{Q^2=0}$.

Another type of theoretical constraint is the use of a theoretical model as done in [3]. The problem here is that

[a] e-mail: arvieux@ipno.in2p3.fr
[b] e-mail: vandewi@ipno.in2p3.fr

Table 1. Constraints on nucleon form-factors at $Q^2 = 0$.

G_M^p	G_E^p	G_M^n	G_E^n
$\frac{G_M^p}{\mu_p} = 1$	1 $< r_p^2 >= -6G'$	$\frac{G_M^n}{\mu_n} = 1$	0 $< r_n^2 >= -6G'$

this induces huge correlations between parameters (up to 500%) so that a numerical analysis, based on estimates of diagonal elements of the error matrix, is incorrect.

The question of the correlation between parameters can be solved by using a Monte Carlo method instead of a purely statistical analysis, but it does not address the question of data selection. We will present in this section three different analyses based on a pure statistical analysis (as in [6]), a pure Monte-Carlo method and a method based on a Monte-Carlo analysis but taking into account, in a quantitative way, the dispersion of the experimental data.

2 Pure statistical analysis

The data are fitted with a functional form:

$$G(Q^2) = \frac{1 + a_1^{(0)}\tau}{1 + b_1^{(0)}\tau + b_2^{(0)}\tau^2 + b_3^{(0)}\tau^3} \qquad (1)$$

where $\tau = Q^2/4{M_p}^2$.

This form ensures that $G(Q^2) \to 1$ at $Q^2 = 0$ (after normalization to the proper charge or magnetic moment). It also ensures that $G(Q^2) \to 1/Q^4$ for large Q (quark counting rule). The values of the coefficients are obtained by minimizing the χ^2 function:

$$\chi^2 \equiv \chi^2_{(0)} = \sum_{i=1}^{N} \frac{[y_i^{(0)} - G(a_1^{(0)}, b_1^{(0)}, b_2^{(0)}, b_3^{(0)}; \tau_i^2)]^2}{\Delta y_i^2} \qquad (2)$$

where $y_i^{(0)}$, $(1 \le i \le N)$ is the experimental form factor.

As an example the values of the $a_1^{(0)}$, $b_1^{(0)}$, $b_2^{(0)}$ and $b_3^{(0)}$ parameters are given in table 2 for G_M^p where they are compared (second line) to the values of Kelly [5]. Below 1 $(GeV/c)^2$the difference between the 2 sets of parameters is less than 0.3 % which was expected as we use the same fitting procedure and same data selection.

For the neutron electric form-factor we use Galster parametrization:

$$G_E^n(Q^2) = \frac{A\tau}{1 + B\tau} G_D(Q^2), \qquad (3)$$

where $G_D = (1 + Q^2/\Lambda^2)^{-2}$ is a dipole form factor and $\Lambda^2 = 0.71(GeV/c)^2$. The fit is strongly constrained at low Q^2 by the mean square radius $< r_n^2 >$.

Table 2. Parameters from the fits eq.(1) for G_M^p . 1^a: our analysis, 1^b: Kelly [5].

Eq.	χ^2/N	$a_1^{(0)}$	$b_1^{(0)}$	$b_2^{(0)}$	$b_3^{(0)}$
1^a	1.055	0.11 ± 0.042	10.97 ± 0.114	18.89 ± 0.263	6.33 ±1.299
1^b	1.060	0.12 ± 0.04	10.97 ± 0.11	18.86 ± 0.28	6.55 ± 1.2

Table 3. Parameters of the fit eq.(3) for the neutron electric form-factor. 3^a: this work, (*): $< r_n^2 >$ not included in the search, 3^b: Kelly [5].

Eq.	χ^2/N	$A^{(0)}$	$B^{(0)}$	$< r_n^2 > (fm^2)$
3^{a*}	0.52	1.518 ± 0.154	2.265 ± 0.709	-0.100 ± 0.010
3^a	1.09	1.770 ± 0.016	3.423 ± 0.285	-0.117 ± 0.001
3^b	0.80	1.700 ± 0.040	3.300 ± 0.320	-0.112 ± 0.003

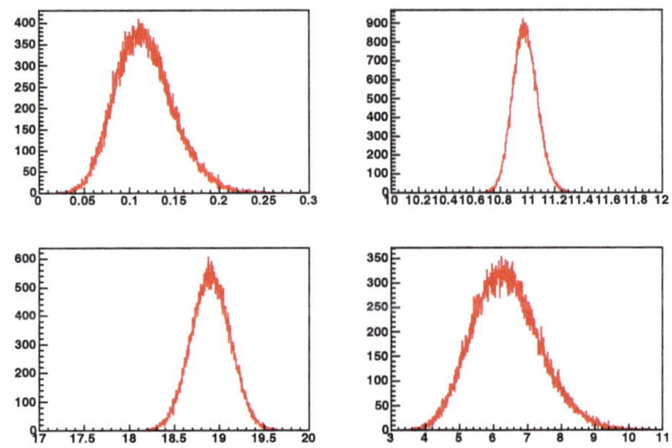

Fig. 1. Distribution of the parameters a_1, b_1, b_2 and b_3 of Eq. 1 for the Monte-Carlo fit of the proton magnetic form factor. This distribution does not follow a Gaussian law

The difference with [5] is larger for G_E^n . It amounts to about 4% at low Q^2, the main reason being a difference of about 5% in $< r_n^2 >$ as shown in table 3. The results for all four form factors resulting from a fit of the world data up to 3 $(GeV/c)^2$are given in table 4, column 2. The statiscal uncertainties are calculated using a standard error propagation method based on the uncertainties of the fit parameters and without taking into account correlations. They are given in table 4, column 3. As the uncertainties on some parameters are large, the correlated non-diagonal terms may be larger than the diagonal term with a negative sign, as it happens e.g. for G_E^p and G_M^n and the calculation of the error on these form factors with correlations is impossible for some values of Q^2. When the calculation is possible, the value is given in brackets.

3 Monte-Carlo analysis

We use the same functional form used in the statistical analysis (and in [5]) but instead of performing only one fit, we calculate the parameter distributions by minimizing

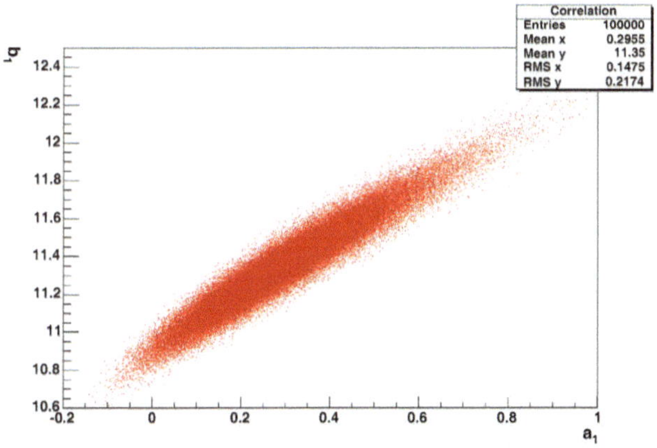

Fig. 2. Correlation between the parameters a_1 and b_1 for G^p_M

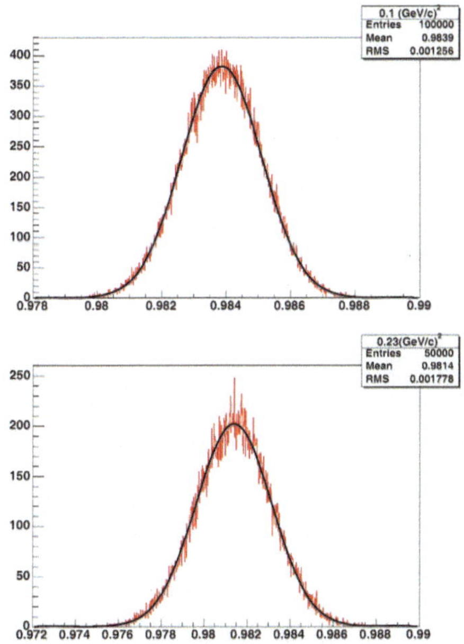

Fig. 3. Distributions for the proton magnetic form factor G^p_M at Q^2= 0.1 (Up) and Q^2= 0.23 (Down) using the parametrisation of Eq. 1

the $\mathcal{N}$ $(1 \leq \nu \leq \mathcal{N})$ $\chi^2_{(\nu)}$ functions:

$$\chi^2_{(\nu)} = \sum_{i=1}^{N} \frac{[y_i^{(\nu)} - G(a_1^{(\nu)}, b_1^{(\nu)}, b_2^{(\nu)}, b_3^{(\nu)}; \tau_i^2)]^2}{\Delta y_i^2} \quad (4)$$

where ν is the drawing index and $y_i^{(\nu)}$ $(1 \leq i \leq N)$ is the experimental form factor obtained in a gaussian distribution with a central value equal to $y_i^{(0)}$ and a variance $\sigma_i = \Delta y_i/1.177$. We do $\mathcal{N}$ minimizations (with $\mathcal{N}$ approximately equal to 50000), and obtain $\mathcal{N}$ solutions: $\{a_1^{(1)}, a_2^{(1)}, \cdots a_n^{(1)}\}$, $\{a_1^{(2)}, a_2^{(2)}, \cdots a_n^{(2)}\}$, $\cdots$, $\{a_1^{(\mathcal{N})}, a_2^{(\mathcal{N})}, \cdots a_n^{(\mathcal{N})}\}$.

As an example, the distributions of the parameters a_1, b_1, b_2 and b_3 for the proton magnetic form factor are displayed in fig.1. This method allows to estimate the correlation functions, as displayed e.g. in fig.2 for the correlation between the a1 and b1 parameters.

Table 4. Comparison of the 3 analysis methods: purely statistical (stat), Monte-Carlo (MC) and "envelop" (env). All values are normalized to G_D. Relative uncertainties are in %.

Q^2	$\frac{G^p_M}{\mu_p}$	$\frac{\Delta G^p_M}{G^p_M}$ (stat)	$\frac{\Delta G^p_M}{G^p_M}$ (MC)	$\frac{\Delta G^p_M}{G^p_M}$ (env)
0.10	0.984	0.32 [0.18]	0.18	0.60
0.23	0.981	0.59 [0.25]	0.26	0.92
0.45	0.994	0.92 [0.23]	0.26	0.99
0.63	1.008	1.16 [0.20]	0.25	0.92

Q^2	$\frac{G^n_M}{\mu_n}$	$\frac{\Delta G^n_M}{G^n_M}$ (stat)	$\frac{\Delta G^n_M}{G^n_M}$ (MC)	$\frac{\Delta G^n_M}{G^n_M}$ (env)
0.10	0.965	8.7 [-]	0.64	0.88
0.23	0.966	17.2 [-]	0.65	1.11
0.45	1.001	29.6 [-]	0.64	1.07
0.63	1.029	40.3 [-]	0.71	1.02

Q^2	G^p_E	$\frac{\Delta G^p_E}{G^p_E}$ (stat)	$\frac{\Delta G^p_E}{G^p_E}$ (MC)	$\frac{\Delta G^p_E}{G^p_E}$ (env)
0.10	0.977	0.88 [0.13]	0.21	0.50
0.23	0.971	1.83 [-]	0.39	0.93
0.45	0.975	3.48 [-]	0.73	1.76
0.63	0.976	5.04 [-]	0.93	2.52

Q^2	G^n_E	$\frac{\Delta G^n_E}{G^n_E}$ (stat)	$\frac{\Delta G^n_E}{G^n_E}$ (MC)	$\frac{\Delta G^n_E}{G^n_E}$ (env)
0.10	0.046	2.86 [1.26]	1.26	4.27
0.23	0.094	1.91 [1.91]	1.94	5.66
0.45	0.157	2.46 [2.97]	3.07	7.72
0.63	0.200	4.12 [3.68]	3.82	9.12

From the parameter distribution, one can deduce a distribution for the form factors themselves as shown in fig.3 for G^p_M at Q^2= 0.1(up) and 0.23(down) $(GeV/c)^2$.

We can extract a variance σ and then make a fit (gaussian or not) of the distribution. The values for G^p_M, G^p_E, G^n_M and G^n_E and their uncertainties at Q^2= 0.1, 0.23, 0.45 and 0.63 are given in table 4, column 4. The great advantage of the Monte-Carlo method is that it gives a direct estimate of the variance which takes correlations at all orders into account. Although they should be more correct, these uncertainties still do not reflect the dispersion of the experimental data.

4 The "envelop" method

In order to quantify the uncertainties due to the dispersion of the experimental points, we divide the data into two subsets: the ones which are above and the ones which are below the theoretical fit as obtained in the preceding

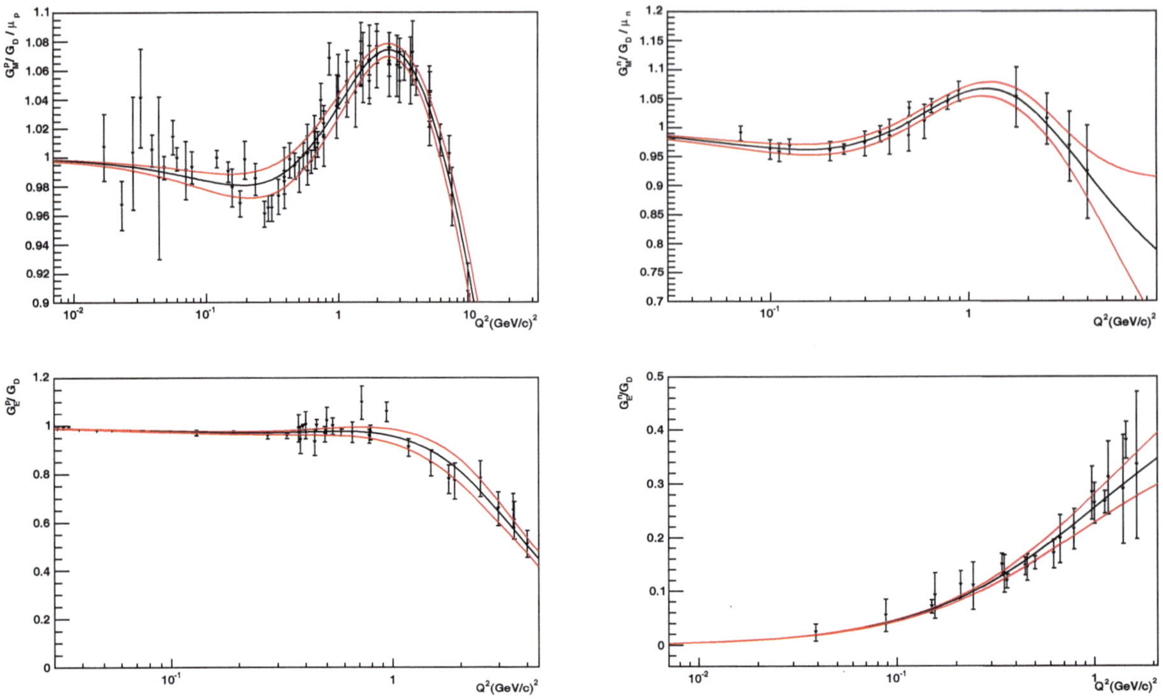

Fig. 4. Form factor uncertainties with the envelop method. Shifted left: Proton form factors. Shifted right: Neutron form factors.

section. We then fit the 2 sets separately with the Monte-Carlo method as described previously. From the two fits, a procedure has been developed to get quantitative estimations on the uncertainties on the form-factors [8]. The results for the four nucleon form factors are shown in Fig.4 and table 4 (column "env").

When comparing the Monte Carlo and envelop method, one can see that the errors are larger by a factor from 2 to 4 in the case of the envelop method which reflects the dispersion of the experimental data. The uncertainties on G^p_M , G^p_E and G^n_M remain small and are less than 3%. In the Q^2 range of interest, the uncertainties on G^n_E are larger but nevertheless they remain smaller than 8-10 %.

5 Summary and conclusions

The results of the present calculations with the 3 methods: purely statistical, Monte-Carlo and envelop methods are summarized in the following table for a few values of Q^2 of experimental interest [9]: 0.1 (SAMPLE, HAPPEX, PV-A4, G^0), 0.23 (PV-A4, G^0), 0.45 (HAPPEX, G^0) and 0.63 (G^0) $(GeV/c)^2$.

In conclusion we have derived a method based on a Monte-Carlo fit of the four nucleon electromagnetic form factors. This method takes into account correlations between parameters at all orders and it allows to quantify the systematic uncertainties due to the dispersion of the experimental data. It has been used here with polynomial functional forms but it can used with any other functional form, as done for example in [3]. The present results should reduce uncertainties in estimating the strange form factors from asymmetries in parity violating electron scattering.

References

1. O. Gayou et al., Phys.Rev.Lett. **88**, (2002) 092301
2. Galster et al., Nucl.Phys. **B32**, (1971) 221
3. J.Friedrich and Th.Walcher, Eur.Phys.J.A. **17**, (2003) 607
4. J. Arrington, Phys.Rev. **C68**, (2003) 034325
5. J.J. Kelly, Phys.Rev. **C70**, (2004) 0688202
6. J.J. Kelly, Phys.Rev. **C66**,(2002) 065203
7. E.J. Brash, A. Kozlov, S. Li and G. Huber, Phys. Rev. **C65**, (2002) 051001(R)
8. Marouan El-Yakoubi, PhD Thesis, Université Paris-Sud, to be published
9. See contributions to this conference and reference therein

Beam normal spin asymmetries: Theory

M. Vanderhaeghen[a]

Department of Physics, College of William and Mary, Williamsburg, VA 23187, USA, and
Thomas Jefferson National Accelerator Facility, Newport News, VA 23606, USA

Paper reprinted from Eur. Phys. J. A with permission

Abstract. The beam normal spin asymmetry in elastic electron-nucleon scattering is discussed. This beam normal spin asymmetry depends on the imaginary part of two-photon exchange processes between electron and nucleon, and measures the non-forward structure functions of the nucleon. After briefly reviewing the theoretical formalism, we discuss calculations in the threshold region, in the resonance region, as well as in the diffractive region, corresponding with high energy and forward angles.

PACS. 25.30.Bf Elastic electron scattering – 24.70.+s Polarization phenomena in reactions

1 Introduction

Elastic electron-nucleon scattering in the one-photon exchange approximation is a time-honoured tool to access information on the structure of hadrons. Experiments with increasing precision have become possible in recent years, mainly triggered by new techniques to perform polarization experiments at the electron scattering facilities. This has allowed to reach a new frontier in the measurement of hadron structure quantities, such as its electroweak form factors, parity-violating effects, nucleon polarizabilities, transition form factors, or the measurement of spin dependent structure functions, to name a few. For example, experiments using polarized electron beams and measuring the ratio of the recoil nucleon in-plane polarization components have profoundly extended our understanding of the nucleon electromagnetic form factors. For the proton, such polarization experiments which access the ratio G_{Ep}/G_{Mp} of the proton's electric (G_{Ep}) to magnetic (G_{Mp}) form factors have been performed out to a momentum transfer Q^2 of $5.6\,\mathrm{GeV}^2$ [1–3]. It came as a surprise that these experiments extracted a ratio of G_{Ep}/G_{Mp} which is clearly at variance with unpolarized measurements [4–6] using the Rosenbluth separation technique.

The understanding of this puzzle has generated a lot of activity recently, and is a prerequisite to use electron scattering as a precision tool. One possible explanation for the discrepancy between the Rosenbluth and polarization methods is the presence of two-photon exchange effects, beyond those which have already been accounted for in the standard treatment of radiative corrections. A general study of two- (and multi)-photon exchange contributions to the elastic electron-proton scattering observables was given in [7]. In that work, it was noted that the interference of the two-photon exchange amplitude with the one-photon exchange amplitude could be comparable in size to the $(G_E^p)^2$ term in the unpolarized cross-section at large Q^2. In contrast, the two-photon exchange effects do not impact the polarization-transfer extraction of G_E/G_M in an equally significant way. Thus a missing and unfactorizable part of the two-photon exchange amplitude at the level of a few percent may well explain the discrepancy between the two methods.

Realistic calculations of elastic electron-nucleon scattering beyond the Born approximation are required in order to demonstrate in a quantitative way that 2γ exchange effects are indeed able to resolve this discrepancy.

Recently, several model calculations of the 2γ exchange amplitude have been performed. In ref. [8], a calculation of the 2γ exchange when the hadronic intermediate state is a nucleon was performed. It found that the 2γ exchange correction with intermediate nucleon can partially resolve the discrepancy between the two experimental techniques. It was found [9] however that the effect is partly cancelled when including the next hadronic intermediate state, the $\Delta(1232)$-resonance. The 2γ exchange contribution to elastic eN scattering has also been estimated at large momentum transfer [10,11], through the scattering off a parton in a proton by relating the process on the nucleon to the generalized parton distributions. This approach effectively sums all possible excitations of inelastic nucleon intermediate states. It was found that the two-photon corrections to the Rosenbluth process indeed can substantially reconcile the two ways of measuring G_E/G_M.

To push the precision frontier further in electron scattering, one needs a good control of 2γ exchange mechanisms and needs to understand how they may or may

[a] e-mail: marcvdh@jlab.org

not affect different observables. This justifies a systematic study of such 2γ exchange effects, both theoretically and experimentally. The real (dispersive) part of the 2γ exchange amplitude can be accessed through the difference between elastic electron and positron scattering off a nucleon. The imaginary (absorptive) part of the 2γ exchange amplitude, on the other hand, can be accessed through a single spin asymmetry (SSA) in elastic electron-nucleon scattering, when either the target or beam spin are polarized normal to the scattering plane, as has been discussed some time ago [12]. As time reversal invariance forces this SSA to vanish for one-photon exchange, it is of order $\alpha = e^2/(4\pi) \simeq 1/137$. Furthermore, to polarize an ultra-relativistic particle in the direction normal to its momentum involves a suppression factor m/E (with m the mass and E the energy of the particle), which typically is of order 10^{-4}–10^{-3} when the electron beam energy is in the 1 GeV range. Therefore, the resulting target normal SSA can be expected to be of order 10^{-2}, whereas the beam normal SSA is of order 10^{-6}–10^{-5}. A measurement of such small asymmetries is quite demanding experimentally. However, in the case of a polarized lepton beam, asymmetries of the order ppm are currently accessible in parity violation (PV) elastic eN scattering experiments. The parity-violating asymmetry involves a beam spin polarized along its momentum. However, the SSA for an electron beam spin normal to the scattering plane can also be measured using the same experimental set-ups. First measurements of this beam normal SSA at beam energies up to 1 GeV have yielded values around -10 ppm [13–15] in the forward angular range and up to an order of magnitude larger in the backward angular range [16]. At higher beam energies, first results for the beam normal SSA in elastic electron-nucleon scattering experiments have also been reported at this meeting [15,17,18].

First estimates of the target normal SSA in elastic electron-nucleon scattering have been performed in [12, 19]. In those works, the 2γ exchange with nucleon intermediate state (so-called elastic or nucleon pole contribution) has been calculated, and the inelastic contribution has been estimated in a very forward-angle approximation. Estimates within this approximation have also been reported for the beam normal SSA [20]. The general formalism for elastic electron-nucleon scattering with lepton helicity flip, which is needed to describe the beam normal SSA, has been developed in [21]. Furthermore, the beam normal SSA has also been estimated at large momentum transfers Q^2 in [21] using a parton model, which was found crucial [10] to interpret the results from unpolarized electron-nucleon elastic scattering, as discussed before. In the handbag model of refs. [10,11,21], the corresponding 2γ exchange amplitude has been expressed in terms of generalized parton distributions, and the real and imaginary parts of the 2γ exchange amplitude are related through a dispersion relation. Hence in the partonic regime, a direct comparison of the imaginary part with experiment can provide a very valuable cross-check on the calculated result for the real part.

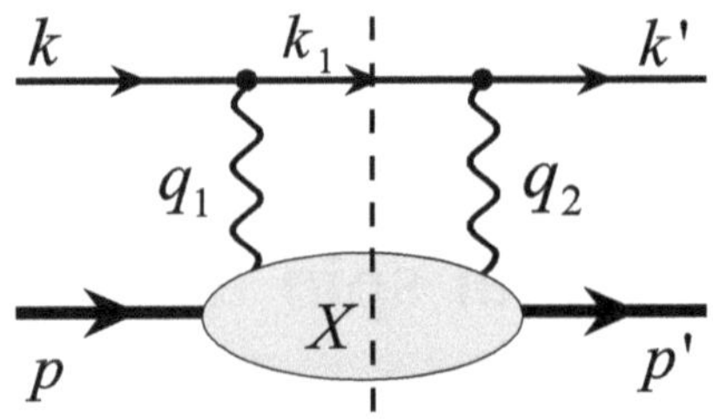

Fig. 1. The two-photon exchange diagram. The filled blob represents the response of the nucleon to the scattering of the virtual photon. In the imaginary part of the two-photon amplitude, the intermediate state indicated by the vertical dashed line is on-shell.

To use the elastic electron-nucleon scattering at low momentum transfer as a high precision tool, such as in present-day PV experiments, one may also want to quantify the 2γ exchange amplitude. To this aim, one may envisage a dispersion formalism for the elastic electron-nucleon scattering amplitudes, as has been discussed some time ago in the literature [22]. To develop this formalism, the necessary first step is a precise knowledge of the imaginary part of the two-photon exchange amplitude, which enters in both the beam and target normal SSA. Using unitarity, one can relate the imaginary part of the 2γ amplitude to the electro-absorption amplitudes on a nucleon, see fig. 1.

The imaginary part of the 2γ exchange has been studied recently in several approaches, both in the threshold region [23], in the resonance region [24], as well as at forward angles in the diffractive region [25–27].

Finally, by understanding the 2γ exchange contributions for the case of electromagnetic electron-nucleon scattering, one may extend this calculation to electroweak processes, where the γZ and W^+W^- box diagrams are in several cases the leading unknown contributions entering electroweak precision experiments.

2 Elastic electron-nucleon scattering beyond the one-photon exchange approximation

In this section, we briefly review the elastic electron-nucleon scattering formalism beyond the one-photon exchange approximation, as has been developed in refs. [7, 21]. For the kinematics of elastic electron-nucleon scattering;

$$e^-(k) + N(p) \to e^-(k') + N(p'), \tag{1}$$

we adopt the definitions

$$P = \frac{p+p'}{2}, \quad K = \frac{k+k'}{2}, \quad q = k - k' = p' - p, \tag{2}$$

and choose

$$Q^2 = -q^2, \qquad \nu = K \cdot P, \tag{3}$$

as the independent invariants of the scattering. The invariant ν is related to the polarization parameter ε of the virtual photon, which can be expressed as (neglecting the electron mass):

$$\varepsilon = \frac{\nu^2 - M^4\tau(1+\tau)}{\nu^2 + M^4\tau(1+\tau)}, \quad (4)$$

where M is the nucleon mass.

For a theory which respects Lorentz, parity and charge conjugation invariance, the general amplitude for elastic scattering of two spin-(1/2) particles can be expressed by 6 independent helicity amplitudes or equivalently by six invariant amplitudes. The total amplitude can be decomposed in general in terms of a lepton spin non-flip and spin flip part:

$$T = T_{non\text{-}flip} + T_{flip}. \quad (5)$$

The non-flip amplitude which conserves the helicity of the electron (in the limit $m_e = 0$) depends upon 3 invariant amplitudes, and has been parametrized in ref. [7] as

$$T_{non\text{-}flip} = \frac{e^2}{Q^2}\,\bar{u}(k')\gamma_\mu u(k) \times \bar{u}(p')\left(\tilde{G}_M\gamma^\mu - \tilde{F}_2\frac{P^\mu}{M} + \tilde{F}_3\frac{\gamma\cdot K P^\mu}{M^2}\right)u(p), \quad (6)$$

where $u(k), u(k')$ and $u(p), u(p')$ are the lepton and nucleon spinors, respectively. The amplitude which flips the electron helicity (*i.e.* is of the order of the mass of the electron, m_e), depends on 3 additional invariants which have been introduced in ref. [21] as

$$T_{flip} = \frac{m_e}{M}\frac{e^2}{Q^2}\left[\bar{u}(k')u(k)\cdot\bar{u}(p')\left(\tilde{F}_4 + \tilde{F}_5\frac{\gamma\cdot K}{M}\right)u(p) + \tilde{F}_6\bar{u}(k')\gamma_5 u(k)\cdot\bar{u}(p')\gamma_5 u(p)\right]. \quad (7)$$

In eqs. (6), (7), $\tilde{G}_M$, $\tilde{F}_2$, $\tilde{F}_3$, $\tilde{F}_4$, $\tilde{F}_5$, $\tilde{F}_6$ are complex functions of ν and Q^2, and the factor e^2/Q^2 has been introduced for convenience. Furthermore in eq. (7), we extracted an explicit factor m_e/M out of the amplitudes, which reflects the fact that for a vector interaction (such as in QED), the electron helicity flip amplitude vanishes when $m_e \to 0$. In the Born approximation, one obtains

$$\begin{aligned}\tilde{G}_M^{Born}(\nu, Q^2) &= G_M(Q^2),\\ \tilde{F}_2^{Born}(\nu, Q^2) &= F_2(Q^2),\\ \tilde{F}_{3,4,5,6}^{Born}(\nu, Q^2) &= 0,\end{aligned} \quad (8)$$

where $G_M(F_2)$ are the proton magnetic (Pauli) form factors, respectively.

3 Single spin asymmetries in elastic electron-nucleon scattering

An observable which is directly proportional to the two- (or multi-) photon exchange is given by the elastic scattering of an unpolarized electron on a proton target polarized *normal* to the scattering plane (or the recoil polarization normal to the scattering plane, which is exactly the same assuming time-reversal invariance). For a target polarized perpendicular to the scattering plane, the corresponding single spin asymmetry, which we refer to as the target normal spin asymmetry (A_n), is defined by

$$A_n = \frac{\sigma_\uparrow - \sigma_\downarrow}{\sigma_\uparrow + \sigma_\downarrow}, \quad (9)$$

where $\sigma_\uparrow$ ($\sigma_\downarrow$) denotes the cross-section for an unpolarized beam and for a nucleon spin parallel (anti-parallel) to the normal polarization vector, defined as

$$S_n^\mu = (0, \boldsymbol{S}_n), \qquad \boldsymbol{S}_n \equiv (\boldsymbol{k}\times\boldsymbol{k}')/|\boldsymbol{k}\times\boldsymbol{k}'|. \quad (10)$$

As has been shown by de Rujula *et al.* [12], the target (or recoil) normal spin asymmetry is related to the absorptive part of the elastic eN scattering amplitude as

$$A_n = \frac{2\,\mathrm{Im}(\sum_{spins} T_{1\gamma}^* \cdot \mathrm{Abs}\,T_{2\gamma})}{\sum_{spins}|T_{1\gamma}|^2}, \quad (11)$$

where $T_{1\gamma}$ denotes the one-photon exchange amplitude. Since the one-photon exchange amplitude is purely real, the leading contribution to A_n is of order $O(e^2)$, and is due to an interference between one- and two-photon exchange amplitudes.

For a beam polarized perpendicular to the scattering plane, one can also define a single spin asymmetry, analogously as in eq. (9) as noted in ref. [20], where now $\sigma_\uparrow$ ($\sigma_\downarrow$) denotes the cross section for an unpolarized target and for an electron beam spin parallel (anti-parallel) to the normal polarization vector, given by eq. (10). We refer to this asymmetry as the beam normal spin asymmetry (B_n). It explicitly vanishes when $m_e = 0$ as it involves an electron helicity flip. Using the general electron-nucleon scattering amplitude of eqs. (6), (7), B_n is given by [21]

$$\begin{aligned}B_n = &\frac{2\,m_e}{Q}\sqrt{2\,\varepsilon\,(1-\varepsilon)}\sqrt{1+\frac{1}{\tau}}\left(G_M^2 + \frac{\varepsilon}{\tau}G_E^2\right)^{-1}\\ &\times\left\{-\tau\,G_M\,\mathcal{I}\left(\tilde{F}_3 + \frac{1}{1+\tau}\frac{\nu}{M^2}\tilde{F}_5\right)\right.\\ &\left.-\,G_E\,\mathcal{I}\left(\tilde{F}_4 + \frac{1}{1+\tau}\frac{\nu}{M^2}\tilde{F}_5\right)\right\} + \mathcal{O}(e^4),\end{aligned} \quad (12)$$

with $\tau \equiv Q^2/(4M^2)$. As for A_n, also B_n vanishes in the Born approximation, and is therefore of order e^2.

4 Imaginary (absorptive) part of the two-photon exchange amplitude

In this section we discuss the relation between the imaginary part of the two-photon exchange amplitude and the absorptive part of the doubly virtual Compton scattering tensor on the nucleon, as shown in fig. 1. In the following we consider the helicity amplitudes for the elastic electron-nucleon scattering, defined in the e^-N c.m. frame, which are denoted by $T(h', \lambda'_N; h, \lambda_N)$. Here $h(h')$ denote the helicities of the initial (final) electrons and $\lambda_N(\lambda'_N)$ denote the helicities of the initial (final) nucleons.

The discontinuity of the two-photon exchange amplitude, shown in fig. 1, can then be expressed as

$$\mathrm{Abs}T_{2\gamma} = e^4 \int \frac{\mathrm{d}^3\boldsymbol{k}_1}{(2\pi)^3 2E_{k_1}} \bar{u}(k', h')\gamma_\mu(\gamma \cdot k_1 + m_e)\gamma_\nu u(k, h) \times \frac{1}{Q_1^2 Q_2^2} \cdot W^{\mu\nu}(p', \lambda'_N; p, \lambda_N), \quad (13)$$

where the momenta are defined as indicated in fig. 1, with $q_1 \equiv k - k_1$, $q_2 \equiv k' - k_1$, and $q_1 - q_2 = q$. Denoting the c.m. angle between initial and final electrons as $\theta_{c.m.}$, the momentum transfer $Q^2 \equiv -q^2$ in the elastic-scattering process can be expressed as

$$Q^2 = \frac{(s - M^2)^2}{2\,s}(1 - \cos\theta_{c.m.}) + \mathcal{O}(m_e^2), \quad (14)$$

with $s = (k + p)^2$. Furthermore, $Q_1^2 \equiv -q_1^2 = -(k - k_1)^2$ and $Q_2^2 \equiv -q_2^2 = -(k' - k_1)^2$ correspond with the virtualities of the two exchanged spacelike photons.

In eq. (13), the hadronic tensor $W^{\mu\nu}(p', \lambda'_N; p, \lambda_N)$ corresponds with the absorptive part of the doubly virtual Compton scattering tensor with two *space-like* photons:

$$W^{\mu\nu}(p', \lambda'_N; p, \lambda_N) = \sum_X (2\pi)^4\, \delta^4(p + q_1 - p_X) \times \langle p', \lambda'_N | J^{\dagger\mu}(0) | X \rangle \langle X | J^\nu(0) | p, \lambda_N \rangle, \quad (15)$$

where the sum goes over all possible *on-shell* intermediate hadronic states X. Note that in the limit $p' = p$, eq. (15) reduces to the forward tensor for inclusive electron-nucleon scattering and can be parametrized by the usual 4 nucleon forward structure functions. In the non-forward case however, the absorptive part of the doubly virtual Compton scattering tensor of eq. (15) which enters in the evaluation of target and beam normal spin asymmetries, depends upon 18 invariant amplitudes [28]. Though this may seem as a forbiddingly large number of new functions, we may use the unitarity relation to express the full non-forward tensor in terms of electroproduction amplitudes $\gamma^* N \to X$. The number of intermediate states X which one considers in the calculation will then put a limit on how high in energy one can reliably calculate the hadronic tensor eq. (15). In the following section, the tensor $W^{\mu\nu}$ will be discussed for the elastic contribution $(X = N)$, in the resonance region as a sum over all πN intermediate states (*i.e.* $X = \pi N$), using a phenomenological state-of-the-art calculation for the $\gamma^* N \to \pi N$ amplitudes, in the diffractive region (corresponding with high-energy, forward scattering) where it can be related to the total photo-absorption cross-section on a proton, as well as in the hard scattering region where it can be related to nucleon generalized parton distributions.

In fig. 2, the full kinematical accessible region for the virtualities Q_1^2, Q_2^2 in the phase space integral of eq. (13) is shown. The full phase space integral is three-dimensional, the third variable represents the invariant mass W of the intermediate state X, defined as $W^2 \equiv p_X^2$. There are special regions in this phase space integral (corresponding with near singularities) which may give important contributions (logarithmic enhancements) under some kinematical conditions.

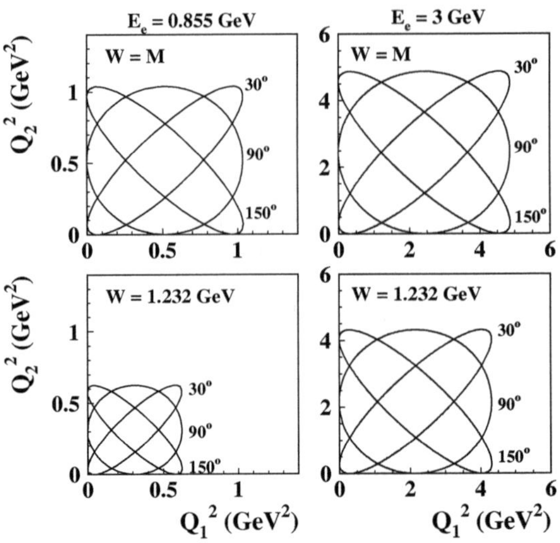

Fig. 2. Kinematical accessible region for the virtualities Q_1^2, Q_2^2 in the phase space integral of eq. (13), for MAMI (left panels) and JLab (right panels) kinematics for different c.m. angles $\theta_{c.m.}$ as indicated. The accessible regions correspond with the inside of the ellipses and are shown both for the elastic (upper panels) and for inelastic (lower panels) intermediate hadronic states. The intersection with the axes correspond with quasi-VCS, whereas the situation at $W = \sqrt{s} - m_e$ where all ellipses shrink to the point $Q_1^2 = Q_2^2 \simeq 0$ corresponds with quasi-RCS.

When the intermediate and initial electrons are collinear, then also the photon with momentum $\boldsymbol{q}_1 = \boldsymbol{k} - \boldsymbol{k}_1$ is collinear with this direction. For the elastic case $(W = M)$ this precisely corresponds with the situation where the first photon is soft (*i.e.* $q_1 \to 0$) and where the second photon carries the full momentum transfer $Q_2^2 \simeq Q^2$. For the inelastic case $(W > M)$ the first photon is hard but becomes quasi-real (*i.e.* $Q_1^2 \sim m_e^2$). In this case, the virtuality of the second photon is smaller than Q^2. An analogous situation occurs when the intermediate electron is collinear with the final electron. These kinematical situations with one quasi-real photon and one virtual photon correspond with quasi-virtual Compton scattering (quasi-VCS). Besides the quasi-VCS singularities, the two-photon exchange amplitude also has a near singularity when the intermediate electron momentum is soft (*i.e.* $|\boldsymbol{k}_1| \to 0$). In this case the first photon takes on the full momentum of the initial electron, *i.e.* $\boldsymbol{q}_1 \to \boldsymbol{k}$, whereas the second photon takes on the full momentum of the final electron, *i.e.* $\boldsymbol{q}_2 \to \boldsymbol{k}'$. This situation occurs when the invariant mass of the hadronic state takes on its maximal value $W_{max} = \sqrt{s} - m_e$. In this case, both photon virtualities become very small and vanish if the electron mass is taken to zero. This kinematical situation with two quasi-real photons corresponds with quasi-real Compton scattering (quasi-RCS).

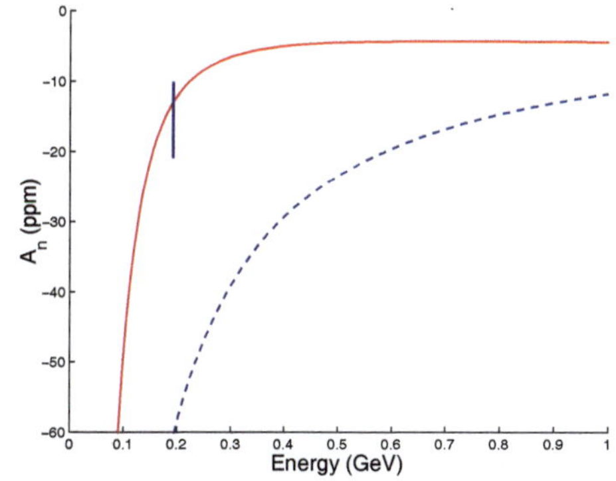

Fig. 3. Beam normal spin asymmetry for $e^{-\uparrow}p \to e^-p$ at fixed c.m. scattering angle $\theta_{cm} = 146.1$ degrees. Calculations and figure from ref. [23]. Dashed (solid) curves are the leading order (full) results. The data point is from the SAMPLE Collaboration [13].

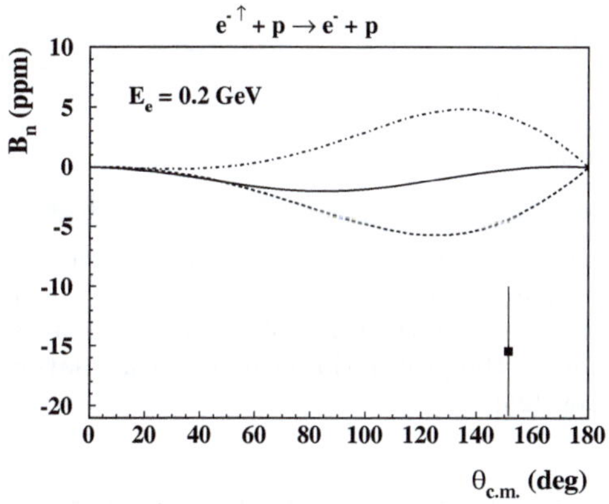

Fig. 4. Beam normal spin asymmetry for $e^{-\uparrow}p \to e^-p$ at a beam energy $E_e = 0.2\,\text{GeV}$ as function of the c.m. scattering angle, for different hadronic intermediate states (X) in the blob of fig. 1: N (dashed curve), πN (dashed-dotted curve), sum of the N and πN (solid curve). Calculations and figure are from ref. [24]. The data point is from the SAMPLE Collaboration [13].

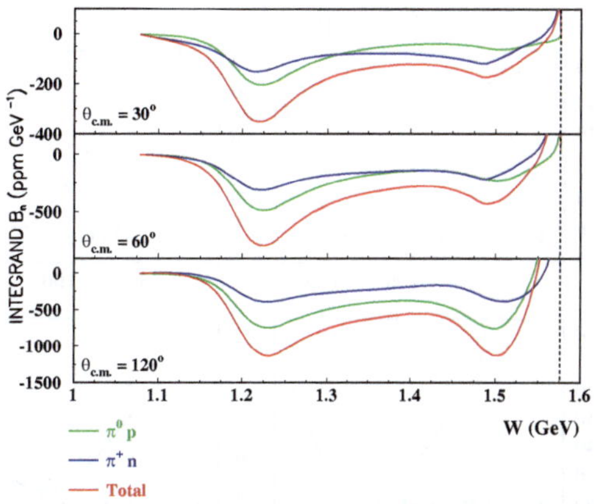

Fig. 5. Integrand in W of B_n for $e^{-\uparrow}p \to e^-p$ at a beam energy of $E_e = 0.855\,\text{GeV}$ and at different c.m. scattering angles as indicated on the figure. The vertical dashed line indicates the upper limit of the W integration, *i.e.* $W_{max} = \sqrt{s} - m_e$. Calculations and figure from ref. [24].

5 Results and discussion

5.1 Threshold region

In ref. [23], the beam normal spin asymmetry was studied at low energies in an effective theory of electrons, protons and photons. This calculation, in which pions are integrated out, effectively corresponds with the nucleon intermediate-state contribution only, expanded to second order in E_e/M_N. To this order, the calculation includes the recoil corrections to the scattering from a point charge, the nucleon charge radius, and the nucleon isovector magnetic moment. One sees from fig. 3 that the theory expanded up to second order in E_e/M_N (indicated by the full results) are able to give a good account of the SAMPLE data point at the low energy $E_e = 0.2\,\text{GeV}$.

When doing the full calculation for the N intermediate state, which is model independent (as it only involves on-shell $\gamma^* NN$ matrix elements), the result is further reduced as seen in fig. 4. Inclusion of threshold pion electroproduction contributions, arising from the πN intermediate states, partly cancels the elastic contributions. Because in this low-energy region, the matrix elements are rather well known, it is not clear how to get a better agreement with the rather large asymmetry measured by SAMPLE [13].

5.2 Resonance region

When measuring the imaginary part of the elastic eN amplitude through a normal SSA at sufficiently low energies, below or around two-pion production threshold, one is in a regime where these electroproduction amplitudes are relatively well known using pion electroproduction experiments as input. As both photons in the 2γ exchange process are virtual and integrated over, an observable such as the beam or target normal SSA is sensitive to the electroproduction amplitudes on the nucleon for a range of photon virtualities. This may provide information on resonance transition form factors complementary to the information obtained from current pion electroproduction experiments.

In ref. [24], the imaginary part of the two-photon exchange amplitude was calculated by relating it through unitarity to the contribution of $X = N$ and $X = \pi N$ intermediate state contributions. For the πN intermediate state contribution, the corresponding pion electroproduction amplitudes were taken from the phenomenological MAID analysis [29], which contains both resonant and non-resonant pion production mechanisms. As an example, the integrand in W of the phase space integral of eq. (13) is shown in fig. 5 at $E_e = 0.855\,\text{GeV}$. The resonance structure is clearly reflected in the integrands for both π^+n and π^0p channels. At forward angles, the quasi-real Compton scattering at the endpoint $W = W_{max}$ only yields a very small contribution, which grows larger when going to backward angles. This quasi-RCS contribution is of opposite sign as the remainder of the integrand, and

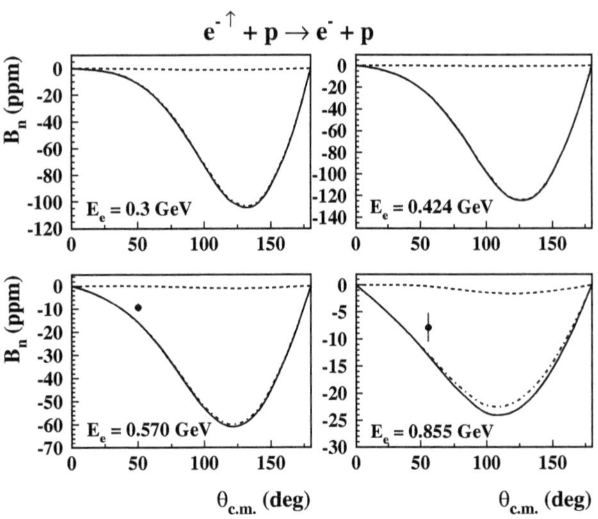

Fig. 6. Beam normal spin asymmetry B_n for $e^{-\uparrow}p \rightarrow e^-p$ as function of the c.m. scattering angle at different beam energies, as indicated on the figure. The calculations are for different hadronic intermediate states (X) in the blob of fig. 1: N (dashed curve), πN (dashed-dotted curves), sum of the N and πN (solid curves). The data points are from the A4 Collaboration (MAMI) [14]. Calculations and figure from ref. [24].

therefore determines the position of the maximum (absolute) value of B_n when going to backward angles.

In fig. 6, the results for B_n are shown at different beam energies below $E_e = 1\,\text{GeV}$. It is clearly seen that at energies $E_e = 0.3\,\text{GeV}$ and higher the nucleon intermediate state (elastic part) yields only a very small relative contribution. Therefore B_n is a direct measure of the inelastic part which gives rise to sizeable large asymmetries, of the order of several tens of ppm in the backward angular range, mainly driven by the quasi-RCS near singularity. First results from the A4 Collaboration for B_n at backward angles (for E_e around $0.3\,\text{GeV}$) shown at this meeting [16] indeed show such a large beam normal spin asymmetry of order $-100\,\text{ppm}$ for θ_{cm} around 150 degrees. At forward angles, the size of the predicted asymmetries is compatible with the first high-precision measurements performed at MAMI, though the model slightly overpredicts (in absolute value) B_n at $E_e = 0.57\,\text{GeV}$ and $0.855\,\text{GeV}$.

5.3 High-energy, forward-scattering (diffractive) region

At very high energies and forward-scattering angles (so-called diffractive limit), it was shown in refs. [25,26] that the beam normal spin asymmetry is dominated by the quasi-real Compton singularity. In this (extreme forward limit) case, the hadronic tensor can be expressed in terms of the total photo-absorption cross-section, allowing to express B_n through the simple analytic expression:

$$B_n = -\frac{m_e\,\sqrt{Q^2}\,\sigma_{tot}^{\gamma p}}{8\pi^2}\,\frac{G_E}{\tau G_M^2 + \varepsilon G_E^2}\left[\log\frac{Q^2}{m_e^2} - 2\right]. \quad (16)$$

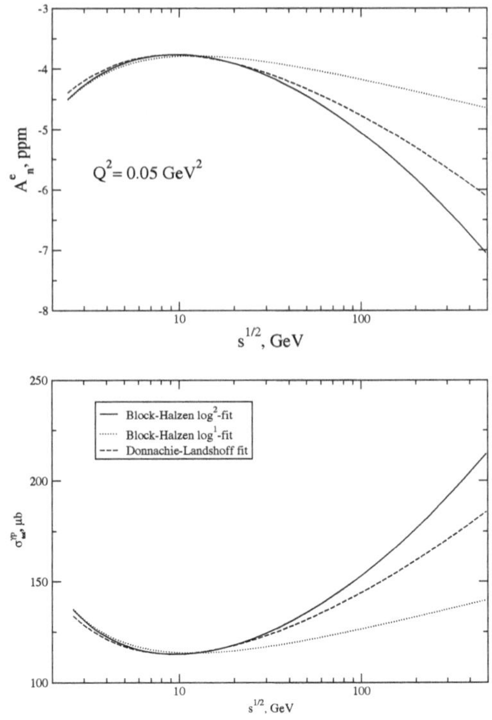

Fig. 7. Beam normal spin asymmetry (top panel) at high energies and very forward angles ($Q^2 = 0.05\,\text{GeV}^2$) using eq. (16) for different parameterizations of the total photo-absorption cross section on the proton (lower panel). Calculations and figure from ref. [25].

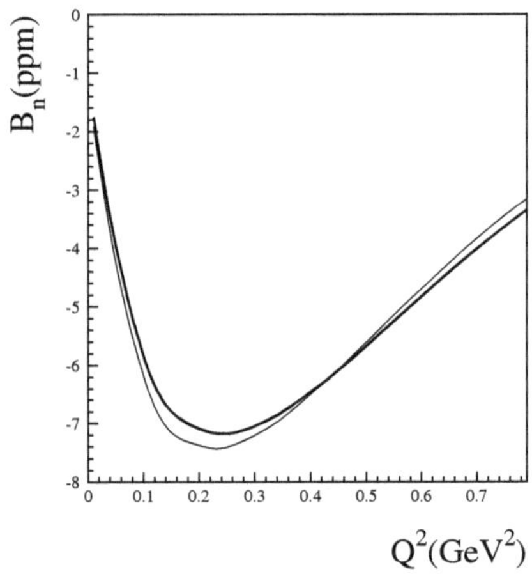

Fig. 8. Beam normal spin asymmetry for elastic e^-p scattering for $E_e = 3\,\text{GeV}$. The thick curve is the leading-Q^2 contribution, whereas the thin line shows the full calculation with subleading terms in Q^2. Calculations and figure from [27].

One notices that the quasi-real Compton singularity gives rise to a (single) logarithmic enhancement factor which is at the origin of the relatively large value of B_n.

In fig. 7, the estimate from ref. [25] based on eq. (16) is shown for different parameterizations of the total photo-absorption cross-section. The beam normal spin asymmetry has been measured at SLAC (E-158) at an energy

Table 1. Forward-angle experimental measurements of the beam normal spin asymmetry.

EXP.	E_e (GeV)	Q^2 (GeV^2)	B_n (ppm)
SAMPLE [13]	0.192	0.10	-16.4 ± 5.9
A4 [14]	0.570	0.11	-8.59 ± 0.89
A4 [14]	0.855	0.23	-8.52 ± 2.31
HAPPEX [15]	3.0	0.11	-6.7 ± 1.5
E-158 [17]	46	0.06	$-3.5 \rightarrow -2.5$

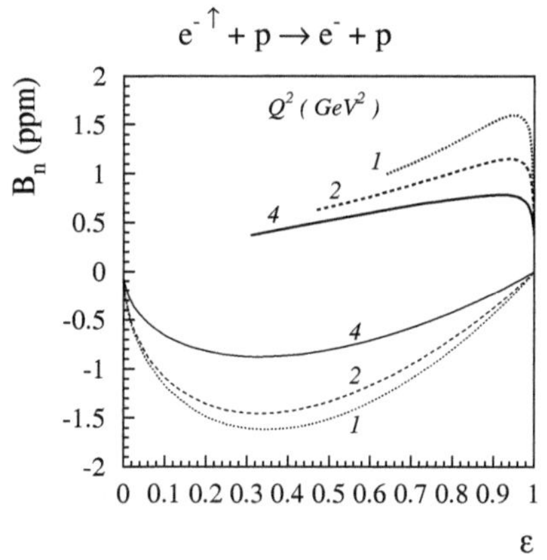

Fig. 9. Beam normal spin asymmetry for elastic e^-p scattering as function of ε at different values of Q^2 as indicated on the figure. The upper thick curves ($B_n > 0$) are the GPD calculations for the kinematical range where $s, -u > M^2$. For comparison, the nucleon pole contribution is also displayed: lower thin curves ($B_n < 0$). Calculations and figure from [21].

$E_e = 46\,\text{GeV}$ ($\sqrt{s} \simeq 9\,\text{GeV}$) and very forward angle ($Q^2 \simeq 0.05\,\text{GeV}^2$). First results, reported at this meeting, indicate a value $B_n \simeq -3.5 \rightarrow -2.5\,\text{ppm}$, confirming the estimate shown in fig. 7.

At intermediate energies, around $E_e \simeq 3\,\text{GeV}$, and forward angles, B_n has also been measured by the HAPPEX and G0 Collaborations. The simple "diffractive" formula of eq. (16) does not rigorously apply any more and one has to calculate corrections due to the deviation from forward scattering. Such calculation has recently been performed in [27] and the result is shown in fig. 8. The resulting asymmetries are in basic agreement with first results reported by HAPPEX [15] and G0 [18].

In table 1, a summary of forward-angle measurements of beam normal spin asymmetries is shown.

5.4 Hard-scattering region

In the hard-scattering region, the normal beam spin asymmetry was estimated in [21] through the scattering off a parton, which is embedded in the nucleon through a generalized parton distribution. Using phenomenological parametrizations for the GPDs, B_n was found to yield values around +1 ppm to +1.5 ppm in the few GeV beam energy range, see fig. 9. In particular, the forward angular range for $e^{-\uparrow}p \rightarrow e^-p$ scattering was found to be a favorable region to get information on the inelastic part of B_n. Because in the handbag calculation, real and imaginary parts are linked, a direct measurement of B_n may yield a valuable cross-check for the real part, which was found crucial in understanding the unpolarized cross section data for $e^-p \rightarrow e^-p$ at large momentum transfer.

References

1. M.K. Jones *et al.*, Phys. Rev. Lett. **84**, 1398 (2000).
2. O. Gayou *et al.*, Phys. Rev. Lett. **88**, 092301 (2002).
3. V. Punjabi *et al.*, Phys. Rev. C **71**, 055202 (2005); 069902 (2005)(E).
4. L. Andivahis *et al.*, Phys. Rev. D **50**, 5491 (1994).
5. M.E. Christy *et al.*, Phys. Rev. C **70**, 015206 (2004).
6. I.A. Qattan *et al.*, Phys. Rev. Lett. **94**, 142301 (2005).
7. P.A.M. Guichon, M. Vanderhaeghen, Phys. Rev. Lett. **91**, 142303 (2003).
8. P.G. Blunden, W. Melnitchouk, J.A. Tjon, Phys. Rev. Lett. **91**, 142304 (2003); Phys. Rev. C **72**, 034612 (2005).
9. S. Kondratyuk, P.G. Blunden, W. Melnitchouk, J.A. Tjon, Phys. Rev. Lett. **95**, 172503 (2005).
10. Y.C. Chen, A. Afanasev, S.J. Brodsky, C.E. Carlson, M. Vanderhaeghen, Phys. Rev. Lett. **93**, 122301 (2004).
11. A.V. Afanasev, S.J. Brodsky, C.E. Carlson, Y.C. Chen, M. Vanderhaeghen, Phys. Rev. D **72**, 013008 (2005).
12. A. De Rujula, J.M. Kaplan, E. de Rafael, Nucl. Phys. B **35**, 365 (1971).
13. SAMPLE Collaboration (S.P. Wells *et al.*), Phys. Rev. C **63**, 064001 (2001).
14. MAMI/A4 Collaboration (F. Maas *et al.*), Phys. Rev. Lett. **94**, 082001 (2005).
15. On behalf of the HAPPEX Collaboration (L. Kaufmann), this conference.
16. On behalf of the A4 Collaboration (L. Capozza), this conference.
17. On behalf of the E-158 Collaboration (K. Kumar), this conference.
18. JLab G0 experiment (E-00-006, E-01-116), spokesperson D. Beck.
19. A. De Rujula, J.M. Kaplan, E. de Rafael, Nucl. Phys. B **53**, 545 (1973).
20. A. Afanasev, I. Akusevich, N.P. Merenkov, hep-ph/0208260.
21. M. Gorchtein, P.A.M. Guichon, M. Vanderhaeghen, Nucl. Phys. A **741**, 234 (2004).
22. J.A. Peñarrocha, J. Bernabéu, Ann. Phys. **135**, 321 (1981); J. Bordes, J.A. Peñarrocha, J. Bernabéu, Phys. Rev. D **35**, 3310 (1987).
23. L. Diaconescu, M.J. Ramsey-Musolf, Phys. Rev. C **70**, 054003 (2004).
24. B. Pasquini, M. Vanderhaeghen, Phys. Rev. C **70**, 045206 (2004).
25. A.V. Afanasev, N.P. Merenkov, Phys. Lett. B **599**, 48 (2004) (arXiv:hep-ph/0407167).
26. M. Gorchtein, Phys. Rev. C **73**, 035213; 055201 (2006).
27. M. Gorchtein, arXiv:hep-ph/0610378.
28. R. Tarrach, Nuovo Cimento A **28**, 409 (1975).
29. D. Drechsel, O. Hanstein, S. Kamalov, L. Tiator, Nucl. Phys. A **645**, 145 (1999).

Transverse single spin asymmetry at the A4 experiment

Preliminary backward-scattering results

L. Capozza[a]

For the A4 Collaboration
Institut für Kernphysik, Johannes Gutenberg Universität, Mainz, Germany

Abstract. The spin asymmetry in the scattering of transversely polarised electrons on unpolarised hydrogen target has been measured by the A4 Collaboration at backward-scattering angle. The square of the four-momentum transfer was $Q^2 = 0.22\,(\mathrm{GeV}/c)^2$ like in one of the previous measurements of the A4 Collaboration at forward angle. In this kinematical configuration the spin asymmetry is of the order of 100 ppm and presents a very clear azimuthal angle dependency. Some preliminary results are presented and are in agreement with theoretical predictions. Also some details about the extraction of the asymmetry are given and the handling of the background is discussed.

PACS. 13.40.Gp Electromagnetic form factors – 11.30.Er Charge conjugation, parity, time reversal, and other discrete symmetries – 13.40.-f Electromagnetic processes and properties – 14.20.Dh Protons and neutrons

1 Introduction

The A4 experiment was thought for studying the parity violation in the elastic electron proton scattering. Longitudinally polarised electrons are scattered off unpolarised protons and the asymmetry in the elastic scattering cross-section between the two beam polarisation states is measured by counting the single scattering events. The experimental apparatus was designed for measuring asymmetries of the order of some parts per million (ppm), which is the magnitude of parity violation effects in the electron-proton scattering.

Things change if the electron spin is turned by 90° to be perpendicular to the beam direction. In this case the asymmetry depends on the azimuthal angle and modulates like the sine of the angle between scattering plane and spin plane, *i.e.* the plane containing the electron spin and their momentum. Instead of being a parity violation electro-weak effect this transverse spin asymmetry (TSA) $A_\perp$ arises from QED contributions to the scattering amplitude beyond the one photon exchange amplitude $\mathcal{M}_\gamma$. At the lowest order in the QED perturbation series $A_\perp$ is proportional to the interference between $\mathcal{M}_\gamma$ and the imaginary part of the two photon exchange amplitude $\mathcal{M}_{\gamma\gamma}$

$$A_\perp \propto \frac{\mathcal{M}_\gamma \,\mathrm{Im}\,\mathcal{M}_{\gamma\gamma}}{|\mathcal{M}_\gamma|^2}\,.$$

This ratio is proportional to the fine-structure constant α. Furthermore the scattering amplitudes are obtained projecting the initial electron spinors to a spin state perpendicular to the momentum direction, which gives a scaling factor m_e/E, being m_e the mass of the electron and E the beam energy, respectively. Such a factor m_e/E is of the order 10^{-4} to 10^{-3} in the A4 experiment. Thus the expected order of magnitude of $A_\perp$ is

$$A_\perp \sim 10^{-6}\text{–}10^{-5}\,,$$

which is measurable with the A4 detector. Moreover the detector is symmetric with respect to the azimuthal scattering angle φ. This feature makes it suited to measure the φ-dependence of the asymmetry as a systematic check.

The A4 collaboration has already measured the transverse single spin asymmetry at forward angle two different Q^2, $0.22\,(\mathrm{GeV}/c)^2$ and $0.1\,(\mathrm{GeV}/c)^2$ [1]. The mean polar scattering angle θ was 35° and the corresponding beam energies 855 MeV and 570 MeV, respectively. After the rearrangement of the detector for measuring at backward angle ($\theta = 145°$) a new measurement of the TSA at $Q^2 = 0.22\,(\mathrm{GeV}/c)^2$ ($E = 315\,\mathrm{MeV}$) has been performed with this new kinematical configuration. The data analysis is ongoing and some preliminary results are presented in the following sections.

2 Experimental setup

The A4 experiment takes place at the MAMI electron accelerator facility in Mainz. The polarised electron beam

[a] e-mail: capozza@kph.uni-mainz.de

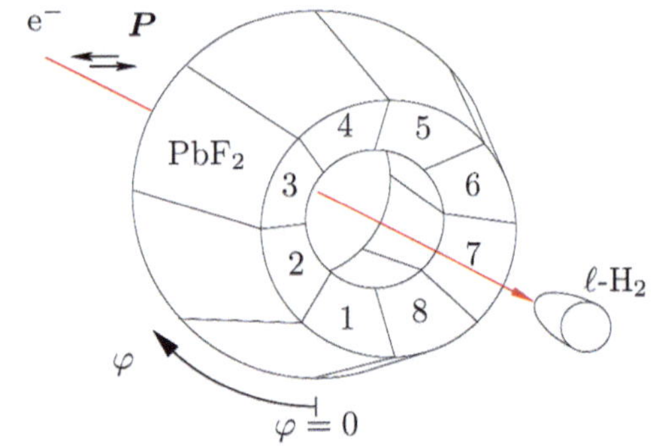

Fig. 1. Schematic view of the experiment for the backward-scattering measurement. The PbF_2 electromagnetic calorimeter covers the polar scattering angle range from 140° to 150°. For the TSA measurement it has been subdivided into 8 sectors in order to see the φ-dependence of the asymmetry.

hits an unpolarised liquid hydrogen target and the number of elastically scattered electrons for both polarisation states $N^{\pm}$ is measured by a counting electromagnetic calorimeter (fig. 1). The beam spin asymmetry is given by $A = (N^+ - N^-)/(N^+ + N^-)$. A detailed description of the experimental setup can be found in [2]. The electron beam current is $20\,\mu$A and using a polarised source based on a strained layer GaAs crystal, a polarisation P of about 80% can be obtained. The spin of the electrons is reversed every 20 ms during the measurement following a randomly selected pattern, $(+--+)$ or $(-++-)$. The polarisation degree of the beam is measured by a Møller polarimeter situated in another experimental hall. A Wien filter located between the electron source and the injector of the accellerator is used for turning the spin of the electrons to a transversal polarisation. The exact angle between electron spin and momentum is measured by a Møller-Mott polarimeter located in the beam dump. Several monitor and stabilisation systems are installed along the accelerator to minimise beam fluctuations that could give false asymmetries. The length of the target was 10 cm for the forward-angle measurement and 23 cm in the backward-angle setup. These lengths correspond to luminosities of $0.5 \cdot 10^{38}\,\mathrm{cm}^{-2}\,\mathrm{s}^{-1}$ and $1.15 \cdot 10^{38}\,\mathrm{cm}^{-2}\,\mathrm{s}^{-1}$, respectively.

The detector system comprises a total absorbing calorimeter constisting of 1022 lead fluoride (PbF_2) crystals ordered in 7 rings and 146 rows obtaining a cylindrical symmetry about the beam line. It covers a solid angle of 0.64 sr given by the complete azimuthal angle φ range (0 to 2π) and the polar angle θ range either from 30° to 40° at forward angle or from 140° to 150° at backward angle.

The most important novelty for the measurements at backward angle was the introduction of a scintillator trigger between the target and the calorimeter. With this new detector it becomes possible to distinguish between charged and neutral particles. Thus the background due to photons, which is quite important at backward angle, can be separated from the signal of elastic scattering electrons.

Table 1. Beam parameters giving instrumental asymmetries.

Param.	Description	Value
X_1	current asymmetry	(-0.2 ± 0.1) ppm
X_2	horizontal position	(-35.1 ± 0.2) nm
X_3	vertical position	(-14.1 ± 0.2) nm
X_4	incidence angle, horizontal	(-4.33 ± 0.03) nrad
X_5	incidende angle, vertical	(-1.56 ± 0.02) nrad
X_6	energy difference	(-0.38 ± 0.05) eV

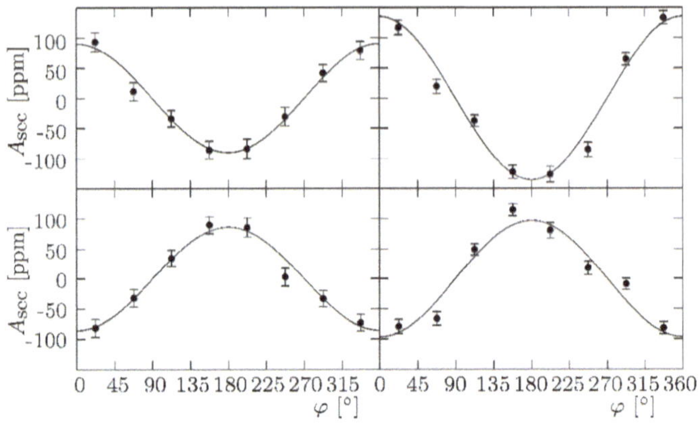

Fig. 2. Preliminary results of the TSA measurement at backward angle. The points are the mean value of the physical asymmetry over each detector sector. The error bars contain only the statistical error. The curves are cosine functions with fitted amplitude. The left plots are extracted from the coicidence spectra, the right ones from the anticoicidence ones. The two rows correspond to two different data samples. For a systematic check, between the two measurements a half-wave plate was inserted into the laser beam at the source changing the overall sign of the electron spin inducing a sign change of the measured asymmetry.

3 Extraction of the physical asymmetry

The origin and the sign of the azimuthal scattering angle φ are defined like shown in fig. 1. Within this convention, defining $\boldsymbol{S}_n$ as the normal to the scattering plane ($\boldsymbol{S}_n = (\boldsymbol{k} \times \boldsymbol{k}')/|\boldsymbol{k} \times \boldsymbol{k}'|$, where $\boldsymbol{k}$ and $\boldsymbol{k}'$ are the incoming and outgoing electron momenta, respectively) and $\boldsymbol{P}$ as the beam polarisation vector, the measured TSA A_{phys} reads

$$A_{\text{phys}} = A_{\perp} \frac{\boldsymbol{S}_n \cdot \boldsymbol{P}}{|\boldsymbol{S}_n||\boldsymbol{P}|} = -A_{\perp} \cos\varphi\,.$$

The physical asymmetry A_{phys} is extracted in the same way as in the parity violation experiment [2]. The relationship between the measured asymmetry A_{exp} and A_{phys} is

$$A_{\text{exp}} = \frac{N^+ - N^-}{N^+ + N^-} = P A_{\text{phys}} + \sum_{i=1}^{6} a_i X_i\,,$$

where the terms in the sum are the so-called instrumental asymmetries. The X_i's are systematic shifts in the beam parameters between the two polarisation states (table 1) and the a_i's are obtained by multilinear regression.

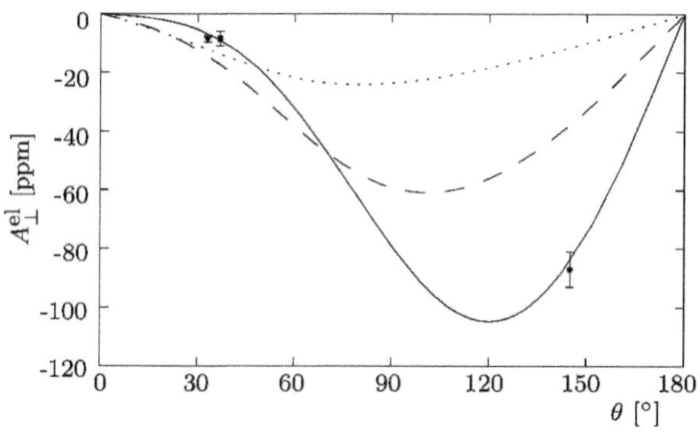

Fig. 3. TSA in elastic scattering as a function of the polar scattering angle. The curves are calculations of [4] for different beam energies E_0. The data points are the A4 measurements at forward angle ($E_0 = 855\,\mathrm{MeV}$: dotted line, square; $E_0 = 570\,\mathrm{MeV}$: dashed line, triangle) and the preliminary result of the backward-scattering measurement ($E_0 = 300\,\mathrm{MeV}$: solid line, circle).

The parity violation asymmetry is obtained averaging over the whole detector. Though, for the TSA measurement this has been segmented into 8 sectors, each of which giving the mean value of $A_{\rm phys}$ over the corresponding φ-range. Such a mean value $A_{\rm sec}$ is related to $A_\perp$ through

$$A_{\rm sec} = \frac{-\int_{\rm sec} \mathrm{d}\varphi A_\perp \cos\varphi}{\int_{\rm sec} \mathrm{d}\varphi} = F_{\rm sec} A_\perp \,,$$

where the $F_{\rm sec}$'s are constant geometrical factors.

3.1 Handling of the background

In the backward-scattering configuration the energy spectrum measured by the calorimeter presents a strong contamination due to high-energy photons produced in the target. In order to separate the electron spectrum from the photon spectrum a plastic scintillator trigger was introduced between target and calorimeter [3]. With this new detector two energy spectra are measured at the same time: one containing the coincidence events, the other one the anticoincidence events, *i.e.* events which are detected by the calorimeter but not by the scintillator. From both spectra a TSA $A_\perp^{\rm c}$ and $A_\perp^{\rm ac}$, respectively, is extracted (fig. 2). Though in the coincidence spectrum the peak of elastic scattering events is well isolated, it still contains a background component coming from γ events where the scintillator triggers, because the photon generates a charged particle along its path from the target to the detector. This can happen if the photon converts inside the vacuum chamber wall, which is situated before the scintillator. The asymmetry $A_\perp^{\rm c}$ is thus a linear combination of the asymmetry $A_\perp^{\rm el}$ in the elastic-scattering cross-section and of the asymmetry $A_\perp^{\rm bg}$ of the background processes. Furthermore the efficiency of the scintillator can be assumed to be 100% in first approximation, that is $A_\perp^{\rm bg} = A_\perp^{\rm ac}$. The elastic scattering asymmetry is then obtained by

$$A_\perp^{\rm el} = \frac{A_\perp^{\rm c} - \epsilon A_\perp^{\rm ac}}{1-\epsilon} \,,$$

where ϵ is the fraction of coincidence spectrum coming from background events. The determination of ϵ requires a careful study but a first estimation is of about 10%.

4 Preliminary backward-scattering results

The data-taking for the measurement of the TSA at backward scattering was done in March 2006 and amounts to 50 hours. The number of elastic-scattering events collected was $1.3 \cdot 10^{11}$. The physical asymmetry as a function of the azimuthal scattering angle both for coincidence and anti-coincidence events is shown in fig. 2. The preliminary result of the measurement of the TSA in the backward elastic scattering at $Q^2 = 0.22\,\mathrm{GeV}^2/c^2$ is $(87 \pm 6_{\rm stat})$ ppm, with the systematic error still to be determined. A comparison of this result and of the previous measurements of the A4 Collaboration with a theoretical calculation of [4] is shown in fig. 3. The calculation models the doubly virtual Compton scattering tensor in the resonance region using unitarity and pion electroproduction amplitudes. Those amplitudes are taken from a phenomenological analysis of corresponding observables. The preliminary result is in agreement with such a calculation.

References

1. F.E. Maas *et al.*, Phys. Rev. Lett. **94**, 082001 (2005).
2. F.E. Maas *et al.*, Phys. Rev. Lett. **93**, 022002 (2004).
3. See the contributions of S. Baunack and B. Gläser in this conference.
4. B. Pasquini, M. Vanderhaeghen, Phys. Rev. C **70**, 045206 (2004).

Transverse beam asymmetries measured from ^{4}He and hydrogen targets

L.J. Kaufman[a]

For the HAPPEX Collaboration
Department of Physics, University of Massachusetts, Amherst, MA 01003, USA

Abstract. The HAPPEX Collaboration at Jefferson Lab has measured the transverse beam spin asymmetries (A_T) for elastic electron scattering from proton and ^{4}He targets. The experiment was conducted using a vertically polarized electron beam of energy $\sim 3\,\text{GeV}$, at a $Q^2 \sim 0.1\,\text{GeV}^2$ and a scattering angle $\theta_{lab} \sim 6°$. The preliminary results are reported here. The ^{4}He measurement is the first measurement of A_T from a nucleus. A_T for ^{4}He is non-negligible; therefore, it will be necessary to make measurements of A_T for future parity-violating experiments using nuclear targets.

PACS. 25.30.Bf Elastic electron scattering – 13.40.Gp Electromagnetic form factors – 29.27.Hj Polarized beams

1 Introduction

The transverse beam spin asymmetry A_T arises from the interference between one-photon (1γ) and two-photon (2γ) exchange in elastic scattering of transversely polarized electrons from an unpolarized target. The HAPPEX experiment measures this asymmetry for vertically polarized electrons scattering from ^{4}He and proton targets. The asymmetry is defined as

$$A_T = \frac{\sigma_\uparrow - \sigma_\downarrow}{\sigma_\uparrow + \sigma_\downarrow}, \tag{1}$$

where $\sigma_\uparrow$ ($\sigma_\downarrow$) is the cross-section for electrons polarized parallel (antiparallel) to the normal of the scattering plane. The asymmetry manifests itself as a spatial asymmetry because

$$A_T \propto \boldsymbol{S}_e \cdot \frac{\boldsymbol{k}_e \times \boldsymbol{k}_e'}{|\boldsymbol{k}_e \times \boldsymbol{k}_e'|}, \tag{2}$$

where $\boldsymbol{S}_e$ and $\boldsymbol{k}_e$ are the spin vector and momentum vector, respectively, of the incoming electron, and $\boldsymbol{k}_e'$ is the momentum vector of the scattered electron. A_T is suppressed by several factors such as the fine-structure constant, α, the Lorentz boost of the incoming electron, and the very-forward scattering angle of the selected kinematics. In the approximation of 1γ exchange or 2γ exchange with contributions only from ground state interactions, $A_T \ll 10^{-6}$.

The extraction of the ratio of the nucleon electric and magnetic form factors is influenced by the real part of the 2γ amplitude. The discrepancy between the unpolarized Rosenbluth separation and polarization transfer data measuring this ratio has motivated interest in studying the effects of 2γ exchange. A_T is only sensitive to the imaginary part of the 2γ exchange and is of interest to parity-violating (PV) experiments because it is a possible background to the PV asymmetry. Studying the 2γ exchange provides a clear and accessible window on the treatment of hadronic intermediate states in box diagrams.

PV experiments are ideal for measuring small asymmetries such as A_T because of their high-luminosity and low-background environment. PV experiments are already set up to control false asymmetries, to rapidly flip the beam polarization in order to cancel effects of slow drifts, and to cancel beam asymmetries with symmetric detector configurations. All PV experiments measure A_T in order to control possible systematic errors. As a result, measurements of A_T exist over a wide range of kinematics [1–3].

2 Experiment

The experiment took place in Hall A at Jefferson Lab (JLab) in Virginia using the standard HAPPEX experimental equipment and techniques. The differences from the PV HAPPEX measurement are the setup of the beam polarization (discussed in sect. 2.1) and the measurement of the beam polarization (discussed in sect. 2.2). The transverse asymmetry was measured as a background to

[a] e-mail: lkaufman@jlab.org; now at University of Maryland, College Park, MD, USA.

the PV HAPPEX measurement in 2004 and 2005 for the proton and ^{4}He, respectively.

The experiment used a 20 cm liquid hydrogen (LH_2) target for the proton measurement and a 20 cm cold, dense ^{4}He gas target for the helium measurement. The scattered electrons were focused onto the focal-plane detectors using the Hall A High Resolution Spectrometers (HRS). A septum magnet located upstream of the HRS bent the 6° scattered electrons into the entrance of the HRS (at 12.5°). The signals from the focal-plane detectors were then integrated in the HAPPEX data acquisition (DAQ).

The beam helicity is chosen pseudorandomly at 15 Hz. The beam remains in this helicity state for 33.3 ms and then is switched to the opposite helicity state for the same amount of time. Each period of constant helicity is referred to as a "window". The asymmetry is calculated for adjacent helicity windows, called a "pair". In addition to the focal plane detectors, the beam current and position monitor signals are also integrated in the DAQ in order to measure and control false asymmetries on the beam. For more detail of the experimental apparatus, see [4,5].

2.1 Setup of the beam polarization

The horizontal acceptance of the Hall A HRS requires vertical polarization for the measurement of A_T. Vertical polarization is not standard at JLab and requires a special setup procedure. The setup of the polarized source optics for the vertical beam polarization, and the helicity-correlated beam properties, are the same as for the PV measurements. The vertical alignment of the electron spin in the injector is discussed below.

When the electrons leave the cathode of the polarized injector, their spins are are aligned parallel or antiparallel to the beam direction. The Wien filter consists of perpendicular electric and magnetic fields (both of which are perpendicular to the electron's trajectory) in order to rotate the electron spin horizontally without deflecting the electron's trajectory. The Wien filter settings are optimized such that after passing through the Wien filter, the electron spin is horizontal to the beam direction.

Downstream of the Wien filter are two solenoids which are counter wound. During normal beam operation, these solenoids are connected in series and are run by the same power supply, such that they provide equal and opposite magnetic fields along the beam direction to provide focusing of the beam. In order to rotate the electron spin to the vertical direction, there needs to be a net magnetic field in the beam direction. This field is achieved by running the solenoids with separate power supplies and thus different currents and magnetic fields. The current settings in each solenoid are optimized such that the magnetic field rotates the electron spin to be fully vertical.

2.2 Measurement of the beam polarization

The beam polarization is measured with two different techniques. The polarization is measured by the Mott polarimeter in the 5 MeV region of the injector and by the Møller polarimeter in Hall A. The standard configuration of the Mott polarimeter is capable of providing a vertical beam polarization measurement. Because vertical polarization does not precess through the accelerator, the Mott polarimeter measures the same polarization as what is present in Hall A. The Hall A Møller polarimeter provides a cross-check to the Mott measurement.

The standard Hall A Møller polarimeter is insensitive to vertical polarization, but new targets were made for the polarimeter so that it would be sensitive to vertical beam polarization. The new targets were tilted vertically by ±20° and polarized along the length of the target. These targets provide an analyzing power in both the beam direction and the vertical direction. The analyzing power is seven times larger along the beam direction compared to the vertical direction. The two targets provide the opposite-sign analyzing-power in the vertical direction, and the use of both targets cancels the effects of the longitudinal target polarization. Although the low analyzing power of the Møller polarimeter allows significant systematic errors, the Møller measurement provides a valuable cross-check of the absolute sign of the polarization. The Mott and Møller polarimeters measured a consistent polarization of $84 \pm 2\%$.

3 Results

The standard PV analysis techniques are used for the analysis of the A_T measurement and are discussed in more detail in [4,5]. The hydrogen data set consists of 3.1×10^5 pairs with a resolution of 610 parts per million (ppm), and the helium data set consists of 9.4×10^5 pairs with a resolution of 1040 ppm. As is typical with PV asymmetry measurements, a half-wave plate (HWP) is used for slow helicity reversal. The data set for each target is split into two approximately equal groups based on the state of the HWP. The two HWP data sets for each target are consistent. The raw detector asymmetry (A_{raw}) corrected for false asymmetries due to the beam (A_{beam}) is defined as $A_{\text{corr}} = A_{\text{raw}} - A_{\text{beam}}$. The correction to the hydrogen detector asymmetry is $A_{\text{beam}} = 0.14$ ppm, and $A_{\text{corr}} = -4.95 \pm 1.09$ (stat) ppm. The correction to the helium detector asymmetry is $A_{\text{beam}} = 0.11$ ppm, and $A_{\text{corr}} = -10.93 \pm 1.07$ (stat) ppm.

The physics result is achieved after correcting A_{corr} for the beam polarization and other normalization factors. The systematic errors are listed in table 1. The three

Table 1. Systematic errors for A_T.

Systematic	Hydrogen (ppm)	Helium (ppm)
False Asymmetry	0.136	0.110
Polarization	0.165	0.300
Linearity	0.070	0.140
Al QE	0.064	0.130
^{4}He QE	–	0.01
Rescattering	0	0.070
Total	0.240	0.370

dominant sources of error are false asymmetries, polarization measurement uncertainty, and the $\sim 1\%$ aluminum quasi-elastic (QE) dilution. The preliminary results are shown below pending final polarization and background analyses. The proton result is

$$A_T^{\mathrm{H}} = -6.58 \pm 1.47\ (\mathrm{stat}) \pm 0.24\ (\mathrm{syst})\ \mathrm{ppm}$$

for a beam energy $E_b = 3\,\mathrm{GeV}$, center-of-mass scattering angle $\theta_{\mathrm{CM}} = 16°$, and 4-momentum transfer $\langle Q^2 \rangle = 0.099\,\mathrm{GeV}^2$. The ^{4}He result is

$$A_T^{\mathrm{He}} = -13.51 \pm 1.34\ (\mathrm{stat}) \pm 0.37\ (\mathrm{syst})\ \mathrm{ppm}$$

for $E_b = 2.75\,\mathrm{GeV}$, $\theta_{\mathrm{lab}} = 6°$, and $\langle Q^2 \rangle = 0.077\,\mathrm{GeV}^2$.

Previous calculations of A_T for the proton are not valid at the HAPPEX kinematics [6,7], but our result agrees well with a recent calculation using the optical theorem to relate A_T to the total photoabsorption cross-section [8]. An exact calculation of A_T for elastic electron-^{4}He scattering predicts A_T to be of the order 10^{-10} assuming that the nucleus remains in its ground state at all times [9]. The ^{4}He A_T result is 5 orders of magnitude larger than this prediction; therefore, excited states of the nucleus must contribute significantly to the asymmetry. The calculation using the optical theorem to relate A_T to the total photabsorption cross-section predicted a value [10] which is in reasonable agreement with our result.

4 Conclusion

We conclude that the hadronic intermediate states are important contributions to the imaginary part of the 2γ-exchange amplitude for elastic electron-proton and electron-nucleus scattering. This is the first measurement of A_T for elastic electron-nucleus scattering which clearly shows that the asymmetry is non-negligible, as was previously thought. It will be especially important to measure A_T for upcoming experiments such as PREX [11], which aim to measure the very small forward-angle PV asymmetry for ^{208}Pb, because of the challenge of reducing experimental backgrounds to acceptably low levels.

References

1. S.P. Wells *et al.*, Phys. Rev. C **63**, 064001 (2001).
2. F.E. Maas *et al.*, Phys. Rev. Lett. **94**, 082001 (2005).
3. K.D. Paschke, *Transverse Asymmetries from HAPPEX and SLAC-E158*, presented at Jefferson Laboratory Annual User's Group Meeting, 2005, unpublished.
4. K.A. Aniol *et al.*, Phys. Rev. Lett. **96**, 022003 (2006).
5. K.A. Aniol *et al.*, Phys. Lett. B **635**, 275 (2006).
6. B. Pasquini, M. Vanderhaeghen, Phys. Rev. C **70**, 045206 (2004).
7. B. Pasquini, M. Vanderhaeghen, Eur. Phys. J. A **24**, 29 (2005).
8. A.V. Afanasev, N.P. Merenkov, Phys. Lett. B **500**, 48 (2004).
9. E.D. Cooper, C.J. Horowitz, Phys. Rev. C **72**, 034602 (2005).
10. Private communication with A.V. Afanasev.
11. R. Michaels, P.A. Souder, G.M. Urciuoli, spokespersons: TJNAF Experiment E06-002.

Overview of hadronic parity violation

B.R. Holstein[a]

Department of Physics-LGRT, University of Massachusetts, Amherst, MA 01003, USA

Abstract. The subject of hadronic parity violation is nearly fifty years old, but a good deal of uncertainty remains, despite many efforts both theoretical and experimental. A brief summary of the field is presented and a plan is proposed for new experimental work which, when combined with a new theoretical tack based on effective field theory, should lead to resolution of the present difficulties.

PACS. 13.75.Cs Nucleon-nucleon interactions (including antinucleons, deuterons, etc.) – 21.30.Fe Forces in hadronic systems and effective interactions – 21.30.-x Nuclear forces

1 Introduction

The subject of hadronic parity violation is an ancient one by particle physics standards. Indeed the field began soon after the 1957 publication of the experiment by Wu *et al.* [1] wherein maximal violation of parity invariance was discovered in nuclear beta decay. The first search for parity violation in the NN interaction was carried out by Tanner in the same year [2]. However, it was not until a decade later that Lobashov *et al.*, using integration techniques, was able to find a $(-6 \pm 1) \times 10^{-6}$ signal among the much larger parity conserving background in radiative neutron capture from ^{181}Ta [3]. Since that time, many different experiments have been conducted and there is no question that such signals exist. Indeed, in at least two cases —photon asymmetry in the decay of a polarized isomer of ^{180}Hf [4]

$$A_\gamma = -(1.66 \pm 0.18) \times 10^{-2} \tag{1}$$

and the cross-section asymmetry for longitudinally polarized neutrons from ^{139}La [5]

$$A_z = (9.55 \pm 0.35) \times 10^{-2} \tag{2}$$

the signals are at the $\sim 10^{-2}$ level and are many standard deviations from zero. However, such large and unambiguous signals are by far the exception. Indeed, the natural size of such effects is

$$\sim G_F m_\pi^2 = 10^{-7}$$

and in order to produce larger effects early experiments tried to use nuclei wherein the presence of closely spaced parity doublets acted as an amplifier. However, interpreting such signals, even in P-shell and S, D-shell nuclei wherein one has believable nuclear wave functions, has proved exceptionally difficult. In this presentation, we present a brief summary of this field, reviewing both past accomplishments (and frustrations!) and we suggest a blueprint for future progress that should allow resolution of the current difficulties within the next decade.

2 Hadronic parity violation: a current view

As outlined above, despite nearly fifty years of experimental work, the understanding of hadronic parity violation remains clouded. A summary of the current state of affairs and of the various experiments can be found in recent review articles by Adelberger and Haxton [6], by Haeberli and Holstein [7], and most recently by Page and Ramsey-Musolf [8].

2.1 Present experiments

On the experimental side there are two classes of (supposedly) interpretable experiments. One involves nearby parity doublets in the S, D shell, for which reasonable wavefunctions exist. These include circular polarization measurements in the photon decay of parity-mixed levels in ^{18}F and ^{21}Ne. In the case of ^{18}F the results of five experiments are in agreement [9]:

$$|P_\gamma(1081)| = \begin{cases} (-7 \pm 20) \times 10^{-4}, & \text{Caltech/Seattle,} \\ (3 \pm 6) \times 10^{-4}, & \text{Florence,} \\ (-10 \pm 18) \times 10^{-4}, & \text{Mainz,} \\ (2 \pm 6) \times 10^{-4}, & \text{Queens,} \\ (-4 \pm 30) \times 10^{-4}, & \text{Florence,} \end{cases} \tag{3}$$

while in the case of ^{21}Ne there exist two consistent measurements [10]:

$$P_\gamma = \begin{cases} (24 \pm 24) \times 10^{-4}, & \text{Seattle/Chalk River,} \\ (3 \pm 16) \times 10^{-4}, & \text{Chalk River/Seattle.} \end{cases} \tag{4}$$

[a] e-mail: `holstein@physics.umass.edu`

Finally, in the case of ^{19}F there exist two consistent measurements of the asymmetry in the decay of the first excited state [11]:

$$A_\gamma = \begin{cases} (-8.5 \pm 2.6) \times 10^{-5}, & \text{Seattle}, \\ (-6.8 \pm 1.8) \times 10^{-5}, & \text{Mainz}. \end{cases} \tag{5}$$

In each of these cases the measurements are at a precision much larger than the canonical 10^{-7} level, but these are significant because of the nuclear amplification effect, whereby the parity admixture in a given nuclear state is enhanced due to an accidental near-degeneracy of nearby levels with identical spin but opposite parity, as is clear from simple first-order perturbation theory:

$$\begin{aligned} |\psi_{J+}\rangle &\simeq |\phi_{J+}\rangle + \frac{|\phi_{J-}\rangle\langle\phi_{J-}|\mathcal{H}_{\rm wk}|\phi_{J+}\rangle}{E_+ - E_-} \\ &= |\phi_{J+}\rangle + \epsilon|\phi_{J-}\rangle, \\ |\psi_{J-}\rangle &\simeq |\phi_{J-}\rangle + \frac{|\phi_{J+}\rangle\langle\phi_{J+}|\mathcal{H}_{\rm wk}|\phi_{J-}\rangle}{E_- - E_+} \\ &= |\phi_{J-}\rangle - \epsilon|\phi_{J+}\rangle. \end{aligned} \tag{6}$$

On the other hand, some experimenters have chosen systems which do not bear the baggage of nuclear complications. However, the price paid in the requirement for experimental precision is a high one. Thus we have measurements of the asymmetry in the scattering of longitudinally polarized protons from both hydrogen [12]

$$\begin{aligned} A_z(45\,\text{MeV}) &= -(1.57 \pm 0.23) \times 10^{-7}, \quad \text{PSI}, \\ A_z(13.6\,\text{MeV}) &= -(0.93 \pm 0.20 \pm 0.05) \times 10^{-7}, \quad \text{Bonn}, \end{aligned} \tag{7}$$

as well as from ^{4}He [13]

$$A_z(46\,\text{MeV}) = -(3.34 \pm 0.93) \times 10^{-7}, \quad \text{PSI} \tag{8}$$

and both results are at the canonical 10^{-7} level and require exceptional control over systematics.

2.2 Present theory

On the theoretical side the first systematic study was carried out in 1964 by F. Curtis Michel [14], who pictured the parity-violating NN interaction as occurring via meson exchange, with the exchanged meson connecting with the nucleon on one side with the usual strong coupling and on the other with phenomenological weak couplings —*cf.* fig. 1.

The basic form of the weak couplings follows from general symmetry arguments and is parametrized as

$$\begin{aligned} \mathcal{H}_{\rm wk} = &\frac{f_\pi}{\sqrt{2}}\bar{N}(\tau\times\pi)_3 N + \bar{N}\Big[h_\rho^0 \tau\cdot\rho^\mu \\ &+ h_\rho^1\rho_3^\mu + \frac{h_\rho^2}{2\sqrt{6}}(3\tau_3\rho_3^\mu - \tau\cdot\rho^\mu)\Big]\gamma_\mu\gamma_5 N \\ &+ \bar{N}(h_\omega^0\omega^\mu + h_\omega^1\tau_3\omega^\mu)\gamma_\mu\gamma_5 N \\ &- h_\rho^{\prime 1}\bar{N}(\tau\times\rho^\mu)_3\frac{\sigma_{\mu\nu}k^\nu}{2M}\gamma_5 N \end{aligned} \tag{9}$$

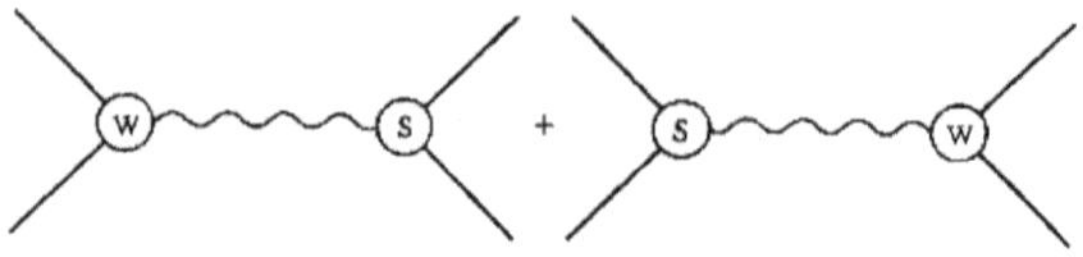

Fig. 1. Meson-exchange diagrams contributing to the PV NN interaction. Here the vertices labelled S, W refer to those involving the strong, weak interactions, respectively.

and this approach leads to the conventional DDH two-body potential:

$$\begin{aligned} V^{\rm PNC}(\boldsymbol{r}) = &\, i\frac{f_\pi g_{\pi NN}}{\sqrt{2}}\left(\frac{\tau_1\times\tau_2}{2}\right)_3(\sigma_1+\sigma_2) \\ &\cdot\left[\frac{\mathbf{p}_1-\mathbf{p}_2}{2M}, f_\pi(r)\right] - g_\rho\left(h_\rho^0\tau_1\cdot\tau_2 + h_\rho^1\left(\frac{\tau_1+\tau_2}{2}\right)_3\right. \\ &\left. + h_\rho^2\frac{(3\tau_1^3\tau_2^3 - \tau_1\cdot\tau_2)}{2\sqrt{6}}\right)\left((\sigma_1-\sigma_2)\cdot\left\{\frac{\mathbf{p}_1-\mathbf{p}_2}{2M}, f_\rho(r)\right\}\right. \\ &\left. + i(1+\chi_V)\sigma_1\times\sigma_2\cdot\left[\frac{\mathbf{p}_1-\mathbf{p}_2}{2M}, f_\rho(r)\right]\right) \\ &- g_\omega\left(h_\omega^0 + h_\omega^1\left(\frac{\tau_1+\tau_2}{2}\right)_3\right)\left((\sigma_1-\sigma_2)\cdot\left\{\frac{\mathbf{p}_1-\mathbf{p}_2}{2M}, f_\omega(r)\right\}\right. \\ &\left. + i(1+\chi_S)\sigma_1\times\sigma_2\cdot\left[\frac{\mathbf{p}_1-\mathbf{p}_2}{2M}, f_\omega(r)\right]\right) \\ &- (g_\omega h_\omega^1 - g_\rho h_\rho^1)\left(\frac{\tau_1-\tau_2}{2}\right)_3(\sigma_1+\sigma_2)\cdot\left\{\frac{\mathbf{p}_1-\mathbf{p}_2}{2M}, f_\rho(r)\right\} \\ &- g_\rho h_\rho^{1\prime} i\left(\frac{\tau_1\times\tau_2}{2}\right)_3(\sigma_1+\sigma_2)\cdot\left[\frac{\mathbf{p}_1-\mathbf{p}_2}{2M}, f_\rho(r)\right], \end{aligned} \tag{10}$$

where $f_V(r) = \exp(-m_V r)/4\pi r$ is the usual Yukawa function.

The theoretical challenge here is to calculate the size of the weak couplings using some sort of first principles technique. Michel employed a simple factorization approach [14], wherein one replaces the sum over a complete set of intermediate states by a single state —the vacuum—

$$\langle NM|V_\lambda A^\lambda|N\rangle \approx \langle M|V_\lambda|0\rangle\langle N|A^\lambda|N\rangle \tag{11}$$

and this paper was soon joined by other works on the subject. One breakthrough came in the late 1960s when it was realized by Tadic, Fischbach, and others that the parity-violating pion coupling could be determined in terms of experimental hyperon decay amplitudes using a simple $SU(3)$ relation [15]:

$$\begin{aligned} \langle\pi^+ n|\mathcal{H}_{\rm wk}^c|p\rangle = &-\sqrt{\frac{2}{3}}\tan\theta_c(2\langle\pi^- p|\mathcal{H}_{\rm wk}|\Lambda^0\rangle \\ &-\langle\pi^-\Lambda^0|\mathcal{H}_{\rm wk}|\Xi^-\rangle). \end{aligned} \tag{12}$$

Unfortunately the signs and magnitudes of the Λ and Ξ^- decay amplitudes are nearly identical so that considerable cancellation takes place. Consequently, the reliability of this estimate is not so clear. Another approach was taken by McKellar and Pick who used the symmetry $SU(6)_W$ to relate vector meson and pion parity-violating amplitudes [16]. However, the seminal contribution in this area remains the 1980 paper bv Desplanques, Donoghue, and Holstein, which has come to be called simply DDH [17]. In this work, the parity-violating NNM couplings were estimated using quark model and $SU(6)$ symmetry ideas, enabling bounds to be placed as well as estimated "best values" for each coupling, as shown in table 1.

However, the bounds are rather generous and the "best values" are really "best guesses." The goal of the field has been to come up with definitive *experimental* values for these quantities. One does this (hopefully) by attempting to correlate findings from the various experiments. However, at the present time this process has *not* produced an internally consistent set of results.

Indeed, without going into detail, the present state of affairs is somewhat discouraging and is summarized in fig. 2. Note that although there exist seven weak couplings in general, the most important by far are the isovector pion coupling, because of its long range, and a linear combination of isoscalar rho and omega couplings. This is the reason why one can represent the experimental situation in terms of a simple two-dimensional plot. The various allowed ranges shown here come from:

i) experiments in ^{18}F which place limits on the isovector pion coupling [9];
ii) experiments in ^{19}F which place limits on a linear combination of the pion and vector meson couplings [11];
iii) *pp* scattering experiments, which measure the size of the vector meson couplings [12];
iv) $p\alpha$ scattering experiments, which measure a linear combination of the pion and vector meson couplings [13];

There are no doubt a number of reasons for this situation, as explored in the above review articles, but the most likely problem is the inadequacy of existing nuclear wavefunctions when dealing with parity-violating observables. A recent paper by Miller is suggestive in this regard [20].

Table 1. Weak NNM couplings as calculated in refs. [17–19]. All numbers are quoted in units of the "sum rule" value $g_\pi = 3.8 \cdot 10^{-8}$.

Coupling	DDH [17] Reasonable range	DDH [17] "Best" value	FCDH [19]
f_π	$0 \to 30$	+12	+7
h_ρ^0	$30 \to -81$	−30	−10
h_ρ^1	$-1 \to 0$	−0.5	−1
h_ρ^2	$-20 \to -29$	−25	−18
h_ω^0	$15 \to -27$	−5	−13
h_ω^1	$-5 \to -2$	−3	−6

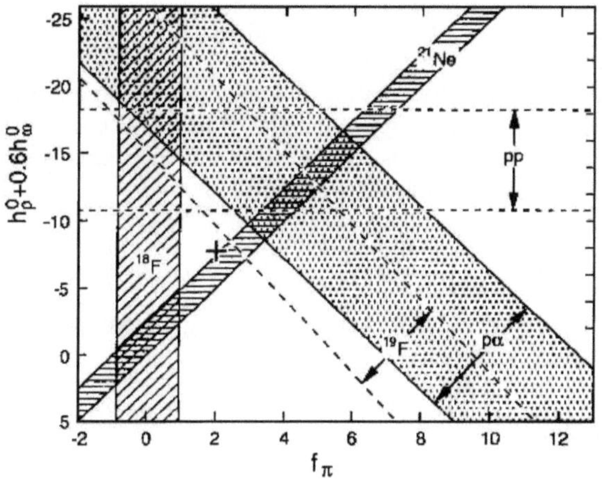

Fig. 2. Restrictions on weak parity-violating pion and isoscalar vector meson couplings (in units of 3.8×10^{-7}) which arise from various particle and nuclear experiments.

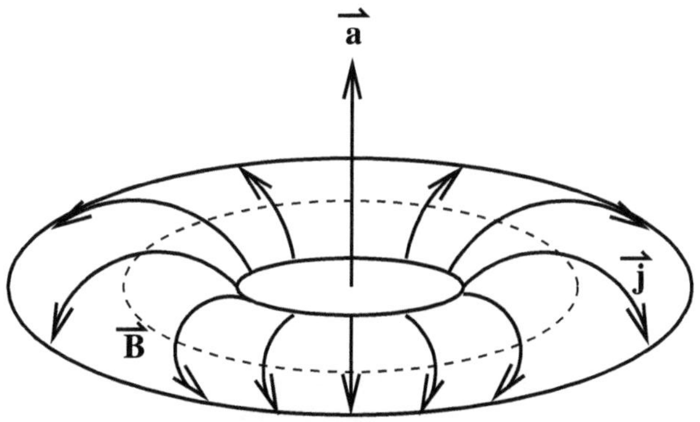

Fig. 3. Toroidal current flow which is allowed in the presence of parity nonconservation and which leads to an anapole moment.

Another intriguing application of hadronic parity violation is its connection to the "anapole" moment. In order to understand this quantity, consider the problem of finding the fields associated with a local charge distribution. In the ordinary (parity-conserving) case, the far-field region can be described in terms of a multipole expansion, in which the multipole characterizes the inverse power law by which the fields fall off. However, in the presence of parity violation, then different current distributions are permitted, such as a toroidal pattern —*cf.* fig. 3— wherein the fields are purely local! The anapole moment characterizes such a field pattern.

Another way to understand the anapole is to consider the problem of writing the most general matrix element of the electromagnetic current between a pair of spin-1/2 states, which is of the form

$$\langle f|V_\mu^{em}|i\rangle = \bar{u}(p_f)\Big[F_1(q^2)\gamma_\mu - F_2(q^2)\frac{i\sigma_{\mu\nu}q^\nu}{2M} + F_3(q^3)\frac{1}{4M^2}(\gamma_\mu\gamma_5 q^2 - q_\mu \not{q}\gamma_5) + F_4(q^2)\frac{i\sigma_{\mu\nu}q^\nu\gamma_5}{2M}\Big]u(p_i). \qquad (13)$$

Here $F_1(q^2)$, $F_2(q^2)$ are usual charge and magnetic form factors, while $F_4(q^2)$ violates both P *and* T and is the

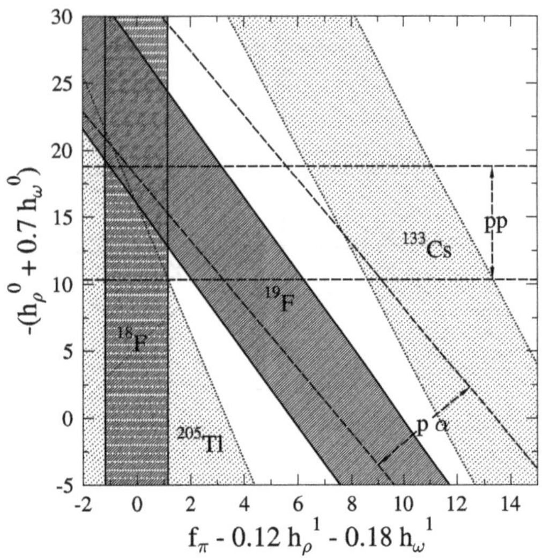

Fig. 4. Restrictions on weak parity-violating pion and isoscalar vector meson couplings (in units of 3.8×10^{-7}) which arise from various particle and nuclear experiments, including the anapole moment limits from ^{133}Cs and Tl.

electric dipole moment. The last form factor —$F_3(q^2)$— violates P but not T and represents the anapole moment. Note that the anapole matrix element is gauge invariant, but vanishes, because of the presence of the factor q^2, in the case of real photon interactions. Nevertheless, the anapole *can* contribute to processes involving virtual photon exchange, and mimics a parity-violating Z^0-exchange process. Note that the factor of q^2 just cancels the $1/q^2$ in the photon propagator, leading to a local spin-dependent and parity-violating interaction

$$\mathcal{H}_w^{eff} = \frac{G_F}{\sqrt{2}}(\kappa_Z + \kappa_a)\boldsymbol{\alpha}_e \cdot \boldsymbol{J}_{nuc}\rho(r), \qquad (14)$$

where $\boldsymbol{J}_{nuc}$ is the nuclear spin, $\boldsymbol{\alpha}_e \equiv \bar{u}(k_f)\gamma u(k_i)$ is the electron matrix element, κ_Z is the direct Z-exchange term, and κ_a is the anapole moment. Wieman *et al.* have managed to measure the anapole moment in a heroic experiment involving detecting the spin dependence using parity violation observed using the hyperfine levels of 6S-7S transitions in ^{133}Cs [21]. The result, using the calculated value $\kappa_Z = 0.013$ for the direct Z-exchange contribution was $\kappa_a = 0.112 \pm 0.016$, which can in turn be interpreted in terms of the DDH parameters as [22]

$$h_\pi - 0.21\left(h_\rho^0 + 0.6h_\omega^0\right) = (0.99 \pm 0.16) \times 10^{-6}. \qquad (15)$$

Unfortunately, inclusion of this result together with that of an upper bound on the anapole moment of ^{205}Tl [23] merely makes the situation of the weak couplings even more confusing, as shown in fig. 4.

3 Hadronic parity violation: new hope!

In order to bring order out of this chaos, a new approach is needed and this has been suggested in recent theoretical work, which proposes approaching the problem from the perspective of effective field theory as well as removing some of the nuclear uncertainty from the problem [24]. In the effective field theory picture the low-energy component of the parity-violating NN interaction is described in terms of its long-range pion exchange component plus a group of short-range contact interactions, without specific reference to vector meson exchange, although the basic form of the operators is similar to that used by DDH. Nevertheless, there is more freedom in choosing the contact couplings here, and this may be of help in sorting out the problem.

In order to remove some of the nuclear uncertainty it is proposed to first utilize a series of experiments on systems which are no heavier than the alpha particle, in order that nuclear effects be completely under control. From general considerations we know that the strangeness conserving effective weak interaction is the product of currents carrying isotopic spin one and zero or isospin 1/2 times 1/2 and therefore can itself carry total isotopic spin zero, one, or two. Also, at the very lowest —near threshold— energies, only mixing between s- and p-states is allowed and it is easy to show from basic symmetry arguments that there can exist only *five* independent weak NN amplitudes:

a) $c_t(k)$: $^3S_1$-$^3P_1$ mixing carrying $\Delta I = 1$,
b) $d_t(k)$: $^3S_1$-$^1P_1$ mixing carrying $\Delta I = 0$,
c) $\lambda_s^{0,1,2}(k)$: $^1S_0$-$^3P_0$ mixing carrying $\Delta I = 0, 1, 2$.

We can characterize each by a scattering amplitude $d_i(k)$, for which the constraint of unitarity requires that

$$d_i(k) = |d_i(k)| \exp i(\delta_S(k) + \delta_P(k)), \qquad (16)$$

where $\delta_S(k)$ and $\delta_P(k)$ are the appropriate strong interaction S- and P-wave phase shifts at momentum k. Of course, at low energy the P-wave phase shift can presumably be neglected with respect to its S-wave counterpart, so Danilov has suggested the parameterization [25]

$$d_i(k) \approx \lambda_i m_i(k), \qquad (17)$$

where

$$m_i(k) = \frac{1}{k}e^{i\delta_S(k)} \sin\delta_S(k) \qquad (18)$$

is the usual parity-conserving S-wave transition amplitude and λ_i is a simple (real) constant. Thus the goal of low-energy experiments is to determine the five constants λ_i. In fact, since at threshold we have

$$\lim_{k\to 0} m_i(s) = a_i, \qquad (19)$$

where a_i is the S-wave scattering length, we can define the five real constants ρ_t, λ_t, $\lambda_s^{0,1,2}$ via

$$\lim_{k\to 0} c_t(k), d_t(k), d_s^{0,1,2}(k) = \rho_t a_t, \lambda_t a_t, \lambda_s^{0,1,2} a_s. \qquad (20)$$

We then need five independent low-energy experiments, and in order to minimize the nuclear uncertainty we shall demand nuclei with $A \leq 4$. Two of these are already in hand:

i) low-energy **p-p** scattering, which has been performed at LANL, PSI, and Bonn [12],
ii) low-energy **p-α** scattering, which has been performed at PSI [13],

but additional results are desperately needed. Two are already on the horizon:

iii) neutron spin rotation in He is an approved experiment at NIST and has been approved for running at the fundamental physics beamline at the SNS,
iv) near threshold radiative capture of polarized neutrons by protons is a LANSCE experiment which is now in a commissioning run running at LANSCE, after which it will be moved to the fundamental physics beamline at the SNS [26]. The capture asymmetry is sensitive to the parity-violating pion coupling h_π.

However, additional experiments are needed in order to bring this program to completion. Some examples include

v) measurement of the asymmetry in low-energy photodisintegration of deuterium by polarized photons, which could be accomplished at HIGS or at the electron accelerator being built at Athens,
vi) neutron spin rotation on hydrogen, which is and approved experiment at the SNS,
vii) polarized neutron capture asymmetry on deuterium, which is also an approved experiment at SNS, and which will presumably run once the $np \to d\gamma$ experiment has completed data taking.

The combination of five such precise experiments should allow the identification of the five independent low-energy weak couplings, and thereby the pieces of the effective parity-violating interaction. Obviously, the fundamental physics neutron beamline at SNS is an *essential* component of any such program.

As a first step in the interpretation of these experiments we can use the approximate calculations of Desplanques and Missimer [27]:

i) **pp** scattering:

$$pp(13.6\,\text{MeV}), \qquad A_L = -0.48M\lambda_s^{pp},$$
$$pp(45\,\text{MeV}), \qquad A_L = -0.82M\lambda_s^{pp},$$

ii) **$p\alpha$** scattering:

$$p\alpha(46\,\text{MeV}),$$
$$A_L = -M\left[0.48\left(\lambda_s^{pp} + \frac{1}{2}\lambda_s^{pn}\right) + 1.07\left(\frac{1}{2}\lambda_t + \rho_t\right)\right],$$

iii) radiative capture —$np \to d\gamma$:

a) circular polarization: $P_\gamma = M(0.63\lambda_t - 0.16\lambda_s^{np})$,

b) photon asymmetry: $A_\gamma = -0.11M\rho_t$,

iv) neutron spin rotation in ^{4}He:

$$\frac{\mathrm{d}\phi^{n\alpha}}{\mathrm{d}z} = \left[0.85\left(\lambda_s^{nn} - \frac{1}{2}\lambda_s^{pn}\right)\right.$$
$$\left.-1.89\left(\rho_t - \frac{1}{2}\lambda_t\right)\right] m_N\,\text{rad/m}.$$

However, it is now time for precision calculation of these connections and this has begun with the work of Carlson *et al.* [28] and of Liu [29].

On the theoretical side it is also important to move beyond the meson exchange approach of DDH and to use a formalism which is model independent. This step has been taken by Zhu *et al.*, who have written out the most general parity-violating interaction in an effective field theoretic approach [24]. At the very lowest energies, one can utilize a pionless theory, wherein the most general interaction is of the form

$$\begin{aligned}
V^{PNC}(\boldsymbol{r}) = \frac{2}{\Lambda_\chi^3}\Bigg\{&\left[C_1 + (C_2 + C_4)\left(\frac{\tau_1 + \tau_2}{2}\right)_3\right.\\
&\left. + C_3\tau_1\cdot\tau_2 + \mathcal{I}_{ab}C_5\tau_1^a\tau_2^b\right]\\
&\times(\boldsymbol{\sigma}_1 - \boldsymbol{\sigma}_2)\cdot\{-i\boldsymbol{\nabla}, g_m(r)\}\\
&+\left[\tilde{C}_1 + (\tilde{C}_2 + \tilde{C}_4)\left(\frac{\tau_1 + \tau_2}{2}\right)_3\right.\\
&\left. + \tilde{C}_3\tau_1\cdot\tau_2 + \mathcal{I}_{ab}\tilde{C}_5\tau_1^a\tau_2^b\right]\\
&\times i\,(\boldsymbol{\sigma}_1 \times \boldsymbol{\sigma}_2)\cdot[-i\boldsymbol{\nabla}, g_m(r)]\\
&+(C_2 - C_4)\left(\frac{\tau_1 - \tau_2}{2}\right)_3\\
&\times(\boldsymbol{\sigma}_1 + \boldsymbol{\sigma}_2)\cdot\{-i\boldsymbol{\nabla}, f_m(r)\}\\
&+C_6 i\epsilon^{ab3}\tau_1^a\tau_2^b(\boldsymbol{\sigma}_1 + \boldsymbol{\sigma}_2)\cdot[-i\boldsymbol{\nabla}, g_m(r)]\Bigg\}
\end{aligned} \tag{21}$$

with

$$\mathcal{I} = \begin{pmatrix} 1 & 0 & 0 \\ 0 & 1 & 0 \\ 0 & 0 & -2 \end{pmatrix} \tag{22}$$

and $g_m(\boldsymbol{r})$ is a function that

i) strongly peaked, with width $\sim 1/m$ about $r = 0$,
ii) approaches $\delta^{(3)}(\boldsymbol{r})$ in zero-width ($m \to \infty$) limit.

such as

$$g_m(r) = \frac{m^2}{4\pi r}\exp(-mr) \tag{23}$$

Also, $\Lambda_\chi = 4\pi F_\pi \sim 1\,\text{GeV}$ is the chiral scale and C_i, $\tilde{C}_i$ are unknown counterterms. We see then that in the pionless theory, there are ten constants. However, at threshold and in the zero-range approximation it is easy to see that they appear only in the five combinations:

$$\begin{aligned}
&\lambda_t \propto (C_1 - 3C_3) - (\tilde{C}_1 - 3\tilde{C}_3),\\
&\lambda_s^0 \propto (C_1 + C_3) + (\tilde{C}_1 + \tilde{C}_3),\\
&\lambda_s^1 \propto (C_2 + C_4) + (\tilde{C}_2 + \tilde{C}_4),\\
&\lambda_s^2 \propto -\sqrt{\frac{8}{3}}(C_5 + \tilde{C}_5),\\
&\rho_t \propto (C_2 - C_4) + 2C_6,
\end{aligned} \tag{24}$$

where we have defined $\bar{\Lambda} = m_N m_\rho^2/\Lambda_\chi^3$. Using finite-size corrections, these relations are somewhat modified. For example, using the results of Desplanques and Benayoun using the Reid soft core potential, we have [30]

$$\begin{aligned}
m_N\rho_t &= -\frac{2}{\bar{\Lambda}^3}\left[B_2\left(\frac{1}{2}C_2-\frac{1}{2}C_4+C_6\right)\right.\\
&\quad\left.+B_3\left(\frac{1}{2}C_2-\frac{1}{2}C_4-C_6\right)\right],\\
m_N\lambda_t &= -\frac{2}{\bar{\Lambda}^3}[B_4(C_1-3C_3+\tilde{C}_1-3\tilde{C}_3)\\
&\quad+B_5(C_1-3C_3-\tilde{C}_1+3\tilde{C}_3)],\\
m_N\lambda_s^0 &= -\frac{2}{\bar{\Lambda}^3}[B_6(C_1+C_3+\tilde{C}_1+\tilde{C}_3)\\
&\quad+B_7(C_1+C_3-\tilde{C}_1-\tilde{C}_3)],\\
m_N\lambda_s^1 &= -\frac{2}{\bar{\Lambda}^3}[B_6(C_2+C_4+\tilde{C}_2+\tilde{C}_4)\\
&\quad+B_7(C_2+C_4-\tilde{C}_2-\tilde{C}_4)],\\
m_N\lambda_s^2 &= \frac{4\sqrt{6}}{\bar{\Lambda}^3}[B_6(C_5+\tilde{C}_5)+B_7(C_5-\tilde{C}_5)]. \qquad (25)
\end{aligned}$$

Using the Reid soft core potential to estimate B_i, we find $B_i = -0.0043, 0.0005, -0.0009, -0.0022, -0.0067, 0.0003$ for $i = 2,\ldots,7$. Thus corrections to the zero-range approximation are relatively small except in the case of λ_t, where they are at the $\sim 30\%$ level.

The connection with DDH can be made by noting that while in the general effective field theoretic analysis C_i, which determines the convective contribution to the PNC potential, and $\tilde{C}_i$, which determines the magnetic contribution, are *independent* parameters. At threshold they contribute in a well-defined combination, as shown above in eq. (24). However, at higher energies they involve different combinations and therefore one has in general *ten* different low-energy constants which must be determined, requiring *ten* independent experiments. However, in the DDH model, due to vector dominance the strength of the convective and magnetic pieces of any given meson exchange channel is *fixed* by vector dominance. Thus we can connect with DDH via the definitions

$$\begin{aligned}
C_1^{DDH} &= -\frac{1}{2}\bar{\Lambda}_\omega^3 g_\omega h_\omega^0, & C_2^{DDH} &= -\frac{1}{2}\bar{\Lambda}_\omega^3 g_\omega h_\omega^1,\\
C_3^{DDH} &= -\frac{1}{2}\bar{\Lambda}_\rho^3 g_\rho h_\rho^0, & C_4^{DDH} &= -\frac{1}{2}\bar{\Lambda}_\rho^3 g_\rho h_\rho^1,\\
C_5^{DDH} &= \frac{1}{4\sqrt{6}}\bar{\Lambda}_\rho^3 g_\rho h_\rho^2, & C_6^{DDH} &= -\frac{1}{2}\bar{\Lambda}_\rho^3 g_\rho h_\rho^{1'}, \qquad (26)
\end{aligned}$$

where g_ρ, g_ω are the strong vector meson coupling constants and

$$\begin{aligned}
\tilde{C}_i^{DDH}/C_i^{DDH} &= 1+\chi_\omega, \quad i=1,2,\\
\tilde{C}_i^{DDH}/C_i^{DDH} &= 1+\chi_\rho, \quad i=3,4,5, \qquad (27)
\end{aligned}$$

which reduces the number of parameters back to five.

Of course, at higher energies —$E \simeq m_\pi$— it is necessary also to introduce the pion as an explicit degree of freedom. In this case the channel corresponding to pion exchange must involve two pieces —a long-range component corresponding the explicit pion exchange and a short-range component which consists of everything else. It turns out that in addition at this order there exists one more parameter, which is a contact term involving two nucleons a pion and a photon. Also, medium-range contributions involving two-pion exchange must also be introduced [24]. This complicates the analysis, but must be done for a completely consistent picture.

At energies $E \simeq m_\pi$ then there must be at least twelve independent experiments which must be done and this is a major program. If one is able to keep well below this energy, then a simpler and much more manageable program is possible. In the case of the experiments involving pp scattering, neutron spin rotation on hydrogen, and the two independent $np \to d\gamma$ parameters this is not a problem, since these results can be obtained very near threshold. However, it is an open question whether the experiments involving ^{4}He —p^4He scattering and neutron spin rotation on ^{4}He— can be treated in this fashion, even near threshold, since Fermi motion cannot be eliminated.

In any case a successful analysis of careful near-threshold experiments by the model-independent EFT potential should allow a rigorous extraction of the Danilov parameters. Of course, these should be confirmed by experiments which overconstrain these results. If this program is successful we will then have model-independent values of basic parity-violating counterterms. A vigorous theoretical program should follow, involving

a) using these now confirmed parameters in order to analyze previous results in heavier nuclei —any discrepancies can then be ascribed to nuclear physics uncertainties;
b) confrontation of the measured numbers with fundamental theory via reliable methods such as lattice gauge theory;
c) use of these numbers to confidently predict "background" effects in future experiments.

At that point, after a half-century of work, we would finally be able to designate the problem of hadronic parity violation as being "solved." It is time to get to work!

This work was supported in part by the National Science foundation under award PHY 02-44801.

References

1. C.S. Wu *et al.*, Phys. Rev. **105**, 1413 (1957).
2. N. Tanner, Phys. Rev. **107**, 1203 (1957).
3. V.M. Lobashov *et al.*, Phys. Lett. B **25**, 104 (1967).
4. K.S. Krane *et al.*, Phys. Rev. Lett. **26**, 1579 (1971); Phys. Rev. C **4**, 1906 (1971).
5. V.W. Yuan *et al.*, Phys. Rev. C **44**, 2187 (1991); V.P. Alfimenko *et al.*, Nucl. Phys. A **398**, 93 (1983); Y. Masuda *et al.*, Nucl. Phys. A **504**, 269 (1989).
6. E.G. Adelberger, W.C. Haxton, Annu. Rev. Nucl. Part. Sci. **35**, 501 (1985).
7. W. Haeberli, B.R. Holstein, in *Symmetries and Fundamental Interactions in Nuclei*, edited by W.C. Haxton, E.M. Henley (World Scientific, Singapore, 1995) p. 17.

8. S. Page, M.J. Ramsey-Musolf, Annu. Rev. Nucl. Part. Sci. **56**, 1 (2006).
9. C.A. Barnes *et al.*, Phys. Rev. Lett. **40**, 840 (1978); M. Bini *et al.*, Phys. Rev. Lett. **55**, 795 (1985); G. Ahrens *et al.*, Nucl. Phys. A **390**, 486 (1982); H.C. Evans *et al.*, Phys. Rev. Lett. **55**, 791 (1965); P.G. Bizetti *et al.*, Lett. Nuovo Cimento **29**, 167 (1980).
10. K.A. Snover *et al.*, Phys. Rev. Lett. **41**, 145 (1978); E.D. Earle *et al.*, Nucl. Phys. A **396**, 221 (1983).
11. E.G. Adelberger, in *Polarization Phenomena in Nuclear Physics*, AIP Conf. Proc. **69**, 1367 (1981); E.G. Adelberger *et al.*, Phys. Rev. C **27**, 2833 (1983); K. Elsener *et al.*, Phys. Lett. B **117**, 167 (1982); Phys. Rev. Lett. **52**, 1476 (1984).
12. S. Kistryn, J. Lang, J. Liechti, Th. Maier, R. Müller, F. Nessi-Tedaldi, M. Simonius, J. Smyrski, S. Jaccard, W. Haeberli, J. Sromicki, Phys. Rev. Lett. **58**, 1616 (1987); P.D. Eversheim, W. Schmitt, S. Kuhn, F. Hinterberger, P. von Rossen, J. Chlebek, R. Gebel, U. Lahr, B. von Przewoski, M. Wiemer, V. Zell, Phys. Lett. B **256**, 11 (1991).
13. J. Lang *et al.*, Phys. Rev. C **34**, 1545 (1986); Phys. Rev. Lett. **54**, 170 (1985).
14. F.C. Michel, Phys. Rev. **133**, B329 (1964).
15. B.H.J. McKellar, Phys. Lett. B **26**, 107 (1967); E. Fischbach, Phys. Rev. **170**, 1398 (1968); D. Tadic, Phys. Rev. **174**, 1694 (1968); W. Kummer, M. Schweda, Acta Phys. Austriaca **28**, 303 (1968).
16. B.H.J. McKellar, P. Pick, Phys. Rev. D **7**, 260 (1973).
17. B. Desplanques, J.F. Donoghue, B.R. Holstein, Ann. Phys. (N.Y.), **124**, 449 (1980).
18. V.M. Dubovik, S.V. Zenkin, Ann. Phys. (N.Y.) **172**, 100 (1986).
19. G.B. Feldman, G.A. Crawford, J. Dubach, B.R. Holstein, Phys. Rev. C **43**, 863 (1991).
20. G.A. Miller, Phys. Rev. C **67**, 042501 (2003).
21. C.S. Wood *et al.*, Science **275**, 1759 (1997).
22. W.C. Haxton, C.-P. Liu, M.J. Ramsey-Musolf, Phys. Rev. C **65**, 045502 (2002).
23. N.H. Edwards *et al.*, Phys. Rev. Lett. **74**, 2654 (1995); P. Vetter *et al.*, Phys. Rev. Lett. **74**, 2658 (1995).
24. S.-L. Zhu *et al.*, Nucl. Phys. A **748**, 435 (2005).
25. G.S. Danilov, Phys. Lett. **18**, 40 (1965); Phys. Lett. B **35**, 579 (1971); Sov. J. Nucl. Phys. **14**, 443 (1972).
26. W.M. Snow *et al.*, Nucl. Instrum. Methods A **440**, 729 (2000).
27. B. Desplanques, J. Missimer, Nucl. Phys. A **300**, 286 (1978).
28. J. Carlson *et al.*, Phys. Rev. C **65**, 035502 (2002).
29. C.-P. Liu, private communication.
30. B. Desplanques, J.J. Benayoun, Nucl. Phys. A **458**, 689 (1986).

Parity violation in neutron spin rotation

W.M. Snow

Indiana University/Indiana University Cyclotron Facility, Bloomington, IN 47408 USA

Abstract. Nonleptonic weak interactions remain one of the most poorly understood sectors of the Standard Model. A quantitative description of the weak NN interaction is needed to understand weak interaction phenomena in atomic, nuclear, and hadronic systems. Measurements with low energy neutrons can lead to significant experimental progress. I describe the phenomenon of parity-odd neutron spin rotation and outline experimental strategies to isolate this phenomenon in light nuclei.

PACS. 11.30.Er charge conjugation, parity – 24.80.+y nuclear tests of fundamental interaction – 28.20.-V neutron physics

1 The weak NN interaction: physics and relation to other systems

Despite nearly 40 years of study, the weak interaction between nucleons is not understood. The NN weak interaction can only be seen through parity-odd (PV) cross-terms in the interference between weak and strong amplitudes. The natural scale for the relative size of weak and strong interaction amplitudes, set by the ratio of the amplitudes for W and Z exchange to those for meson exchange between nucleons, is extremely small (10^{-7}). Reviews of the subject conducted over the last 2 decades[1][2][3][4] observe that, with the exception of one amplitude in the p-p system, the weak NN amplitudes are unknown. It is important to understand low energy weak NN amplitudes for a number of reasons:

(1) NN weak interactions are in principle a new probe of strong QCD. Quark-quark charged and neutral currents in the regime of perturbative QCD obey the Standard Model[5] so any deviations at low energy must be due to strong QCD effects. The range of the weak interaction between quarks is much smaller than the nucleon size, and the strong NN repulsion at short distances means that weak interaction dynamics between nucleons must involve meson exchange and soft QCD physics. The weak NN interaction thus samples both short-distance quark-quark correlations in QCD and its chiral symmetry-dominated long-range properties: it is an "inside-out" probe of QCD.

(2) Nonleptonic weak interactions in strangeness changing processes such as nonleptonic kaon decay and hyperon decay are poorly understood. Observed weak amplitudes can differ from their natural size by almost an order of magnitude. Should these phenomena be blamed only on the strange quark, or might the weak interactions among u and d quarks in hadrons also give rise to unexpected results? If strong interaction theory is unable to describe the effects of a very small, well-understood, short-range perturbation such as the weak interaction, then we would have something qualitatively new to learn about the strongly interacting limit of QCD.

(3) Knowledge of weak NN couplings can make possible a quantitative interpretation of many PV phenomena in measurements at nuclear and atomic scales, including (a) PV observables in shell model nuclei[6], (b) PV in neutron reactions in heavy nuclei[7][8], analyzed using ideas from quantum chaos and nuclear statistical spectroscopy, (c) the contribution to PV in the ^{133}Cs atom[9] arising from the nuclear anapole moment[10][11], and (d) to PV electron scattering from the virtual photon coupling to an axial current among quarks in the nucleon[12][13][14][15]. Finally, weak NN matrix elements in nuclei bear many similarities with the matrix elements that must be calculated to interpret limits on neutrino masses from double beta decay searches[16]. In all cases, the PV observables open a new window into specific features of many-body states that can only be interpreted if the NN weak interaction amplitudes are known.

(4) Theoretical advances in the description of PV in the NN and few nucleon systems now make much better contact with QCD. Previous model-dependent approaches, such as the valence quark model used by Desplanques, Donoghue, and Holstein (DDH)[17], employed a weak pion coupling constant f_π and six other meson couplings h_ρ^0, h_ρ^1, $h_\rho^{1'}$, h_ρ^2, h_ω^0, and h_ω^1 in a meson exchange model. f_π has been calculated using QCD sum rules[18][19] and in an SU(3) Skyrme model[20], and a calculation in a chiral quark model is in progress[21].

Recently a systematic analysis of the weak NN interaction using an effective field theory (EFT) approach and chiral perturbation theory has appeared[22][23][4]. 6 constants appear in this hybrid EFT approach for low-energy (< 40 MeV) NN processes[24], which takes the wave functions from strong interaction phenomenology and the weak NN operator structure from EFT. 5 are closely related to the five independent parity-odd S-P NN elastic scattering amplitudes first emphasized by Danilov[25]: $m_N\lambda_s^{pp}$, $m_N\lambda_s^{np}$, and $m_N\lambda_s^{nn}$ for the three independent $^1S_0 \rightarrow ^3P_0$ transitions (pp,pn,nn, ΔI=0,1,2), $m_N\lambda_t$ for the $^3S_1 \rightarrow ^1P_1$ transition (np, ΔI=0), and $m_N\rho_t$ for the $^3S_1 \rightarrow ^3P_1$ transition (np, ΔI=1). A sixth parameter C_6^π describes long-range weak pion exchange and is related to the DDH weak pion coupling f_π. In parallel with this EFT development, preparations have been made to calculate the weak NN couplings using lattice gauge theory in the partially quenched approximation[26]. Taken together, these theoretical efforts set the stage for eventual quantitative predictions of the weak NN interaction directly from the Standard Model and QCD.

It is not practical in the immediate future to conduct enough PV measurements in the NN system to constrain all of these couplings. At the moment only one constant, $m_N\lambda_s^{pp}$, is determined from experiment. However the wave functions of nuclear few body systems are now understood well enough that PV measurements in these systems can be cleanly interpreted in terms of weak NN couplings. In the case of the strong NN interaction, the recent development of Green's Function Monte Carlo (GFMC) techniques, coupled with the extensive strong NN interaction database, has allowed the energy levels of light ($A < 6$) nuclei to be calculated to 1% accuracy[27][28][29]. With the strong interaction contribution to nuclear wave functions known, the effects of the weak NN interaction can be added in as perturbations to see how they contribute to PV observables in finite nuclei. PV observables in np and pp systems have been calculated[30][24] and those in few-nucleon systems can be calculated[31][32] in terms of weak couplings.

2 Parity-odd neutron spin rotation: the phenomenon

A transverse rotation of the neutron spin about its momentum vector as it passes through isotropic, unpolarized matter describes a corkscrew in space. This effect manifestly violates parity[33] in an aesthetically pleasing, direct manner which is obvious even to a nonspecialist. From a neutron optical viewpoint, this phenomenon is caused by the presence of a helicity-dependent neutron index of refraction. The index of refraction n of a medium in terms of the coherent forward scattering amplitude $f(0)$ for a low-energy neutron is $n = 1 - \frac{2\pi\rho f(0)}{k^2}$ where $\mathbf{k}$ is the incident neutron wave vector and ρ is the number density of scatterers in the medium. At low energy in an unpolarized medium, $f(0)$ is the sum of two terms: a parity-conserving term f_{PC} dominated by the strong interaction and consisting of s-waves at low energy, and a parity violating term f_{PV} that contains only weak interactions and is dominated by a p-wave contribution at low energy. f_{PV} is proportional to $\sigma_\mathbf{n} \cdot \mathbf{k_n}$ ($\sigma_\mathbf{n}$ is the neutron spin vector) and so has opposite signs for the positive and negative helicity neutron spin states. As a neutron moves a distance z in the medium, the two helicity states accumulate different phases: $\phi_\pm = \phi_{PC} \pm \phi_{PV}$. ϕ_{PV} causes a relative phase shift of the two neutron helicity components and therefore a rotation of the neutron polarization about its momentum. Since the parity-odd amplitude is proportional to k, the rotary power per unit length $d\phi/dz = 4\pi\rho f_{PV}/k$ tends to a constant for low energy neutrons[34].

Parity violation in the neutron-electron interaction also contributes to ϕ_{PV}. This amplitude can be calculated from the Standard Model in the low energy limit as $d\phi_{ne}/dz = \frac{G_F}{\sqrt{2}} g_A(4\sin^2\theta_W - 1)Z\rho$. It is accidentally small given the size of the Weinberg angle[35] and is a negligible contribution for light nuclei in comparison to the NN weak interaction.

An order-of-magnitude estimate leads one to expect weak rotary powers in the $10^{-6} - 10^{-7}$ rad/m range. In the case of parity violation in compound resonances in neutron-nucleus reactions there are amplification mechanisms[36] which can enhance parity-odd observables by factors as large as 10^5. These amplification mechanisms depend in an essential way on the complexity of the states involved and the reaction dynamics. Part of the amplification comes from the decrease in the spacing between levels as the number of nucleons increases, which brings opposite parity states closer together and increases their weak mixing amplitudes [37][38], and for low energy neutron reactions in heavy nuclei it leads to a generic amplification of order 10^2 in parity-odd amplitudes. In addition, for low energy neutron-nucleus interactions the resonances are mainly $l = 0$ and $l = 1$ with the scattering amplitudes in s-wave resonances larger than for p-wave resonances by a factor of order 10^2 to 10^3. At an energy close to a p-wave resonance, the weak interaction mixes in an s-wave component that is typically much larger, and this factor also amplifies the asymmetry. These amplification mechanisms for p-wave resonances were predicted theoretically [39][40] before they were measured [41][42].

Parity-odd neutron spin rotation has been observed in ^{117}Sn[43], Pb[44], and ^{139}La[45], and has been searched for in ^{133}Cs, Rb, and ^{81}Br[46]. All of the nonzero PV spin rotation observations so far seen with meV energy neutron beams come from the tails of a higher-energy p-wave resonance. In the case of Pb there is still some controversy about which isotope is responsible[47][48][49][50].

3 Parity-odd neutron spin rotation: experimental strategy

To measure the small parity-odd rotation, a neutron polarimeter is used to measure the $\hat{y}$ polarization component of a neutron beam initially polarized along the $\hat{z}$ axis and traveling in the $\hat{x}$ direction. The challenge is to distinguish small PV rotations from rotations that arise from residual magnetic fields. Simply flipping the neutron polarization as is commonly done in many PV asymmetry measure-

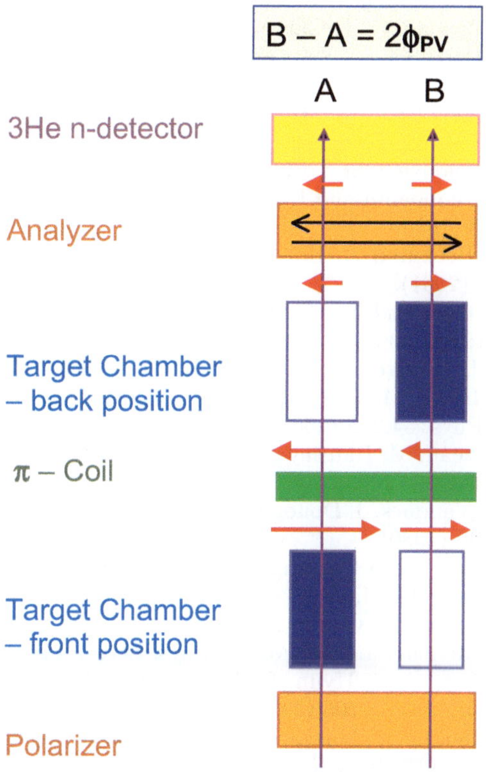

Fig. 1. Conceptual diagram (top view) illustrating the strategy to isolate the PV neutron spin rotation signal in the presence of a large background from residual longitudinal magnetic fields. The neutron beam and target chambers are split into two parallel sections. The arrows show the projection of the neutron polarization vector onto the plane of the figure at different points along the apparatus. Neutrons polarized out of the plane pass through a segmented target region. The upstream-downstream target segmentation in combination with the precession of the neutron polarization in the π-coil and the motion of the liquid allows the PV component of spin rotation signal from the liquid to be separated from the much larger parity-even component due to residual magnetic fields. The left-right segmentation allows for the removal of common-mode noise from nonstatistical neutron source intensity fluctuations and the chosen upstream-downstream pattern produces opposite PV components of the signal from the liquid. Although in practice the net rotation from residual magnetic fields will be larger in magnitude than the PV component, in the illustration we have chosen a particular example with $\int \mathbf{B}_{\text{front}} \cdot \mathbf{dl} + \int \mathbf{B}_{\text{back}} \cdot \mathbf{dl} = 0$ to reduce visual clutter.

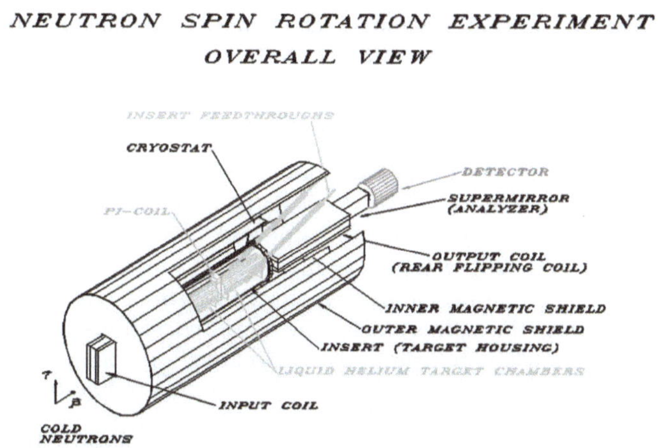

Fig. 2. Overview of an apparatus to measure PV neutron spin rotation in liquid helium. Polarized neutrons are guided into a magnetically-shielded region with the target and π-coil. A polarization analyzer and oscillating guide field converts the in-plane polarization component into an intensity difference measured by a segmented current-mode ion chamber.

ments does not work since the sense of the rotation of the corkscrew through the target does not change. Figure 1 explains the technique proposed to extract the PV component of the spin rotation signal and Figure 2 shows the apparatus for the liquid helium experiment.

To make this separation two target positions are used, one in front and the other behind a central solenoid called the π-coil. Vertically polarized neutrons enter the front target region and rotate in the $y - z$ plane. As they pass through the π-coil the spin precesses by π radians about the $\hat{z}$ axis. The π-coil precession reverses the $\hat{y}$ component of the neutron spin for rotations that occur in the front target region. By moving the liquid from the front target position to the rear, the PV rotation from the liquid reverses sign while rotations that arise from residual magnetic fields are unchanged (to the extent that the liquid does not change the trajectories or the dwell time of the neutrons in the field). The neutrons leave the target, pass through an analyzer, and are counted in a segmented ^{3}He ionization chamber[51]. To further suppress systematic errors and noise, the polarimeter and target chamber are split into right and left halves so that the PV components of the spin rotation are opposite. Segmentation of the ion chamber along the beam axis allows the rotation angle to be measured as a function of neutron energy: the PV rotation angle is independent of neutron energy while rotations arising from magnetic fields depend on the neutron energy.

4 Parity violating spin rotation in H, D, and ^{4}He

The only PV spin rotation experiment on a few-body nucleus that has been attempted is ^{4}He. A Washington/NIST group achieved a sensitivity of 14×10^{-7} rad/m[52], about a factor of two below the precision needed to provide new information about the NN weak interaction. No systematic effects were seen at the 2×10^{-7} rad/m level. A new experiment at NIST[53] will attempt to reach a precision of 3×10^{-7} rad/m. Estimates of systematic effects from diamagnetism in liquid helium and a number of scattering effects which can change the neutron dwell time in the residual magnetic field in a target-dependent manner are estimated at the 10^{-8} rad/m level. A number of auxiliary measurements to amplify certain classes of potential systematic error must be performed to verify these estimates experimentally.

PV spin rotation in n-^{4}He has not yet been calculated in the new hybrid EFT framework: the question of whether or not the internal momenta of the nucleons in ^{4}He are too large for the s-p dominance assumed in the 6-parameter EFT analysis needs to be addressed. An old calculation done in the DDH framework gives $d\phi_{He}/dz = -0.97 f_\pi - 0.32 h^0_\rho + 0.11 h^1_\rho - 0.22 h^0_\omega + 0.22 h^1_\omega)$[54] in n-^{4}He. Within the isoscalar/isovector subspace of NN weak amplitudes, PV spin rotation in n-^{4}He constrains a linear combination which is approximately orthogonal to already-measured PV effects in the isospin-conjugate p-^{4}He system[55] and in ^{133}Cs[9].

One can imagine attempting a measurement of PV neutron spin rotation in liquid parahydrogen[56] since there is no spin flip scattering until the neutron energy reaches 15 meV and therefore the neutron polarization is preserved. PV spin rotation in hydrogen has been calculated recently in the hybrid EFT approach to be $d\phi_H/dz = 0.30 C^\pi_6 + 2.50 m_N \lambda^{np}_s - 0.57 m_N \lambda_t + 1.41 m_N \rho_t \text{rad/m}$. Rough estimates of the expected rotation angle using DDH best values give $5-7\times10^{-7}$ rad/m. A PV neutron spin rotation measurement in deuterium may or may not be practical because of the depolarization of neutrons. A measurement of the neutron energy dependence of the depolarization of cold neutrons in a 4 cm long orthodeuterium target in both the liquid and solid phases was recently performed at the FUNSPIN polarized cold neutron beam facility at PSI[57]. Analysis of the data is in progress. There is no theoretical calculation of PV spin rotation in n-D.

PV neutron spin rotation measurements are an essential part of a program to determine the weak NN interaction[4][58]. In combination with other measurements in NN and few-body nuclei, such as parity-odd gamma asymmetries in polarized neutron capture, gamma circular polarization from unpolarized neutron capture, and PV helicity dependence in deuteron photodisintegration, it is possible to determine and perhaps overconstrain the 6 low energy parameters in the new hybrid EFT description of the weak NN interaction. Success would provide essential information for the understanding of PV phenomena in atoms and nuclei and may provide new insight into strong QCD dynamics.

This work is supported in part by NSF PHY-0457219.

References

1. E.G. Adelberger and W.C. Haxton, Ann. Rev. Nucl. Part. Sci. 35, 501 (1985).
2. W. Haeberli and B. R. Holstein, in *Symmetries in Nuclear Physics*, eds. W. C. Haxton and E. Henley (1995).
3. B. Desplanques, Phys. Rep. 297, 1 (1998).
4. M. J. Ramsey-Musolf and S. Page, hep-ph/0601127, submitted to Ann. Rev. Nucl. Part. Sci. (2006).
5. G. Arnison, *et al.*, Phys. Lett. 166B, 484 (1986).
6. E. G. Adelberger, J. Phys. Soc. Jpn. 54, 6 (1985).
7. J.D. Bowman, G. T. Garvey, and M. B Johnson, Ann. Rev. Nucl. Part. Sci. 43, 829 (1993).
8. S. Tomsovic, M. B. Johnson, A. Hayes, and J. D. Bowman, Phys. Rev. C62, 054607 (2000).
9. C.S. Wood *et al.*, Science 275, 1759 (1997).
10. Y. B. Zeldovich, Sov. Phys. JETP 6, 1184 (1957).
11. V. V. Flambaum and I. B. Khriplovich, Sov. Phys. JETP 52, 835 (1980).
12. D. H. Beck and B. Holstein, Int. Journal of Mod. Phys. E10, 1 (2001).
13. D. H. Beck and R. D. McKeown, Ann. Rev. Nucl. Part. Sci. 51:189 (2001)
14. D. H. Beck, Prog. Nucl. Part. Phys. 50:429 (2003)
15. E. J. Beize, M. L. Pitt, and D. T. Spayde, Prog. Nucl. Part. Phys. 54:289 (2005)
16. G. Prezeau, M. Ramsay-Musolf, and P. Vogel, Phys. Rev. D68, 034016 (2003).
17. B. Desplanques, J. Donoghue, and B. Holstein, Ann. Phys. 124, 449 (1980).
18. E. Henley, W.-Y. Hwang, and L. Kisslinger, Phys. Lett. B271, 403 (1998).
19. G. A. Lobov, Phys. At. Nuclei 65:534 (2002)
20. U. G. Meissner and H. Weigel, Phys. Lett. B447, 1 (1999).
21. H. J. Lee, C. H. Hyun, C. H. Lee, and H. C. Kim, Eur. Phys. J A24, s2, 105 (2005).
22. S. L. Zhu, C. M. Maekawa, B. R. Holstein, M. J. Ramsey-Musolf, and U. van Kolck, Nucl. Phys. A748, 435 (2005).
23. B. Holstein, Fizika B14, 165 (2005).
24. C. P. Liu, nucl/th-0609078 (2006).
25. G. S. Danilov, Phys. Lett. 18, 40 (1965).
26. S. R. Beane and M. J. Savage, Nucl. Phys. B636, 291 (2002).
27. J. Carlson, R. Schiavilla, Rev. Mod. Phys. 70, 743 (1998).
28. S. C. Pieper and R. B. Wiringa, Ann. Rev. Nucl. Part. Sci. 51, 53 (2001).
29. S. C. Pieper, R. B. Wiringa, and J. Carlson, Phys. Rev. C70, 054325 (2004).
30. R. Schiavilla, J. Carlson, and M. Paris, Phys. Rev. C67, 032501 (2003), Phys. Rev. C70, 044007 (2004).
31. J. Carlson, R. Schiavilla, V. R. Brown, and B. F. Gibson, Phys. Rev. C65, 035502 (2002).
32. J. Carlson and M. Paris, Eur. Phys. J A24, s1, 123 (2005).
33. F.C. Michel, Phys. Rev. B133, B329 (1964).
34. L. Stodolsky, Nucl. Phys. B197, 213 (1982).
35. L. Stodolsky, Phys. Lett. B50, 353 (1974).
36. M. Forte, in Physics with Reactor neutrons and neutrinos, ed. T. Von Egidy, Inst. Phys. Conf. Series 42 (1978).
37. J. B. French, V. K. B. Kota, A. Pandey, and A. Tomsovic, Ann. Phys. 181:198 (1988)
38. J. B. French, V. K. B. Kota, A. Pandey, and A. Tomsovic, Ann. Phys. 181:235 (1988)
39. O. P. Sushkov and V. V. Flambaum, JETP Lett. 32:352 (1980)
40. V. E. Bunakov and V. P. Gudkov, JETP Lett. 36, 328 (1982).
41. V. P. Alfimenkov *et al.*, JETP Lett. 35:51 (1982)
42. V. P. Alfimenkov *et al.*, Nucl. Phys. A 398:93 (1983)
43. M. Forte *et al.*, Phys. Rev. Lett. 45, 2088 (1980).
44. B. R. Heckel *et al.*, Phys. Lett. B 119:298 (1982)
45. B. R. Heckel *et al.*, Phys. Rev. C29, 2389 (1984).
46. S. Saha, PhD thesis, University of Washington (1990).
47. Bolotsky VP *et al.*, Phys. At. Nucl. 59:1808 (1996)

48. G. A. Lobov, Phys. At. Nucl. 63:1387 (2000)
49. R. Golub, I. L. Karpikhin, P. A. Krupchitsky, S. K. Lamoreaux, and V. V. Vasiliev, Phys. At. Nucl. 65,795 (2002)
50. J. Andrzejewski *et al.*, Phys. At. Nucl. 67:1233 (2004)
51. S. Penn *et al.*, Nucl. Inst. Meth. A457, 332 (2001).
52. D.M. Markoff, PhD thesis, University of Washington (1997), unpublished.
53. C. D. Bass *et al.*, J. Res. NIST 110, 189 (2005).
54. V. Dmitriev, V. V. Flambaum, O. P. Shuskov, and V. B. Telitsin, Phys. Lett. 125, 1 (1983).
55. J. Lang *et al.*, Phys. Rev. C34, 1545 (1986), Phys. Rev. Lett. 54, 170 (1985).
56. D.M. Markoff, J. Res. Nat. Inst. Stand. Technol. 110, 209 (2005).
57. W. M. Snow *et al.*, Measurement of Cold Neutron Depolarization in Liquid and Solid Deuterium, PSI experiment R-05-02 (2005).
58. W. M. Snow, J. Res. NIST 110, 189-194 (2005).

Two-pion-exchange parity-violating potential and np → dγ

C.H. Hyun[1,a], S. Ando[1], and B. Desplanques[2]

[1] Department of Physics and Institute of Basic Science, Sungkyunkwan University, Suwon 440-746, Korea
[2] Laboratoire de Physique Subatomique et de Cosmologie (UMR CNRS/IN2P3-UJF-INPG), F-38026 Grenoble Cedex, France

Abstract. We calculate the parity-violating nucleon-nucleon potential in heavy-baryon chiral perturbation theory up to next-to-next-to-leading order. The one-pion exchange enters in leading order and the next-to-next-to-leading order consists of two-pion-exchange and the two-nucleon contact terms. In order to investigate the effect of the higher-order contributions, we calculate the parity-violating asymmetry in $np \to d\gamma$ at threshold. The one-pion-exchange contribution dominates the physical observable, while the two-pion-exchange contribution is about or less than 10% of the one-pion-exchange one.

PACS. 21.30.Fe Forces in hadronic systems and effective interactions – 12.15.Ji Applications of electroweak models to specific processes

1 Introduction

We employ the idea of effective field theory (EFT) to derive the parity-violating (PV) nucleon-nucleon (NN) potential and apply it to the calculation of the PV photon asymmetry A_γ in $np \to d\gamma$ at threshold. The PV NN potential is obtained by replacing a parity-conserving (PC) vertex in the strong NN potential with a PV vertex. Most of the low-energy PV calculations have been relying on a one-meson exchange potential, the so-called DDH potential [1]. Some literature [2–4] has investigated the PV two-pion-exchange potential (TPEP), and it was revived quite recently in the light of EFT [5]. In ref. [5], a thorough derivation of the PV potential is performed up to the next-to-next-to-leading order (NNLO) in both pionless and pionful EFTs. Since the PV asymmetry in $np \to d\gamma$ is sensitive to the pion-exchange contribution, we adopt heavy-baryon chiral perturbation (HBχPT) to obtain the potential relevant to $np \to d\gamma$ up to NNLO.

Theoretical estimates of A_γ have been extensively worked out with the DDH potential [6–9]. The results with various strong-interaction models turn out basically identical. A_γ is dominated by the PV one-pion-exchange potential (OPEP), while the heavy-meson contribution is negligible. Thus, measurements of A_γ could provide us an opportunity to determine the weak pion-nucleon coupling constant h_π^1. On the other hand, the possibility of a 10% effect from the PV TPEP has been discussed [10].

In this work we present a first estimate of the intermediate-range contribution to A_γ. We employ the Argonne v18 potential for the PC potential and Siegert's theorem for the current operators. The PV potential relevant to $np \to d\gamma$ in our study comprises OPEP and TPEP components. There are various terms other than the TPEP one that appear at NNLO: the two-nucleon contact term, higher-order corrections of PC and PV πNN vertices, etc. In the present work, we retain only the TPEP one and neglect the remaining NNLO contributions for simplicity. The effect of the remaining terms will be considered elsewhere [11].

The NNLO calculation will allow us to estimate the order and the magnitude of higher-order corrections, which will be important in pinning down the value of h_π^1 and its uncertainty. At the same time, the NNLO contribution will provide a criterion for the validity of the EFT approach to PV phenomena.

2 Formalism

2.1 PV potential

The PV potential relevant to our estimate of A_γ in momentum space has the form

$$\tilde{V}_i(\boldsymbol{q}) = i(\boldsymbol{\tau}_1 \times \boldsymbol{\tau}_2)^z (\boldsymbol{\sigma}_1 + \boldsymbol{\sigma}_2) \cdot \boldsymbol{q}\ \tilde{v}_i(q), \tag{1}$$

where $q \equiv |\boldsymbol{q}|$ and $\boldsymbol{q} = \boldsymbol{p}_1 - \boldsymbol{p}_2$. OPE (fig. 1(a)) and TPE (fig. 1(b)-(d)) terms are obtained as

$$\tilde{v}_{1\pi}(q) = -\frac{g_A h_\pi^1}{2\sqrt{2} f_\pi} \frac{1}{q^2 + m_\pi^2}, \tag{2}$$

$$\tilde{v}_{2\pi}(q) = \sqrt{2}\pi \frac{h_\pi^1}{\Lambda_\chi^3} \left\{ g_A \tilde{L}(q) - g_A^3 \left[3\tilde{L}(q) - \tilde{H}(q) \right] \right\}, \tag{3}$$

[a] e-mail: hch@meson.skku.ac.kr

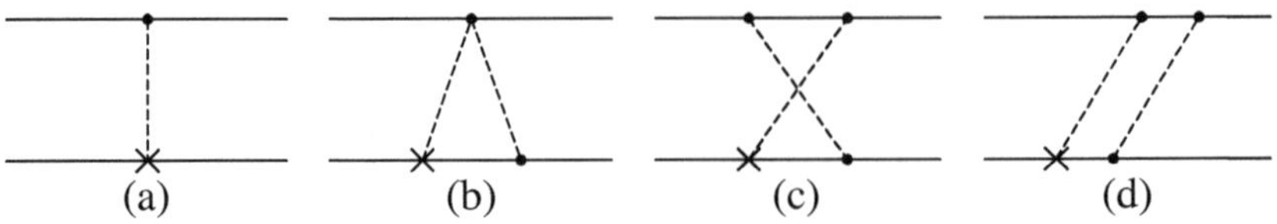

Fig. 1. Diagrams for PV OPE and TPE potentials. Diagram (a) is for the OPE term, while diagrams (b)-(d) are for the TPE ones. Lines (dashed lines) denote nucleons (pions), vertices with a dot represent PC vertices, vertices with an "×" represent the PV vertex proportional to h_π^1.

with

$$\tilde{L}(q) = \frac{\sqrt{q^2+4m_\pi^2}}{q} \ln\left(\frac{\sqrt{q^2+4m_\pi^2}+q}{2m_\pi}\right), \qquad (4)$$

$$\tilde{H}(q) = \frac{4m_\pi^2}{q^2+4m_\pi^2}\tilde{L}(q), \qquad (5)$$

where g_A is the axial coupling constant, f_π the pion decay constant and $\Lambda_\chi = 4\pi f_\pi$. The potential of eq. (1) in coordinate space can be written as

$$V_i(\boldsymbol{r}) = \int \frac{\mathrm{d}^3\boldsymbol{q}}{(2\pi)^3}\tilde{V}_i(\boldsymbol{q})\,\mathrm{e}^{-i\boldsymbol{q}\cdot\boldsymbol{r}} = i(\boldsymbol{\tau}_1\times\boldsymbol{\tau}_2)^z(\boldsymbol{\sigma}_1+\boldsymbol{\sigma}_2)\cdot[\boldsymbol{p},\, v_i(r)], \qquad (6)$$

where $\boldsymbol{p}$ is the conjugate momentum of the relative coordinate $\boldsymbol{r} \equiv \boldsymbol{r}_1 - \boldsymbol{r}_2$. For easier numerical calculation, we cast eqs. (4),(5) in dispersion relations as

$$\tilde{L}(q) = -\int_{4m_\pi^2}^{\infty}\frac{\mathrm{d}t'}{2\sqrt{t'}}\sqrt{t'-4m_\pi^2}\left(\frac{1}{t'+q^2}-\frac{1}{t'-4m_\pi^2}\right), \qquad (7)$$

$$\tilde{H}(q) = \frac{4m_\pi^2}{2}\int_{4m_\pi^2}^{\infty}\frac{\mathrm{d}t'}{\sqrt{t'}}\frac{1}{\sqrt{t'-4m_\pi^2}}\frac{1}{t'+q^2}. \qquad (8)$$

Furthermore, we introduce a monopole form factor of the type $\Lambda^2/(\Lambda^2+q^2)$ in the Fourier transform of eq. (1). The roles of the form factor and the cutoff are i) to make the numerical calculation more easier and efficient, and ii) to cut away the high-momentum region where the dynamics are essentially unknown and irrelevant to the low-energy processes. With the form factor, we rewrite the potential in coordinate space as

$$V_i^\Lambda(\boldsymbol{r}) = i(\boldsymbol{\tau}_1\times\boldsymbol{\tau}_2)^z(\boldsymbol{\sigma}_1+\boldsymbol{\sigma}_2)\cdot[\boldsymbol{p},\, v_i^\Lambda(r)], \qquad (9)$$

where

$$v_{1\pi}^\Lambda = \frac{g_A h_\pi^1}{2\sqrt{2}f_\pi}\frac{\Lambda^2}{\Lambda^2-m_\pi^2}\frac{1}{4\pi r}(\mathrm{e}^{-m_\pi r}-\mathrm{e}^{-\Lambda r}), \qquad (10)$$

$$v_{2\pi}^\Lambda = \sqrt{2}\pi\frac{h_\pi^1}{\Lambda_\chi^3}\left\{g_A L^\Lambda(r) - g_A^3\left[3L^\Lambda(r)-H^\Lambda(r)\right]\right\} \qquad (11)$$

with

$$L^\Lambda(r) = \frac{\Lambda^2}{8\pi r}\int_{4m_\pi^2}^{\infty}\frac{\mathrm{d}t'}{\sqrt{t'}}\sqrt{t'-4m_\pi^2} \times\left(\frac{\mathrm{e}^{-\sqrt{t'}r}-\mathrm{e}^{-\Lambda r}}{\Lambda^2-t'}-\frac{\mathrm{e}^{-\Lambda r}}{t'-4m_\pi^2}\right), \qquad (12)$$

$$H^\Lambda(r) = \frac{m_\pi^2\Lambda^2}{2\pi r}\int_{4m_\pi^2}^{\infty}\frac{\mathrm{d}t'}{\sqrt{t'}}\frac{1}{\sqrt{t'-4m_\pi^2}}\frac{\mathrm{e}^{-\sqrt{t'}r}-\mathrm{e}^{-\Lambda r}}{\Lambda^2-t'}. \qquad (13)$$

2.2 Asymmetry in np → dγ

The photon asymmetry in $np \to d\gamma$, A_γ, is defined from the differential cross-section of the process as

$$\frac{\mathrm{d}\sigma}{\mathrm{d}\Omega} \propto 1 + A_\gamma\cos\theta, \qquad (14)$$

where θ is the angle between the neutron polarization and the outgoing photon momentum. Non-zero A_γ values arise from the interference of opposite-parity transition amplitudes, *e.g.* $M1$ and $E1$.

At the thermal energy where the process occurs, lowest-order EM operators may suffice, therefore we consider the $E1$ operator,

$$\boldsymbol{J}_{\mathrm{E1}} = -i\frac{\omega_\gamma}{4}(\tau_1^z-\tau_2^z)\,\boldsymbol{r}, \qquad (15)$$

where ω_γ is the energy of the outgoing photon. At leading order of h_π^1, A_γ is proportional to h_π^1, and we can write A_γ as

$$A_\gamma = a_\gamma h_\pi^1 \qquad (16)$$

with

$$a_\gamma = -2\frac{\mathrm{Re}\,(\mathcal{M}_1\mathcal{E}_1^*)}{|\mathcal{M}_1|^2}, \qquad (17)$$

where $\mathcal{E}_1$ and $\mathcal{M}_1$ are matrix elements of the $E1$ and $M1$ transitions, respectively. Analytic forms of these amplitudes can be found in [9].

3 Result

Figure 2 shows $v_{1\pi}^\Lambda(r)$ (eq. (10)) and $v_{2\pi}^\Lambda(r)$ (eq. (11)) in the range $0 < r < 5\,\mathrm{fm}$ as functions of the cutoff Λ. The curve denoted by "Yukawa" in the OPEP corresponds to an infinite cutoff value. In the long-range region (right panels), both the OPEP and TPEP hardly depend on Λ. The magnitude of the TPEP at large r-values is smaller than the OPEP by an order of magnitude, which confirms the dominance of OPEP in the long-range region. For small r-values, *e.g.*, $r < 1\,\mathrm{fm}$, a substantial cutoff dependence is observed. The potential converges to the limiting case (infinite cutoff) with increasing Λ. The change with the cutoff value is simple for the OPEP, but the TPEP shows more diversity in its dependence on the cutoff value. The TPEP is a decreasing function at small r but the sign of the derivative changes at a certain value of r. The value of r at which the derivative of the TPEP becomes zero is sensitive to Λ, and it has a significant effect on a_γ.

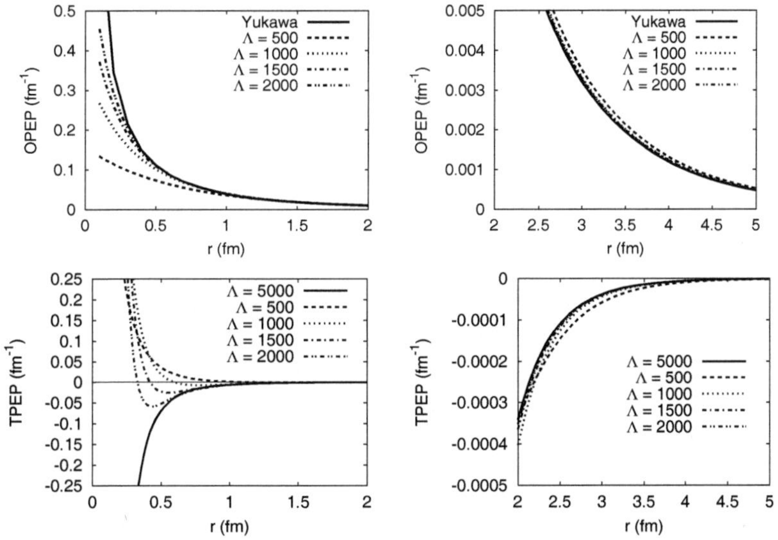

Fig. 2. $v_{1\pi}^{\Lambda}(r)$ (upper row) and $v_{2\pi}^{\Lambda}(r)$ (lower row) in the short-intermediate (left column) and long (right column) ranges.

Table 1. OPE and TPE contributions to the asymmetry as functions of cutoff Λ.

Λ (MeV)	500	1000	1500	2000
a_γ (OPE)	-0.1074	-0.1125	-0.1126	-0.1124
a_γ (TPE)	-0.0022	0.0073	0.0117	0.0133

In table 1, we show the numerical results for a_γ defined in eq. (16). With the OPEP, a_γ is stable against a variation of the cutoff value, with at most a 5% fluctuation for the cutoff values considered in the present work. The TPEP contribution to a_γ, on the other hand, varies significantly in magnitude and even a sign change occurs. From the behavior of $v_{1\pi}^{\Lambda}(r)$ and its contribution to a_γ, it can be deduced that a decreasing function gives a negative contribution to a_γ. With a larger Λ, the value of r at which $\partial_r v_{2\pi}^{\Lambda}(r) = 0$, becomes smaller, and the decreasing region of $v_{2\pi}^{\Lambda}(r)$ gets narrower. This means that the negative contribution becomes smaller in magnitude and a more positive contribution comes from the TPEP. This explains well the sign and magnitude behavior of a_γ for the TPEP.

4 Summary

We have calculated the PV two-pion-exchange NN potential from EFT, and applied it to the calculation of the PV asymmetry in $np \to d\gamma$ at threshold. In the intermediate- and long-range region, the TPEP is smaller than the OPEP by an order of magnitude, while the TPEP is comparable to the OPEP for $r < 0.5$ fm. A sizable cutoff dependence of the potential appears at $r \leq 1$ fm for both the OPEP and the TPEP. The asymmetry also shows a non-negligible cutoff dependence. For the asymmetry with the OPEP, the dependence is relatively weak and the result shows convergent behavior with larger cutoff values. On the contrary, the asymmetry with the TPEP depends strongly on the cutoff value, even accompanied by a sign change, and a convergence of the result is not clearly seen. As a result, the TPEP contribution to the asymmetry varies from -2% to 12% of the OPEP in the cutoff range 500 to 2000 MeV. Since 2000 MeV is a sufficiently large cutoff value, the contribution of TPEP, though uncertain, lies in the range similar to or less than 10 of the OPEP. This is a significant correction to the OPEP result compared to the one-heavy-meson exchange, which is less than 1% of the OPEP [9], but it is good to justify that the perturbative expansion of EFT works well for the PV NN potential. The low-energy constant terms at NNLO, which are not considered in this work, could give a similar size correction.

References

1. B. Desplanques, J.F. Donoghue, B.R. Holstein, Ann. Phys. (N.Y.) **124**, 449 (1980).
2. B. Desplanques, Phys. Lett. B **41**, 461 (1972).
3. H.J. Pirner, D.O. Riska, Phys. Lett. B **44**, 151 (1973).
4. M. Chemtob, B. Desplanques, Nucl. Phys. B **78**, 139 (1974).
5. S.-L. Zhu, C.M. Maekawa, B.R. Holstein, M.J. Ramsey-Musolf, U. van Kolck, Nucl. Phys. A **748**, 435 (2005).
6. B. Desplanques, Phys. Lett. B **512**, 305 (2001).
7. C.H. Hyun, T.-S. Park, D.-P. Min, Phys. Lett. B **516**, 321 (2001).
8. R. Schiavilla, J. Carlson, M.W. Paris, Phys. Rev. C **70**, 044007 (2004); **67**, 032501(R) (2003).
9. C.H. Hyun, S.J. Lee, J. Haidenbauer, S.W. Hong, Eur. Phys. J. A **24**, 129 (2005).
10. B. Desplanques, Phys. Rep. **297**, 1 (1998).
11. C.H. Hyun, S. Ando, B. Desplanques, nucl-th/0611018, to be published in Phys. Lett. B.

Theoretical overview of atomic parity violation

Recent developments and challenges

A. Derevianko[1,a] and S.G. Porsev[2]

[1] Physics Department, University of Nevada, Reno, NV 89557, USA
[2] Petersburg Nuclear Physics Institute, Leningrad district, Gatchina, 188300, Russia

Abstract. Recent advances in interpreting the most accurate-to-date measurement of atomic parity violation in Cs are reviewed. The inferred nuclear weak charge, $Q_W(^{133}\text{Cs}) = -72.65(28)_{\text{expt}}(36)_{\text{theor}}$, agrees with the prediction of the standard model at 1σ level. Further improved interpretation is limited by an accuracy of solving the basic correlation problem of the atomic structure. We report on our progress in solving this problem within the relativistic coupled-cluster formalism. We include single, double and triple electronic excitations in the coupled-cluster expansion. Numerical results for energies, electric-dipole matrix elements, and hyperfine-structure constants of Cs are presented.

PACS. 32.80.Ys Weak-interaction effects in atoms – 31.15.Dv Coupled-cluster theory – 32.10.Fn Fine and hyperfine structure – 32.70.Cs Oscillator strengths, lifetimes, transition moments

1 Introduction

This year, 2006, marks 50 years of the fall of parity conservation, one of the hallmark discoveries of the 20th century physics. The field has started with the seminal Lee and Yang paper [1] and the discovery of parity violation in the nuclear β-decay [2]. Soon after this discovery, Zel'dovich [3] contemplated the possibility of observing the parity-nonconserving (PNC) signal in atoms. He concluded that the effect was too small to be of experimental significance. In 1974, however, M.-A. Bouchiat and C. Bouchiat [4] realized that the PNC is amplified in *heavy* atoms. They showed that the relevant PNC amplitude scales steeply with the nuclear charge Z. In atomic physics, the first P-violating signal has been observed in 1978 by Barkov and Zolotorev [5] in the Bi atom. Over the following decades the experiments were refined, with PNC signal observed in several atoms. So far the most accurate measurement has been carried out in ^{133}Cs by the Boulder group [6].

Rich history of atomic parity violation is examined in a number of review articles, *e.g.*, refs. [7,8] and a book [9]. Here, due to the space limitation, we restrict our attention to recent developments in interpreting P-violation in the Cs atom and report on the progress towards a more accurate interpretation of atomic PNC in this atom.

The PNC measurements are interpreted in terms of the weak nuclear charge Q_W, which quantifies the strength of the electroweak coupling between atomic electrons and quarks in the nucleus. The relation between Q_W and the PNC amplitude, E_{PNC}, can be represented as

$$E_{\text{PNC}} = k\, Q_W\,, \tag{1}$$

where k is an atomic-structure factor. Apparently, the interpretation requires atomic-structure calculations of k with an accuracy that matches the experimental uncertainty in E_{PNC}. In particular, in Cs, the present theoretical uncertainty dominates over the experimental error bar resulting in an overall 0.6% uncertainty in the inferred value of $Q_W(^{133}\text{Cs})$.

The underlying theory of the electroweak interactions is well established and its predictions have been verified in a variety of experiments. Much of the present efforts are driven by searches for "new physics" beyond the standard model (SM). We may distinguish between two approaches to such searches: low- and high-energy experiments. Atomic parity violation probes the low-energy electroweak sector of the SM. While the relevant momentum transfer is just $\sim 30\,\text{MeV}$, the exquisite accuracy of the interpretation constrains new physics at much higher energies. Indeed, following arguments of ref. [10], we find that the present 0.6% determination of the $Q_W(^{133}\text{Cs})$ probes the new physics at a mass scale of

$$\left\{8\sqrt{2}\pi\frac{1}{G_F}\left|\frac{Q_W}{\delta Q_W}\right|\right\}^{1/2} \approx 20\,\text{TeV},$$

where G_F is the Fermi constant. Apparently by reducing the uncertainty in the determination of the weak charge,

[a] e-mail: andrei@unr.edu

Table 1. Present status of the 2.5σ deviation of the inferred $Q_W(^{133}\mathrm{Cs})$ from the prediction of the standard model. In the bulk of the table, we summarize the recent theoretical progress on computing "small" sub-1% corrections to the 6s-7s PNC amplitude. These corrections directly affect the inferred value of the weak charge of Cs. For illustrative purposes all the corrections (pulls) are tabulated in terms of the original Bennet and Wieman [16] uncertainty, $\sigma = \{\sigma^2_{\mathrm{expt}} + \sigma^2_{\mathrm{theor}}\}^{1/2} \approx 0.5\%$.

Deviation (1999) Bennet and Wieman [16]	2.5σ	Based on calculations by Dzuba *et al.* (1989) [14] and Blundell *et al.* (1990) [15]
Theoretical correction	Pull	Reference
Breit interaction	-1.2σ	Derevianko (2000) [20] Dzuba *et al.* (2001) [21] Kozlov *et al.* (2001) [22]
Vacuum polarization ($+0.8\sigma$) Vertex/self-energy (-1.3σ)	-0.5σ	Johnson *et al.* (2002) [23] Milstein *et al.* (2002) [24,25] Kuchiev and Flambaum (2002) [26] Sapirstein *et al.* (2003) [27] Shabaev *et al.* (2005) [28]
Neutron skin	-0.4σ	Derevianko (2002) [29]
Updated correlated E_{PNC} and β	$+0.7\sigma$	Dzuba *et al.* (2002) [30]
e-e P-violating interaction, Renormalization $Q \to 0$, Giant nuclear resonance	-0.08σ	Sushkov and Flambaum (1978) Milstein *et al.* (2002) [24]
Total deviation (agreement)	1σ	

$|\delta Q_W|$, we may access even higher mass scales. Such a reduction in the uncertainty of Q_W is the goal of our present efforts outlined in sect. 3.

While discussing the impact of atomic PNC on particle physics, it is worth noting that colliders are blind to certain "new physics" scenarios [10]. For example, atomic PNC is uniquely sensitive to extra Z bosons predicted in grand unified theories, technicolor models, SUSY, and string theories. Generally, atomic PNC is one of a few probes of electroweak coupling below the Z-pole. Competing and at the same time complementary experiments include determination of the weak charge of the electron at SLAC [11] and ν-nucleus deep inelastic scattering by the NuTeV Collaboration [12]. With an upgrade at Jefferson Lab, an even more accurate constraint on electroweak coupling is expected to come from a P-violating e-p scattering experiment (Q-weak Collaboration) [13]. Still, the atomic PNC will continue to play an important complementary role, as the atomic PNC is sensitive to electron-neutron couplings, while the Q-weak experiment will probe electron-proton coupling, *i.e.*, a different linear combination in terms of the up and down quark-electron couplings.

2 Deviation from the standard model and recent theoretical progress

Parity violation in Cs has been the subject of a recent controversy. To infer the value of the weak charge, one needs to combine measurements with atomic-structure calculations. The uncertainty in the value of the inferred Q_W is usually determined by summing experimental and theoretical uncertainties in quadrature,

$$\sigma = \left\{\sigma^2_{\mathrm{expt}} + \sigma^2_{\mathrm{theor}}\right\}^{1/2} . \quad (2)$$

In 1997, the most accurate to date (0.35%) measurement of PNC has been carried out by the Boulder group [6]. At that time, the accuracy of calculations [14,15] has been estimated to be 1%. In 1999, Bennet and Wieman [16], in light of new lifetime data which improved theory-experiment agreement, have reduced the theoretical error bar to 0.4%. While compared with the prediction of the SM, the resulting value of Q_W differed by 2.5σ, one of the largest deviations in the recent history. The deviation has prompted a substantial interest from the particle physics community (see, *e.g.*, [10,17–19]). At the same time the reduced theoretical uncertainty raised the questions whether some "small" sub-1% atomic-structure effects could be the reason for the deviation.

Over the last few years, there has been an important progress in understanding "small" corrections, such as Breit, radiative, and neutron-skin corrections. These advances are summarized in table 1.

Taking into account the tabulated "small" corrections, the revised value of vector transition polarizability and the somewhat improved value of E_{PNC} [30], we arrive at the following revised value of the weak charge (the assigned theoretical uncertainty of 0.5% is consistent with an estimate of ref. [30] and the central value with that of ref. [28]):

$$Q_W(^{133}\mathrm{Cs}) = -72.65(28)_{\mathrm{expt}}(36)_{\mathrm{theor}} \,. \qquad (3)$$

This value agrees with the SM value [19] of $-73.19(13)$ at 1σ level.

The present theoretical uncertainty in PNC calculations is estimated to be 0.5%. Since the overall error is obtained with quadrature (2), reducing σ_{theor} below the experimental uncertainty of 0.35% will either reveal new physics or set important constraints on competing extensions to the SM. In addition, it will set a stage for further experimental efforts, such as those by the Paris group [31, 32]. Other experimental efforts which will benefit from the improved calculations are PNC measurements with Ba^+ [33,34], Tl [35], and Fr [36].

3 High-accuracy atomic-structure calculations

The present accuracy in the determination of the weak charge of Cs is limited by the accuracy of solving the basic correlation problem. The many-body problem is hard. Even classically, the three-body problem cannot be solved in closed form. While adequate numerical approaches (*e.g.*, configuration interaction and variational methods) were developed for few-electron atoms, accurate solution of the many-body problem for atoms with many electrons still remains a challenge.

In a particular case of the PNC problem in Cs, we need to evaluate the PNC amplitude for the $6S_{1/2} \to 7S_{1/2}$ transition

$$E_{\mathrm{PNC}} = \sum_n \frac{\langle 7S|D|nP_{1/2}\rangle\langle nP_{1/2}|H_W|6S\rangle}{E_{6S} - E_{nP_{1/2}}} + \sum_n \frac{\langle 7S|H_W|nP_{1/2}\rangle\langle nP_{1/2}|D|6S\rangle}{E_{7S} - E_{nP_{1/2}}} \,. \qquad (4)$$

Here D and H_W are the electric-dipole and the weak interaction matrix elements, and E_i are the atomic energy levels. The effective weak interaction, averaged over quarks is accumulated inside the nucleus

$$H_W = -\frac{G_F}{\sqrt{8}} Q_W \, \gamma_5 \, \rho(\mathbf{r}) \,, \qquad (5)$$

where $\rho(\mathbf{r})$ is the neutron-density distribution. One could easily recast eqs. (4) and (5) in terms of the structure factor k of eq. (1).

We wish to evaluate accurately the sum (4). Let us comment on the pre-requisites for such calculations. i) Due to the particular characters of the involved operators in summation (4), the approximate wave functions have to reproduce the short-range (close to the nucleus) and long-range atomic properties simultaneously. The accurate short-range description requires *ab initio* relativistic approach, as the ratio of the velocity of the electron to the speed of light near the nucleus is $\sim \alpha Z \sim 0.4$ for Cs. ii) With respect to the accuracy, we notice that simple Dirac-Hartree-Fock (DHF) calculations for the hyperfine-structure (HFS) constant of the ground state are about 50% off from the experimental value. This constant describes the strength of the coupling of the electron to the nuclear magnetic moment and its short-distance nature mimics the behavior of the weak matrix elements. Therefore we arrive at the conclusion that so-called correlation corrections (those beyond the DHF approximation) have to be addressed. We treat the correlations within the systematic and successful methods of many-body perturbation theory.

How do we estimate the accuracy of calculating E_{PNC}? Ultra-precise experimental data are available for Cs. These data have accuracy better than our anticipated theoretical uncertainty thus allowing us to calibrate our calculations. Atomic energies are known to many significant figures. The Ultra-precise (0.01% accurate) value of the lifetime of the $6P_{3/2}$ state of Cs has been recently inferred from photoassociation spectroscopy [37,38] with ultracold atoms. The HFS constants for Cs are also measured to a good accuracy (the HFS constant A_{6S} of $^{133}\mathrm{Cs}$ is known *exactly* by the definition of the second).

To reiterate, to further improve the interpretation of atomic parity violating signals, we need a high-accuracy *ab initio* relativistic many-body method capable of reaching the accuracy level of 0.1% for Cs.

4 Relativistic coupled-cluster method

Many-body perturbation theory (MBPT) provides a systematic way of treating correlation corrections, the critical issue for an improved interpretation of atomic PNC. First, we review main ideas of MBPT and the all-order coupled-cluster (CC) method. Then, we describe our present CC-inspired computational scheme and illustrate it with numerical results for the basic atomic properties of the Cs atom.

4.1 Generalities

In MBPT the atomic Hamiltonian is partitioned as

$$H = \left(\sum_i h_{\mathrm{nuc}}(\mathbf{r}_i) + \sum_i U_{\mathrm{DHF}}(\mathbf{r}_i) \right) + \left(\frac{1}{2} \sum_{i \neq j} \frac{1}{r_{ij}} - \sum_i U_{\mathrm{DHF}}(\mathbf{r}_i) \right) ,$$

where h_{nuc} includes the kinetic energy of an electron and its interaction with the nucleus, U_{DHF} is the DHF potential, and the last term represents the residual Coulomb interaction between electrons. In MBPT the first part of the Hamiltonian is treated as the lowest-order Hamiltonian H_0 and the residual Coulomb interaction as a perturbation. The perturbative expansion is built in powers of residual interaction and the derivations typically involve methods of second quantization and diagrammatic techniques.

One of the mainstays of practical applications of MBPT is the assumption of the convergence of series in powers of the residual interaction. Sometimes the convergence is poor and then one sums certain classes of diagrams to "all orders" using the iterative techniques. One of the most popular all-order methods is the coupled-cluster (CC) formalism [39,40]. It is widely employed in atomic and nuclear physics, and quantum chemistry [41]. The *relativistic* atomic-structure CC-type calculations were carried out, for example, in refs. [42–47].

The Hamiltonian in the second quantization (based on the DHF basis) reads

$$H = H_0 + G = \sum_i \varepsilon_i \left\{a_i^\dagger a_i\right\} + \frac{1}{2}\sum_{ijkl} g_{ijkl}\left\{a_i^\dagger a_j^\dagger a_l a_k\right\}, \quad (6)$$

where H_0 is the lowest-order Hamiltonian and the residual Coulomb interaction G is treated as a perturbation. The operators a_i and $a_i^\dagger$ are annihilation and creation operators, and $\{\ldots\}$ stands for a normal product of operators with respect to the core quasivacuum state $|0_c\rangle$. In the lowest order the atomic wave function with the valence electron in an orbital v reads $|\Psi_v^{(0)}\rangle = a_v^\dagger|0_c\rangle$. For example, v can represent the ground state $6s$ orbital of the Cs atom. Formally, we can introduce a wave operator Ω that produces the exact many-body wave function, $|\Psi_v\rangle = \Omega\,|\Psi_v^{(0)}\rangle$.

Central to the CC method is the realization that the wave operator can be represented via the exponential ansatz [48]

$$\Omega = \{\exp(K)\} = 1 + K + \frac{1}{2!}\{K^2\} + \ldots. \quad (7)$$

The operator K can be compellingly separated into cluster operators combining simultaneous excitations of core and valence electrons from the reference state $|\Psi_v^{(0)}\rangle$ to all orders of MBPT,

$$K = S + D + T + \cdots, \quad (8)$$

i.e., K is separated into single (S), double (D), triple (T), and higher-rank excitations. For the univalent systems it is convenient to subdivide cluster operators into core and valence classes

$$K = K_c + K_v\,. \quad (9)$$

Clusters K_c involve excitations from the core orbitals only, while K_v describe simultaneous excitations of the core and valence electrons. Then $S = S_c + S_v$, $D = D_c + D_v$, etc. The cluster operators satisfy so-called Bloch equations [48].

While the CC formulation is exact, in practice the full cluster operator K is truncated at a certain level of excitations. If we restrict it to single and double excitations,

$$K \equiv K^{(\mathrm{SD})} \approx S_c + D_c + S_v + D_v\,, \quad (10)$$

we arrive at the widely employed coupled-cluster single-double (CCSD) method.

The linearized version of the CCSD method was employed by the Notre Dame group for high-accuracy calculations of various atomic properties [42,43,46,47]. In this approximation, one discards nonlinear terms in the expansion of the exponent in eq. (7), $\Omega \approx 1+K^{(\mathrm{SD})}$. We will refer to this approximation as the singles-doubles (SD) method. The resulting SD equations are written out in ref. [42]. A typical *ab initio* accuracy attained for properties of heavy alkali-metal atoms is at the level of 1%.

Since the present goal is to reduce theoretical uncertainties to the level of 0.1–0.2% we have to go beyond the SD approach. A systematic step in improving the SD method would be an additional inclusion of triple and nonlinear double excitations. However, considering the present state of available computational power, the full incorporation of triples (specifically, core triples) seems to be unmanageable for heavy atoms. For instance, for Cs storing and manipulating core triple amplitudes would require ~ 100 Gb of memory.

To motivate next-generation formalism, we have explicitly computed 1648 fourth-order diagrams for matrix elements that appear due to triple excitations and nonlinear terms (*i.e.*, those omitted in the SD method) [49, 50]. We observe from numerical results for electric-dipole matrix elements in Na [50] and Cs [51] that the contributions from *valence* triples T_v and nonlinear doubles D_{nl} are much larger than those from *core* triples T_c. This leads to our present level of approximation: we discard core triples and core nonlinear terms and incorporate the *valence triples* and *valence nonlinear terms* into the SD formalism. The resulting approximation will be referred to as CCSDvT method.

4.2 Driving equations in the CCSDvT approximation

Below we write down the CC equations for cluster amplitudes in the CCSDvT approximation. Here we present the topological structure of the equations only. A detailed tabulation of the formulas can be found in our paper [52]. The equations in the SD approximation are presented in explicit form in ref. [42]. The CCSDvT equations for the core cluster amplitudes S_c and D_c are the same as in the SD approximation.

For valence triple amplitudes we obtain symbolically

$$-[H_0, T_v] + \delta E_v T_v \approx T_v[D_c] + T_v[D_v]. \quad (11)$$

Here $[H_0, T_v]$ is a commutator, and δE_v is the correlation valence energy defined as

$$\delta E_v = \delta E_{\mathrm{SD}} + \delta E_{\mathrm{CC}} + \delta E_{\mathrm{vT}}\,, \quad (12)$$

where the correction δE_{SD} is obtained within the SD approach, the correction δE_{CC} comes from nonlinear CC contributions and δE_{vT} is due to valence triples. Contributions $T_v[D_c]$ and $T_v[D_v]$ denote the effect of core and valence doubles on valence triples, respectively. At present we include only these effects omitting the effect of valence and core triples on valence triples ($T_v[T_v]$ and $T_v[T_c]$) and nonlinear CC contributions. These are higher-order effects which computationally are much more demanding.

The topological structure of the valence singles equation is

$$\begin{aligned}&-[H_0,S_v]+\delta E_v S_v \approx \mathrm{SD}\\&+S_v[S_c\otimes S_v]+S_v[S_c\otimes S_c]\\&+S_v[S_c\otimes D_v]+S_v[S_v\otimes D_c]+S_v[T_v].\end{aligned} \tag{13}$$

Here $S_v[S_c \otimes S_v]$ stands for a contribution from the excitations of core and valence electrons resulting from a product of clusters S_c and S_v. All other terms are defined in a similar fashion.

Finally, equation for valence doubles can be symbolically represented as

$$\begin{aligned}&-[H_0,D_v]+\delta E_v D_v \approx \mathrm{SD}\\&+D_v[S_c\otimes S_v]+D_v[S_c\otimes S_c]\\&+D_v[S_c\otimes D_v]+D_v[S_v\otimes D_c]+D_v[S_c\otimes D_c]\\&+D_v[D_c\otimes D_v]+D_v[S_c\otimes T_v]\\&+D_v[S_v\otimes T_c]+D_v[T_v].\end{aligned} \tag{14}$$

Solution of the above equations provides us with the cluster amplitudes and correlation energies. Numerical results for the energies will be presented in sect. 4.4. At this point, with the obtained wave functions we proceed to evaluating matrix elements.

4.3 Matrix elements

The SD method has already proven to be successful in calculations of various atomic properties. For heavy alkali-metal atoms the attained level of agreement with experimental data for the hyperfine constants is at 5% and the accuracy of a similar calculation for the electric-dipole amplitudes is 0.5% (see, *e.g.*, [47]). At the same time the accuracy required for our goals should be at the level of 0.1–0.2%. In order to improve the overall accuracy we develop the technique of relativistic calculations of matrix elements beyond the SD approach.

Given two computed CCSDvT wave functions, we may evaluate matrix elements of one-electron operator Z as

$$Z_{wv} = \frac{\langle \Psi_w | \sum_{ij} z_{ij}\, a_i^\dagger a_j | \Psi_v \rangle}{\sqrt{\langle \Psi_w | \Psi_w \rangle \langle \Psi_v | \Psi_v \rangle}}\,. \tag{15}$$

The explicit expressions are given in ref. [52]. Compared to the SD approximation, we include contribution of valence triples T_v. They contribute both directly via explicit contributions to matrix element formula and indirectly through the modification of the SD amplitudes.

It is worth pointing out, that the importance of the valence triples has been realized earlier by the Notre Dame group [43]. They have shown that at the SD level, the error for the HFS constants is as large as 5% for Cs. To rectify this problem, they proposed and implemented a scheme that approximates the effect $S_v[T_v]$, *i.e.*, the effect of valence triples on valence singles. While improving the agreement for the HFS constants, their approximation leads to a poorer agreement (compared to SD method) for the dipole matrix elements. The advantage of the Notre Dame scheme is that it avoided expensive storing of triple excitations. Due to improved computational resources, we are able to store triples. Accounting for the triples in a rigorous fashion leads to a better agreement between theory and experiment.

Compared to the Notre Dame approximation we also include dressing of matrix elements based on the CC ansatz. The idea of our method [51] is as follows. When the CC exponent is expanded in eq. (15), we encounter an infinite number of terms. We devised a method of partial summation (dressing) of the resulting series. Our formalism is built upon an expansion of the product of cluster amplitudes into a sum of n-body insertions. We considered two types of insertions: the particle (hole) line insertion (line "dressing") and the two-particle (two-hole) random-phase-approximation-like insertion. We demonstrated how to "dress" these insertions and formulated iterative equations.

Another formal improvement over Notre Dame calculations comes from including the CC nonlinear terms in the equations for valence singles (13) and doubles (14). We also include the contribution of the core triples to matrix elements from a direct fourth-order calculation.

4.4 Numerical results

Our developed numerical CCSDvT code is an extension of the relativistic SD code [46] which employs a B-spline basis set. This basis numerically approximates a complete set of single-particle atomic states. Here we use 35 out of 40 positive-energy basis functions. Basis functions with $l_{\max} \leq 5$ are used for singles and doubles. For triples we employ a more limited set of basis functions with $l_{\max}(T_v) \leq 4$. Excitations from core sub-shells $[4s, \ldots, 5p]$ are included in the calculations of triples while excitations from sub-shells $[1s, \ldots, 3d]$ are discarded.

Computed removal energies of $6s$, $6p_{1/2}$, and $6p_{3/2}$ states of atomic cesium are presented in table 2. The dom-

Table 2. Contributions to removal energies of $6s$, $6p_{1/2}$, and $6p_{3/2}$ states for Cs in cm^{-1} in various approximations. $\delta E^{\rm tot}_{\rm extrapolated}$ correction is obtained by computing SD properties with increasingly larger basis sets and interpolating them to $l = \infty$ [53]. A comparison with experimental values is presented in the lower panel.

	$6s$	$6p_{1/2}$	$6p_{3/2}$
$E_{\rm DHF}$	27954	18790	18389
$\delta E_{\rm SD}$	3869	1611	1623
$\delta E_{\rm CCSDvT}$	3350	1387	1220
$E^{\rm tot}_{\rm CCSDvT}$	31304	20178	19608
QED[1]	18	−0.4	0
$\delta E^{\rm tot}_{\rm extrapolated}$	30	20	20
$E^{\rm tot}_{\rm final}$	31352	20198	19628
$E_{\rm experim}$[2]	31407	20228	19675

[1] Reference [54].
[2] Reference [55].

Table 3. Magnetic-dipole hyperfine structure constants A (in MHz) and matrix elements of the electric-dipole moment (in a.u.) for ^{133}Cs. Results of calculations and comparison with experimental values are presented. See text for the explanation of entries.

	$A(6s)$	$A(6p_{1/2})$	$\langle 6p_{1/2}\|D\|6s\rangle$	$\langle 6p_{3/2}\|D\|6s\rangle$
DHF	1425.4	160.94	5.2777	7.4264
SD	2438.0	310.71	4.4829	6.3075
Δ(CCSDvT)	−136.9	−20.92	0.0256	0.0363
Complementary corrections				
Line dressing	−12.5	−2.16	0.0094	0.0107
Vertex dressing	4.3	0.29	−0.0067	−0.0088
MBPT-IV (core triples, ...)	7.8	1.14	0.0001	0.0001
Breit + QED	−6.5[1]		0.0024[2]	
Extrapolation for $l = \infty$[3]	5.0	0.37	−0.004	−0.006
Final CCSDvT + corrections	2299.2	289.43	4.5097	6.3398
Experiment	2298.2	291.89(8)[4]	4.5049(17)[5]	6.3404(3)[6]

[1] Reference [56].
[2] References [20,54].
[3] Reference [53].
[4] Reference [57].
[5] References [58,38].
[6] Reference [38].

inant contribution to the energies comes from the DHF values. The remaining (correlation) contribution is given by eq. (12). We computed this correlation correction in SD and CCSDvT approximations. As it follows from the table the agreement with experiment is at the level of 0.1–0.2% for all considered states. We anticipate that including other corrections missed at this stage (*e.g.*, non-linear corrections to core amplitudes and core triples) can further improve the agreement with the experimental results.

With the computed wave functions of the $6s$, $6p_{1/2}$ and $6p_{3/2}$ states we determine matrix elements. Numerical results for magnetic-dipole hyperfine-structure constants A and electric-dipole transition amplitudes are presented in table 3. This table is organized as follows. First we list the DHF and SD values, and the differences between CCSDvT and SD values, $\Delta(\text{CCSDvT}) = \text{CCSDvT} - \text{SD}$. We base our final *ab initio* results on the most sophisticated CCSDvT values. These values also include all-order dressing, and corrections due to core triples, computed in the the fourth order of MBPT. QED corrections are included where available. The results for the HFS constants include finite nuclear size (Born-Weisskopf) effect.

We find an excellent, 0.1% level, agreement for dipole matrix elements and the HFS constant of the ground state. The agreement for the the HFS constant of the $6p_{1/2}$ is only at 1% level. We are presently working on testing the sensitivity of this constant to higher-order effects.

We would like to emphasize that presently the correlation corrections at the level of a few 0.1% are comparable to radiative corrections. In this regard it would be useful to carefully compute them to unmask the remaining many-body effects.

5 Summary and outlook

Atomic parity violation plays an important role in testing the low-energy electroweak sector of the standard model. The interpretation of experiments in terms of nuclear weak charge requires calculations challenging the capabilities of modern atomic theory. Over the last few years, we have witnessed a substantial advance in evaluating corrections to parity-violating amplitudes in heavy atoms. These small (sub-1%), but important corrections include Breit, radiative (vacuum polarization, self-energy, and vertex) and neutron skin corrections. As a result of this progress, the most accurate-to-date measurement of atomic parity violation in Cs has been brought into substantial agreement with the prediction of the standard model.

Presently, the theoretical interpretation is clouded by uncertainties in solving the basic correlation problem of the atomic structure. In this paper we outlined our next-generation many-body formalism for solving this problem. We tested our coupled-cluster-inspired method by computing basic atomic properties of the Cs atom. All the computed properties are important for quantifying the accuracy of the calculations of parity-violating amplitudes. We find an agreement at 0.1% for the ground-state hyperfine structure constant, $E1$ transition amplitudes, and energies. However, a relatively poor 1% agreement of the HFS constant A for the $6P_{1/2}$ state with experiment requires further improvements of the method. The advantage of the employed coupled-cluster method is that it allows for such systematic improvements. It is anticipated that the further theoretical progress will refine constraints on new physics beyond the standard model and enable the next round of experimental studies.

This work was supported in part by the US National Science Foundation, by the US NIST precision measurement grant, by the Russian Foundation for Basic Research under Grant Nos. 07-02-00210-a and 05-02-16914-a, and by Saint Petersburg State Scientific Center. This manuscript was completed with a partial support by NSF through a grant for the Institute for Theoretical Atomic, Molecular, and Optical Physics at Harvard University and Smithsonian Astrophysical Observatory.

References

1. T.D. Lee, C.N. Yang, Phys. Rev. **104**, 254 (1956).
2. C.S. Wu, E. Ambler, R.W. Hayward, D.D. Hoppes, R.P. Hudson, Phys. Rev. **105**, 1413 (1957).
3. Ya.B. Zel'dovich, Zh. Eksp. Teor. Fiz. **36**, 964 (1959) [Sov. Phys. JETP **9**, 681 (1959)].
4. M.A. Bouchiat, C. Bouchiat, J. Phys. **35**, 899 (1974).
5. L.M. Barkov, M.S. Zolotorev, Pis'ma Zh. Eksp. Teor. Fiz. **27**, 379 (1978) [JETP Lett. **27**, 357 (1978)].
6. C.S. Wood, S.C. Bennett, D. Cho, B.P. Masterson, J.L. Roberts, C.E. Tanner, C.E. Wieman, Science **275**, 1759 (1997).
7. M.A. Bouchiat, C. Bouchiat, Rep. Prog. Phys. **60**, 1351 (1997).
8. J.S.M. Ginges, V.V. Flambaum, Phys. Rep. **397**, 63 (2004).
9. I.B. Khriplovich, *Parity non-conservation in atomic phenomena* (Gordon and Breach, New York, 1991).
10. M.J. Ramsey-Musolf, Phys. Rev. C **60**, 015501 (1999).
11. P.L. Anthony, R.G. Arnold, C. Arroyo, K. Bega, J. Biesiada, P.E. Bosted, G. Bower, J. Cahoon, R. Carr, G.D. Cates *et al.*, Phys. Rev. Lett. **95**, 081601 (2005).
12. G.P. Zeller, K.S. McFarland, T. Adams, A. Alton, S. Avvakumov, L. de Barbaro, P. de Barbaro, R.H. Bernstein, A. Bodek, T. Bolton *et al.*, Phys. Rev. Lett. **88**, 091802 (2002).
13. Homepage of the Q-weak Collaboration at Jefferson Lab, `www.jlab.org/qweak/`.
14. V.A. Dzuba, V.V. Flambaum, O.P. Sushkov, Phys. Lett. A **141**, 147 (1989).
15. S.A. Blundell, W.R. Johnson, J. Sapirstein, Phys. Rev. Lett. **65**, 1411 (1990); Phys. Rev. D **45**, 1602 (1992).
16. S.C. Bennett, C.E. Wieman, Phys. Rev. Lett. **82**, 2484 (1999).
17. R. Casalbuoni, S. de Curtis, D. Dominici, R. Gatto, Phys. Lett. B **460**, 135 (1999).
18. J.L. Rosner, Phys. Rev. D **61**, 016006 (2000).
19. J.L. Rosner, Phys. Rev. D **65**, 073026 (2002).
20. A. Derevianko, Phys. Rev. Lett. **85**, 1618 (2000).
21. V.A. Dzuba, C. Harabati, W.R. Johnson, M.S. Safronova, Phys. Rev. A **63**, 044103 (2001).
22. M.G. Kozlov, S.G. Porsev, I.I. Tupitsyn, Phys. Rev. Lett. **86**, 3260 (2001).
23. W.R. Johnson, I. Bednyakov, G. Soff, Phys. Rev. Lett. **87**, 233001 (2001).
24. A.I. Milstein, O.P. Sushkov, I.S. Terekhov, Phys. Rev. Lett. **89**, 283003 (2002).
25. A. Milstein, O. Sushkov, I. Terekhov, Phys. Rev. A **67**, 62103 (2003).
26. M. Kuchiev, V. Flambaum, Phys. Rev. Lett. **89**, 283002 (2002).
27. J. Sapirstein, K. Pachucki, A. Veitia, K.T. Cheng, Phys. Rev. A **67**, 052110 (2003).
28. V.M. Shabaev, K. Pachucki, I.I. Tupitsyn, V.A. Yerokhin, Phys. Rev. Lett. **94**, 213002 (2005).
29. A. Derevianko, Phys. Rev. A **65**, 012106 (2002).
30. V.A. Dzuba, V.V. Flambaum, J.S.M. Ginges, Phys. Rev. D **66**, 076013 (2002).
31. J. Guena, D. Chauvat, P. Jacquier, E. Jahier, M. Lintz, S. Sanguinetti, A. Wasan, M.A. Bouchiat, A.V. Papoyan, D. Sarkisyan, Phys. Rev. Lett. **90**, 143001 (2003).
32. J. Guena, M. Lintz, M.A. Bouchiat, Phys. Rev. A **71**, 042108 (2005).
33. N. Fortson, Phys. Rev. Lett. **70**, 2383 (1993).
34. T.W. Koerber, M.H. Schacht, W. Nagourney, E.N. Fortson, J. Phys. B **36**, 637 (2003).
35. P.A. Vetter, D.M. Meekhof, P.K. Majumder, S.K. Lamoreaux, E.N. Fortson, Phys. Rev. Lett. **74**, 2658 (1995).
36. E. Gomez, L.A. Orozco, G.D. Sprouse, Rep. Prog. Phys. **69**, 79 (2006).
37. C. Amiot, O. Dulieu, R.F. Gutterres, F. Masnou-Seeuws, Phys. Rev. A **66**, 052506 (2002).
38. N. Bouloufa, A. Crubellier, O. Dulieu, private communications.
39. F. Coester, H.G. Kümmel, Nucl. Phys. **17**, 477 (1960).
40. J. Čižek, J. Chem. Phys. **45**, 4256 (1966).
41. R.F. Bishop, H.G. Kümmel, Physics Today **3**, 52 (1987).
42. S.A. Blundell, W.R. Johnson, Z.W. Liu, J. Sapirstein, Phys. Rev. A **40**, 2233 (1989).
43. S.A. Blundell, W.R. Johnson, J. Sapirstein, Phys. Rev. A **43**, 3407 (1991).
44. E. Eliav, U. Kaldor, Y. Ishikawa, Phys. Rev. A **50**, 1121 (1994).
45. E.N. Avgoustoglou, D.R. Beck, Phys. Rev. A **57**, 4286 (1998).
46. M.S. Safronova, A. Derevianko, W.R. Johnson, Phys. Rev. A **58**, 1016 (1998).
47. M.S. Safronova, W.R. Johnson, A. Derevianko, Phys. Rev. A **60**, 4476 (1999).
48. I. Lindgren, J. Morrison, *Atomic Many-Body Theory*, 2nd ed. (Springer-Verlag, Berlin, 1986).
49. A. Derevianko, E.D. Emmons, Phys. Rev. A **66**, 012503 (2002).
50. C.C. Cannon, A. Derevianko, Phys. Rev. A **69**, 030502(R) (2004).
51. A. Derevianko, S.G. Porsev, Phys. Rev. A **71**, 032509 (2005).
52. S.G. Porsev, A. Derevianko, Phys. Rev. A **73**, 012501 (2006).
53. M.S. Safronova, PhD Thesis, University of Notre Dame (2000).
54. V.V. Flambaum, J.S.M. Ginges, Phys. Rev. A **72**, 052115 (2005).
55. C.E. Moore, *Atomic energy levels*, Vol. **III** (National Bureau of Standards, Washington, D.C., 1958).
56. J. Sapirstein, K.T. Cheng, Phys. Rev. A **67**, 022512 (2003).
57. R.J. Rafac, C.E. Tanner, Phys. Rev. A **56**, 1027 (1997).
58. R.J. Rafac, C.E. Tanner, Phys. Rev. A **58**, 1087 (1998).

Pump-probe measurement of atomic parity violation in cesium with a precision of 2.6%

M. Lintz[a], J. Guéna[b], and M.-A. Bouchiat

Laboratoire Kastler Brossel, Département de Physique de l'Ecole Normale Supérieure[c], 24 Rue Lhomond, 75231 Paris cedex 05, France

Abstract. We present the atomic parity violation measurements made in Cs vapour using a pump-probe scheme. After pulsed excitation of the $6S$-$7S$ forbidden transition in the presence of a longitudinal electric field, a laser beam resonant with one of the $7S$-$6P$ transitions stimulates the $7S$ atom emission for a duration of 20 ns. The polarisation of the amplified probe beam is analysed. A seven-fold signature allows discrimination of the parity-violating linear dichroism, and real-time calibration by a similar, known, parity-conserving linear dichroism. The zero-field linear dichroism signal due to the magnetic dipole transition moment is observed for the first time, and used for *in situ* determination of the electric field. The result, $\mathrm{Im}\, E_1^{pv} = (-808\pm21)\times10^{-14} ea_0$, is in perfect agreement with the corresponding, more precise measurement obtained by the Boulder group. A transverse field configuration with large probe amplification could bring atomic parity violation measurements to the 0.1% accuracy level.

PACS. 32.80.Ys Weak-interaction effects in atoms – 11.30.Er Charge conjugation, parity, time reversal, and other discrete symmetries – 33.55.Be Zeeman and Stark effects – 42.50.Gy Effects of atomic coherence on propagation, absorption, and amplification of light; electromagnetically induced transparency and absorption

1 Introduction

In this paper we present the parity violation (PV) measurements we have performed in Cs [1], by making use of a probe beam superposed with the beam that drives the $6S$-$7S$ forbidden transition. After the probe beam has been amplified by stimulated emission of the excited vapour, its polarisation is analysed so as to extract the PV asymmetry, calibrated against a parity-conserving signal. This experimental scheme has the potential for high-precision PV measurements, since the PV asymmetry to be measured is an increasing function of the excited atom number density. The quantity of metal necessary for vapour cell operation is orders of magnitude smaller than in an atomic beam experiment, and would be acceptable for an experiment with the long-lived isotope ^{135}Cs.

[a] *Present address:* Observatoire de la Côte d'Azur, BP 4229, F-06304 Nice, France; e-mail: `michel.lintz@oca.eu`.

[b] *Present address:* LNE-SYRTE, 61 avenue de l'Observatoire, F-75014 Paris, France; e-mail: `jocelyne.guena@obspm.fr`.

[c] Unité de Recherche de l'Ecole Normale Supérieure et de l'Université Pierre et Marie Curie, associée au CNRS (UMR 8552).

2 Motivations

Since the emergence of the field [2], atomic parity violation (APV) measurements in heavy atoms have evolved and diversified [3]. These difficult experiments are motivated by the fact that APV measurements provide unique inputs for testing the Standard Model (SM). Measurements at the pole of the Z^0 boson mass energy have reached an impressive precision, but testing, for instance, the variation of the value of the weak mixing angle with energy requires data taken at different energy scales, such as the recent measurement of the PV asymmetry in Møller scattering [4], and the APV measurements in cesium. The corresponding momentum transfers are about 160 MeV/c and a few MeV/c, respectively, to be compared with 100 GeV/c at the Z^0 pole. Besides the energy scale, the information extracted from a determination of the Cs weak charge Q_W is a test in the hadronic sector of the SM, and a test different from those extracted, for instance, from the inelastic-scattering processes exploited in [5]. In a model-independent analysis of the results, the corresponding allowed regions, in the plane spanned by the elementary weak charges of the u and d quarks, turn out to be nearly orthogonal.

Relevance to physics beyond the SM is one more motivation. APV experiments are particularly sensitive, among the P-violating processes, to those possibly mediated by particles with a mass in the MeV range, while such a "light" boson could escape detection in high-energy experiments [6]. The U boson predicted by some supersymmetric extensions of the SM is a candidate to explain, through dark-matter recombination, the astrophysical observation that the bulge of the galaxy shows a significant emission at 511 keV [7]. However, to account for the moderate intensity of the emission line, one has to assume [6] that the axial coupling of the U boson (mass ≈ 10 MeV) to electrons is very small. Indeed, the present agreement between APV measurements in Cs and the SM [8] excludes such an axial coupling at the 10^{-6} level. Some of the Kaluza-Klein models of space-time also make predictions of effects observable in APV without counterparts in high-energy experiments [9].

The Boulder measurement [10], with two 0.5% precision measurements on two hyperfine (HF) components of the Cs forbidden transition, is the only experiment that could achieve a precision test of the standard model at very low energy. Controlling the systematics [11] has been a difficult aspect of this experiment, and a cross-check of this measurement would be highly valuable, in regard of the implications of APV experiments. The experimental scheme developed in Paris and presented in this paper combines the advantages i) of the "Stark" (electric-field enhanced excitation) experiments: very well-defined signature, removing the need for scanning the excitation wavelength to identify the PV signal, and hence the risk associated with the "wavelength-dependent analysis angle" of the optical-rotation experiments in heavier atoms; and ii) of those latter "transmission" experiments: use of the probe beam allows to detect the whole excited vapour column with high efficiency.

Most importantly, our experimental scheme has only two features in common with those of the Boulder experiment: the choice of the $6S_{1/2}$-$7S_{1/2}$ forbidden transition and the application of a Stark field, both suggested in [2]. As a result the systematic effects are very different, a crucial point since the measurements presented here aim at a cross-check of the Boulder experiment. Some of the many differences will be highlighted in the text. The most original feature here is the increase of the detected asymmetries with the applied electric field. Our proposal [12] to best take advantage of this feature will be presented briefly, keeping in mind the goal of APV measurements at the 0.1% level.

3 Principles of the experiment

A pulsed, 539 nm laser beam excites the $6S_{1/2}, F = 3 \rightarrow 7S_{1/2}, F = 4$ HF transition of Cs while a longitudinal electric field is applied, for a duration of only 100 ns not to trigger discharges in the 8 cm long cell [13]. After this powerful (1.5 mJ in 15 ns) excitation of the forbidden transition, a weak (1 mW), infrared beam stimulates emission on the $7S_{1/2}, F = 4 \rightarrow 6P_{3/2}, F = 4$ allowed transition, for 20 ns. The pump and probe beams are superposed with the same direction $\hat{k}$ and the same linear polarisations $\hat{\epsilon}_{exc} \parallel \hat{\epsilon}_{pr}$. The PV signal is the change of the probe polarisation associated to the excitation of the forbidden transition.

An effective dipole operator [2] can be used to describe the excitation of the forbidden transition:

$$-\hat{\epsilon}_{exc} \cdot \boldsymbol{d}_{eff} = \hat{\epsilon}_{exc} \cdot (i \operatorname{Im} E_1^{pv} \boldsymbol{\sigma} - M_1^{'} \boldsymbol{\sigma} \wedge \hat{k} + i\beta \boldsymbol{\sigma} \wedge \boldsymbol{E} + \alpha \boldsymbol{E}). \tag{1}$$

The first term is the PV electric dipole amplitude ($\boldsymbol{\sigma}$ is the spin Pauli operator, and $i \operatorname{Im} E_1^{pv}$ is purely imaginary), the second one is the magnetic dipole contribution, and the two last ones are the Stark-induced, respectively vector and scalar, contributions. To give orders of magnitude, the corresponding partial $7S_{1/2}$ lifetimes are about 12 million years for the PV electric dipole amplitude, 12 days for the magnetic dipole amplitude, 300 s and 3 s, respectively, for the vector and scalar Stark-induced transitions in a 2 kV/cm $\boldsymbol{E}$ field. However, in a longitudinal field $\boldsymbol{E}_l = E_l \hat{z}$, the $\alpha \boldsymbol{E} \cdot \hat{\epsilon}_{exc}$ term is absent since the excitation polarisation is transverse to $\hat{k} \parallel \hat{z}$. Also important is the fact that, due to the choice of the linearly polarised excitation, the real term $M_1^{'} \boldsymbol{\sigma} \wedge \hat{k}$ is out of phase with the $i\beta \boldsymbol{\sigma} \wedge \boldsymbol{E}_l$ purely imaginary term. This is a crucial difference with the Boulder experiment, where these two terms interfere, which contributes to several systematics [11]. The magnetic term will be omitted throughout, except in sect. 8 where it is exploited for calibration.

In the absence of the PV term, the effective transition operator (eq. (1)) would be $\hat{\epsilon}_{exc} \cdot (i\beta \boldsymbol{\sigma} \wedge \boldsymbol{E}_l) = -i\beta E_l \sigma_x$, assuming that $\hat{\epsilon}_{exc} = \hat{y}$. Adding the PV contribution changes this term into $-i\beta E_l \{\sigma_x - (\operatorname{Im} E_1^{pv} / \beta E_l) \sigma_y\}$. Indeed, the PV term has the same, geometrical consequence as a tilt of the excitation polarisation

$$\hat{\epsilon}_{exc} \rightarrow \hat{\epsilon}_{exc} + \theta^{pv} \hat{z} \wedge \hat{\epsilon}_{exc} \tag{2}$$

by an angle $\theta^{pv} = -\operatorname{Im} E_1^{pv} / \beta E_l$. As a result, the eigenaxes of the excited-state density matrix $\rho_{7S} \propto (\beta E_l \sigma_x)^2 \propto (\beta E_l F_x)^2$ will be tilted as well. Consequently, the vapour will act on the probe as an anisotropic amplifier *with eigen-axes tilted*, by the angle θ^{pv}, *with respect to the symmetry planes* imposed by the direction of the linear polarisation $\hat{\epsilon}_{exc}$. The tilt angle θ^{pv} is small, 10^{-6} radians at $E_l \approx 1.6$ kV/cm, but is odd in E_l reversal, a clear-cut signature. Hence, the PV signal will be a change of the linear polarisation of the amplified probe beam, odd under E_l reversal, manifesting the presence, in the probe gain matrix, of a chiral contribution associated with the pseudoscalar $(\hat{\epsilon}_{exc} \cdot \hat{\epsilon}_{pr})(\hat{\epsilon}_{exc} \wedge \hat{\epsilon_{pr}} \cdot \boldsymbol{E}_l)$.

Although the expected probe polarisation signal can be calculated [14], an efficient calibration has been obtained by deliberately tilting $\hat{\epsilon}_{exc}$ by a known angle θ_{cal} and measuring the corresponding probe polarisation signal in rigorously identical conditions. For θ_{cal} in the milliradian range, proportionality is good enough [15] to allow linear extraction of θ^{pv} from the PV signal. As compared to the Boulder experiment, this calibration procedure is obtained in real time, and is free from any lineshape correction.

4 Short review of the systematic effects

Systematics can be classified as rank-1 or rank-2, according to the number of defects involved.

Rank-1: i) A tilt of $\hat{\epsilon}_{exc}$, if it were *odd* in the reversal of the applied electric field, would give rise to a systematic effect: this is looked for and estimated by real-time monitoring of $\hat{\epsilon}_{exc}$. ii) Remembering that a longitudinal magnetic field induces Larmor rotation of the eigen-axes of the vapour, an E_l-odd B_z field would give rise to a systematic. Although the symmetry of the experiment forbids even an even B_z field, a significant fraction of the data acquisition is devoted to the measurement of B_z^-, using atomic signals (see sect. 6).

Rank-2: i) An offset between the pump and probe polarisations, in the presence of an *imperfect reversal* of the applied $\boldsymbol{E}_l$ field, could give rise to a systematic. However, the data processing (see sect. 6) is made in a way that rejects their coupling. In addition, these two imperfections monitored using specific signals are kept below 10^{-4} and 10^{-3}, respectively. ii) Defects that break cylindrical symmetry. Ideally, simultaneous rotation of the pump and probe polarisations (and the corresponding polarimeters) should leave the experiment unchanged, except for possible *transverse* electric or magnetic fields, or a misalignment between the excitation and probe beam [16]. The coupling of a transverse electric field to a transverse magnetic field gives rise, in the calibrated probe polarisation signal, to a contribution,

$$2\omega_{F'}\tau\frac{E_t}{E_l}(\hat{B}_t\cdot\hat{E}_t-(\hat{B}_t\cdot\hat{\epsilon}_{exc})(\hat{E}_t\cdot\hat{\epsilon}_{exc})), \tag{3}$$

in which $\omega_{F'}\tau$ is the average Larmor precession under the magnetic field $\boldsymbol{B}_t$ during the time spent by the atoms in the excited state. This term mimics the PV E_l-odd term if the $\boldsymbol{B}_t$ and $\boldsymbol{E}_t$ fields are *both* odd, or *both* even, in $\boldsymbol{E}_l$ reversal. When averaged over two polarisations at 90° of each other, this systematic is reduced by a factor of 1/2 ("class 1" systematic effect). On the other hand, the angular dependence, $\hat{B}_t\cdot(\hat{E}_l\wedge\hat{\epsilon}_{exc})\cdot(\hat{B}_t\cdot\hat{\epsilon}_{exc})$, of the systematic due to the second-order perturbation by a magnetic field completely cancels in the 90° switch polarisation average: "class 2" systematic effect. In ref. [16] can be found the inventory of both class-1 (the most important ones) and class-2 systematic effects.

5 Implementation of the experiment

We briefly recall the experimental setup, more extensively described in [17,1], where the absolute frequency stabilisation of the laser sources is presented.

The excitation and probe beams are recombined using a dichroic mirror (see fig. 1). Beforehand, the polarisation of each of the two beams has been prepared in a "polariser unit" made of a calcite Glan polariser that defines the vertical polarisation, and several half-wave plates that are mechanically inserted to work with horizontal, or ±45° polarisation. The excitation polariser unit also includes i) a wave plate chosen to be precisely $1\times\lambda_{exc}$ at *normal incidence*. When appropriately tilted, it allows to compensate

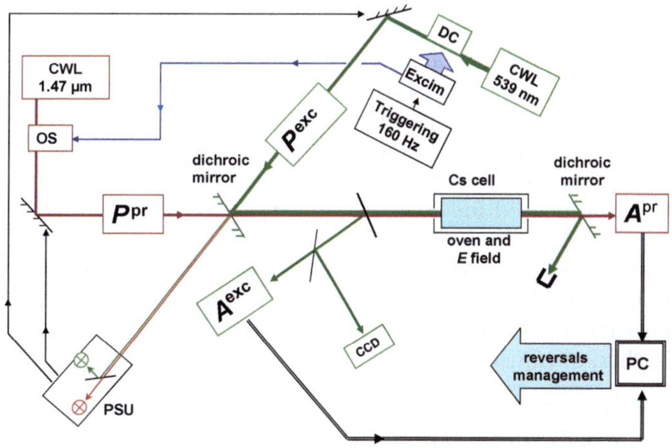

Fig. 1. Main parts of the experimental set-up. CWL: single-frequency, continuous-wave laser. OS: sub-nanosecond optical switch. P_{pr}, P_{exc}, A_{pr}, A_{exc}, polariser and analyser units. Excim: excimer laser. DC: dye cells amplifier unit. PSU: beam-position stabilisation unit. CCD: camera for excited-beam profile monitoring. PC: computer for data acquisition, and management of the parameter reversals. The two-stage oven allows separate control of the Cs cell reservoir and wall temperatures.

the birefringence in the input Cs cell window; ii) a Faraday modulator, providing the $\pm\theta_{cal}$ tilts required for the calibration procedure. The probe polariser unit includes a tilted glass plate to compensate for the linear dichroism in the recombination mirror. The "analyser units" are based on a two-channel, balanced-mode polarimeter [18], preceded by several half-wave plates, two of which are inserted synchronously with the half-wave plates of the corresponding polariser unit, so as to operate the polarimeter always in the same conditions. The third half-wave plate has its axes at 45° to the axes of the polarimeter. Inserting this "cleaner" half-wave plate allows to reject the polarimeter signals that are not due to a beam polarisation change [19]. The probe analyser records two kinds of signal: the first 20 ns probe shot, immediately after excitation of the forbidden transition, which contains the excited-state contribution, and, one ms later, a burst of four probe shots, which serves as a reference for the probe polarisation signals.

Sapphire appeared as an appropriate material for the cell body, due to its very low surface conductivity in the presence of cesium vapour [20]. This should have suppressed the B_z^- field. However, a considerable electron charge (and corresponding current) was observed at the anode window of the sapphire cells, until we could mechanically suppress the multiplication, due to secondary emission, of the electrons emitted at the cathode window [21]. Then the B_z^- field was observed to be at most a few tens of μG, corresponding to a false effect of a fraction of the PV signal (11% on average), for which we can correct practically in real time. Measurement of B_z^- uses the (large) optical rotation signal on the $F=4$ to $F=5$ probe transition [22].

6 Measurement procedure

The basic element of the data acquisition is the simultaneous recording of the two signals of the two-channel probe polarimeter, to form the imbalance $D^{ampl} \equiv (S_1 - S_2)/(S_1 + S_2)$ with the *amplified* probe beam. After the reference probe pulses have been detected, the corresponding imbalance is subtracted: $D_{at} \equiv D^{ampl} - D^{ref}$, and this is repeated for 30 shots (duration 200 ms) before θ_{cal} is reversed. The next reversal is that of the electric field (period 800 ms). Insertion and removal of the "cleaner" half-wave plate is repeated every 7 s. The input probe polarisation undergoes a 90° switch every 14 s. A value θ^{pv} is then obtained by

$$[\theta^{pv}]_{\hat{\epsilon}_{exc}} = \theta_{cal} \times \left\langle \sigma_E \left[\frac{\langle \sigma_{clean} D_{at}(\{\sigma_j\})\rangle_{\sigma_{clean}\sigma_{cal}}}{\langle \sigma_{clean}\sigma_{cal} D_{at}(\{\sigma_j\})\rangle_{\sigma_{clean}\sigma_{cal}}} \right] \right\rangle_{\sigma_E \sigma_{90^\circ pr}}, \quad (4)$$

where $\sigma_i = \pm 1$ indicates the state of reversal (or plate insertion) i. Beside this PV linear dichroism, other linear combinations are monitored for diagnosis of possible drifts or systematics. The probe optical rotation, obtained by inserting a factor $\sigma_{90^\circ pr}$ after the σ_E factor in the equation above, is used to measure the B_z^- field. In order to suppress systematics (see sect. 4), two "isotropic values" of θ^{pv} are obtained, every 5 min, after $\hat{\epsilon}_{exc}$ has been switched by 90°, then ±45°:

$$S_{xy} \equiv \frac{1}{2}([\theta^{pv}]_x + [\theta^{pv}]_y) \quad \text{and} \quad S_{uv} \equiv \frac{1}{2}([\theta^{pv}]_u + [\theta^{pv}]_v). \quad (5)$$

The recording of isotropic values goes on for about 90 min, after which a last reversal is performed, concerning the cell orientation with respect to the light beams. Operating at normal incidence at the cell windows gives rise to an excess noise on the polarimetric measurement, due to an etalon, interference effect between the cell windows and optical surfaces inside the polarimeter. To prevent this, the cell is tilted by an angle $\psi = 3 \times 10^{-3}$ rad. However, the atoms excited by the 539 nm beam reflected at the exit cell also contribute. This beam is tilted by an angle 2ψ with respect to the probe beam, and it can be shown that the rank-2 systematic associated with this tilt coupled to a transverse electric field is linear in ψ [16,1]. It is suppressed by i) reversing the angle ψ, and ii) reducing the window reflection (see sect. 7).

Finally, a PV data-taking run starts and closes with the measurement of the optical rotation on the $4 \rightarrow 5$ probe transition to determine the value of the stray B_z^- field and correct for the associated contribution. This lengthens data acquisition by about 60%. On the other hand, the measurement of the stray, transverse electric and magnetic fields, made by application of large, controlled, transverse magnetic fields, does not increase significantly data-taking duration, although it is regularly repeated throughout the data acquisition. In the results presented below, the systematics are controlled and kept at the 1% level. The 2.6% total error bar on the average is mainly statistical, and includes the slight contribution due to the B_z^- systematic correction.

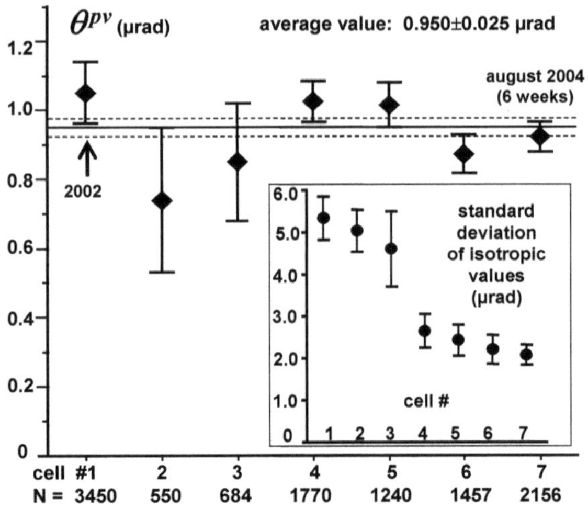

Fig. 2. The values of θ^{pv} measured in the seven successive cells, at an electric field of 1.62 kV/cm. Inset: the standard deviation, as a function of cell number. Adapted from ref. [1] with permission from APS.

7 The PV results and their improvement

Seven cells have been used successively for the measurements. The measured values of θ^{pv} are displayed in fig. 2. For each cell the number N of isotropic values recorded is indicated. The first measurement took place in 2002 [23] and confirmed the validity of the method. Later, efforts devoted to the reduction of the noise and of the systematics have improved the statistics, as is obvious from the inset of fig. 2: the standard deviation of the isotropic values recorded in cell #7 is 2.5 times smaller than that in cell #1.

This has resulted from implementing:

- sapphire cells with highly parallel windows [24]. The fraction of the excitation beam reflected at each window can be made very small (10^{-3} instead of 5% in cell #1) by choosing a temperature such that the reflection is cancelled by interference. This further suppresses the systematic effect associated with the tilted reflected beam (see sect. 6). It also suppresses the interference between the beams reflected at the input and exit windows.
- a "polarisation magnifier" [25]. This device, made of 4 or 6 plates at Brewster incidence, enhances the angle of the polarisation tilt to be measured by the polarimeter. Since it attenuates the beam, it allows higher probe power without changing the polarimeter preamplifier chain, improving the signal/noise ratio when the photon shot noise contribution is significant.
- a better extinction for the probe optical switch. The probe intensity that leaks through the closed optical switch can contribute to the noise.
- long-term absolute frequency stabilisation of the excitation [1].

A statistical analysis [1] shows that the values obtained in the different cells agree. Indeed the cells were not identical: the material of the cell tube (alumina ceramics *vs.*

monocrystalline sapphire), the polishing/origin/tilt of the windows, the filling of the cells (unexpected foreign gas observed in cell #3) etc. were different. The agreement provides confidence in the absence of a bias from cell imperfections. The fact that the averages of the S_{xy} and S_{uv} values agree to about 0.7% also gives confidence in the cylindrical symmetry of the experiment.

8 Calibration aspects; measurement of the E_l field

First, the measurements presented above rely on the determination of the Faraday rotation angle used for calibration of the PV linear dichroism signal. This was done by different methods, in particular by measuring the mechanical rotation of the polariser that compensates for the Faraday rotation, with an accuracy of about 0.5%, that could be improved if need be.

Second, the determination of $\theta^{pv} = -\operatorname{Im} E_1^{pv}/\beta E_l$ can be exploited only if E_l is known with enough accuracy *inside the vapour*, *i.e.*, measured by the atoms themselves. This can be done by measuring the Stark parity-conserving alignment, proportional to $\theta_{cal}\beta^2 E_l^2$, and then measuring the zero-field alignment, proportional to $M_1^{'2}$. The proportionality factors are the same, except for the influence of saturation by the probe beam: in the presence of the applied, $\approx 1.6\,\mathrm{kV/cm}$ electric field, the amplification of the probe beam is non-negligible, while it is very small, and buried in noise and background, in the absence of electric field. An accurate calibration of the electric field can be obtained by the comparison of these two, similar, linear dichroism signals, both measured *vs.* probe beam intensity I_{pr} and extrapolated to $I_{pr} = 0$:

$$E_l^{exp} = \frac{M_1^{'}}{\beta}\left(\sqrt{\frac{\ln\left[1+D_{at}(E=E_l,\theta=\pm\theta_{cal})/2\theta_{cal}\right]}{(1+\epsilon)D_{at}(E=0,\theta=\pm\pi/4)}}-1\right).$$

Here, the small ϵ quantity expresses the deviation of the exact result, calculated numerically [14], with respect to the simple one assuming an exponential-type amplification. In our experimental conditions [1], we obtain $\epsilon = 0.100$, with a resulting uncertainty of 0.3% on E_l^{exp}. Presently, the uncertainty in E_l^{exp} is negligible as compared to the 2.5% uncertainty in the determination of θ^{pv}.

9 Conclusions and perspectives

For the measured field $E_l^{exp} = 1.619\,\mathrm{kV/cm}$ we obtain $\theta^{pv}_{exp} = (0.950 \pm 0.025)\,\mu\mathrm{rad}$, in excellent agreement with the value of 0.962 ± 0.005 obtained from the Boulder measurement [10] on the same $6S, F = 3 \rightarrow 7S, F = 4$ hyperfine transition. Using the value of the β polarisability published in [26], the corresponding value of the PV transition dipole, $\operatorname{Im} E_1^{pv}(3 \rightarrow 4) = -0.808 \times 10^{-11}|e|a_0$, is obtained with a $2.1 \times 10^{-13}|e|a_0$ absolute uncertainty.

Our experimental method is very different from that used by the Boulder group [11], with many consequences as regards calibration, or systematics. It makes use of the amplification of a probe beam, the higher the gain, the larger the asymmetry to be measured. Hence, the PV asymmetry is an *increasing* function of the applied electric field. One important consequence is the considerable enhancement of sensitivity that can be foreseen in an experiment with a *transverse* electric field and a longer interaction length [12]. Despite the transverse-field configuration, a special multi-electrode design can "restore cylindrical symmetry", which helps in tracking and rejecting systematic effects [27]. With reasonably moderate values of the probe gain (still far from spontaneous superradiance), a sensitivity of 0.1% looks a realistic objective, although the regime of higher gains (*triggered* superradiance) probably deserves specific attention. The suggested experiment involves the excitation in the presence of transverse magnetic and electric fields, with two counter-propagating excitation beams. It is further motivated by the recent progress in the atomic-physics calculations of the factor that links the nuclear spin-independent value of E_1^{pv} to the Cs nucleus weak charge, now aiming at the 0.1% accuracy [28].

References

1. J. Guéna *et al.*, Phys. Rev. A **71**, 042108 (2005).
2. M.A. Bouchiat, C. Bouchiat, J. Phys. (Paris) **35**, 899 (1974).
3. See B. Frois, M.A. Bouchiat (Editors), *Parity violation in atoms and polarized electron scattering* (World Scientific, Singapore, 1999); J. Guéna *et al.*, Mod. Phys. Lett. A **20**, 375 (2005); D. Budker, this conference.
4. E158 Collaboration (P.L. Antony *et al.*), Phys. Rev. Lett. **95**, 081601 (2005).
5. C.Y. Prescott *et al.*, Phys. Lett. B **77**, 347 (1978).
6. C. Bouchiat, P. Fayet, Phys. Lett. B **608**, 87 (2005).
7. C. Boehm *et al.*, Phys. Rev. D **69**, 101302 (2004).
8. V.M. Shabaev *et al.*, Phys. Rev. Lett. **94**, 213002 (2005); see also the review by J.S.M. Ginges, V.V. Flambaum, Phys. Rep. **397**, 63 (2004).
9. A. Delgado *et al.*, JHEP **01**, 030 (2000).
10. C.S. Wood *et al.*, Science **275**, 1759 (1997).
11. C.S. Wood *et al.*, Can. J. Phys. **77**, 7 (1999).
12. J. Guéna *et al.*, J. Opt. Soc. Am. B **22**, 21 (2005).
13. E. Jahier *et al.*, Eur. Phys. J. D **13**, 221 (2001).
14. C. Bouchiat, M.A. Bouchiat, Z. Phys. D **36**, 105 (1996).
15. M.A. Bouchiat *et al.*, Opt. Commun. **77**, 374 (1990).
16. M.A. Bouchiat *et al.*, Eur. Phys. J. D **28**, 331 (2004).
17. J. Guéna *et al.*, Quantum Semiclass. Opt. **10**, 733 (1998).
18. J. Guéna *et al.*, Opt. Commun. **71**, 6 (1989).
19. M.A. Bouchiat *et al.*, Opt. Commun. **119**, 403 (1995).
20. M.A. Bouchiat *et al.*, Appl. Phys. B **68**, 1109 (1999), J. Guéna *et al.*, Rev. Sci. Instrum. **76**, 053108 (2005).
21. J. Guéna *et al.*, Appl. Phys. B **75**, 739 (2002).
22. M.A. Bouchiat *et al.*, Z. Phys. D **33**, 89 (1995).
23. J. Guéna *et al.*, Phys. Rev. Lett. **90**, 143001 (2003).
24. E. Jahier *et al.*, Appl. Phys. B **71**, 561 (2000).
25. M. Lintz *et al.*, Rev. Sci. Instrum. **76**, 043102 (2005).
26. S.C. Bennett, C.E. Wieman, Phys. Rev. Lett. **82**, 2484 (1999).
27. M.A. Bouchiat *et al.*, (in French) unpublished.
28. A. Derevianko, this conference.

Program of parity nonconservation measurements in francium

G. Gwinner[1], E. Gomez[2], A. Perez Galvan[2], D. Sheng[2], Y. Zhao[2], L.A. Orozco[2], G.D. Sprouse[3], J.A. Behr[4], K.P. Jackson[4], M.R. Pearson[4], V.V. Flambaum[5], and S. Aubin[6]

[1] University of Manitoba, Department of Physics & Astronomy, Winnipeg, MB R3T 2N2, Canada
[2] University of Maryland, Department of Physics, College Park, MD 20742, USA
[3] Stony Brook University, Department of Physics & Astronomy, Stony Brook, NY 11794, USA
[4] TRIUMF, 4004 Wesbrook Mall, Vancouver, BC V6T 2A3, Canada
[5] University of New South Wales, School of Physics, Sydney, NSW 2052, Australia
[6] University of Toronto, Department of Physics, Toronto, ON M5S 1A7, Canada

Abstract. A program to study parity nonconservation in francium at TRIUMF is presented. The essential part of the experiment will be a collection of $\approx 10^6$ trapped atoms on-line at the ISAC radioactive beam facility. The atoms can be excited through a parity-forbidden E1 transition either in the microwave or in the optical regime. The rate of excitation gives information on the weak nucleon-nucleon interactions through the anapole moment of the Fr nucleus.

PACS. 32.80.Ys Weak-interaction effects in atoms – 24.80.+y Nuclear tests of fundamental interactions and symmetries

1 Introduction

Parity nonconservation (PNC) in atoms comes from two types of interaction: Nuclear spin independent and nuclear spin dependent [1]. Nuclear spin dependent PNC occurs in three ways [2,3]: An electron interacts weakly with a single valence nucleon (electron vector current and nucleon axial-vector current, $A_n V_e$), the nuclear chiral current created by weak interactions between nucleons (anapole moment), or the combined action of the hyperfine interaction and the spin-independent Z^0 exchange interaction from nuclear vector currents ($V_n A_e$) [4–6].

Zel'dovich first proposed the existence of the nuclear anapole moment in 1957 [7]. Vetter *et al.* obtained a limit for its value in thallium [8], and it was measured with an accuracy of 15% through the hyperfine dependence of atomic parity nonconservation (PNC) in cesium by Wood *et al.* [9,10]. In this paper, we discuss the feasibility of studying weak nucleon-nucleon interactions by measuring a nuclear anapole moment through parity-violating electric dipole transitions in atoms. Bouchiat and Bouchiat studied atomic PNC in a series of seminal papers [11,12] and showed a significant enhancement in the effects for heavy atoms (roughly a Z^3 scaling, where Z is the charge of the nucleus). We focus our program on francium, the heaviest of the alkali atoms, but also a radioactive nucleus.

The systematic study of the atomic properties of francium has advanced significantly with the availability of atom traps on-line with accelerators (see review by Gomez *et al.* [13]). It has reached comparable quantitative understanding to that of the stable alkalis both theoretically and experimentally. Recent work related to proposed time-reversal invariance and parity nonconservation tests in traps [14–16] points to the many potential advantages of combining traps with tests of fundamental symmetries, but also highlights possible systematic errors present in such measurements.

The FrPNC collaboration is embarking on a program of PNC experiments in francium at TRIUMF. The effort combines precision measurement techniques from atomic and optical physics for the elucidation of the weak interaction in a chain of Fr isotopes, where the effect is predicted to be one order of magnitude larger than in Cs. This paper reviews the possible avenues for atomic PNC measurements in francium.

2 Atomic parity nonconservation

The exchange of weak neutral currents between electrons and nucleons constitutes the main source of parity violation in atomic transitions. The currents are of two kinds, depending on whether the electron or the nucleon enters as the axial vector current. The Hamiltonian for an infinitely heavy nucleon without radiative corrections is [17]

$$H = \frac{G}{\sqrt{2}}(\kappa_{1i}\gamma_5 - \kappa_{nsd,i}\,\sigma_n \cdot \alpha\,\delta(\mathbf{r})), \qquad (1)$$

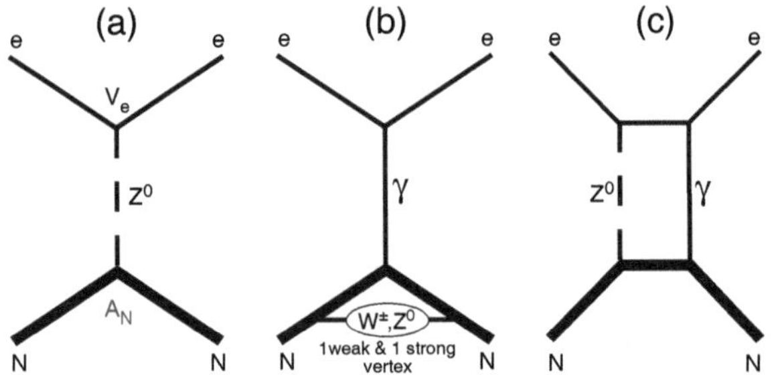

Fig. 1. Nuclear spin dependent PNC processes; (a) standard model tree level $V_e A_N$ Z exchange; (b) electron-nuclear anapole interaction, PNC stems from vertex corrections due to weak hadronic interactions; (c) combination of hyperfine interaction and Z exchange.

where $G = 10^{-5}$ m_p^{-2} is the Fermi constant, m_p is the proton mass, γ_5 and α are Dirac matrices, σ_n are Pauli matrices, and κ_{1i} and $\kappa_{nsd,i}$ are constants of the interaction with $i = p, n$ for a proton or a neutron and *nsd* = nuclear spin dependent. The standard model tree level values for these constants with $\kappa_{nsd,i} = \kappa_{2i}$ are

$$\kappa_{1p} = \frac{1}{2}(1 - 4\sin^2\theta_W), \kappa_{1n} = -\frac{1}{2},$$
$$\kappa_{2p} = -\kappa_{2n} = \kappa_2 = -\frac{1}{2}(1 - 4\sin^2\theta_W)\eta, \qquad (2)$$

with $\sin^2\theta_W \sim 0.23$ the Weinberg angle and $\eta = 1.25$. κ_{1i} (κ_{2i}) represents the coupling between nucleon and electron currents when the electron (nucleon) is the axial vector.

In an atom, the contribution from Eq. 1 for all the nucleons must be added. For the nuclear spin independent part (*nsi*), we obtain

$$H^{nsi}_{\mathrm{PNC}} = \frac{G}{\sqrt{2}}\frac{Q_W}{2}\gamma_5\,\delta(\mathbf{r}). \qquad (3)$$

This contribution is independent of the nuclear spin and is proportional to the weak charge

$$Q_W = 2(\kappa_{1p}Z + \kappa_{1n}N), \qquad (4)$$

with N the number of neutrons. Because of the strong cancellation in κ_{1p} the standard model value for the weak charge is almost equal to $-N$. The theoretical uncertainty present in all the extractions of weak interaction parameters from atomic PNC comes from the the calculation of the matrix element γ_5 as the experiment is not sensitive to the weak charge itself but to the product as Eq. 3 states.

The second term of Eq. 1 is nuclear spin dependent (*nsd*), and due to the pairing of nucleons, its contribution has a weaker dependence on Z. In a shell model description with a single valence nucleon of unpaired spin, Flambaum and Murray obtained [18]

$$H^{nsd}_{\mathrm{PNC}} = \frac{G}{\sqrt{2}}\frac{K\boldsymbol{I}\cdot\boldsymbol{\alpha}}{I(I+1)}\kappa_{nsd,i}\,\delta(\mathbf{r}), \qquad (5)$$

where $K = (I + 1/2)(-1)^{I+1/2-l}$, l is the valence nucleon orbital angular momentum, and I is the nuclear spin. The terms proportional to the anomalous magnetic moment of the nucleons and the electrons have been neglected.

The interaction constant is given by [18]

$$\kappa_{nsd,i} = \kappa_{a,i} - \frac{K - 1/2}{K}\kappa_{2,i} + \frac{I+1}{K}\kappa_{Q_W}, \qquad (6)$$

with $\kappa_{2,i}$ given by Eq. 2 corresponding to the tree level approximation, and two corrections, the effective constant of the anapole moment $\kappa_{a,i}$, and κ_{Q_W} generated by the nuclear spin independent part of the electron-nucleon interaction together with the hyperfine interaction (see Fig. 1). Flambaum and Murray show that [18]

$$\kappa_{a,i} = \frac{9}{10}g_i\frac{\alpha\mu_i}{m_p\tilde{r_0}}\mathcal{A}^{2/3},$$
$$\kappa_{Q_W} = -\frac{1}{3}Q_W\frac{\alpha\mu_N}{m_p\tilde{r_0}\mathcal{A}}\mathcal{A}^{2/3}, \qquad (7)$$

where α is the fine structure constant, μ_i and μ_N are the magnetic moment of the external nucleon and of the nucleus, respectively, in nuclear magnetons, $\tilde{r_0} = 1.2$ fm is the nucleon radius, $\mathcal{A} = Z + N$, and g_i gives the strength of the weak nucleon-nucleus potential with $g_p \sim 4$ for a proton and $0.2 < g_n < 1$ for a neutron [17]. The interaction is stronger in heavier atoms since both $\kappa_{a,i}$ and κ_{Q_W} scale as $\mathcal{A}^{2/3}$ ($Q_W/\mathcal{A} \sim 1/2$ in κ_{Q_W}). The anapole moment is the dominant contribution to the interaction in heavy atoms, for example in ^{209}Fr, $\kappa_{a,p}/\kappa_{Q_W} \approx 15$. As a result, nuclear spin dependent atomic PNC in heavy atoms is best suited to determine nuclear anapole moments by correcting the measured value for the small, calculated contributions from the κ_2 and κ_{Q_w} terms.

The anapole moment of a nucleus is a parity nonconserving, time reversal conserving moment that arises from weak interactions between the nucleons (see the review by Haxton and Wieman [4]). It can be detected in a PNC electron-nucleus interaction and reveals itself in the spin dependent part of the PNC interaction. Wood *et al.* [9,10] measured the anapole moment of ^{133}Cs by extracting the dependence of atomic PNC on the hyperfine energy levels involved, and consequently nuclear spin. The measurement shows that atomic PNC is a unique probe for neutral weak interactions inside the nucleus, which otherwise remain hidden by much larger electromagnetic charged currents [19].

The anapole moment is defined classically by (see reference [7])

$$\mathbf{a} = -\pi\int d^3r\,r^2\mathbf{J}(\mathbf{r}), \qquad (8)$$

with $\mathbf{J}$ the electromagnetic current density. The anapole moment in francium arises mainly from the weak interaction between the valence nucleons and the core. It is possible to think of it as a weak radiative correction that is detectable only with an electromagnetic interaction. Flambaum, Khriplovich, and Sushkov [3], by including weak interactions between nucleons in their calculation of the nuclear current density, estimate the anapole moment from Eq. 8 for a single valence nucleon to be

$$\mathbf{a} = \frac{1}{e}\frac{G}{\sqrt{2}}\frac{K\boldsymbol{j}}{j(j+1)}\kappa_{a,i} = C^{an}_i\boldsymbol{j}, \qquad (9)$$

where j is the nucleon angular momentum. These values correspond to the nuclear values for the case of a single valence nucleon. The calculation is based on the shell model for the nucleus, under the assumption of homogeneous nuclear density and a core with zero angular momentum leaving the valence nucleon carrying all the angular momentum.

3 The anapole moment measurement

There is interest in measuring the anapole moment in francium. Gomez *et al.* have proposed inducing microwave RF E1 transitions between the hyperfine levels of the francium ground state [20]. Alternatively, it can be extracted from the optical measurement of $7s \to 8s$ transitions involving different hyperfine states. The transition that is forbidden by the selection rules becomes allowed due to the weak interactions that create the anapole moment. The measurement of the anapole moment gives information on the weak nucleon-nucleon interactions inside the nucleus. A measurement of the anapole moment in a chain of isotopes should provide information to separate the anapole moment due to the valence proton from that of the neutron. The constraints that could be obtained for the proton and the neutron weak anapole moments depend on the theoretical model, but in a 'naive' picture of valence-only anapoles, they are almost orthogonal in the weak meson-nucleon coupling space [4,20].

The anapole moment induces a small mixing of electronic states of opposite parity. The effect of the anapole moment Hamiltonian on the ground state hyperfine levels according to first order perturbation theory is

$$\overline{|sFm\rangle} = |sFm\rangle + \sum_{F'm'} \frac{\langle pF'm'|\mathbf{H_a}|sFm\rangle}{E_p - E_s}|pF'm'\rangle, \quad (10)$$

where E_p, E_s are the energies of the p and s states respectively and

$$\mathbf{H_a} = |e|\, \boldsymbol{\alpha}\cdot\mathbf{a}\,\delta(\mathbf{r}), \quad (11)$$

is the anapole moment Hamiltonian from Eq. 5 with $\mathbf{a}$ the anapole moment given by Eq. 9. The matrix element in Eq. 10 gives [17]

$$\langle pF'm'|\ \mathbf{H_a}|sFm\rangle = i\frac{\xi Z^2 R}{(\varrho_s\varrho_p)^{3/2}}\frac{2\gamma+1}{3}\frac{(I+1/2)\kappa_a Ry}{I(I+1)}$$
$$\times(F(F+1) - I(I+1) - 3/4)\delta_{F,F'}\delta_{m,m'}, \quad (12)$$

with $\xi = Gm_e^2\alpha^2/\sqrt{2}\pi = 3.651 \times 10^{-17}$, m_e the electron mass, ϱ_s and ϱ_p the effective principal quantum number for the s and p electronic states, $\gamma = \sqrt{(J+1/2)^2 - Z^2\alpha^2}$, J the electron total angular momentum, and Ry the Rydberg constant. The relativistic enhancement factor R is given by

$$R = 4(a_0/2Zr_0)^{2-2\gamma}/\Gamma^2(2\gamma+1), \quad (13)$$

with a_0 the Bohr radius and $r_0 = \tilde{r_0}\mathcal{A}^{1/3}$ the size of the nucleus.

The anapole moment mixes only states with the same F and m and the mixing grows as $Z^{8/3}R$. For the ^{209}Fr ground state, we obtain

$$\overline{|sFm\rangle} = |sFm\rangle\ -i5.9\times10^{-13}\kappa_a$$
$$\times(F(F+1) - 25.5)|pFm\rangle. \quad (14)$$

The mixing coefficient is imaginary due to time reversal symmetry. In practice, the mixing would be measured through the E1 transition amplitude A_{E1} it induces between the perturbed initial and final states. The effect in francium is 11 times larger than in cesium [21,22].

3.1 The microwave/RF experiment

We propose a measurement of the nuclear anapole moment by direct excitation of the microwave E1 transition between or within the ground hyperfine states of an alkali atom. The transition is parity-forbidden, but is allowed by the anapole-induced mixing of opposite parity states. The general approach has been suggested and followed for atomic PNC measurements in the past [23–29]. We modify the idea of Fortson [19] for atomic PNC with an ion placed at the anti-node of a standing optical wave by placing many trapped atoms at the anti-node of a standing microwave field. The stability requirements for the field and the interrogated sample are relaxed given the microwave wavelengh. The experiment requires $\approx 10^6$ trapped atoms localized around the anti-node of the electric standing wave in a Fabry-Perot resonator (microwave frequency $\nu_m \sim 45$ GHz and wavelength $\lambda_m \sim 6.6$ mm for francium) as shown in Fig. 2. In this location, corresponding to a node of the magnetic field, the allowed M1 transition between the same states as the much weaker PNC-induced E1 ($|A_{E1}/A_{M1}| \approx 10^{-9}$) is greatly suppressed (further reduction is required and discussed in detail in [20]). This placement could be accomplished by confining the atoms with a blue-detuned optical dipole trap which would minimize the perturbation of the atoms by the light field. The signal, PNC transitions, would be amplified by interfering it with a parity-allowed Raman transition in the presence of a static magnetic field. Raman transitions are a convenient optical way to prepare a coherent superposition of the two levels in the hyperfine manifold, by an off-resonance 2-photon stimulated absorption-emission laser pulse using an intermediate excited state.

The fields present define the system of coordinates for the experiment. The combination of electric and magnetic fields gives the observable $i(\mathbf{E_M} \times (\mathbf{E_1} \times \mathbf{E_2})) \cdot \mathbf{B}$, with $\mathbf{E_M}$ the microwave electric field, $\mathbf{E_1}$ and $\mathbf{E_2}$ the Raman fields, $\mathbf{B}$ the static magnetic field and the i is present in accordance with time reversal symmetry.

Francium atoms can be initially trapped in a high-efficiency magneto-optical trap (MOT) [30], and then transferred to a second MOT in a separate chamber by means of a push beam. An optical dipole trap, located at the electric field anti-node of the standing wave microwave cavity, loaded from the second MOT would hold the atoms in place for the measurement. Given all the constraints for

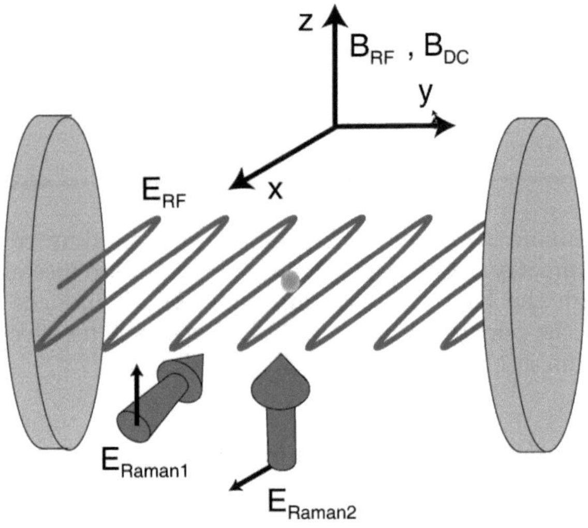

Fig. 2. Field configuration in the Fabry-Perot cavity.

preparation and transfer of the francium atoms, a Fabry-Perot configuration for the microwave cavity seems most apropriate.

Once the atoms are in the dipole trap, the atoms have to be optically pumped into a single Zeeman sublevel. The measurement sequence would then start, in the presence of a right handed system of coordinates, with the preparation of a coherent superposition of the hyperfine ground states with a Raman pulse of duration t_R. The E1 transition will be driven with the cavity microwave field for a fixed time t_{E1}, and then the sequence ends with the determination of the population in the upper hyperfine level. The sequence would repeat under a left handed system of coordinates and the PNC signal is the difference between the two measurements. The signal-to-noise ratio is proportional to the matrix element and to the duration of the Raman pulse. The signal also grows with the square root of the number of atoms (see Ref. [20]).

While measurements in francium benefit from a large expected effect (11 times larger than in Cs [21]); large atomic samples of other alkalis are easily prepared. We could obtain the same signal-to-noise ratio in a cesium sample with 100 times more atoms and the same-strength driving field. While the fundamental signal-to-noise ratio indicates the inherent trade-offs between different alkali species, technical noise specific to the instruments dedicated to the measurement must also be considered. For a discussion of technical noise in the cesium PNC Boulder experiment see Ref. [10].

3.2 The optical experiment

An optical atomic PNC experiment can address both the spin independent part (Eq. 3) and the spin dependent (Eq. 5) parts of the weak interaction. Although different approaches to optical measurements have been used (see Ref. [1]), they all interfere the PNC amplitude with an allowed transition to enhance the signal. A possible avenue for francium has been suggested in the contribution by Orozco to Ref. [31]. It follows closely the technique used by the Boulder group in cesium (see article by C. Wieman in reference [32]), but particular consideration has to be given to the specific values of the atomic structure of francium, such as its ionization potential.

The accumulation and preparation of the atomic sample would be the same as for the microwave/RF technique, with the atoms confined in an optical dipole trap but instead of a microwave Fabry-Perot, they would be in an optical Fabry-Perot tuned to the $7s \rightarrow 8s$ transition of Fr. A static electric field $\mathbf{E}$, a static magnetic field $\mathbf{B}$, and the Poynting vector $\mathbf{S}$ of the excitation field in the Fabry Perot define a handed system of coordinates such that the observable of the experiment is proportional to $\mathbf{B}\cdot(\mathbf{S}\times\mathbf{E})$. The static electric field also induces a Stark shift that allows a parity conserving amplitude between the $7s$ and $8s$ levels of Fr. It is then possible to interfere this electromagnetic term with the weak-interaction amplitude giving rise to a left-right asymmetry with respect to the handedness of the coordinate system

The very high intensities available in a standing wave will exert a repelling force that will tend to move the cold atoms to a region of low intensity. FM modulation at integers of the free spectral range of the cavity can create a slowly moving travelling envelope to solve this problem as already suggested by the Boulder group. A successful excitation of the PNC transition can be monitored by the characteristic decay fluorescence via the $7p$ state. Another possible complication is the ionization of the Fr atoms that absorb a second photon from the cavity while excited in the $8s$ state. A careful balance between intracavity power and signal loss will be needed to find the optimal operating point. Redundancy in the reversal of the coordinates would suppress systematic errors. More thorough experimental and theoretical studies will be required to confirm that the trap environment does not adversely affect the PNC measurement.

The signal-to-noise ratio, assuming a shot-noise-limited experiment, depends directly on the size of the effect and has a square root dependence on the intensity of the excitation light and the number of atoms [31]. Increasing the intensity does not gain as much as having a larger effect. The optimization of the experiment will require a careful understanding of all the sources of technical noise that modify the shot noise assumption.

4 Outlook and conclusions

The prospects for carrying out a series of PNC measurements in a chain of Fr isotopes at TRIUMF are very good. So far, ISAC has demonstrated high yields for many light, short-lived isotopes. With the development of the actinide target, ISAC will have Fr yields two or three orders of magnitude larger than those obtained at the superconducting linear accelerator at Stony Brook [33], where much of the preliminary work for this program has been carried out.

The systematic change in the number of neutrons would allow for careful studies of the nuclear structure dependence of the PNC signal. Data from numerous isotopes on both sides of the $N = 126$ shell closure together with nuclear structure calculations should allow to test the validity of and improve upon the valence-nucleon picture of the anapole moment.

Significant work remains to study first theoretically and then experimentally all possible systematic effects. The experiments are difficult, but the combination of precision measurement techniques from atomic and optical physics should allow us to study the weak interaction in the francium atom.

We would like to thank D. DeMille for many useful comments on this measurement strategy. This work has been supported by NSF (USA), NSERC, and NRC (Canada).

References

1. M.-A. Bouchiat and C. Bouchiat, Rep. Prog. Phys. **60**, 1351 (1997).
2. Y. B. Zel'dovich, Sov. Phys.-JETP **9**, 682 (1959).
3. V. V. Flambaum, I. B. Khriplovich, and O. P. Sushkov, Phys. Lett. B **146**, 367 (1984).
4. W. C. Haxton and C. E. Wieman, Annu. Rev. Nucl. Part. Sci. **51**, 261 (2001).
5. W. R. Johnson, M. S. Safronova, and U. I. Safronova, Phys. Rev. A **67**, 062106 (2003).
6. J. S. M. Ginges and V. V. Flambaum, Phys. Rep. **397**, 63 (2004).
7. Y. B. Zel'dovich, Sov. Phys.-JETP **6**, 1184 (1958).
8. P. A. Vetter, D. M. Meekhof, P. K. Majumder, S. K. Lamoreaux, and E. N. Fortson, Phys. Rev. Lett. **74**, 2658 (1995).
9. C. S. Wood, S. C. Bennett, D. Cho, B. P. Masterson, J. L. Roberts, C. E. Tanner, and C. E. Wieman, Science **275**, 1759 (1997).
10. C. S. Wood, S. C. Bennett, J. L. Roberts, D. Cho, and C. E. Wieman, Can. J. Phys. **77**, 7 (1999).
11. M. A. Bouchiat and C. Bouchiat, J. Phys. (Paris) **35**, 899 (1974).
12. M. A. Bouchiat and C. C. Bouchiat, Phys. Lett. B **48**, 111 (1974).
13. E. Gomez, L. A. Orozco, and G. D. Sprouse, Rep. Prog. Phys. **66**, 79 (2006).
14. M. V. Romalis and E. N. Fortson, Phys. Rev. A **59**, 4547 (1999).
15. C. Chin, V. Leiber, V. Vuletic, A. J. Kerman, and S. Chu, Phys. Rev. A **63**, 033401 (2001).
16. S. Sanguinetti, J. Guéna, M. Lintz, P. Jacquier, A. Wasan, and M.-A. Bouchiat, Euro. Pys. J. D **25**, 3 (2003).
17. I. B. Khriplovich, *Parity Non-Conservation in Atomic Phenomena* (Gordon and Breach, New York, 1991).
18. V. V. Flambaum and D. W. Murray, Phys. Rev. C **56**, 1641 (1997).
19. N. Fortson, Phys. Rev. Lett. **70**, 2383 (1993).
20. E. Gomez, S. Aubin, L. A. Orozco, G. D. Sprouse, and D. DeMille, arXiv:physics/0412124 (2004).
21. V. A. Dzuba, V. V. Flambaum, and O. P. Sushkov, Phys. Rev. A **51**, 3454 (1995).
22. M. S. Safronova and W. R. Johnson, Phys. Rev. A **62**, 022112 (2000).
23. C. E. Loving and P. G. H. Sandars, J. Phys. B **10**, 2755 (1977).
24. V. G. Gorshkov, V. F. Ezhov, M. G. Kozlov, and A. I. Mikhailov, Sov. J. Nucl. Phys. **48**, 867 (1988).
25. D. Budker, in *Physics Beyond the Standard Model*, edited by P. Herczeg, C. M. Hoffman, and H. V. Klapdor-Kleingrothaus (World Scientific, Singapore, 1998).
26. V. E. Balakin and S. I. Kozhemyachenko, JETP Lett. **31**, 297 (1980).
27. V. N. Novikov and I. B. Khriplovich, JETP Lett. **22**, 74 (1975).
28. E. A. Hinds and V. W. Hughes, Phys. Lett. B **67**, 487 (1977).
29. E. G. Adelberger, T. A. Traino, E. N. Fortson, T. E. Chupp, D. Holmgren, M. Z. Iqbal, and H. E. Swanson, Nuc. Instr. and Meth. **179**, 181 (1981).
30. S. Aubin, E. Gomez, L. A. Orozco, and G. D. Sprouse, Rev. Sci. Instrum. **74**, 4342 (2003).
31. L. A. Orozco, in *Trapped Particles and Fundamental Physics, Les Houches 2000*, edited by S. N. Atutov, R. Calabrese, and L. Moi (Kluwer Academic Publishers, Amsterdam, 2002).
32. *Precision Tests of the Standard Electroweak Model*, edited by P. Langacker (World Scientific, Singapore, 1995).
33. J. S. Grossman, L. A. Orozco, M. R. Pearson, J. E. Simsarian, G. D. Sprouse, and W. Z. Zhao, Phys. Rev. Lett. **83**, 935 (1999).

Towards measuring nuclear-spin-dependent and isotopic-chain atomic parity violation in ytterbium

K. Tsigutkin[1], J.E. Stalnaker[1,a], D. Budker[1,2], S.J. Freedman[1,2], and V.V. Yashchuk[3]

[1] Department of Physics, University of California, Berkeley, CA 94720-7300, USA
[2] Nuclear Science Division, Lawrence Berkeley National Laboratory, Berkeley CA 94720, USA
[3] Advanced Light Source Division, Lawrence Berkeley National Laboratory, Berkeley CA 94720, USA

Abstract. We discuss experiments aimed at measurements of atomic parity nonconservation (PNC) effects in the $6s^2\ ^1S_0 \rightarrow 5d6s\ ^3D_1$ transition (408 nm) in atomic Ytterbium (Z=70). According to theoretical predictions, the PNC-induced E1 amplitude of this transition is ~ 100 times larger than the analogous amplitude in Cs. Such an experiment will determine differences in PNC effects between different hyperfine components for odd-neutron-number Yb isotopes and, thereby, will allow measurements of the nuclear anapole moments in nuclei with unpaired neutrons. In addition, measurements of PNC in different isotopes would give information on neutron distributions within the nuclei. The apparatus designed and built for this experiment is described, and results of measurements towards understanding of systematic effects influencing the accuracy, and the current status of the ongoing PNC measurements are presented.

PACS. 32.80.Ys Weak-interaction effects in atoms – 32.70.-n Intensities and shapes of atomic spectral lines – 32.60.+i Zeeman and Stark effects

1 Introduction

Atomic ytterbium (Yb) was proposed as a system for measuring the effects of parity nonconservation (PNC) by D. DeMille [1] who pointed out a rather non-obvious enhancement of the PNC transition amplitude arising due to configuration mixing in addition to relative proximity of the excited $5d6s\ ^3D_1$ and $6s6p\ ^1P_1$ states of opposite nominal parity (Fig. 1). The PNC-induced $6s^2\ ^1S_0 \rightarrow 5d6s\ ^3D_1$ transition amplitude has been calculated [1–3] to be two orders of magnitude larger than the PNC amplitude in cesium where high-precision measurements have been carried out [4,5].

While the accuracy of the theoretical calculations of PNC effects in Yb is at a $\sim$ 20% level [2,3], there is little hope that it will ever reach sub-one-percent levels achieved for Cs and other single-electron systems (see, for example, review [6]). Therefore, the motivation for PNC experiments in Yb is not testing the Standard Model via a comparison of high-precision theory and experiment, but rather probing low-energy nuclear physics by comparing PNC effects on either a chain of naturally occurring Yb isotopes (^{168}Yb-0.13%, ^{170}Yb-3.04%, ^{171}Yb-14.28%, ^{173}Yb-16.13%, ^{174}Yb-31.83%, ^{176}Yb-12.76%), or in different hyperfine components of the same odd-neutron-number isotope (^{171}Yb-nuclear spin $I = 1/2$ or ^{173}Yb-$I = 5/2$). The former would yield information on the mean-square neutron radii in the chain of isotopes, while the latter would probe nuclear-spin-dependent PNC, which is sensitive to the nuclear *anapole moments* (see, for example, reviews [6,7]) that arise from weak interactions be-

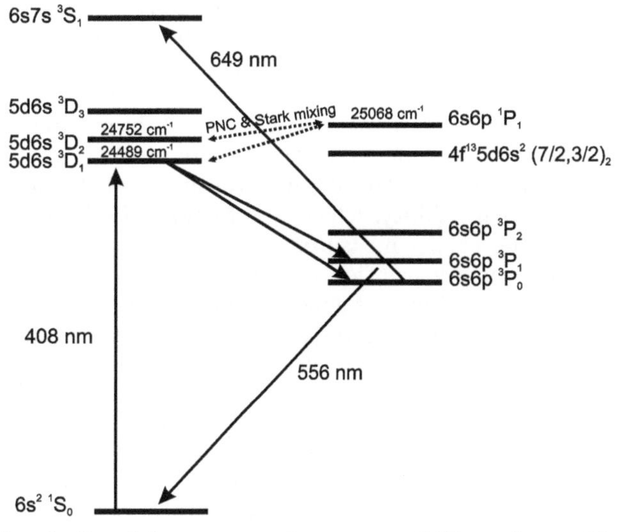

Fig. 1. Low-lying energy eigenstates of Yb and transitions relevant to the PNC experiment.

[a] *Present address:* National Institute of Standards and Technology, 325 S. Broadway Boulder, CO 80305-3322

tween the nucleons. Both of these types of measurements probe nuclear physics that is exceedingly difficult to access by other means, and neither relies on high-precision theory.

2 Preliminary experiments

Experimental work towards an Yb-PNC measurement has started at Berkeley several years ago, and, in its initial phase, was directed towards measurement of various spectroscopic properties of the Yb system of direct relevance to the PNC measurement, including determination of radiative lifetimes [8], measurement of the Stark-induced amplitudes, hyperfine structure, isotope shifts, and dc-Stark shifts of the $6s^2\ ^1S_0 \rightarrow 5d6s\ ^3D_{1,2}$ transitions, and the quadrupole moment of the latter transition ($\lambda = 404$ nm) [9], and the forbidden magnetic-dipole (M1) amplitude of the former transition ($\lambda = 408$ nm) [10]. For the measurement of the forbidden magnetic-dipole transition amplitude ($\approx 10^{-4}\ \mu_B$), we used the M1-(Stark-induced)E1 interference technique. A simple atomic system where transition amplitudes and interferences are well understood has proven useful for gaining insight into curious Jones dichroism effects that had been studied in condensed-matter systems at extreme conditions and whose origin had been a matter of debate (see Ref. [11] and references therein).

Most recently [12], we reported on an experimental and theoretical study of the dynamic (ac) Stark effect on the $6s^2\ ^1S_0 \rightarrow 5d6s\ ^3D_1$ forbidden transition. A general framework for parameterizing and describing off-resonant ac-Stark shifts was presented. A model was developed to calculate spectral line shapes resulting from resonant excitation of atoms in an intense standing light-wave in the presence of off-resonant ac-Stark shifts. A bi-product of this work was an independent determination (from the saturation behavior of the 408-nm transition) of the Stark transition polarizability, which was found to be in agreement with our earlier measurement [9].

The present incarnation of the Yb PNC experiment (described below) involves a measurement using an atomic beam. An alternative approach would involve working with a heat-pipe-like vapor cell. Various aspect of such an experiment were investigated, including measurements of collisional perturbations of relevant Yb states [13], nonlinear optical processes in a dense Yb vapor with pulsed UV-laser excitation [14], and an altogether different scheme for measuring PNC via optical rotation on a transition between excited states [15].

3 Experimental technique for the PNC measurement

The general idea of the present Yb-PNC experiment is to excite the forbidden 408-nm transition (Fig. 1) with resonant laser light in the presence of a static electric field. The purpose of the electric field is to provide a reference

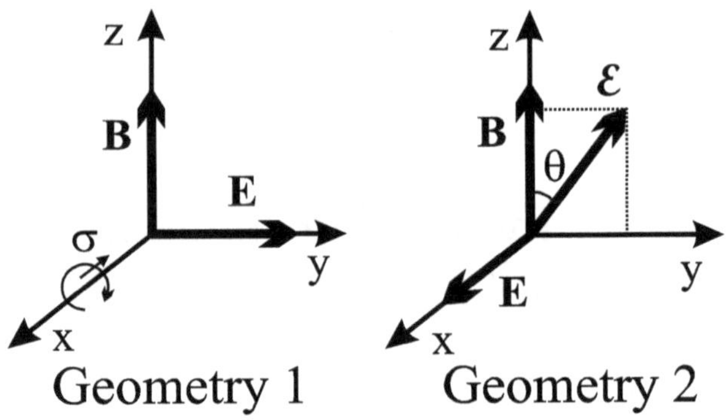

Fig. 2. Possible orientation of fields for PNC-Stark interference experiment. In both cases, light is applied collinearly with x.

Stark-induced transition amplitude, much larger than the PNC amplitude. In such a Stark-PNC interference method [16,17], one is measuring the part of the transition probability that is linear in both the reference Stark-induced amplitude and the PNC amplitude. In addition to enhancing the PNC-dependent signal, employing the Stark-PNC interference technique provides for all-important reversals of involved vector quantities, allowing one to separate the PNC effects from various systematics due to their associated sign changes.

We begin the discussion of the experiment by considering the rotational invariant to which the PNC-Stark-interference term is proportional. Since the effect violates parity and conserves time-reversal invariance, we require the rotational invariant to be P-odd and T-even. The two configurations which have been used for Stark-interference experiments of this type are shown in Fig. 2.

Geometry One was used in the Cs Stark-interference experiment [18] and Geometry Two was used in an early Stark-interference experiment in Tl [19]. The rotational invariant for Geometry One is given by

$$\sigma \cdot (\mathbf{E} \times \mathbf{B}), \tag{1}$$

where σ is the helicity of the excitation light. For Geometry Two, the invariant is

$$(\mathcal{E} \cdot \mathbf{B})([\mathbf{E} \times \mathcal{E}] \cdot \mathbf{B}), \tag{2}$$

where $\mathcal{E}$ is the electric field of the (linearly polarized) light.

The two geometries have very different characteristics and are sensitive to different sets of systematics. Neither geometry is clearly superior. One advantage of Geometry Two is an additional suppression of spurious M1-Stark interference, one of the most important systematic effects in precision PNC-Stark interference experiments. In both Geometries One and Two, this effect can be minimized by the use of a power-build-up cavity to generate a standing light wave. Since a standing wave has no net direction of propagation, $\mathbf{k}$, any transition rate which is linear in the M1 amplitude (proportional to $\mathbf{k} \times \mathcal{E}$) will cancel out. However, this is the only source of suppression of the M1-Stark interference in Geometry One. This can be seen by replacing σ in the rotational invariant (1) with $\mathbf{k}$.

This is a T-even, P-even rotational invariant which arises from M1-Stark interference. Geometry Two, on the other hand, is such that the M1 and Stark amplitudes are out of phase. Thus, the M1 and Stark amplitudes do not interfere and therefore do not produce additional interference terms which can change with the changing fields and mimic the PNC-Stark interference. This geometry can even be used in the absence of a power-build-up cavity (and in fact was in the work of Ref. [19]). Thus, the M1-Stark interference term requires a misalignment of the fields along with a net direction of propagation resulting from an imperfect standing wave. This additional reduction of M1-Stark interference has led us to choose Geometry Two for the PNC-Stark experiment.

The rotational invariant (2) highlights some of the salient features of our experiment: the PNC effect reverses with dc electric field (**E**), and does not reverse with the magnetic field (**B**); however, the presence of the magnetic field is necessary. In fact, the magnetic field should be strong enough so that the transitions to different Zeeman components of the upper state are resolved (because the PNC effect averaged over all Zeeman components is zero). Another reversal is flipping the axis of the light polarization with respect to the E-B plane, which also changes the sign of the PNC effect.

4 PNC signatures for even and odd isotopes

Here we address briefly the effect of the weak mixing between $^3D_1$ and $^1P_1$ states on the $6s^2\ ^1S_0 \rightarrow 5d6s\ ^3D_1$ transition. In the present experiment, the forbidden transition rate is enhanced by applying a DC electric field to Yb atoms which mixes the $^3D_1$ and $^1P_1$ states. In order to discriminate the weak mixing contribution to the transition rate, a magnetic field is applied leading to Zeeman splitting of the transition into several components. In the case of the Geometry 2 (Fig. 2), for the even ($I = 0$) Yb isotopes, the transition is split onto three components. For the odd isotopes, due to the hyperfine splitting of the $^3D_1$ state, the magnetic field produces different Zeeman patterns for upper levels with different values of the total angular momentum F', but in each case, there is an even number of Zeeman components.

A Stark-induced transition amplitude is generally expressed in terms of real scalar (α), vector (β), and tensor (γ) transition polarizabilities [16,9], however, for the case of a $J = 0 \rightarrow J' = 1$ transition, only the vector transition polarizability contributes. The transition amplitude between states with total angular momenta and projections F, M and F', M' takes the form[1]

$$A^{Stark}_{FMF'M'} = i\beta_{FF'}(-1)^q \left(\mathbf{E} \times \mathcal{E}\right)_q \langle F, M, 1, -q|F', M'\rangle\ , \quad (3)$$

where $q = M - M'$ labels the spherical component and $\langle F,\ M,\ 1,\ -q|F',\ M'\rangle$ is a Clebsch-Gordan coefficient. The

[1] The $(-1)^q$ factor was inadvertently omitted in some of our earlier publications. This has not affected any of the quoted results.

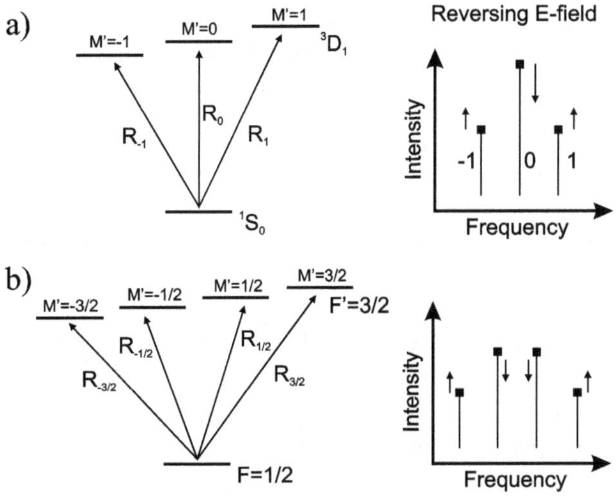

Fig. 3. (a) Level diagram and Zeeman-component spectrum for $^1S_0 \rightarrow ^3 D_1$ transition for even Yb isotopes. (b) Same for $^1S_0\ F = 1/2 \rightarrow ^3 D_1\ F = 3/2$ for ^{171}Yb. The small arrows indicate the effect of the Stark-PNC interference term – a small relative enhancement of some components and a suppression of others – that reverses with **E** and θ.

transition amplitude arising due to PNC can be expressed as

$$A^{PNC}_{FMF'M'} = i\xi_{FF'}(-1)^q \mathcal{E}_q \langle F, M, 1, -q|F', M'\rangle\ . \quad (4)$$

Here $\xi_{FF'}$ characterizes the PNC-induced electric-dipole transition amplitude.

Assuming that the magnetic field is perpendicular to the electric field and is strong enough to resolve the Zeeman components of the transition, that the light propagates parallel to the electric field, and selecting the quantization axis along the magnetic field, we obtain the following results for the transition rates. For isotopes with zero nuclear spin, the transition rates to Zeeman sublevels $M' = 0, \pm 1$ are:

$$\mathcal{R}_0 = \frac{8\pi}{c}\mathcal{I}\left[\beta^2 E^2 \sin^2\theta + 2\xi\,\beta E \sin\theta\cos\theta\right]\ , \quad (5)$$

$$\mathcal{R}_{\pm 1} = \frac{4\pi}{c}\mathcal{I}\left[\beta^2 E^2 \cos^2\theta - 2\xi\,\beta E \sin\theta\cos\theta\right]\ , \quad (6)$$

where $\mathcal{I}$ is the light intensity. In these expressions, we neglect the term in the transition rate which is quadratic in PNC mixing. For an arbitrary angle of the light polarization θ, there are generally three components of the transition as shown in Fig. 3a. Neglecting Stark shifts, the distance between adjacent peaks corresponds to an energy difference

$$\Delta E \approx g\mu_0 M' B\ , \quad (7)$$

where g is the Landé factor, μ_0 is the Bohr magneton, and B is the magnetic field. The effect of the Stark-PNC interference is to change the relative strength of the two extreme components of the transition with respect to the central component. Expressions (5,6) show explicitly that the PNC effect reverses with both **E** and θ.

As an example of a transition for an odd isotope, consider $F = 1/2 \rightarrow F' = 3/2$ of ^{171}Yb (Fig. 3b). Since the magnetic moment of the ground $^1S_0$ $F = 1/2$ state is solely due to the nucleus, the Zeeman splitting of this state is three orders of magnitude smaller than that of the $^3D_1$ $F = 3/2$ state and is unresolved at the magnetic-field strength of interest. Therefore, the total rates for the transitions to the upper-state Zeeman components M' are sums of the respective rates. The vector transition polarizability $\beta_{FF'}$ is related to that for an isotope with zero nuclear spin (β) according to

$$\beta_{FF'} = \beta(-1)^{I+F}\sqrt{(2F'+1)(2F+1)}\begin{Bmatrix} 1 & F' & I \\ F & 0 & 1 \end{Bmatrix}, \quad (8)$$

where the brackets denote a $6j$-symbol [20]. The distance between peaks is determined by

$$\Delta E \approx g_{F'}\mu_0 M'B, \quad (9)$$

where $g_{F'}$ is given by

$$g_{F'} = g \cdot \frac{F'(F'+1) + J'(J'+1) - I(I+1)}{2F'(F'+1)}. \quad (10)$$

In calculating the PNC amplitude, one needs to take into account both the nuclear-spin-independent effect (as in the case of the even isotopes), as well as the nuclear spin-dependent contribution, including the effect of the nuclear anapole moment. The PNC signature for this transition is an asymmetry between the two pairs of Zeeman components of the transition (Fig. 3b) which, once again, reverses with **E** and θ.

In addition, we note that the $F = 1/2 \to F' = 1/2$ transition of ^{171}Yb is not influenced by the PNC effect under the discussed conditions. This transition is split by the magnetic field into two Zeeman components, whose intensities must be invariant under the B-field reversal. Thus, the PNC asymmetry cancels. The total rates for these two Zeeman components turn out to be equal and independent of the polarization angle θ. Observation of this transition gives us an additional handle to study systematics.

Using the theoretical value of $\xi \simeq 10^{-9}\ ea_0$ [2,3] combined with the measured $\beta \simeq 2\times 10^{-8}\ ea_0/(\mathrm{V/cm})$ [9,12], we can estimate the relative asymmetry in the line shapes. For example, for $\mathcal{R}_0$ and $\theta = \pi/4$, and for $E = 1$ kV/cm, the asymmetry is $\simeq 10^{-4}$, with the nuclear spin-dependent and isotope-dependent effects expected at a few percent of that (depending on a specific model).

5 Apparatus and signals

A schematic of the Yb-PNC apparatus is shown in Fig. 4. A beam of Yb atoms is produced (inside of a vacuum chamber with a residual pressure of $\approx 5\times 10^{-6}$ Torr) with an effusive source (not shown), which is a stainless-steel oven loaded with Yb metal. The oven is outfitted with a multi-slit nozzle providing initial collimation of the atomic beam. The oven is operated with the front ≈ 100°C hotter than the rear to avoid clogging of the nozzle. The typical operating temperatures are 500°C in the rear and 600°C in the front. In order to reduce the Doppler width of the 408-nm transition and the overlap between its various hyperfine and isotopic components, a downstream external vane collimator was installed. The collimator provides a collimation angle of $\approx 1°$ in the horizontal direction. This reduces the Doppler width of the 408-nm transition to ≈ 12 MHz. The transparency of the collimator is estimated to be $\approx 90\%$. The vane collimator is mounted on a movable platform, allowing precise alignment of its angle relative to the atomic beam during the experiment, as well as its removal from the beam.

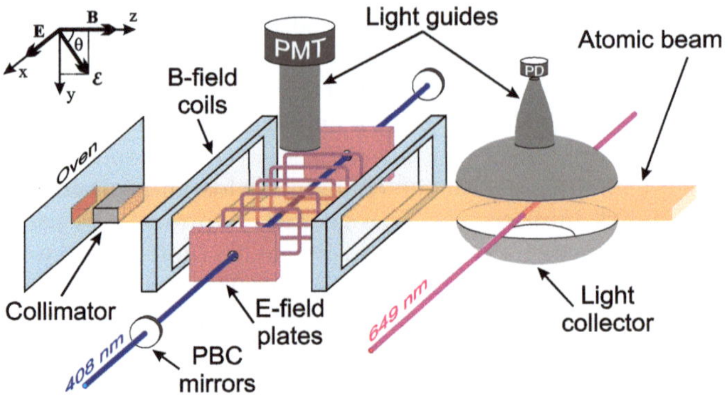

Fig. 4. Schematic of the present Yb-PNC apparatus. Not shown are the atomic-beam oven and the vacuum chamber containing all the depicted elements, except the photomultiplier (PMT), photodiode (PD), and the photodiode light-guide. PBC–power buildup cavity.

Downstream from the collimator, the atoms enter the main interaction region where the Stark- and PNC-induced transition takes place. The light at the transition wavelength of 408.345 nm is produced by doubling the output of a Ti:Sapphire laser (see schematic of the optical arrangement, Fig. 5). A Coherent 899 Ti:Saphire laser is pumped by ≈ 12 W from a Spectra Physics 2080 argon-ion laser operating on all lines. The Ti:Sapphire laser produces ≈ 1.2 W of light at 816 nm. This light is frequency doubled using a commercial bow-tie resonator with a Lithium-Triborate (LBO) crystal (Laser Analytical Systems Wavetrain cw). The output of the frequency doubler is ≈ 80 mW.

The 408-nm light is coupled into a power buildup cavity (PBC). We operate the cavity in the confocal regime (mirror radii of curvature and the mirror separation are 25 cm) in order to maintain a relatively large diameter of the light beam by simultaneously exciting many degenerate transverse modes. This is important for our ability to control the ac-Stark shifts without compromising the size of the signals. Precision optical mounts are used for PBC mirrors with micrometer adjustments for the horizontal and vertical angles and the pivot point of the mirror face. The mirrors are attached with RTV adhesive to thin aluminum rings. It was hoped that the thin wall aluminum frame and RTV adhesive would reduce birefringent-induced stress resulting from the adhesion as well as from thermal expansion and contraction. The alu-

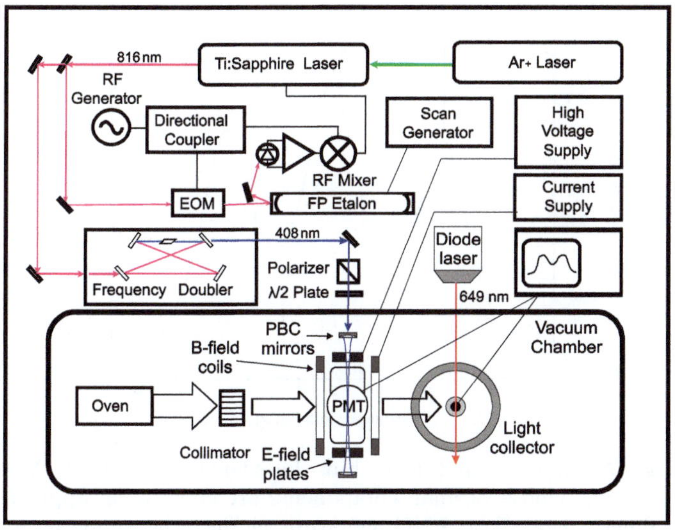

Fig. 5. Schematic of the optical setup and the frequency-locking systems. EOM–electro-optical modulator, FP-Fabry-Perot. The "diode laser" producing 649-nm light consists of a commercial master laser (New Focus Vortex) and a home-made injection-locked diode amplifier.

minum frame of one of the mirrors is attached to a piezo-electric ceramic which is attached to the optical mount. The three micrometer adjustments of one of the optical mounts are effected with precision, vacuum compatible pico-motors (New Focus 8302-v) to allow for adjustment of the cavity alignment while in vacuum. The finesse of the PBC was measured using the cavity-ring-down method to be $\mathcal{F} = 4240(70)$. The fraction of the incident light power that could be coupled through the cavity was typically 10-18%, limited to a large extent by the losses in the mirrors (measured at about 240 ppm per bounce). We note that the losses for the mirrors presently used in this experiment are ≈ 40 times larger than those for the mirrors used in some other PBCs (e.g., those in Refs. [18,21]). A reason for such relatively high losses is that our wavelength is significantly shorter. (The losses are expected to be significantly lower in the state-of-the-art mirrors presently on order).

The Coherent 899 Ti:Sapphire laser is rated to have a short-term instability (effective line width) of 500 kHz. In order to reduce the line width of the laser (which is necessary for efficient coupling into the PBC) an electro-optic modulator (EOM) was placed inside the laser cavity. Changing the voltage across the EOM changes the index of refraction of the crystal and therefore the optical-path length. The EOM is capable of extremely fast response with a ≈ 5 MHz bandwidth, limited by the speed of the electronics. The EOM is a double-crystal assembly (LINOS Optics PM 25 IR); the crystals are cut at Brewster's angle to minimize loss and are compensated to prevent walk off of the beam. It is possible to insert the EOM into the cavity and achieve lasing with minimal realignment of the laser. The output of the laser drops $\approx 5-10\%$ when the EOM is added. The laser is locked to the PBC using the fm sideband technique [22].

We found that using the PBC as the "master" cavity leads to frequency oscillations at acoustic frequencies. In order to remove these oscillations, the resonant frequency of the cavity is locked to a more stable confocal Fabry-Perot etalon, once again using the fm sideband locking technique. Thus, the stable (scannable) cavity provides the master frequency, with the power-build-up cavity serving as the secondary master for the laser.

The relative frequency of the laser light is determined using a homemade spherical-mirror Fabry-Perot interferometer operating at 408 nm (not shown in Fig. 5). Since the experiment relies on a detailed understanding of the spectral line shape of the transition over a region of ≈ 100 MHz, a frequency reference with closely spaced frequency markers is needed. To this end, the interferometer is operated with the mirror spacing chosen so that the transverse cavity modes overlap at frequency intervals of $\Delta\nu_{res} = c/(2\,N\,L)$, with $N = 7$ (c is the speed of light, and L is the mirror separation) [23]. This allows one to achieve relatively closely spaced frequency markers without making an excessively long cavity. The interferometer consists of two mirrors each of which has a radius of curvature of $R = 50$ cm and the mirrors are separated a distance 38.9 cm. This results in a spacing between the cavity resonances of 55.12 MHz. The cavity is made of invar and placed in an evacuated, passively thermally stabilized enclosure. The frequency drifts of the cavity resonances are ≈ 1 MHz/min.

For the PNC measurements, we plan to take most of the data with an electric-field magnitude in the interaction region of 1.5 kV/cm. The electric field is generated with two copper electrodes with dimensions 4.4x2.5x0.4 cm^3 separated by a 6.8-cm gap. In the inter-electrode gap, nine equidistant copper frames are placed for producing a homogeneous voltage drop along the optical axis of the cavity. These frames are connected between each other and the electrodes through 10-MΩ 0.1%-accuracy high-voltage resistors forming a voltage divider. A 10-kV voltage is supplied to the E-field plates by a SPELLMAN CZE1000R power supply modified to allow computer controlled switching of the voltage polarity. During preliminary diagnostic experiments where higher electric field was desirable (see, for example, the data in Figs. 6 and 7), the inter-electrode frames were removed and the gap was reduced to 1 cm. Fields of 12 kV/cm can be reliably obtained without breakdown.

The magnetic field is generated by a pair of rectangular coils having dimensions: 12(width)x3.2(height)x1(depth) cm^3 and separated by a 1.4-cm gap. The coils are designed to produce a uniform (1% non-uniformity) axial magnetic field up to 100 G within the region where the atomic beam intersects the waist of the power-buildup cavity (see Fig. 4). The coils are powered by a computer-controlled power supply allowing modulation and reversal of the magnetic field.

Light emitted from the interaction region at 556 nm (Fig. 1) resulting from the second step of the fluorescence cascade following excitation to the 5d6s $^3D_1$ state is collected with a light guide and detected with a photomulti-

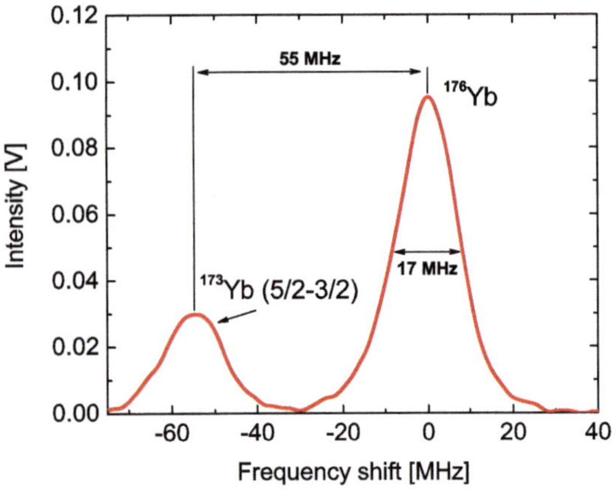

Fig. 6. An example of the 556-nm fluorescence signal detected in the interaction region recorded as the frequency of the 408-nm light was swept across resonances corresponding to two neighboring isotopic an hyperfine components of the Stark-induced transition. E=12 kV/cm, B=0.

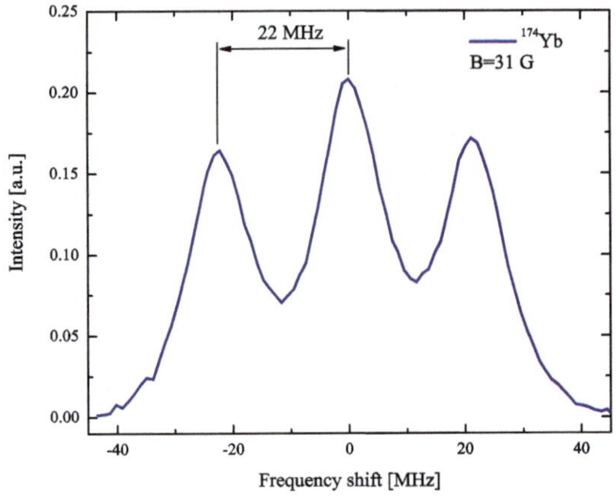

Fig. 7. Same as in Fig. 6, but the scan is over the ^{174}Yb resonance whose Zeeman components are split by an applied magnetic field. The PNC signature is the relative suppression/enhancement of the central peak with respect to the two side peaks that is odd in E and even in B.

plier tube. Because of the field requirements in the interaction region, it is difficult to achieve efficient light detection. In order to improve the detection efficiency for the PNC experiment a separate downstream detection region, dedicated to monitoring the number of atoms making a transition is used. We utilize the fact that 65% of the atoms excited to the 5d6s $^3D_1$ state decay to the metastable 6s6p $^3P_0$ state (Fig. 1). These atoms are probed by resonantly exciting them (with 649-nm light) out of the 6s6p $^3P_0$ state to the 6s7s $^3S_1$ state. The subsequent fluorescence is detected in a region free of other experimental components.

The 649-nm light is derived from a single-frequency diode laser (New Focus Vortex) producing $\approx$ 1.2 mW of cw output that is amplified in home-made injection-locking system to $\approx$ 7 mW of light sent into the detection region.

The detection region consists of two optically polished curved aluminum reflectors. One reflector, which has hemispherical shape with a radius of curvature of 8.9 cm, covers the upper hemisphere with respect to the intersection of the 649-nm laser beam with the atomic beam. The second reflector is parabolic and is located below the beam intersection. This reflector is positioned in such a way that the focal point of the parabola coincides with the center of curvature of the upper reflector and with the beam intersection. Fluorescent light from the atoms excited with the 649-nm light is collected onto a plastic light guide through a 5-cm diameter opening in the upper reflector. The light-guide is a "Winston cone" [24] designed to efficiently condense the light to the output diameter of the light guide of 1 cm matched to the active area of a silicon photodiode. The use of a photodiode with a higher quantum efficiency than a photomultiplier tube (afforded by larger signals in the detection region) further improves the detection efficiency. We estimate a detection efficiency of $\approx$ 40% for the atoms in the 6s6p $^3P_0$, corresponding to $\approx$ 26% of the atoms excited to the 5d6s $^3D_1$.

Examples of signals obtained with the present apparatus are shown in Figs. 6 and 7.

6 Statistical sensitivity

Based on various parameters achieved in the present apparatus, one can estimate the statistical sensitivity of the PNC experiment. One straightforward way to do it is to note that the Stark-interference technique is a tool for raising the signal above the background and controlling systematic effects. It does not, however, improve the statistical sensitivity for a shot-noise-limited experiment, but merely allows one to achieve the shot-noise limit. Because of this, one can ignore the Stark-interference nature of the experiment in the estimate of the statistical sensitivity of the PNC measurement, and consider instead the statistical sensitivity for direct excitation of the PNC amplitude without any static electric field.

With $\approx$ 5 W of circulating power in the power-build-up cavity, using a detection efficiency of 26% as discussed above, and a quantum efficiency of the photodiode of $\approx$ 90%, we arrive at the number of detected transitions per second of $\approx$ 0.7. Thus, for a shot-noise-limited experiment, the fractional precision is $\delta E1_{\rm PNC}/E1_{\rm PNC} \approx 1/\sqrt{\tau(s)}$, where τ is the total integration time of the experiment. A realistic integration time of 10^4 s, gives a fractional precision of $\approx$ 1%.

Achieving this statistical sensitivity seems quite realistic given the current apparatus. A preliminary analysis of systematic effects indicates that systematic errors can be controlled at this level as well. Pushing the precision of the experiment to a level beyond this point will require a significant amount of effort, most notably a dramatic improvement in the power-build-up cavity in order to achieve the necessary statistical sensitivity. While it is difficult to project the ultimate sensitivity of a PNC experiment in

Yb, the current status of the experiment gives us reason to hope that a high precision experiment is possible.

7 Analysis of systematics

The current Yb-PNC apparatus has been designed to minimize systematic effects, and the majority of the systematic effects we anticipate will result only in a combination of multiple apparatus imperfections. Several reversals are available (electric and magnetic fields, and light-polarization angle) to control systematics. An additional helpful feature is a specific characteristic light-detuning dependence of the PNC asymmetry.

Our general approach to dealing with systematics is similar to that employed, for example, in Refs. [25,26]. We begin by making a list of all reasonable apparatus imperfections (field misalignments, inhomogeneities, imperfect reversals, etc.), and write out all possible P,T-even rotational invariants that produce the same signature as the PNC effect upon reversals. We can then artificially impose exaggerated combinations of imperfections in order to measure (and, where possible, minimize) the remaining imperfections. A detailed analysis of the systematics will be given elsewhere. At present, we are confident that initially, systematics should be under control at a $\sim 1\%$ level, sufficient for reliable measurements of the isotopic and hyperfine PNC effects.

8 Conclusions and acknowledgements

We have reported on the progress in the experiment measuring parity-nonconservation effects in ytterbium, which takes advantage of an extraordinary anticipated enhancement of atomic PNC effects (by two orders of magnitude), compared to, for example, Cs. The Yb-PNC experiment relies on proven, although challenging, experimental techniques of atomic physics. The atomic structure of Yb is vastly different from the structure of atoms which have previously been studied in atomic PNC experiments, so the specific features of the experiment are quite different from the earlier work, although there are many features drawn from earlier PNC experiments [19,4,25]. The seven stable isotopes of Yb will allow for the first measurement of atomic PNC effects in a chain of isotopes. Assuming the weak-charge variation between the isotopes as predicted by the Standard Model, this will provide a unique way of measuring the variation of neutron r.m.s. radii in an isotopic chain. Conversely, if the neutron radii are known, one can measure the isotopic variation of the nuclear weak charge independent of uncertainties associated with atomic theory. Measurements of the PNC effect on different hyperfine components of the transition of atoms with non-zero nuclear spin nuclei (^{171}Yb and ^{173}Yb) will allow for the first measurements of the nuclear anapole moment in nuclei with unpaired neutrons.

The authors thank A. Dilip, B. P. Das, and M. G. Kozlov for useful discussions. This work has been supported by NSF (Grant 0457086), and by the Director, Office of Science, Office of Basic Energy Sciences, Nuclear Science Division, of the U.S. Department of Energy under contract DE-AC03-76SF00098.

References

1. D. DeMille, Physical Review Letters **74**(21), 4165 (1995)
2. S.G. Porsev, G. Rakhlina Yu, M.G. Kozlov, JETP Letters **61**(6), 459 (1995)
3. B.P. Das, Physical Review A **56**, 1635 (1997)
4. C.S. Wood, S.C. Bennett, J.L. Roberts, D. Cho, C.E. Wieman, Canadian Journal of Physics **77**(1), 7 (1999)
5. J. Guena, M. Lintz, M.A. Bouchiat, Physical Review A (General Physics) **71**, 042108 (2005)
6. J.S.M. Ginges, V. Flambaum, Physics Reports **397**(2), 63 (2004)
7. W.C. Haxton, C.E. Wieman, in *Annual Review of Nuclear and Particle Science* (2001), Vol. 51, pp. 261–293
8. C.J. Bowers, D. Budker, E.D. Commins, D. DeMille, S.J. Freedman, A.T. Nguyen, S.Q. Shang, M. Zolotorev, Physical Review A **53**(5), 3103 (1996)
9. C.J. Bowers, D. Budker, S.J. Freedman, G. Gwinner, J.E. Stalnaker, D. DeMille, Physical Review A **59**(5), 3513 (1999)
10. J.E. Stalnaker, D. Budker, D.P. DeMille, S.J. Freedman, V.V. Yashchuk, Physical Review A **66**(3), 31403 (2002)
11. D. Budker, J.E. Stalnaker, Physical Review Letters **91**(26), 263901/1 (2003)
12. J.E. Stalnaker, D. Budker, S.J. Freedman, J.S. Guzman, S.M. Rochester, V.V. Yashchuk, Physical Review A **73**, 043416 (2006)
13. D.F. Kimball, D. Clyde, D. Budker, D. DeMille, S.J. Freedman, S. Rochester, J.E. Stalnaker, M. Zolotorev, Physical Review A **60**(2), 1103 (1999)
14. B. DeBoo, D.F. Kimball, C.H. Li, D. Budker, Journal of the Optical Society of America B-Optical Physics **18**(5), 639 (2001)
15. D.F. Kimball, Physical Review A **63**, 052113 (2001)
16. M.A. Bouchiat, C. Bouchiat, Journal de Physique I **36**(6), 493 (1975)
17. R. Conti, P. Bucksbaum, S. Chu, E.D. Commins, L. Hunter, Physical Review Letters **42**(6), 343 (1979)
18. C.S. Wood, S.C. Bennett, D. Cho, B.P. Masterson, J.L. Roberts, C.E. Tanner, C.E. Wieman, Science **275**(5307), 1759 (1997)
19. P.S. Drell, E.D. Commins, Physical Review A **32**, 2196 (1985)
20. I.I. Sobelman, *Atomic Spectra and Radiative Transitions*, Springer Series on Atoms and Plasmas, 2nd edn. (Springer, New York, 1992)
21. C.J. Hood, H.J. Kimble, J. Ye, Physical Review A **64**(3), 033804/1 (2001)
22. R.W.P. Drever, J.L. Hall, F.V. Kowalski, J. Hough, G.M. Ford, A.J. Munley, H. Ward, Applied Physics B-Photophysics and Laser Chemistry **B31**(2), 97 (1983)
23. D. Budker, S.M. Rochester, V.V. Yashchuk, Review of Scientific Instruments **71**(8), 2984 (2000)
24. R. Winston, W.T. Welford, J.C. Miano, P. Bentez, *Non-imaging optics* (Elsevier Academic Press, Amsterdam ; Boston, Mass., 2005)
25. A.T. Nguyen, D. Budker, D. DeMille, M. Zolotorev, Physical Review A **56**(5), 3453 (1997)
26. B.C. Regan, E.D. Commins, C.J. Schmidt, D. DeMille, Physical Review Letters **88**(7), 071805 (2002)

Prospects for parity violation measurements in cold francium atoms

S. Sanguinetti[3], S. N. Atutov[1,a], R. Calabrese[1], L. Corradi[2], A. Dainelli[2], A. Khanbekyan[4], E. Mariotti[4], C. de Mauro[4], P. Minguzzi[3], L. Moi[4], G. Stancari[1], L. Tomassetti[1], and S. Veronesi[4]

[1] Dipartimento di Fisica, Università di Ferrara and INFN, I–44100 Ferrara, Italy
[2] INFN, Laboratori Nazionali di Legnaro, I–35020 Legnaro, Italy
[3] Dipartimento di Fisica, Università di Pisa, I–56127 Pisa, Italy
[4] Dipartimento di Fisica, Università di Siena, I–53100 Siena, Italy

Abstract. We present the Trap-rad experiment, which in the long term aims to an atomic parity violation (APV) measurement in francium isotopes. We describe the present status of the experiment, the francium production and the magneto-optical trap (MOT). Some details are given about our new CCD detection system. We explore then the possibility to perform APV measurements with francium. The low Trap-rad francium production rates should allow to perform preliminary tests. With higher rates attainable in other accelerator facilities, it should be possible to begin a francium APV experiment.

PACS. 32.80.Pj Optical cooling of atoms; trapping – 32.80.Ys Weak-interaction effects in atoms

1 Introduction

In the last years, laser cooling and trapping techniques opened the way to high precision spectroscopic measurements of radioactive species. The possibility to collect and accumulate atoms in a small volume allowed researchers to study many radioactive elements [1].

Among them francium, the heaviest alkali, is expected to be an excellent candidate for atomic parity violation (APV) measurements. As for cesium [2], the simple electronic structure should allow refined atomic calculations, necessary to extract the weak charge information and to test the Standard Model. The heavy nucleus should lead to a large APV effect, about 18 times the cesium one [3]. Also the possibility to perform measurements on different francium isotopes seems attractive.

One of the main difficulties would be the limited quantity of Fr atoms available for such a measurement: up to now, the maximum number obtained in a MOT was about 250000, at Stony Brook [4]. The signal to noise ratio directly depends on the quantity of francium. We should then adapt the APV experiments carried out for cesium to a situation with a very small number of cold atoms.

In the following we present the Trap-rad experiment for the production and trapping of francium and we make some considerations about the possible schemes for an APV measurement with francium.

2 Francium production and trapping

The Trap-rad experiment at the INFN *Laboratori Nazionali di Legnaro* aims at preparing a cold francium sample in order to initiate parity violation studies for this element. With our relatively low production rates, we do not expect to be able to perform a complete APV measurement. However, we should gather precious information to understand what can be the best scheme for such a measurement at a high francium production rate facility.

We produce francium via a nuclear fusion-evaporation reaction, between a 100 MeV ^{18}O beam coming from a tandem accelerator and a ^{197}Au target. Thanks to beam energy fine tuning, we can maximize the rate for a specific isotope, from 208 to 211 atomic mass. Francium is extracted in its ionic form from the target and conveyed to a pyrex cell, onto an yttrium neutralizer. With $1.5 \cdot 10^{12}$ ^{18}O/s, we typically obtain a $1 \cdot 10^6$ ^{210}Fr ions/s production rate. The francium atoms released by the neutralizer can be captured by the magneto-optical trap setup. More details about the production, the transport and the neutralization of francium can be found in Ref. [5].

The MOT is formed by six trapping laser beams at 718 nm, six repumping beams at 817 nm and a magnetic-field gradient (~7 G/cm) [6]. The trapping beams, red shifted with respect to the D_2 transition, cool the atoms

[a] *Permanent address:* Institute of Automation and Electrometry, Novosibirsk 90, Russia.
Correspondence to: Dr. Stefano Sanguinetti, Dipartimento di Fisica E. Fermi, Largo B. Pontecorvo 3, I–56127 Pisa, Italy; e-mail: sanguine@df.unipi.it.

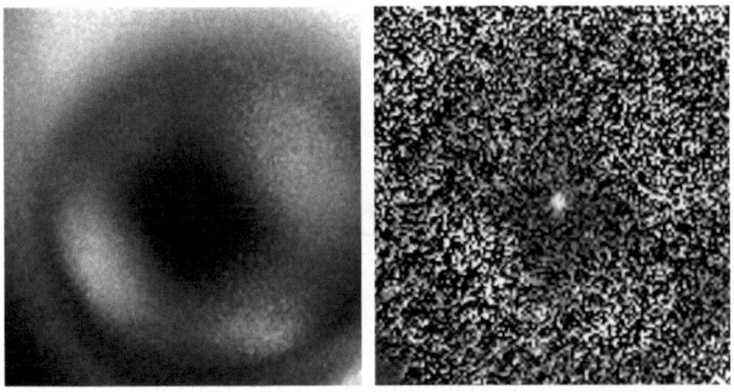

Fig. 1. Images of about 80 trapped Rb atoms: raw image (left) and background-subtracted image (right).

in the three cartesian directions; the repumping light prevents the accumulation of atoms in the wrong hyperfine ground state. Then the atoms accumulate in the zero magnetic field region.

3 MOT detection

Since we wanted to be able to detect a small number of trapped atoms, we recently set up a highly sensitive detection system by using a cooled CCD camera. The possibility to spatially discriminate the MOT signal from the background stray light was very attractive. Indeed the main contribution to the noise comes from the laser light scattered on the cell walls. Thanks to the large depth of field of our imaging system, we could also individuate a dark area corresponding to a small valve placed in front of the CCD camera: we aligned our system in such a way that the atomic cloud appears exactly in front of this area.

In order to further improve the signal to noise ratio, in the photograms analysis we perform a differential detection which reduces the noise coming from the stray light fluctuations. The detailed procedure is described in Ref. [7]. After calibration of our signals, we saw that we are able to detect as few as 50 Fr atoms (Fig. 1).

4 Parity violation measurements with francium

We here describe a possible scheme for an APV measurement with francium, after Ref. [8]. We do not claim to find the best solution: nevertheless the following considerations should be a good starting point for such a complex experiment.

4.1 Measurements on the highly forbidden 7S–8S transition

The mixing between opposite parity states due to the weak neutral currents can be detected with laser spectroscopy. Following the example of APV measurements in cesium, we plan to perform measurements on the 7S–8S highly

Table 1. Magnetic dipole, polarizabilities and PV amplitude calculated for the 7S–8S transition in Fr, after Refs. [3,9].

M_1	$-46\ (1 \pm 16\%) \cdot 10^{-5}\ \lvert\mu_B/c\rvert$
α	$-375.3(3.6) \cdot a_0^3$
β	$74.3(7) \cdot a_0^3$
$\Im m(E_1^{PV})$	$-1.48(3) \cdot 10^{-10}\ e\ a_0\ (^{223}\mathrm{Fr})$
	$-1.33(2) \cdot 10^{-10}\ e\ a_0\ (^{210}\mathrm{Fr})$

forbidden transition in francium at 506 nm. In the presence of a static electric field **E**, the effective dipole for this transition is given by

$$\mathbf{d}^{\mathrm{eff}}_{7S,F-8S,F'} = -\alpha\mathbf{E} - i\beta\boldsymbol{\sigma}\wedge\mathbf{E} + M_1'\boldsymbol{\sigma}\wedge\hat{\mathbf{k}} - i\Im m(E_1^{PV})\boldsymbol{\sigma}, \quad (1)$$

where σ_i are the Pauli matrices, $\hat{\mathbf{k}}$ the wave vector of the excitation laser, α and β are respectively the scalar and vectorial polarizabilities, M_1' is the magnetic dipole moment and E_1^{PV} the parity violating amplitude due to the weak neutral currents (Table 1). F and F' are the hyperfine quantum numbers for the ground and the excited state. Since we shall consider $F \neq F'$, the scalar polarizability term turns out to be zero.

For a transverse applied electric field, circularly polarized light and francium atoms with a polarization $\mathbf{P}_e$, we have an excitation rate N_{8S} proportional to the following expression:

$$N_{8S} \sim \beta^2E^2 - \chi\,(M_1' + \xi\,\Im m E_1^{PV})\,\beta\mathbf{E}\wedge\hat{\mathbf{k}}\cdot\mathbf{P}_e, \quad (2)$$

with ξ the helicity of the excitation laser and χ a known angular factor.

Note the circular dichroism ($\xi\,\Im m E_1^{PV}$) term: parity violation can be detected by reversing the helicity and detecting the change in the excitation rate. The $\Im m E_1^{PV}/M_1'$ measurement directly gives a real-time calibration of the APV effect. The expression (2) also shows the interesting possibility to measure $M_1'/\beta E$: since it is similar but easier with respect to the APV measurement, it could be an interesting preliminary test.

4.2 The cold francium sample

We already discussed the possibility to accumulate francium atoms in a MOT. Because of the presence of magnetic fields in this kind of trap, we do not plan to measure APV in a MOT. There are then two possibilities: either to transfer the atoms to a second kind of trap, for example a dipole trap, or to perform the APV experiment on a cold atomic beam. In a dipole trap we could take advantage of long storage times to accumulate even more francium atoms than in the MOT. However, we expect the APV measurement to heat the atoms and eventually kick them out of the trap: to avoid losses we should periodically cool the trapped atoms before they escape. Furthermore, we presently do not know if the far detuned light from the

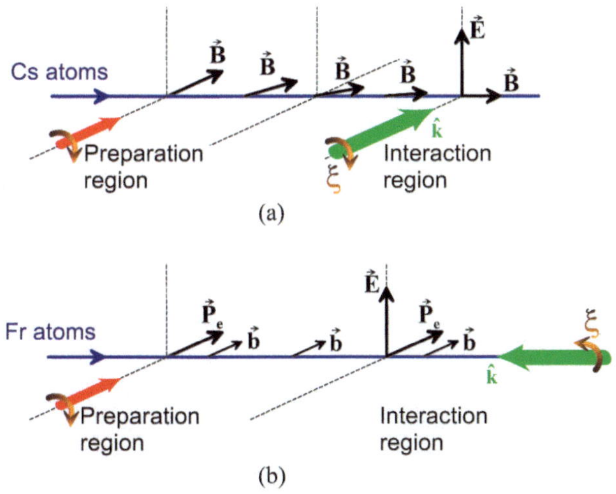

Fig. 2. Scheme for a PV measurement with (a) a thermal Cs beam (Boulder experiment); (b) a cold Fr beam.

dipole trap can disturb the APV measurement. For the cold beam situation, we expect to have a smaller quantity of francium in the interaction region, but the situation is much more definite: in principle the atoms can be conveyed to a safe area, away from unwanted magnetic fields. For this reason we treat here the example of a cold beam for the APV measurement in francium.

4.3 APV experiment with a cold beam

We report in Fig. 2(a) a possible scheme for our APV experiment. The arrangement is similar to the Boulder experiment, performed on a thermal cesium beam (Fig. 2(b)) [10]. In the preparation region an infrared laser transversally polarizes the atoms. However, our scheme takes advantage of the cold sample: the laser excitation is longitudinal to the atomic motion, so as to involve as many atoms as possible. We also plan to apply a magnetic field **b**, much smaller than the cesium case, only to maintain the atomic polarization on the transverse axis: as a consequence the detected pseudoscalar would be $\mathbf{E}\cdot\xi\hat{\mathbf{k}}\wedge\mathbf{P}_e$ instead of $\mathbf{E}\cdot\xi\hat{\mathbf{k}}\wedge\mathbf{B}$. Note that in this configuration we do not need to rotate the magnetic field along the atomic path.

Let us now make some considerations about the magnitude of the expected signal. We shall consider a $4\cdot 10^9$ Fr ions/s production rate, easily attainable at the ISOLDE facility at CERN. We can reasonably expect that after efficient neutralization and laser cooling (for example Zeeman slowing) we shall have 10^8 cold Fr atoms/s. Given typical geometrical parameters for cold beams (velocity about 90 m/s, 2 mrad divergence angle, radius less than 1 mm over 60 cm), we should have $N = 7\cdot 10^5$ atoms in the interaction region.

For a 100 mW excitation laser with a 1 mm waist and a 1000 V/cm electric field, the excitation rate R would be about 0.3 s^{-1}.

With an $\eta = 10\%$ fluorescence detection efficiency, we expect then a signal to noise ratio

$$S/N = \frac{\Im m E_1^{PV}}{\beta E}\sqrt{\eta R N} = 1.3\cdot 10^{-3}\ \sqrt{\mathrm{Hz}}. \qquad (3)$$

In order to be able to reach 1% precision for the APV measurement in an acceptable time, we should enhance the signal to noise ratio. It is indeed possible to increase the available laser power by using a Fabry-Pérot cavity, as in the Boulder cesium experiment. If we consider a 10000 enhancement factor, we have

$$S/N = 0.13\ \sqrt{\mathrm{Hz}}. \qquad (4)$$

Even with the Fabry-Pérot cavity, the excitation rate should be low enough to avoid unwanted heating effects on the cold beam. In a 160 hours data acquisition time the 1% precision should be attainable.

5 Conclusion

We have set up a francium production and trapping experiment in Legnaro. Last tests show that our CCD detection system is more sensitive with respect to our previous traditional fluorescence detection: it will allow to improve the francium trap optimization.

An APV experiment with francium is possible with the high production rates attainable at the ISOLDE facility. Even in the case of the cold francium beam, less favorable than the dipole trap for the signal to noise ratio, it should be possible to reach a good precision. With the lower production rates in Legnaro, we should be able to perform preliminary spectroscopic measurements on the highly forbidden 7S–8S transition.

References

1. J. A. Behr *et al.*, Phys. Rev. Lett. **79**, (1997) 375; Z.-T. Lu *et al.*, Phys. Rev. Lett. **72**, (1994) 3791; G. Gwinner *et al.*, Phys. Rev. Lett. **72**, (1994) 3795; R. Guckert *et al.*, Phys. Rev. A **58**, (1998) R1637; J. E. Simsarian *et al.*, Phys. Rev. Lett. **76**, (1996) 3522; Z.-T. Lu *et al.*, Phys. Rev. Lett. **79**, (1997) 994.
2. M. A. Bouchiat, C. Bouchiat, J. Phys. France **35**, (1974) 899; J. Guéna *et al.*, Mod. Phys. Lett. A **20**, (2005) 375.
3. V. A. Dzuba *et al.*, Phys. Rev. A **51**, (1995) 3454; M. S. Safronova, W. R. Johnson, Phys. Rev. A **62**, (2000) 022112.
4. S. Aubin *et al.*, Rev. Sci. Instr. **74**, (2003) 4342.
5. S. N. Atutov et al., J. Opt. Soc. Am. B **20**, (2003) 953; G. Stancari et al., Nucl. Instrum. Methods A **557**, (2006) 390.
6. H. J. Metcalf, P. van der Straten, *Laser cooling and trapping of atoms* (Springer, New-York 1999).
7. S. N. Atutov *et al.*, *LNL annual report 2005* (2006) 205.
8. S. Sanguinetti *et al.*, Eur. Phys. J. D **25**, (2003) 3.
9. I. M. Savukov *et al.*, Phys. Rev. Lett. **83**, (1999) 2914; M. S. Safronova *et al.*, Phys. Rev. A **60**, (1999) 4476.
10. C. S. Wood *et al.*, Science **275**, (1997) 1759.

Electroweak physics at LHC

J. Erler

Instituto de Física, Universidad Nacional Autónoma de México, 04510 México D.F., Mexico

Abstract. The prospects for electroweak physics at the LHC are reviewed focusing mainly on precision studies. This includes projections for measurements of the effective Z pole weak mixing angle, $\sin^2\theta_W^{\text{eff.}}$, of top quark, W boson, and Higgs scalar properties as well as new physics searches.

PACS. 12.15.-y Electroweak interactions – 13.85.-t Hadron-induced high- and super-high-energy interactions

1 Introduction

The Large Hadron Collider (LHC) is well on its way to produce first collisions in 2007. Initial physics runs are scheduled for 2008 with several fb^{-1} of data and the precision program can be expected to take off in 2009. The low luminosity phase with about 10 fb^{-1} of data (corresponding to 150 million W bosons, 15 million Z bosons, and 11 million top quarks) per year and experiment [1] will already allow most precision studies to be performed. Some specific measurements, most notably competitive results on $\sin^2\theta_W^{\text{eff.}}$, will probably have to wait for the high luminosity phase with $\mathcal{O}(100\ \text{fb}^{-1})$ per year and experiment. The determination of the Higgs self-coupling would even call for a luminosity upgrade by another order of magnitude.

Good knowledge of the lepton and jet energy scales will be crucial. Initially these will be known to 1% and 10%, respectively, but with sufficient data one can use the Z boson mass for calibration, allowing 0.02% and 1% determinations. Furthermore, a 2% measurement of the luminosity and 60% b-tagging efficiency can be assumed [1].

To give a point of reference, a combination of all currently available precision data yields for the Higgs mass,

$$M_H = 88^{+34}_{-26}\ \text{GeV},$$

and for the strong coupling, $\alpha_s(M_Z) = 0.1216 \pm 0.0017$. The fit value for the top quark mass, $m_t = 172.5 \pm 2.3$ GeV, is dominated by and coincides with the Tevatron combination [2]. The χ^2/d.o.f. at the minimum of the global fit is 47.4/42 with a probability for a larger χ^2 of 26%. The 90% CL range for M_H is 47 GeV $< M_H <$ 146 GeV, where upon inclusion of direct search results from LEP 2 [3] the 95% CL upper limit increases to 185 GeV. Besides the notorious list of 1.5 to 3 σ deviations, the electroweak Standard Model (SM) remains in very good shape. One of the largest discrepancies is the NuTeV result [4] on deep inelastic neutrino scattering (ν-DIS) which comes with several challenging theory issues [5].

When discussing future improvements for the key observables, m_t, $\sin^2\theta_W^{\text{eff.}}$, and the W boson mass, M_W, it is useful to keep some benchmark values in mind. An increase of M_H from 100 to 150 GeV (distinguishing between these values provides a rough discriminator between minimal supersymmetry and the SM) is equivalent to a change in M_W by $\Delta M_W = -25$ MeV. On the other hand, this 25 MeV decrease can be mimicked by $\Delta m_t = -4$ GeV, and also by an increase of the fine structure constant at the Z scale, $\Delta\alpha(M_Z) = +0.0014$. We know $\alpha(M_Z)$ an order of magnitude better than this — despite hadronic uncertainties in its relation to the fine structure constant in the Thomson limit. On the other hand, improving m_t will be important. The same shift in M_H is also equivalent to $\Delta\sin^2\theta_W^{\text{eff.}} = +0.00021$, which in turn can be mimicked by $\Delta m_t = -6.6$ GeV or by $\Delta\alpha(M_Z) = +0.0006$. Once the Higgs boson has been discovered and its mass determined kinematically, these observables are then free to constrain heavy new particles which cannot be produced or detected directly. As an example serves the mass of the heavier top squark eigenstate in the minimal supersymmetric standard model at certain parameter values [6].

2 High precision measurements

LEP and SLC [7] almost completely dominate the current average Z pole weak mixing angle, $\sin^2\theta_W^{\text{eff.}} = 0.23152 \pm 0.00016$. Via measurements of leptonic forward-backward (FB) asymmetries, the Tevatron Run II is expected to add another combined ± 0.0003 determination [8], competitive with the most precise measurements from LEP (the FB asymmetry for $b\bar{b}$ final states) and SLD (the initial state polarization asymmetry). Having $p\bar{p}$ collisions are a crucial advantage here. At the LHC, by contrast, one has

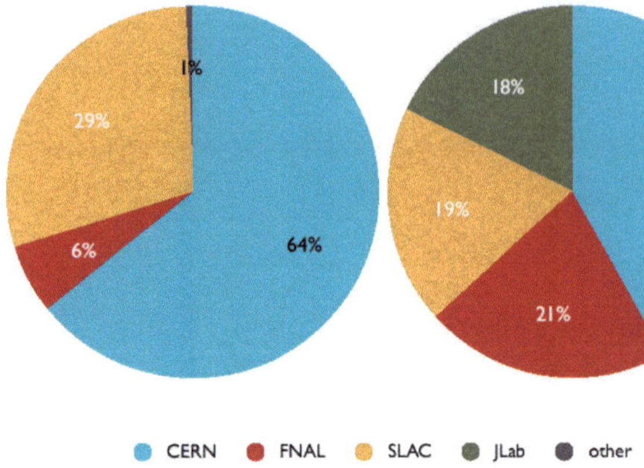

Fig. 1. Precision (square of inverse error) weighted contributors to our knowledge of $\sin^2\theta_W^{\text{eff.}}$ by laboratory. Shown is the status (left) and a projection (right) for the Tevatron Run IIA (2 fb^{-1} per experiment) along with 4% *ep* and 3.25% *ee* cross section asymmetries from JLab.

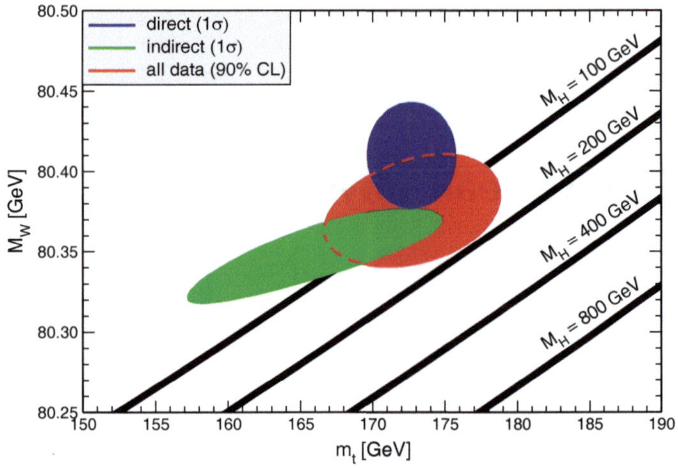

Fig. 2. 1 σ constraints and the 90% CL allowed region by all precision data in the $M_W - m_t$ plane. The lines show the SM predictions for various Higgs masses, their widths indicating the theory uncertainty from $\alpha(M_Z)$ in the prediction.

to focus on events with a kinematics suggesting that a valence quark was involved in the collision and which proton provided it (Z rapidity tag). This will be possible for a small fraction of events only, requiring high luminosity running. Furthermore, sufficient rapidity coverage of $|\eta| < 2.5$ will be necessary for even a modest ± 0.00066 determination [9]. Incidentally, a similar precision is expected from fixed target elastic proton scattering by the Qweak experiment [10] using the polarized electron beam at JLab. A breakthrough measurement at the LHC with an error as small as ± 0.00014 [8,9] will require a much more challenging rapidity coverage of $|\eta| < 4.9$ for jets and missing transverse energy. Thus, it is presently unclear what the impact of the LHC on $\sin^2\theta_W^{\text{eff.}}$ will be. On the other hand, as has been covered at this meeting, there may be further opportunities at JLab after the 12 GeV upgrade of CEBAF in parity violating deep inelastic scattering (DIS-Parity) [11] building on the current 6 GeV DIS-Parity effort [12], and most notably, from an improved measurement of polarized Møller scattering (e2ePV) [13], reducing the error of the E 158 experiment at SLAC [14] by about a factor of four (see Fig. 1).

Our current knowledge regarding M_W and m_t is summarized in Fig. 2. The height of the uppermost (blue) ellipse is the average, $M_W = 80.410 \pm 0.032$ GeV, of final UA2 [15] and Tevatron Run I [16–18], as well as preliminary LEP 2 [19] results. With the exception of a less precise threshold determination at LEP 2, all these results are based on direct reconstruction[1]. The other (green) 1 σ ellipse is from all data excluding M_W and the Tevatron m_t. Its elongated shape arises because one combination is tightly constrained by $\sin^2\theta_W^{\text{eff.}}$ while the orthogonal one is from less precise measurements including low energy observables and the partial Z decay width into $b\bar{b}$ pairs [7] (with a very different m_t dependence than other neutral current observables). Fig. 2 demonstrates that the direct and indirect contours in the $M_W - m_t$ plane are consistent with each other and independently favor small Higgs masses, $M_H \lesssim 150$ GeV. All channels and experiments combined, the Tevatron Run II will likely add another ± 30 MeV constraint. The huge number of W bosons will enable the LHC to provide further ± 30 MeV measurements per experiment and lepton channel (e and μ) for a combined ± 15 MeV uncertainty (it is assumed here that the additional precision that can be gained by cut optimization is compensated approximately by common systematics). The measurements are limited by the lepton energy and momentum scales, but these can be controlled using leptonic Z decays. With the even larger data samples of the high luminosity phase, one may alternatively consider the W/Z transverse mass ratio, opening the avenue to a largely independent measurement with an error as low as ± 10 MeV [8], for a combined uncertainty about three times smaller than our benchmark of ± 25 MeV.

The uncertainty in the world average $\sin^2\theta_W^{\text{eff.}}$ after the LHC era will only be half of our benchmark of ± 0.00021 at best, but it is less effected by m_t than M_W. More generally, $\sin^2\theta_W^{\text{eff.}}$ and M_W can be used to constrain the so-called oblique parameters, describing vector boson self-energies, such as S and T [20] shown in Fig. 3. The complementarity between M_W and $\sin^2\theta_W^{\text{eff.}}$ (asymmetries and E 158) can be appreciated from the different slopes. Similarly, Z decay properties (other than asymmetries), ν-DIS, and the weak charges of heavy elements, Q_W, from atomic parity violation, all yield different slopes and shapes illustrating the power of having a wide variety of observables available.

The total W decay width, Γ_W, represents another observable of relevance to oblique parameters, but its sen-

[1] Frequently, ν-DIS results are also represented as measurements of M_W, since this accounts to a good approximation for the m_t dependence in the SM. This is, however, a coincidence and, in general, ν-DIS is affected differently by new physics than M_W, and ν and $\bar{\nu}$ scattering actually provide two independent observables, although $\bar{\nu}$-DIS is usually less accurate.

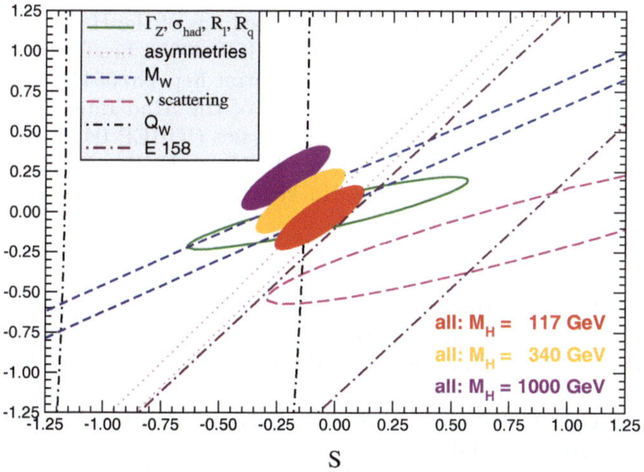

Fig. 3. 1 σ constraints and the 90% CL regions (for various reference Higgs masses) in S and T allowed by all precision data. They are defined to vanish in the SM, so that any significant deviation from zero may signal the presence of new physics.

sitivity to new physics and its complementarity to and correlation with other quantities depends on how it is obtained. It can be extracted indirectly through measurements of cross section ratios,

$$\left[\frac{\sigma(pp \to Z \to \ell^+\ell^- X)}{\sigma(pp \to W \to \ell\nu X)}\right]_{\rm exp.} \times \left[\frac{\sigma(pp \to W)}{\sigma(pp \to Z)}\right]_{\rm th.} \times \frac{\Gamma_{\rm SM}(W \to \ell\nu)}{\mathcal{B}_{\rm LEP}(Z \to \ell^+\ell^-)},$$

(CDF currently quotes $\Gamma_W = 2.079 \pm 0.041$ GeV [21]) but the leptonic W decay width has to be input from the SM. More interesting is therefore the direct method using the tail of the transverse mass distribution. An average of final Tevatron Run I and preliminary DØ II [22] and LEP 2 [19] results gives, $\Gamma_W = 2.103 \pm 0.062$ GeV. The final Tevatron Run II is expected to contribute ± 50 MeV measurements for each channel and experiment. Detailed studies for the LHC are not yet available, but historically the absolute error in Γ_W at hadron colliders has traced roughly the one in M_W. If this trend carries over to the LHC, a $\pm 0.5\%$ error of Γ_W may be in store.

Some Tevatron Run II results are already included in the current ± 2.3 GeV [2] uncertainty in m_t, and with the expected total of 8 fb^{-1} the error may decrease by another factor of two. The LHC is anticipated to contribute a ± 1 GeV determination from the lepton + jets channels alone [23]. The cleaner but lower statistics dilepton channels may provide another ± 1.7 GeV determination, compared with ± 3 GeV from the systematics limited all hadronic channel [23]. The combination of these channels (all dominated by the b jet energy scale) would yield an error close to the additional irreducible theoretical uncertainty of ± 0.6 GeV from the conversion from the pole mass (which is approximately what is being measured [24]) to a short distance mass (such as $\overline{\rm MS}$) which actually enters the loops. Folding this in, the grand total may give an error of about ± 1 GeV, so that the parametric uncertainty from m_t in the SM prediction for M_W would be somewhat smaller than the anticipated experimental error in M_W.

3 Other electroweak physics

With 30 fb^{-1} of data, the LHC will also be able to determine the CKM parameter, V_{tb}, in single top quark production to $\pm 5\%$ [25] (one expects $\pm 9\%$ from the Tevatron Run II although no single top events have been observed so far). Anomalous flavor changing neutral current decays, $t \to Vq$ (where V is a gluon, photon, or Z boson, and $q \neq b$), can be searched for down to the $10^{-4} - 10^{-5}$ level [23]. This is a sensitivity gain by three orders of magnitude over current HERA bounds [26], and relevant, *e.g.* to extra W' bosons. Measuring $t\bar{t}$ spin correlations at the 10% level [23] will allow to establish the top quark as a spin 1/2 particle, to study non-standard production mechanisms (*e.g.* through resonances), and to discriminate between W^+b and charged Higgs (H^+b) decays.

If the Higgs boson exists, its production at the LHC will proceed primarily through gluon fusion, $gg \to H$, and/or vector boson fusion, $qq' \to Hqq'$. Higgs couplings can generally be determined to $10 - 30\%$ [8]. The top Yukawa coupling is best studied in associated production, $pp \to t\bar{t}H$, to $20-30\%$ precision [23]. Most difficult proves the Higgs self coupling, λ, whose measurement would need a luminosity upgrade. With 3 ab^{-1} of data, λ can be measured to $\pm 20\%$, for 150 GeV $< M_H <$ 200 GeV, while only $\pm 70\%$ precision would be possible for a lighter (and weaker coupled) Higgs boson [8].

The LHC is, of course, primarily a discovery machine with the outstanding task to find the Higgs boson or else to rule out its existence [28]. As an example for a potential discovery beyond the SM, an extra Z' (W') boson would reveal itself through a high dilepton invariant ($\nu\ell$ transverse) mass peak. Current Z' limits (which depend on the nature of the Z') ranging from 650 to 850 GeV from CDF [27] and from 434 GeV to 1.8 TeV from LEP 2 [19] can be extended to 4.2–5 TeV with 100 fb^{-1} of data, while 1 ab^{-1} from an upgraded LHC would add another TeV to the reach [29].

In some cases one can turn things around and use electroweak physics to understand the LHC. For example, by computing W and Z cross sections and comparing them to LHC production rates one can extract the luminosity of the machine. This assumes knowledge of the relevant parton density functions (PDFs) which will probably be available with 2% uncertainties reflecting the limitation of the method [25]. In turn, one can obtain information on u and d quark PDFs by measuring the $W^\pm$ charge asymmetry, defined as the differential (with respect to the $e^\pm$ rapidity) cross section asymmetry [8].

4 Conclusions

The LHC is posed to achieve breakthrough discoveries in the electroweak symmetry breaking sector. As for precision measurements, one can expect particularly great improvements in M_W, while a competitive measurement of $\sin^2\theta_W^{\rm eff.}$ will be possible with large rapidity coverage only. Measurements of m_t, Γ_W, Yukawa and Higgs self-couplings will be performed and the top quark will be subjected to detailed studies.

At a meeting on parity violation these measurements should be put into context with low energy precision measurements which will remain important complements even with the LHC in operation. This is because (i) they are capable of contributing results on $\sin^2\theta_W^{\rm eff.}$ which can compete both with Z pole factories and hadron colliders (see Fig 1), (ii) they are subject to entirely different experimental and theoretical issues, and (iii) they are generally affected quite differently by beyond the SM physics.

It is a pleasure to thank the organizers of PAVI06 for the invitation to a very enjoyable meeting. This work was supported by CONACyT (México) contract 42026–F and by DGAPA–UNAM contract PAPIIT IN112902.

References

1. ATLAS Collaboration: P. Pralavorio *et al.*, PoS **HEP2005** (2006) 294.
2. Tevatron Electroweak Working Group and CDF and DØ Collaborations: E. Brubaker *et al.*, e-print hep-ex/0603039.
3. LEP Working Group for Higgs boson searches and ALEPH, DELPHI, L3, and OPAL Collaborations: R. Barate *et al.*, Phys. Lett. **B565** (2003) 61–75.
4. NuTeV Collaboration: G. P. Zeller *et al.*, Phys. Rev. Lett. **88** (2002) 091802.
5. T. Londergran, this conference.
6. J. Erler, S. Heinemeyer, W. Hollik, G. Weiglein, and P. M. Zerwas, Phys. Lett. **B486** (2000) 125–133.
7. ALEPH, DELPHI, L3, OPAL, and SLD Collaborations, LEP Electroweak Working Group and SLD Electroweak and Heavy Flavour Groups: S. Schael *et al.*, Phys. Rept. **427** (2006) 257.
8. U. Baur, to appear in the Proceedings of the Hadron Collider Physics Symposium 2005 (HCP 2005), Les Diablerets, Switzerland, July 4–9, 2005, e-print hep-ph/0511064.
9. W. Quayle, in the Proceedings of the 32nd International Conference on High-Energy Physics (ICHEP 04), Beijing, China, Aug 16–22, 2004, ed. H. Chen, D. Du, W. Li, and C. Lu (World Scientific, Hackensack 2005) Vol. 1, 531–534.
10. S. Page, this conference.
11. P. Souder, this conference.
12. X. Zheng, this conference.
13. D. Mack, this conference.
14. K. Kumar, this conference.
15. UA2 Collaboration: J. Alitti *et al.*, Phys. Lett. **B276** (1992) 354–364.
16. CDF Collaboration: A. A. Affolder *et al.*, Phys. Rev. **D64** (2001) 052001.
17. DØ Collaboration: V. M. Abazov *et al.*, Phys. Rev. **D66** (2002) 012001.
18. CDF and DØ Collaborations: V. M. Abazov *et al.*, Phys. Rev. **D70** (2004) 092008.
19. ALEPH, DELPHI, L3, and OPAL Collaborations and LEP Electroweak Working Group: J. Alcaraz *et al.*, e-print hep-ex/0511027.
20. M. E. Peskin and T. Takeuchi, Phys. Rev. **D46** (1992) 381–409.
21. CDF II Collaboration: D. Acosta *et al.*, Phys. Rev. Lett. **94** (2005) 091803.
22. Tevatron Electroweak Working Group and CDF and DØ Collaborations: B. Ashmanskas *et al.*, e-print hep-ex/0510077.
23. J. Womersley, PoS **TOP2006** (2006) 038.
24. M. C. Smith and S. S. Willenbrock, Phys. Rev. Lett. **79** (1997) 3825–3828.
25. K. Mazumdar, AIP Conf. Proc. **792** (2005) 587–590.
26. ZEUS and H1 Collaborations: J. Ferrando *et al.*, Eur. Phys. J. **C33** (2004) S761–S763.
27. CDF Collaboration: A. Abulencia *et al.*, Phys. Rev. Lett. **96** (2006) 211801.
28. M. Escalier, AIP Conf. Proc. **792** (2005) 623–626.
29. S. Godfrey, in the Proceedings of the APS/DPF/DPB Summer Study on the Future of Particle Physics (Snowmass 2001), ed. N. Graf, eConf **C010630** (2001) P344, e-print hep-ph/0201093.

The E158 experiment

K.S. Kumar[a]

University of Massachusetts, Amherst, MA 01003, USA

Paper reprinted from Eur. Phys. J. A with permission

Abstract. We have carried out a precision measurement of the parity-violating asymmetry A_{PV} in the scattering of longitudinally polarized electrons off electrons in a liquid-hydrogen target. The measurement was performed with the 50 GeV beam line at SLAC. The final result with the full data set collected in three production runs is $A_{PV} = -131 \pm 14$ (stat) ± 10 (syst) parts per billion. The result leads to new limits on possible contact interactions at the TeV scale. We discuss future prospects for more precise measurements.

PACS. 12.15.Mm Neutral currents – 11.30.Er Charge conjugation, parity, time reversal, and other discrete symmetries

1 Introduction

Precision measurements of weak neutral current (WNC) interactions, mediated by the Z-boson, play a central role in tests of the electroweak theory and in the search for new dynamics at very high-energy scales. One class of WNC experiments involves measurements of the fractional difference in the cross-section for longitudinally polarized electrons scattering off unpolarized nuclear targets. A non-zero asymmetry (A_{PV}) is a signature of parity violation and arises from the interference between the WNC and electromagnetic amplitudes [1].

WNC measurements can test the standard model by comparing the extracted value of the weak mixing angle $\sin^2\theta_W$ in each case (evolved to the same energy scale) to the precise value obtained in high-energy collider measurements. A deviation would be a signature of new contact interactions at the TeV scale. Since fixed target WNC measurements are carried out at $Q^2 \ll M_Z^2$, such new physics amplitudes can interfere with the electromagnetic amplitude, unlike the case of WNC measurements at the Z-resonance [2]. It is necessary, however, to measure $\sin^2\theta_W$ to fractional accuracy better than 1% [3].

Prior to the E158 experiment, two published measurements achieved such sensitivity: the weak-charge measurement in ^{133}Cs [4] and the NuTeV neutrino deep-inelastic-scattering measurement [5]. Here, we discuss a measurement of A_{PV} in electron-electron (Møller) scattering [6], a purely leptonic reaction with little theoretical uncertainty [7,8]. The E158 experiment at the Stanford Linear Accelerator Center (SLAC) was designed to use the longitudinally polarized 50 GeV electron beam to measure A_{PV} to a relative accuracy of about 10%.

2 Experimental overview

Target electrons, in a 1.54 m long cell of liquid hydrogen (10.5 gm/cm^2), were bombarded by a 48 GeV electron beam, the longitudinal polarization of which was changed pseudo-randomly. Møller electrons (beam electrons scattering off target electrons), were isolated by a forward magnetic spectrometer consisting of a 3-dipole "chicane" and 4 quadrupole magnets. Møller electrons of interest in the full range of the azimuth (spanning the polar angular range 4.5 mrad $< \theta_{lab} <$ 8 mrad) traversed through the bores of the quadrupoles and were brought to a ring focus on a calorimeter 60 m downstream of the target.

The experimental asymmetry was measured by averaging the fractional difference in the cross-section over complementary pairs of beam pulses of opposite helicity. In order to achieve the desired statistical precision of 10 parts per billion (ppb) in a reasonable length of time, the intergrated signal of more than 20 million electrons were detected for each beam pulse. The calorimeter provided both radial and azimuthal segmentation. The bulk of the electron-proton (ep) scattered flux was intercepted by the outermost ring. The radiative tail of the ep flux was the main background in the inner Møller rings, totalling $\approx 8\%$. Other background was less than 1%.

Data were collected over 3 run periods in 2002 and 2003. Rapid reversal of the polarization of the electron beam, generated via photoemission on a GaAs photocathode by circularly polarized laser light, suppressed spurious asymmetries. Additional suppression was achieved by pas-

[a] Representing the E158 Collaboration. Funded by US Department of Energy Grant No. DE-FG02-88R40415.A018;
e-mail: kkumar@physics.umass.edu

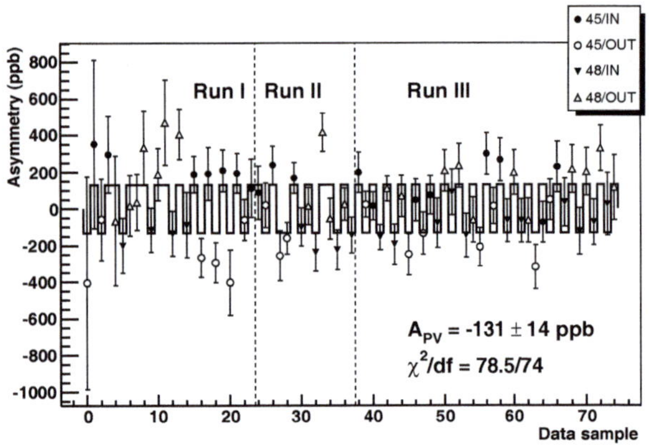

Fig. 1. $A_{\rm PV}$ for 75 data samples.

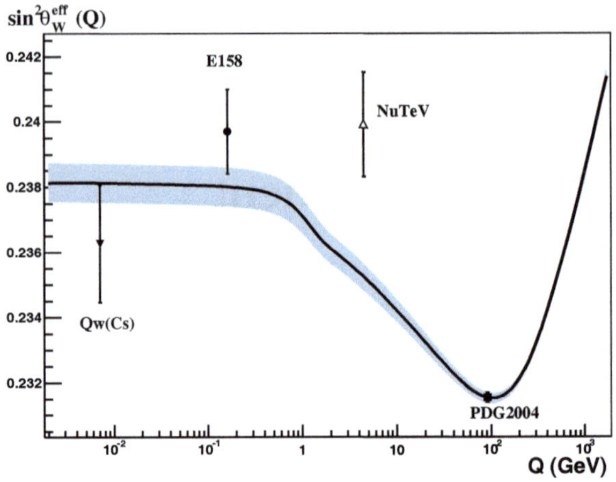

Fig. 2. $\sin^2\theta_{\rm W}$ as a function of Q^2.

sively reversing the sign of the asymmetry by two independent methods. First, the state of a half-wave plate in the laser line was toggled each day. Second, spin precession in the 24.5° bend after beam acceleration created opposite helicity orientation at 45 GeV and 48 GeV beam energies. Roughly equal statistics were accumulated with opposite signs of the measured asymmetry.

3 Results and implications

Figure 1 shows the parity-violating asymmetry as a function of data-set number. Each data-set constitutes about two days of data, after which either the beam energy or the state of the half-wave plate were changed to flip the sign of the measured asymmetry. The grand average result for the parity-violating asymmetry in Møller scattering at $Q^2 = 0.03\,\mathrm{GeV}^2$ was found to be

$$A_{\rm PV} = -131 \pm 14(\mathrm{stat}) \pm 10(\mathrm{syst}) \quad (\mathrm{ppb}). \tag{1}$$

From this measurement, the value of $\sin^2\theta_W$ can be extracted within the context of the standard model. Using a definition which reproduces the effective leptonic couplings at the Z pole, we determine:

$$\sin^2\theta_{\rm W}^{\rm eff} = 0.2397 \pm 0.0010(\mathrm{stat}) \pm 0.0008(\mathrm{syst}). \tag{2}$$

Figure 2 shows the E158 result, which establishes the "running" of $\sin^2\theta_{\rm W}$ [9] by more than 6 standard deviations. Also shown are the two other precise low-energy $\sin^2\theta_{\rm W}$ determinations mentioned earlier. It can be seen that the ^{133}Cs result and the E158 result are consistent with the standard model expectation. The deviation of the NuTeV result thus implies that either there are new contact interactions specific to neutrino interactions or that there are additional strong interaction effects that are unaccounted for. One leading candidate is charge symmetry violation in the parton distribution functions [10].

The E158 measurement can be used to set limits on the size of possible new contributions beyond the standard model. Assuming a new contact interaction scale [11] characterized by $\Lambda_{\rm LL}$, the 95% C.L. limit is 7 TeV or 16 TeV depending on the sign of the contact interaction term.

4 Future prospects

The figure of merit to measure the WNC amplitude in Møller scattering rises with incident beam energy. It is therefore possible to contemplate improved new measurements of $A_{\rm PV}$ at future facilities [12]. One possibility is to carry out a new measurement at Jefferson Laboratory after it is upgraded to 12 GeV, where it is potentially possible to achieve a factor of 5 improvement over the reported measurement. The ultimate measurement could be carried out at the proposed International Linear Collider (ILC), using the electron beam downstream of the primary collider interaction region. More than an order-of-magnitude improvement is possible [13], which competes very favorably with future collider determinations of the weak mixing angle and measurements of the W-boson mass.

References

1. Ya.B. Zel'dovich, Sov. Phys. JETP **94**, 262 (1959).
2. K.S. Kumar *et al.*, Mod. Phys. Lett. A **10**, 2979 (1995).
3. M.J. Ramsey-Musolf, Phys. Rev. C **60**, 015501 (1991).
4. S.C. Bennett, C.E. Wieman, Phys. Rev. Lett. **82**, 2484 (1999).
5. G.P. Zeller *et al.*, Phys. Rev. Lett. **88**, 0918802 (2002).
6. A. Czarnecki, W.J. Marciano, Phys. Rev. D **53**, 1066 (1996).
7. P.L. Anthony *et al.*, Phys. Rev. Lett. **92**, 181602 (2004).
8. P.L. Anthony *et al.*, Phys. Rev. Lett. **95**, 081601 (2005).
9. A. Czarnecki, W.J. Marciano, Int. J. Mod. Phys. A **15**, 2365 (2000); J. Erler, A. Kurylov, M.J. Ramsey-Musolf, Phys. Rev. D **68**, 016006 (2003); A. Ferroglia, G. Ossola, A. Sirlin, Eur. Phys. J. C **34**, 165 (2004); J. Erler, M.J. Ramsey-Musolf, hep-ph/0409169 (2004); F.J. Petriello, Phys. Rev. D **68**, 033006 (2003).
10. J.T. Londergan, A.W. Thomas, hep-ph/0407247 (2004) and references therein.
11. E.J. Eichten, K.D. Lane, M.E. Peskin, Phys. Rev. Lett. **50**, 811 (1983).
12. K.S. Kumar, Eur. Phys. J. A **24**, s2, 191 (2005).
13. K.S. Kumar, in *DPF/DPB Summer Study on New Directions in High Energy Physics*, econf **C960625**, NEW168 (1996).

The Q^p_{weak} experiment at Jefferson Laboratory

S.A. Page (for the Q_{weak} Collaboration)[a]

Department of Physics & Astronomy, University of Manitoba, Winnipeg MB Canada R3T 2N2

Abstract. A major new experiment is being prepared at Jefferson Laboratory to measure the proton's weak charge via the parity violating asymmetry in elastic electron-proton scattering at very low momentum transfer. The Standard Model makes a firm prediction of the proton's weak charge, $Q^p_w = 1 - 4\sin^2\theta_W$, based on the running of the weak mixing angle $\sin^2\theta_W$ from the Z^0 pole down to low energies, corresponding to a 10 σ effect in our experiment. Our ultimate goal is to determine the proton's weak charge with 4% combined statistical and systematic errors, which in turn leads to a 0.3% measurement of $\sin^2\theta_W$. The experiment is currently under construction; installation in Hall C at Jefferson Lab followed by data taking is planned for 2009.

PACS. 12.15.-y Electroweak interactions – 11.30.Er Charge conjugation, parity, time reversal, and other discrete symmetries – 13.60.Fz Elastic and Compton scattering

1 Introduction

Precision measurements of parity violation have traditionally played an important role in guiding our understanding of the electroweak interaction. A key prediction of the Standard Model is the variation of $\sin^2\theta_W$ with momentum transfer Q^2, referred to as the "running of $\sin^2\theta_W$". As with the QED and QCD couplings, $\alpha(\mu^2)$ and $\alpha_s(\mu^2)$, the running of $\sin^2\theta_W(\mu^2)$ is an effective parameter defined at a scale $\mu^2 \sim Q^2$ at which a given experiment is performed. The μ-dependence arises from loop corrections to the electroweak gauge couplings, and thus reflects the content of the Standard Model beyond tree-level. Physics beyond the Standard Model influences the running of $\sin^2\theta_W$ via the incorporation of new diagrams in corrections to the electroweak gauge couplings.

Thus far, measurements at the Z^0 pole have constrained the weak mixing angle $\sin^2\theta_W$ to impressive precision at that energy scale, and a few experiments have attempted to establish the expected running of $\sin^2\theta_W$ to lower energies, with less certain results. Higher precision experiments to determine $\sin^2\theta_W$ at low energy are needed to complement the Z^0 pole measurements and constrain new physics beyond the Standard Model.

The Standard Model running of $\sin^2\theta_W$ predicts a shift of $+0.007$ at low Q^2 with respect to the Z^0 pole best fit value of 0.23113 ± 0.00015. Figure 1 shows a calculation by Erler and Ramsey-Musolf [1] for $\sin^2\theta_W$ together with existing and proposed world data. The expected evolution of $\sin^2\theta_W$ corresponds to a 10 standard deviation effect in the planned Q_{weak} experiment at Jefferson Laboratory, including both experimental and theoretical systematic errors. Any significant deviation from the Standard Model prediction would be a signal of new physics, whereas agreement would place new and significant constraints on possible Standard Model extensions.

It should be noted that high energy studies which determine $\sin^2\theta_W$ at the Z^0 pole are complementary to the precision low energy experiment described here. Indeed, the precise Z^0 pole measurements set the overall magnitude of the $\sin^2\theta_W$ curve, while additional experiments at low energy are required to establish the running with energy scale. Low energy values of $\sin^2\theta_W$ have been extracted from parity violating Möller scattering [6], atomic parity violation [2] and neutrino deep inelastic scattering experiments [3], but significant uncertainties in the theoretical interpretation of the latter [4] limits the impact of the results. In contrast, a precision measurement of the weak charge of the proton, $Q^p_w = 1 - 4\sin^2\theta_W$, under construction at Jefferson Laboratory [5], addresses similar physics issues but is free of many-body theoretical uncer-

[a] The Q_{weak} Collaboration: D. Armstrong, T. Averett, J. Birchall, P. Bosted, T. Botto, A. Bruell, C. Capuano, R.D. Carlini, G. Cates, S. Chattopadhyay, S. Covrig, C. Davis, K. Dow, J. Dunne, D. Dutta, R. Ent, J. Erler, W. Falk, W.M. Finn, T. Forest, W. Franklin, D. Gaskell, M. Gericke, K. Grimm, F. Hersman, M. Holtrop, K. Johnston, R. Jones, K. Joo, C. Keppel, P. King, M. Kohl, E. Korkmaz, S. Kowalski, L. Lee, Y. Liang, A. Lung, D. Mack, S. Majewski, J. Mammei, R. Mammei, J. Martin, D. Meekins, H. Mkrtchyan, N. Morgan, K. Myers, A. Opper, J. Pan, K. Paschke, M. Pitt, M. Poelker, T. Porcelli, Y. Prok, W. Ramsay, M. Ramsey-Musolf, J. Roche, N. Simicevic, G. Smith, T. Smith, P. Souder, R. Suleiman, E. Tsentalovich, W. van Oers, P. Wang, S. Wells, S. Wood, R. Young, H. Zhu, C. Zorn, T. Zwart.

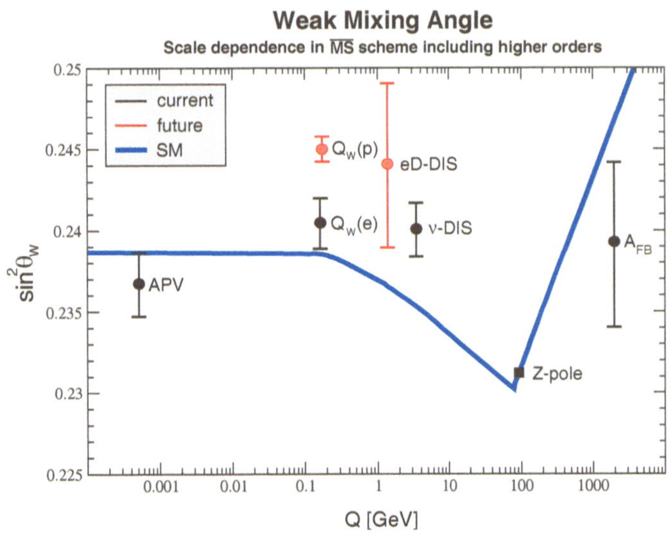

Fig. 1. Calculated running of the weak mixing angle in the Standard Model, as defined in the modified minimal subtraction scheme [1]. The black error bars show the current situation, while the red error bars refer to the proposed 4% Q^p_w measurement and other possible future measurements.

tainties and is designed for significantly better precision. The dominant hadronic effects that must be accounted for in extracting Q^p_w from the data are contained in form factor contributions which can be sufficiently constrained by the current program of parity violating electron scattering measurements without reliance on theoretical nucleon structure calculations. The Q_{weak} experiment is thus a crucial element of a program of very sensitive low energy tests of the Standard Model, as it will be both more precise than and complementary to previous efforts such as the recently completed SLAC E158 experiment [6], which carried out parity violating asymmetry measurements at a similar Q^2 in the purely leptonic sector.

2 Overview of the experiment

The proton's weak charge, $Q^p_w = 1 - 4\,\sin^2\theta_W$ will be evaluated from a precision measurement of the parity violating asymmetry in elastic electron-proton scattering at very low momentum transfer. The parity-violating asymmetry A is the ratio of the helicity dependent part of the cross section for *ep* elastic scattering to the helicity average cross section:

$$A = \frac{\sigma_+ - \sigma_-}{\sigma_+ + \sigma_-}, \qquad (1)$$

which, expressed in terms of Sachs electromagnetic form factors G^γ_E, G^γ_M, weak neutral form factors G^Z_E, G^Z_M and the neutral weak axial form factor G_A, has the form:

$$A = \left[\frac{-G_F Q^2}{4\pi\alpha\sqrt{2}}\right] \times$$

$$\left[\frac{\varepsilon G^\gamma_E G^Z_E + \tau G^\gamma_M G^Z_M - (1-4\sin^2\theta_W)\varepsilon' G^\gamma_M G^Z_A}{\varepsilon (G^\gamma_E)^2 + \tau (G^\gamma_M)^2}\right] \qquad (2)$$

where

$$\varepsilon = \frac{1}{1+2(1+\tau)\tan^2\frac{\theta}{2}}, \quad \varepsilon' = \sqrt{\tau(1+\tau)(1-\varepsilon^2)} \qquad (3)$$

are kinematical quantities, Q^2 is the four-momentum transfer, $\tau = Q^2/4M^2$ where M is the proton mass, and θ is the laboratory electron scattering angle. It was shown in [7] that for forward-angle scattering where $\theta \to 0$, $\epsilon \to 1$, and $\tau << 1$, the asymmetry can be written as:

$$A = \left[\frac{-G_F}{4\pi\alpha\sqrt{2}}\right]\left[\,Q^2 Q^p_w + F^p(Q^2,\theta)\right]$$

$$\to \left[\frac{-G_F}{4\pi\alpha\sqrt{2}}\right]\left[\,Q^2 Q^p_w + Q^4 B(Q^2)\right] \qquad (4)$$

where F^p is a form factor. Neglecting radiative corrections, the leading term in equation 4 is the proton's weak charge: $Q^p_w = 1 - 4\sin^2\theta_W$. The quantity $B(Q^2)$ is the leading term in the nucleon structure defined in terms of neutron and proton electromagnetic and weak form factors. These structure contributions can be reduced by carrying out the asymmetry measurements at lower momentum transfer, but at the expense of reduced sensitivity to Q^p_w. The value of $B(Q^2)$ can be determined experimentally by extrapolation from the ongoing program of forward angle parity-violating experiments at higher Q^2. The optimum value of Q^2 for the Q_{weak} experiment is near 0.03 $(GeV/c)^2$ based on our estimate of the anticipated final precision of the various HAPPEx, HAPPEx II, G^0, and Mainz A4 measurements. At this momentum transfer, the parity violating asymmetry is expected to be $A = -0.3$ *ppm*, and we must achieve a total uncertainty of 2% in the measurement of A in order to meet our precision goal of 0.3% in $\sin^2\theta_W$. The contributions of the weak charge and structure dependent terms to the parity-violating asymmetry at 0.03 $(GeV/c)^2$ are expected to be:

$$A(0.03(GeV/c)^2) = A(Q^p_w) + A(Had_V) + A(Had_A)$$

$$= -0.19\,ppm - 0.09\,ppm - 0.01\,ppm \qquad (5)$$

where the hadronic structure contributions are separated into vector and axial vector components.

In order to achieve the required precision in the Q^p_w measurement, a number of key requirements must be met. Clearly, a high statistics data sample is required, implying high luminosity, a large detector acceptance, and an integrating low-noise detector system. In addition, the measured asymmetry is PA where P is the beam polarization, implying that the beam polarization must be high and precisely measured. The hadronic structure contribution $B(Q^2)$ must be independently known, and its contribution must be accurately subtracted from the physics asymmetry A in order to isolate the weak charge contribution; to do so requires independent and precise knowledge of the detector-response-weighted averages $< Q^2 >$ and $< Q^4 >$, which should be measured in ancillary experiments. On the systematics side, helicity correlated systematic errors

Table 1. Total error estimate for the Q_{weak} experiment. The contributions to both the physics asymmetry and the extracted Q^p_w are given. In most cases, the error magnification due to the 39% hadronic dilution is a factor of 1.64. The enhancement for the Q^2 term is somewhat larger.

Source of error	Contribution to $\Delta A_{phys}/A_{phys}$	Contribution to ΔQ^p_w / Q^p_w
Counting Statistics	1.8%	2.9%
Hadronic structure	—	2.2%
Beam polarimetry	1.0 %	1.6%
Absolute Q^2	0.7%	1.1%
Backgrounds	0.5%	0.8%
Helicity-correlated beam properties	0.5%	0.8%
TOTAL:	2.3%	4.3%

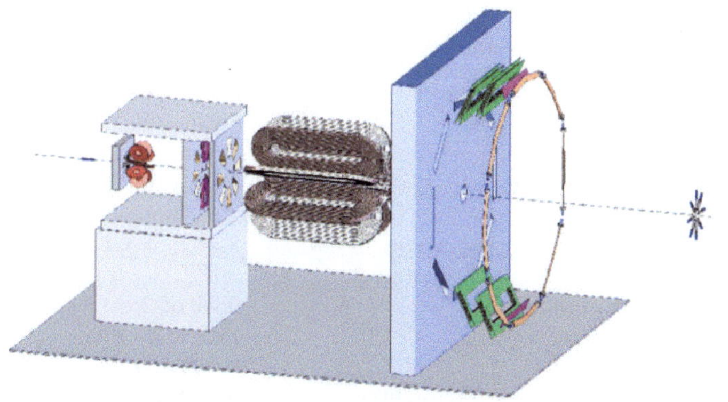

Fig. 2. Layout of the Q_{weak} apparatus. The beam and scattered electrons travel through the target, the first collimator, the Region 1 GEM detectors, the mini-torus, the two-stage second precision collimator which surrounds the region 2 drift chambers, the toroidal magnet, the shielding wall, the region 3 drift chambers, the trigger scintillators and finally through the quartz Čerenkov detectors. The tracking system chambers and trigger scintillators will be retracted during high current running when Q_{weak} asymmetry data are acquired. The Q_{weak} luminosity monitor, which will be used to monitor target fluctuations and to provide a sensitive null asymmetry test, is located downstream of the apparatus very close to the beam pipe.

must be reduced to a negligible level: the Q_{weak} experiment has set a goal of 6×10^{-9} or less for helicity correlated systematic error contributions to the asymmetry.

Table 1 contains a brief summary of the key uncertainties and error budgets for this experiment. These have been refined through extensive simulations since the Q_{weak} proposal was first submitted. The experiment model now fully accounts for the effects of all Bremsstrahlung losses, including those inside the target flask. The simulation also contains details of a realistic collimator/detector system. A 2200 hour measurement of the parity violating asymmetry in elastic electron-proton scattering at a momentum transfer of $Q^2 = 0.03$ (GeV/c)^2 employing 180 μA of 85% polarized beam on a 35 cm liquid hydrogen target will determine the proton's weak charge with 4% combined statistical and systematic errors; this in turn implies a determination of $\sin^2\theta_W$ at the $\pm0.3\%$ level at low energy. The results of previous experiments in parity violating electron-proton scattering will be used to constrain hadronic corrections to the data; a recently published analysis by Young et. al [8] of published SAMPLE, PVA4, HAPPEx and G0 data confirms the expected hadronic structure uncertainty listed in Table 1 for the Q_{weak} experiment. We note that a stand-alone measurement of $\sin^2\theta_W$, the Q_{weak} experiment is competitive with any channel measured in the recently completed SLD and LEP programs at the Z resonance.

The layout of the experiment is given in Figure 2. A longitudinally polarized electron beam, a liquid hydrogen target, a room temperature 8-fold symmetric toroidal magnetic spectrometer, and a set of detectors for the scattered electrons at forward angles, are the key elements of the experimental apparatus. The toroidal magnetic field will focus elastically scattered electrons onto a set of 8, rectangular quartz Čerenkov detectors coupled to photomultiplier tubes, which will be read out in current mode to achieve the high statistical precision required for the measurements. Inelastically scattered electrons are bent out of the detector acceptance by the spectrometer and make only a minimal contribution to the Čerenkov signal.

Basic parameters of the experiment are summarized in Table 2. The main technical challenges result from the small expected asymmetry of approximately -0.3 ppm; we will measure this asymmetry to $\pm1.8\%$ statistical and $\pm1.4\%$ systematic errors. The optimum kinematics corresponds to an incident beam energy of $E_0 = 1.165$ GeV, scattered electron polar angles $\theta_e = 8.4 \pm 3$ degrees, and azimuthal detector acceptance as large as possible (8 electron detectors with acceptance $\Delta\phi_e = 24$ degrees each, totalling 53% of 2π). Fixing $Q^2 = 0.03$ (GeV/c)^2 limits nucleon structure contributions which increase with Q^2, and avoids very small asymmetries where corrections from helicity correlated beam parameters begin to dominate the measurement uncertainty. With these constraints applied, the figure-of-merit is relatively insensitive to the primary beam energy; using a higher beam energy would result in a physically longer experiment with stronger magnetic field requirements, smaller scattering angles, and the possibility of opening new secondary production channels that might contribute to backgrounds.

Radiation hardness, background insensitivity, uniformity of response, and low intrinsic noise are criteria that are optimized by the choice of synthetic quartz Čerenkov bars (Spectrosil 2000, n = 1.47) for the main detectors. Extensive GEANT simulations were used to optimize the detector thickness, which necessitates a compromise between high photoelectron yield and low shower noise. We expect a tolerable 3.5% excess noise in current mode operation compared to counting statistics. Nonlinearity in the (detector + electronics) system is expected to be less than 10^{-4}; the integrating electronics custom designed

Table 2. Basic parameters of the Q^p_{weak} experiment.

Parameter	Value
Incident Beam Energy	1.165 GeV
Beam Polarization	85%
Beam Current	180 μA
Target Thickness	35 cm ($0.04X_0$)
Running Time	2200 hours
Nominal Scattering Angle	8.4°
Scattering Angle Acceptance	±3°
ϕ Acceptance	53% of 2π
Solid Angle	$\Delta\Omega = 45$ msr
Average Q^2	0.030 $(GeV/c)^2$
Average Physics Asymmetry	-0.288 ppm
Average Expt'l Asymmetry	-0.24 ppm
Integrated Cross Section	3.9 μb
Integrated Rate (all sectors)	6.4 GHz
Statistical Error on the Asymmetry	1.8%
Statistical Error on Q^p_W	2.9%

and built at TRIUMF will contribute negligibly to the excess noise, and will permit beam-off null asymmetry tests to be made in a few hours of acquisition time. The final quartz detectors will be 2 m long, 18 cm high and 1.25 cm thick along the path of the elastically scattered electrons. A shielding hut will protect the Čerenkov detectors from room backgrounds which are expected to be significant during the asymmetry measurement; the front wall of the shielding hut will have an aperture for each octant, and the apertures will be sufficiently large so as to not define the acceptance of the detectors.

It is essential to maximize the fraction of the detector signal arising from the electrons of interest, and to measure this fraction experimentally. Background levels in the bars due to photons are expected to be on the order of 1% of the signal from the elastic electrons. The asymmetry due to background must be corrected for, and we must measure both the detector-signal-weighted $< Q^2 >$ and $< Q^4 >$ – the latter in order to subtract the appropriate hadronic form factor contribution – in order to be able to extract a precise value for Q^p_w from the measured asymmetry. The Q^2 definition will be optimized by ensuring that the entrance aperture of the main collimator will define the acceptance for elastically scattered events.

The Q^2 acceptance of the system will be extracted from ancillary measurements at low beam current, in which the quartz Čerenkov detectors are read out in pulse mode and individual particles are tracked through the spectrometer system – thus, we will be able to measure the light yield in the Čerenkov detectors for a given Q^2 of each scattered electron. The Čerenkov detector front end electronics are designed to operate in both current mode and pulse mode for compatibility with both the parity measurements and the ancillary Q^2 calibration runs. The tracking system will be capable of mapping the Q^2 acceptance to ±0.7% in two opposing octants simultaneously; the tracking chambers will be mounted on a rotating wheel assembly so that the entire system can be mapped in 4 sequential measurements. A small quartz scanning detector, modelled after its SLAC E158 counterpart, will be positionable directly behind the main detector bars and used as part of the acceptance mapping and linearity testing at high and low currents.

The Q_{weak} tracking system consists of 3 sets of chambers. The upstream Region 1 chambers are based on GEM (Gas Electron Multiplier) technology for their fast response time and excellent position resolution. The Region 2 chambers at the entrance of the spectrometer are horizontal drift chambers; a small "mini-toroid" magnet will be installed downstream of the first collimator to sweep low energy Möller electrons out of the acceptance of the region 2 chambers during the low-current Q^2 calibration runs. The Region 3 chambers, a pair of vertical drift chambers (VDCs), will be located just upstream of the focal plane where the quartz Čerenkov detectors are mounted. The Region 3 chambers will be used to momentum-analyze the events, as well as to provide a complete characterization of the particle trajectories entering the Čerenkov detectors. Finally, trigger scintillators will be positioned between the Region 3 VDCs and the Čerenkov bars in order to provide a trigger to the electronics and a timing reference. The chambers plus trigger scintillator system will be retracted during normal Q_{weak} data taking at high current.

The combined beam current and target length requirements lead to a cooling requirement of approximately 2.5 kW, considerably above the present capacity of the JLab End Station Refrigerator (ESR). This will require us to draw additional refrigeration capacity from the central helium liquefier (CHL), providing a cost effective solution for the required target cooling power. Target density fluctuations must be minimized via careful design of the target flask and recirculation system; in addition, we plan to minimize their impact by taking advantage of a dramatic reduction in density fluctuations that occurs at higher frequency. Based on measurements of the noise frequency spectrum for solid and liquid targets in Hall A at JLab, we anticipate a suppression by 1-2 orders of magnitude by operating at a spin flip frequency above 150 Hz. The JLab source group has already implemented a 250 Hz spin reversal mechanism, and the Q_{weak} detector readout electronics have been designed accordingly.

The Q_{weak} apparatus also includes a luminosity monitor consisting of an array of quartz Čerenkov detectors located downstream of the Q_{weak} experiment at a very small scattering angle. The detectors will be instrumented with radiation hardened vacuum photodiodes with external current-to-voltage converters. The high rate (29 GHz/octant integrating mode) and the resulting small statistical error in the luminosity monitor signals will enable us to use this device for removing our sensitivity to target density fluctuations. In addition, the luminosity monitor will provide a valuable null asymmetry test, since it is expected to have a negligible physics asymmetry as compared to the main detector.

Experimental systematic errors are minimized by construction of a symmetric apparatus, optimization of the target design and shielding, utilization of feedback loops

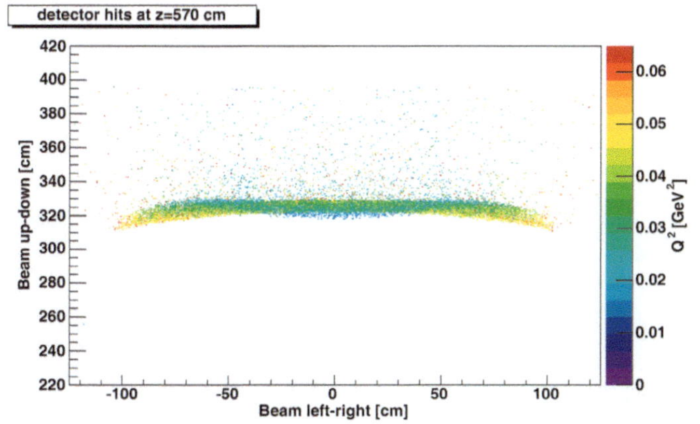

Fig. 3. Elastic event distribution by Q^2 incident on a Čerenkov detector at the focal plane of the Q_{weak} spectrometer. The detector bars are $200 \times 18 \times 1.25$ cm^3.

in the electron source to null out helicity correlated beam excursions, and careful attention to beam polarimetry. A program of ancillary measurements will be carried out to determine the system response to helicity correlated beam properties and background terms.

Systematic effects have been extensively modelled with the Q_{weak} GEANT package. Particular attention has been given to a systematic optimization of the collimator geometry so as to maximize figure of merit while obtaining as clean as possible an image of the elastic event distribution on the Čerenkov bars and an acceptably low inelastic rate. The toroidal magnetic field focuses elastic electrons in the θ direction and defocuses in ϕ, with the finite coil geometry resulting in an elastic event distribution that is contained within the quartz detector bar acceptance but is not perfectly rectangular. As seen in figure 3, the elastic event distribution 'droops' slightly at the edges of the detector bars, and the radiative tail of the electron distribution extends off the top of the Čerenkov bars. Sensitivities to beam motion, size, and angle of incidence on the target are enhanced by the tails of the elastic event distribution, and have been extensively modelled in Monte Carlo simulations. Coupled with realistic tolerances on the alignment and uniformity of the detector bars and magnet coils, these studies result in a set of helicity correlated beam property requirements summarized in Table 3.

The electron beam polarization must be measured with an absolute uncertainty at the 1% level. At present, this can be achieved in Hall C using an existing Möller polarimeter, which operates at beam currents below 8 μA – the current limit is due to heating and subsequent magnetization loss of the Fe target foils. An R&D program to push the operating range of the Möller polarimeter to higher beam currents has been underway for the past few years; a scheme involving a very thin target that is normally out of the beam and a fast kicker magnet to steer the beam onto the foil for brief periods has demonstrated good performance at average beam currents up to 40 μA. A major effort to design and build a Compton polarimeter

Table 3. Helicity correlated beam property requirements for the Q_{weak} experiment. The symbol x_0 refers to the d.c. beam position relative to the symmetry axis of the apparatus, δx refers to h.c. modulation x; r is the distance from the beam axis, and D is the beam diameter. These requirements should ensure that individual sources of systematic error produce false scattering asymmetries no greater than 6×10^{-9}.

H.C. Modulation	Error goes as	Requirement DC condition	H.C. limit
Position	$x_0 r^2 \delta x$	$x_0 < 3$ mm	$\delta x = 20$ nm
Size	$D_0^3 \delta D$	$D_0 = 4$ mm	$\delta D < 0.7$ μm
Direction	$\theta_0 \delta\theta$	$\theta_0 = 60$ μrad	$\delta\theta < 0.3$ μrad
Energy	δE	$E = 1.165$ GeV	$\frac{\delta E}{E} < 6 \times 10^{-9}$

in Hall C at Jefferson Lab is also underway as part of the laboratory's support of this and other experiments where precise beam polarimetry is an issue; the Compton polarimeter will provide a continuous on-line measurement of the beam polarization at full current (180 μA).

3 Conclusion and outlook

The Q_{weak} experiment is a major new initiative being prepared at Jefferson Laboratory to measure the proton's weak charge and hence test the Standard Model prediction of the running of $\sin^2\theta_W$ to high precision. Construction of the apparatus is currently underway, with plans to mount the experiment in Hall C during 2009. Extensive simulations of the experiment coupled with a vigorous R&D program and analysis of hadronic form factor contributions that are determined from previous experiments, confirm that our goal of a 4% measurement of the proton's weak charge at low momentum transfer, yielding a 0.3% measurement of $\sin^2\theta_W$, should be well within reach.

Work supported in part by the US DOE, NSF, NSERC (Canada), Jefferson Laboratory, and TRIUMF.

References

1. J. Erler, A. Kurylov & M.J. Ramsey-Musolf, Phys. Rev. D **68**, (2003) 016006.
2. S.C. Bennett & C.E. Wieman, Phys. Rev. Lett. **82**, (1999) 2484; C.S. Wood *et al.*, Science **275**, (1997) 1759.
3. G.P. Zeller & *et al.* (NuTeV collaboration), Phys. Rev. Lett. **88**, (2002) 091802.
4. T. Londergan, this conference.
5. R.D. Carlini *et al.*, The Q_{weak} Experiment, Proposal JLab E02-020, Thomas Jefferson National Accelerator Facility (2004), http://www.jlab.org/qweak/.
6. P. L. Anthony *et al.* [SLAC E158 Collaboration], Phys. Rev. Lett. **95**,(2005) 081601.
7. M. J. Musolf *et al.*, Phys. Rep. **239**, (1994) 1.
8. R.D. Young *et al.*, Phys. Rev. Lett. **97**, (2006) 102002.

Outlook for an improved measurement of parity violation in Moeller scattering at Jefferson Laboratory

The e2ePV experiment

D.J. Mack

TJNAF, Newport News, VA, 23606, USA

Abstract. Jefferson Laboratory has the potential to make a dramatically improved measurement of parity violation in Moeller scattering ($e + e \rightarrow e + e$). In the context of the Standard Model, the measurement would yield the best determination of $\sin^2 \theta_W$ at low energy ($\Delta \sin^2 \theta_W = \pm 0.00025$), and one of the best at any energy scale. As a new physics search via the running of the weak mixing angle, the experiment would have unparalleled sensitivity to new parity-violating $e-e$ interactions, probing electron substructure to 29 TeV (95% CL). In terms of specific models, pulls of 6σ are allowed in R-parity violating SUSY, about 5σ in E6 Z', and almost 3σ in R-parity conserving SUSY. The latter makes an improved Moeller measurement complementary to searches for SUSY loop-induced Electric Dipole Moments. Interpretability limits are well below the projected experimental error. A conceptual design for a 12 GeV JLab experiment is presented.

PACS. 11.30.Er Charge conjugation, parity, time reversal, and other discrete symmetries – 11.30.Pb Supersymmetry – 12.15.-y Electroweak interactions – 13.60.Fz Elastic and Compton scattering

1 Introduction

The SLAC E158 collaboration recently completed the first measurement of parity violation in Moeller scattering ($e+e \rightarrow e+e$) [1]. This pioneering experiment, summarized by K. Kumar at this conference, was not only an impressive technical achievement but is the clearest confirmation of the predicted running of the weak mixing angle, $\sin^2\theta_W$. It also sets the tightest limits on compositeness in the $e-e$ sector. However, the final error bar was too large to play a significant role in determining the global average of $\sin^2\theta_W$ or to provide significantly improved constraints on most models of new $e-e$ interactions. A greatly improved measurement could take place at the upgraded Jefferson Laboratory which would possess discovery-class sensitivity by measuring $\sin^2 \theta_W$ at low energy with an error on par with the best individual collider measurements.

2 Motivation for the new experiment

2.1 An important new $\sin^2\theta_W$ measurement

The Z^0 pole precision data for $\sin^2 \theta_W$ are shown in Figure 1 [2]. Since statistical weights are proportional to the inverse *square* of the error, the present world value of $\sin^2 \theta_W$ is largely determined by the two apparently inconsistent points with the smallest errors: one leptonic from SLD, the other semileptonic from LEP. The SLD experiment suggests a low value of the Higgs mass, while the LEP experiment suggests a much higher value of the Higgs mass. While there is only a small chance that the 3σ discrepancy is due to statistics, global analyses nevertheless average these data, yielding the small error bar seen in the figure. If one of the two measurements has a significantly underestimated systematic error, then the uncertainty on $\sin^2 \theta_W$ at the Z^0 pole is larger than this naive weighted average implies. This has implications for Higgs mass searches, as well as for new physics searches via the apparent running of $\sin^2 \theta_W$. Additional data of high precision are needed, and Fermilab is expected to make a significant contribution in the next few years [3].

Low energy experiments can also determine $\sin^2 \theta_W$. The first two entries in Table 1 are existing data from the weak charge of Cs and the electron. The remaining entries in the table are for hypothetical experiments. While the error from a potential nuclear elastic $0^+ \rightarrow 0^+$ measurement would be far too large to be useful, future precise measurements of the proton and the electron weak charge could make significant improvements in the low energy situation due to their proportionality to the suppressed quantity $1 - 4\sin^2 \theta_W$.

In both the electron and proton cases, the small asymmetry makes these measurements not only systematically challenging but statistically challenging. The case of the proposed 4% Jefferson Lab Q^p_W measurement, currently under construction and reviewed at this conference by S. Page, has been studied in great detail and appears to be

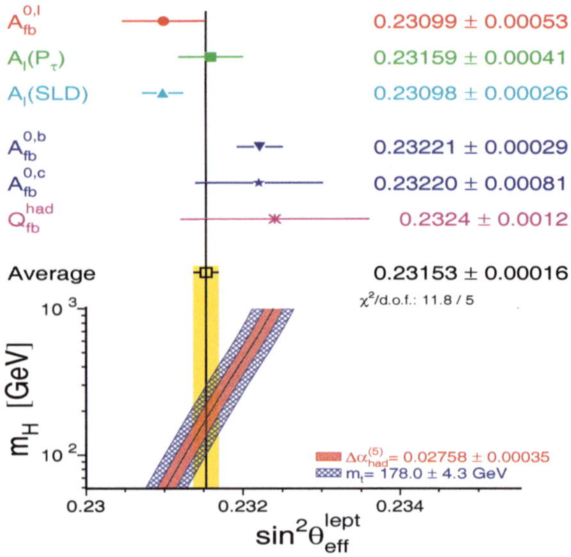

Fig. 1. The Z^0 pole values of $\sin^2\theta_W(eff)$ from purely leptonic and semi-leptonic reactions [2].

Table 1. Summary of existing and potential low energy $\sin^2\theta_W$ experiments. For a generic tree-level weak charge, $Q_W = A + B\sin^2\theta_W$, the magnification from the relative error on the weak charge to the relative error on $\sin^2\theta_W$ is $\frac{A+B\sin^2\theta_W}{B\sin^2\theta_W}$.

Experiment	$\frac{\Delta Q_W}{Q_W}$	Error Mag.	$\frac{\Delta\sin^2\theta_W}{\sin^2\theta_W}$	$\Delta\sin^2\theta_W$
data:				
Q_W^{Cs}[4]	0.5%	1.4	0.7%	0.0016
Q_W^e[1]	13.1%	0.041	0.5%	0.0013
future:				
Q_W^e	2.5%	0.041	0.1%	0.00025
Q_W^p[5]	4%	0.078	0.3%	0.00072
$Q_W(0^+ \to 0^+)$	1%	0.93	0.9%	0.0022

both feasible [5] and interpretable [6]. However, due to hadronic dilutions, and the fact that statistical and systematic errors are already well-matched, it is not possible to significantly reduce the error in Q_W^p. The situation for Q_W^e has not been examined in the same detail, but assuming similar systematic errors, a 2.5% measurement of the electron's weak charge appears possible. Due to differences in electroweak radiative corrections which cause the electron asymmetry to be even further suppressed than the proton asymmetry, the error magnification is also a factor of 2 smaller for the electron. Thus, the electron measurement can yield an error on $\sin^2\theta_W$ which 3 times smaller than the proton measurement, on par with the best individual measurements at the Z^0 pole.

To better understand how JLab is able to mount a competitive Moeller scattering experiment, it is helpful to examine the dependence of the statistical figure of merit on beam energy. Taking $\sigma A^2 It$ as our statistical figure of merit, where the cross section $\sigma \propto 1/E$, the analyzing power $A \propto E$, where E is the beam energy, I is the beam current, and t is the counting time, one finds the figure of merit is proportional to EIt or the total integrated beam power[1]. Although JLab and SLAC can both produce O(1) MWatt beams, SLAC invested only about 0.07 MWatt-Years in the E158 experiment, while a JLab Moeller experiment would require 0.5 MWatt-Years. Significantly smaller statistical errors than the 2.5% JLab measurement under discussion here would require either a new facility with multi-MWatt beams (*e.g.* the measurement K. Kumar has proposed for the exhaust beam at the NLC[7]) or much longer integration times. However, the fixed target Moeller scattering technique requires an absolute measurement of a PV asymmetry, and is ultimately limited by a systematic error floor.

Given the large facility investment that will be required to achieve errors on $\sin^2\theta_W$ of ± 0.00025, and the potential difficulties in reducing the systematic errors by more than a factor of several, it seems likely any further breakthroughs will eventually have to come from non-accelerator experiments. One technique for reducing uncertainties in atomic parity violation experiments is to cancel the correlated errors by measuring isotope ratios [8]. From a low-order expansion of the atomic weak charge, one can easily show the seemingly paradoxical result that isotope ratios experiments measure the *proton* weak charge [9]. The same expansion reveals the factor $\frac{<N>}{\Delta N} \simeq O(10)$ which magnifies any remaining uncorrelated errors. Fortunately, laser quanta are cheap, so high statistical precision is possible. There is also an effort at Jefferson Laboratory to measure the neutron radius of ^{208}Pb, reviewed at this conference by R. Michaels and J. Piekarewicz, to help calibrate relevant models. Finally, D. Budker at this conference reported that his group's decade-long program of preparatory measurements on Yb is complete, and that they are ready to begin searching for the predicted enhancement of parity violation in selected transitions of that atom.

2.2 Constraining new e-e interactions

Strangely enough, even though $\sin^2\theta_W$ is a free parameter in the Standard Model, a precision measurement at low energies can constrain new interactions beyond the Standard Model. In a neutral weak measurement, after the known Standard Model reaction- and hadronic structure-dependent electroweak box diagrams are regressed out, the value of $\sin^2\theta_W$ still "runs" from low to high energy scales due to the γZ mixing diagram. (See the solid curve in Figure 2 [13].) Because any parity violating interactions beyond the Standard Model must also contribute to the virtual particle spectrum of the vacuum, they may reveal themselves through a deviation from this expected running. Note that the theoretical uncertainty on the running

[1] For fixed beam power, if the energy drops too low, then the target refrigeration requirements become excessive. It is assumed here that we can provide the 5+ kWatts of target refrigeration needed at 12 GeV.

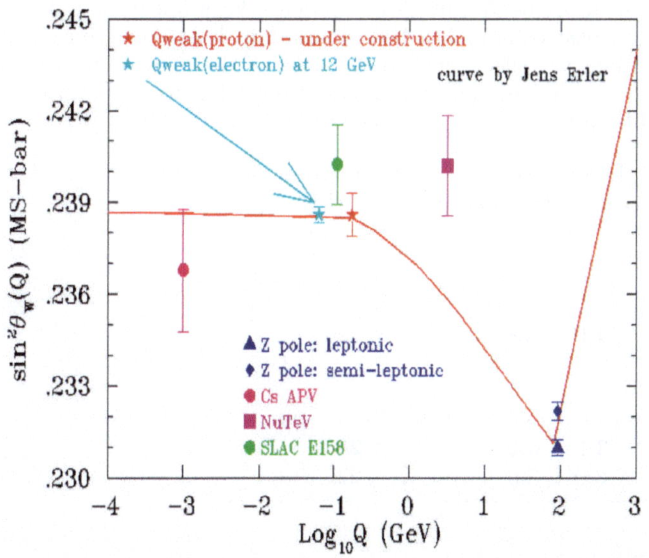

Fig. 2. The running of $\sin^2\theta_W$ curve from reference [13]. The normalization is defined by the experimental data at the Z^0 pole. The scale dependence is a property of the vacuum and is due to the γZ mixing diagrams. As in the case of the running of the electromagnetic fine structure constant α, at low energies the strength increases with increasing energy due to reduced screening. However, the strength decreases again once W^+W^- threshold is crossed.

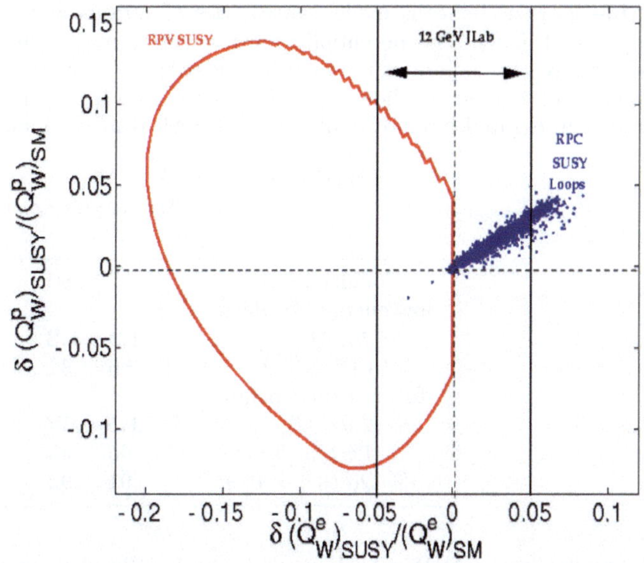

Fig. 3. The correlation between the allowed shifts of the electron and proton weak charges due to R-parity violating SUSY (red contour) and R-Parity conserving SUSY (ellipse of dots) [16]. The Standard Model corresponds to the origin at (0,0). If the 2.5% JLab Q^e_W measurement were to agree exactly with the Standard Model, the parameter space outside the vertical lines would be excluded at 95% confidence level.

has recently been reduced to only ±0.00007 [14], hence the interpretability is currently limited by the normalization of the curve at the Z^0 pole which is arguably as small as the ±0.00016 shown in Fig. 1.

Existing data and projected uncertainties are also shown in Figure 2. The Q_W^{Cs} and E158 results are both consistent with the predicted SM running. However, because the E158 result is so highly interpretable and so obviously inconsistent with the Z^0 pole value, it provides the single most convincing demonstration of the predicted running of $\sin^2\theta_W$. The NuTeV [15] result was reviewed by T. Londergan at this conference; since the extraction of $\sin^2\theta_W$ result is at best ambiguous in this case, one could argue that the NuTeV result probably does not belong on this plot. In contrast to the rapid resolution by the atomic theory community of problems in the interpretability of the Q_W^{Cs} experiment, it may take decades to clear the hadronic interpretability cloud which hangs over the NuTeV value for the weak mixing angle.

It is probably worth noting that each separate low energy reaction on Figure 2 is complementary in that it is sensitive to new physics in a different manner. Thus, the future JLab Q^p_W experiment is complementary to Q_W^{Cs} because they measure different isospin combinations of $e-q$ couplings. The Q^e_W experiments are complementary in that they alone are sensitive to new $e-e$ interactions.

Such complementarities are better reflected in specific models, a few of which are listed in Table 2. Although one has to be cautious about interpreting the value of an experiment in terms of a small subset of models, the importance of the Q_W^{Cs} experiment in constraining new $e-q$ interactions (particularly new $e-d$ quark interactions) is very impressive. This measurement needs to be confirmed, yet we were told by M. Lintz at this conference that the Paris effort, after a great deal of groundwork, is no longer funded. The E158 result is without parallel as a constraint on electron compositeness, but does not exceed collider limits for E6 Z' models. By the end of the decade, the Q^p_W experiment should nicely complement the existing Q_W^{Cs} result, yielding roughly balanced constraints on new $e-u$ and $e-d$ interactions.

In the contact interaction formalism of reference [6], because the sensitivity to the new physics mass-to-coupling ratio Λ/g is proportional to $1/\sqrt{\Delta Q_W}$, the E158 error bar would have to be reduced by a factor of 4 to bring $e-e$ constraints into line with anticipated $e-q$ constraints. In a mature program of measurements, this level of improvement in experimental technique could take a whole generation. Fortunately, these measurements are still in the pioneering stage, and we believe a JLab measurement with an error of 2.5% is possible. By the end of the JLab weak charge program, compositeness for electrons and the two lightest quarks would be probed to about 28 TeV (at 95% confidence level).[2]

Figure 3 shows the correlation between the allowed phase space for SUSY perturbations on Q^p_W and Q^e_W [16]. The electron weak charge has the potential to place very

[2] As for a potential 1% $Q_W(0^+ \rightarrow 0^+)$ measurement, it would nicely complement the Cs and proton measurements since it is sensitive to an intermediate isospin combination and has essentially no axial contributions.

Table 2. Miscellaneous model sensitivities of various weak charge measurements (all 95% confidence level lower mass limits in units of TeV). The potentially most sensitive measurement in a given column has been marked in bold. Sensitivities of low energy PV experiments are from reference [9] with some rescaling to reflect the latest values of the final or projected experimental errors. Collider limits for new Z's are from reference [10] while leptoquark limits are from reference [11]. With its present integrated luminosity of 2 fb^{-1}, FNAL Run II should be able to raise $e-q$ compositness limits to 6-10 TeV [12].

Experiment	Z' $M(Z_X)$	$M(Z_{LR})$	Leptoquarks $M_{LQ}(up)$	$M_{LQ}(down)$	Compositeness (LL) e-q	e-e
Collider Limits	.67	.80	"1.5"	"1.5"	2.5-3.7	2.2-2.4
low energy PV data:						
0.5% Q_W^{Cs}[4]	1.2	**1.3**	**4.0**	3.8	**28**	—
13.1% Q_W^e[1]	.66	.34	—	—	—	13
future experiments:						
2.5% Q_W^e	**1.5**	.77	—	—	—	**29**
4% Q_W^p[5]	.95	.45	3.1	**4.3**	**28**	—
1% $Q_W(0^+ \to 0^+)$	.91	.92	3.0	3.0	22	—

strong limits on RPV SUSY models, and pulls of 6σ are allowed. Effects due to RPC SUSY loops would be manifested more subtly, but pulls of almost 3σ are still allowed. The cancellation of SUSY loop pulls between u and d quarks in hadronic targets leads to small net pulls which may be only significantly addressable in the future by APV isotope ratios. Thus Q_W^e, along with $(g-2)_\mu$ and potentially APV isotope ratios, may be complementary to searches for Electric Dipole Moments which require CP violation.

The fact that precision electroweak measurements set higher limits than present day colliders will seem quaint in a few years when the LHC acquires the ability to materialize TeV-scale particles, yet low energy measurements will still remain important. If the LHC does not find anything in this region, precision measurements will continue to probe masses above LHC limits albeit in the model space of larger couplings. If the LHC does observe a heavy neutral particle, precision electroweak measurements may help to identify it.

Table 3. Nominal parameters for the e2ePV experiment.

Parameter	12 GeV JLab
E	12 GeV
E'	3-6 GeV
Beam polarization	85%
$< Q^2 >$	.008 $(GeV/c)^2$
θ_{cm}	90°-120°
θ_{lab}	0.53°-0.92°
$< A_{pv} >$	40 ppb
$< A_{expt} >$	34 ppb
σ	$64 \cdot 10^{-29}$ cm^2
Current	100 μA
Target Length	150 cm
Luminosity $cm^{-2}sec^{-1}$	$4.4 \cdot 10^{39}$
Time	4000 hours
Luminosity*Time	$6.4 \cdot 10^{46}$ cm^{-2}
Rate	282 GHz (35 GHz/octant)
Counting Statistics	23 ppm/pair at 300 Hz reversal
$\Delta A(stat) = \frac{1}{P}\frac{1}{\sqrt{N}}$	0.58 ppb
$\frac{\Delta A}{A}(stat)$	1.7%

3 Conceptual design

The 13% error on Q_W^e from E158 needs to be reduced by a factor of 4-5 to have a significant impact on the world $\sin^2\theta_W$ database or to raise $e-e$ compositeness limits to the same level as $e-q$ limits expected by the end of the decade. Table 3 lists parameters for a potential JLab Q_W^e experiment which would yield a counting statistical error of 1.7%. Assuming an approximate matching of statistical and systematic errors, and some allowance for excess noise in the detectors, a final error of about 2.5% on Q_W^e appears possible. A 4000 hour integration period corresponds to a year of JLab operations, assuming 32 weeks of operation per year and an overall accelerator and endstation equipment efficiency of 75%. Of course, a detailed examination will find efficiencies to be slightly lower, and additional time would be needed for installation and commissioning, so the experiment would realistically have to occupy one of JLab's high luminosity endstations for 2-3 years.

One concern is the target, which will consist of a 150 cm long liquid H_2 cell with over 5 kW of cooling power. Assuming the cooling power is made available, then density fluctuations will be managed by a combination of rapid beam polarization reversal and normalization via a low noise luminosity monitor. The Q_W^p experiment, which will employ a shorter target with a 2 kW heat load, already plans to use a faster $\simeq$ 300 Hz reversal to help "freeze" target density fluctuations. Another concern is false beam asymmetries since we have to make a precision measurement of an asymmetry of 40 parts per billion! This will require even tighter control of helicity-correlated changes in the polarized source and perhaps a new generation of beam diagnostics to measure changes in the beam spot size, etc. However, qualitatively similar problems are presented by the Q_W^p and ^{208}Pb radius experiments currently under construction (with somewhat larger asymmetries of roughly 250 ppb and 500 ppb, respectively). The success

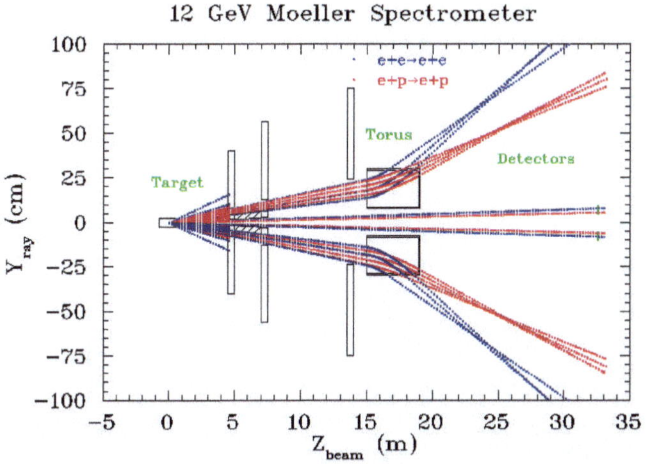

Fig. 4. Conceptual design for a toroidal, Moeller-focusing spectrometer. (Note the change in units between horizontal and vertical axes.) In addition to the Moeller focus at roughly 75 cm radius, there is a good quality $e+p$ elastic focus at about 55 cm.

of the ongoing 3rd generation of JLab parity violation experiments will make possible the 4th generation Moeller experiment.

My view of lessons from E158 is that it would be helpful to have a spectrometer which achieves a better focus for Moeller electrons to minimize backgrounds. The detector should also have finer binning in position to allow a more detailed comparison between simulation and experimental yields and asymmetries. Bending a primary electron beam after it has passed through a 17% radiation length target, as was done in the E158 spectrometer, will inevitably produce severe backgrounds and so should be avoided. The desire to keep a minimum of 2 bounces between the target and detectors is also not controversial, but this admittedly requires field integrals which are much easier to produce at 12 GeV than at 48 GeV.

The 12 GeV conceptual design consists of a toroidal spectrometer with 8 resistive coils and a focal plane instrumented with Position Sensitive Ion Chamber (PSIC) detectors. The $1/R$ field profile of a toroidal spectrometer is a natural way to produce the θ_{scatt}-dependent deflection needed to focus the Moeller electrons. Initial studies have already confirmed that a 2-bounce system with a cm-scale Moeller focus can be obtained over the large required momentum bite (at least on the mid-plane between adjacent coils as shown in Figure 4). Well-designed PSIC detectors with 4.5 radiation length pre-radiators should allow highly linear, low noise, and extremely radiation resistant operation with better than cm-scale resolution. However, much more work needs to be done. For example, azimuthal defocusing in the toroidal spectrometer must be brought under control, and trim coils probably need to be designed which would allow a robust tuning capability in the face of coil fabrication errors and small misalignments between the target and collimator and toroidal spectrometer. PSIC detectors must also be thoroughly tested for sensitivity to soft backgrounds [17].

We are in the process of assembling an international collaboration to tackle this measurement. Other conceptual designs are possible, and the collaboration will have to select one as a reference design. Once JLab calls for 12 GeV proposals for non-standard equipment, we will have about one year to write our Letter of Intent.

I would like to acknowledge Jens Erler and William Marciano for recent discussions which were crucial to refining the physics focus, the other members of the e2ePV working group for extensive input and encouragement, and the PAVI06 organizers and support staff for a tremendously stimulating and well organized conference on the beautiful island of Milos. The Southern Universities Research Association (SURA) operates the Thomas Jefferson National Accelerator Facility for the United States Department of Energy under contract DE-AC05-84ER40150.

References

1. P.L. Anthony *et al.*, Phys. Rev. Lett. **95**, (2005) 081601-1-5.
2. D. Abbanev *et al.*, Precision Electroweak Measurements on the Z Resonance, 7 Sept 2005, hep-ex/0509008, or http://www.cern.ch/LEPEWWG.
3. U. Baur *et al.*, hep-ph/0202001 v1 1Feb 2002.
4. S.C. Bennett and C.E. Wieman, Phys. Rev. Lett. **82**, 2484-7 (1999). But see V.M. Shabaev *et al.* Phys. Rev. A **72**, 062105 (2005) for a comprehensive overview of theoretical corrections which yield a combined theoretical and experimental error of 0.6% and only 1.1σ deviation from the Standard Model.
5. R. Carlini *et al.*, Qweak proposal, TJNAF E05-008.
6. J. Erler *et al.*, Phys. Rev. D **68** (2003) 016006.
7. K. Kumar, "Fixed Target Moeller Scattering at the NLC", Proceedings of the 1996 DPF/DPB Summer Study on New Directions in High Energy Physics (Snowmass 96). SLAC-R-732:C960625
8. D. DeMille, Phys. Rev. Lett. **74**, 4165 (1995).
9. M.J. Ramsey-Musolf, Phys. Rev. C **60**, 015501 (1999).
10. J. Erler and P. Langacker, "Electroweak Model and Constraints on New Physics", Phys. Lett. **B592**, (2004), 114, and hep-ph/0407097.
11. M. Kuze and Y. Sirois, Progress in Particle and Nuclear Physics **50** (2003) 1-62.
12. S. Jain *et al*, Phys. Rev. D **62** (2000) 095003, and see M. Titov hep-ex/0512006 v1 for a recent update.
13. J. Erler, private communication.
14. J. Erler and M.J. Ramsey-Musolf, Phys. Rev. D **72**, (2005) 073003.
15. G. Zeller *et al.*, Phys. Rev. Lett. **88** (2002) 091802.
16. The original figure from A. Kurylov *et al.*, Phys. Rev. D **68**, (2003) 035008 is updated here by Shufang Su (U. Arizona) to incorporate constraints from SLAC E158. Changes were small.
17. P. Souder, private communication.

Parity violation and the nature of charges

B. Desplanques[a]

LPSC, Université Joseph Fourier 1, CNRS/IN2P3, INPG, F-38026 Grenoble Cedex, France

Abstract. The origin of parity violation in physics is still unknown. At present it is introduced in the theory by requiring that the $SU(2)$ subgroup entering the description of interactions involves the left-handed components. In the present contribution, one elaborates on a suggestion made by Landau that particles and antiparticles could be like "stereo-isomeric" molecules, which would naturally provide parity violation. Particles and antiparticles could thus be combinations of the parity doublets associated with a chiral symmetry realized in the Wigner-Weyl mode. Consequences of such a description and the possible problems it could raise are examined.

PACS. 11.30.Er Charge conjugation, parity, time reversal and other discrete symmetries

1 Introduction

Thinking about the $\theta-\tau$ puzzle, Lee and Yang were led to suggest, 50 years ago, that the parity symmetry could be violated in weak interactions [1]. This was confirmed a few months later by Wu *et al.* who observed a preferential direction for the β emission in the decay of oriented ^{60}Co [2]. Since then, the knowledge of the weak interaction has considerably increased. Together with strong and electromagnetic interactions, it is embedded in the standard model, which is based on the $SU(3)_c \otimes SU(2)_L \otimes U(1)_Y$ gauge group with spontaneous symmetry breaking. Parity violation is ascribed to the appearance of the left-handed components associated with the $SU(2)$ weak isospin subgroup ($SU(2)_L$). It is sometimes considered that this property could be a low-energy one and that the standard model should be completed by another $SU(2)$ subgroup, involving this time the right-handed components ($SU(2)_R$) [3]. Implying further higher-mass gauge bosons, the parity symmetry could thus be restored in high-energy processes.

When trying to answer a question, it is not rare that one has better to enlarge the problem. Another question, which is also a fundamental one, is the nature of charges or, for our concern here, what makes a particle different from its antiparticle. At present, particles and antiparticles appear as solutions of the same equation, the Dirac one for instance, and they are essentially characterized by some numbers, reflecting their "charges".

Could it be that parity violation and the nature of charges are related to each other? A hint is provided by the approximate conservation of the PC symmetry, which suggests that the two operations P and C have a deep relationship. Which one however?

Some ideas along the above lines have been briefly proposed by Landau [4], soon after the suggestion of the parity violation by Lee and Yang. He considered that the difference between particles and antiparticles, as far as the space symmetry is concerned, "is no greater than that due to chemical stereo-isomerism". Their "charges" of interest here are especially the leptonic and baryonic ones.

In the present paper, we want to elaborate upon these ideas. In sect. 2, we make a couple of observations that relativize to some extent the origin of parity violation in physics. Section 3 is devoted to describing a framework accounting for particles and antiparticles. The fourth section is concerned with the origin of parity violation in this framework and the emergence of the charge conjugation as a parity operation on an internal structure with a chiral character. Some results from a toy model are presented in sect. 5. An outlook is given in sect. 6.

2 Some observations

When asked about the origin of parity violation in physics, an answer often given is that nature is essentially left-handed. This is built into the theory which relies on the gauge subgroup $SU(2)_L$. Had the theory been written in terms of the charge conjugate fields (ψ describes the creation of a particle and the destruction of the antiparticle, instead of the inverse), the gauge subgroup would be $SU(2)_R$. The left-handedness would have transformed into a right-handedness. The handedness thus depends on what we refer to as particles or antiparticles. Considering both particles and antiparticles on the same footing,

[a] e-mail: desplanq@lpsc.in2p3.fr

there would not be any preferred handedness. Actually, this symmetry is nothing but the PC symmetry which we assume to hold here. It suggests that the charge conjugation operation, C, could have something to do with the usual parity operation.

Another observation concerns the mathematical definition of a parity operation. As noticed by Lee and Wick [5], if P is a parity operation, the product of P and any unitary operator U that leaves the interaction invariant, is also a parity operation. This can be applied to the case where one takes for U the charge conjugation operator C, which leaves invariant the strong and electromagnetic interactions. Thus, the product PC could be considered as a parity operation. This definition is quite advantageous as the corresponding symmetry essentially holds for all interactions. Again, this feature suggests that charge conjugation could have something to do with the usual parity operation.

Following Landau in anticipation of the next sections, we introduce the product PC as a combined parity operation. We will denote it as $PC = \mathcal{P}$, to remind the reader that the operations P and C originate from a unique parity operation, acting however on different degrees of freedom.

3 Framework for particles and antiparticles

Models with chiral symmetry and parity symmetry have received a lot of attention in the strong-interaction domain (QCD). In the simplest case and quite generally, they are characterized by a conserved current, $J^A_\mu(x)$, and an anticommutation relation between the axial charge $Q^A = \int \mathrm{d}\boldsymbol{x}\, J^A_0(x)$, and the (genuine) parity operator, $\mathcal{P}$:

$$[\mathcal{P}, Q^A]_+ = \mathcal{P}\, Q^A + Q^A\, \mathcal{P} = 0. \tag{1}$$

A first realization of the symmetry, which stems from the above equation in the case $Q^A|\text{state} \neq 0> \neq 0$, assumes that particles appear in the form of doublets: two equal-mass states of opposite parity or, equivalently, two states with opposite chiral charge, transforming into each other by a parity operation (Wigner-Weyl mode). It supposes $Q^A|0> = 0$. A second realization (Nambu-Jona-Lasinio mode) supposes $Q^A|0> \neq 0$ and implies the existence of a Goldstone boson. In the absence of parity doublets, this is the realization that has been retained in the QCD case where the Goldstone boson is known to be the pion. While looking for parity doublets, it has been assumed that the charges of these particles should be the same. Little attention has been given to the fact that they could be different and have a chiral character. Thus, particles and antiparticles could be associated with the states with opposite chiral charges of an interaction exhibiting chiral symmetry or as combinations of the parity doublets it implies. It is evident that this chiral symmetry has nothing to do with the QCD one but its existence makes plausible the existence of another one in a different sector of the interaction.

4 Origin of parity violation and C operation

In atomic or nuclear physics, most of the observed parity-violating effects are ascribed to some parity admixture in the systems one is dealing with. The question arises whether it could also be so at a more fundamental level. In the present case, an important observation about parity conservation stems from eq. (1). The parity operation $\mathcal{P}$ and the chiral charge Q^A do not commute; one cannot therefore generally have eigenstates of the parity and the chiral charge simultaneously. As most experiments involve "charged" particles in the initial or final states, one should expect to see some apparent parity violation, though parity is conserved at the interaction level. The observed parity violation would reflect the fact that these particles are not, intrinsically, eigenstates of the parity.

While parity violation could be traced back to charges with a chiral nature, the problem would rather be to explain why it is so strong in some cases (100% in weak interactions) and absent in other cases (strong and electromagnetic interactions). An answer would require to consider some dynamics, perhaps unknown at the present time. A qualitative understanding can nevertheless be proposed. It supposes that the spin and momentum properties of the chiral substructure decouple from those of the particle as a whole. In this decoupling limit, it is conceivable that the genuine parity operation $\mathcal{P}$ factorizes into a part involving the external degrees of freedom (the usual P operation) and a part involving the internal degrees of freedom (the usual C operation). Of course, the decoupling cannot be complete and some parity violation reflecting the chiral nature of "charges" could show up. It is noticed that some experiments do not necessarily involve "charged" particles, like the neutral kaon decay into two mesons (the $\pi^+\pi^-$ state is globally considered here as a neutral one). Parity violation in this case could be observed too. Actually, this would be a true parity violation, which is known as a PC violation [6].

The above explanation ascribing parity violation to an intrinsic chiral structure goes beyond Landau's proposal, which was likely to be only suggestive. Relying on some underlying chiral symmetry like here makes it more systematic, as implied by observation and accounted for by the standard model.

Developments presented in this section have a rather general character. They do not, however, provide any realistic description of the physical world. Actually, without entering much into details, many questions may be raised. The first one concerns the nature of the chiral structure that could underly some charges and their quantization (possibly approximate). A second non-trivial question has to do with the conservation of the axial current that stems from the chiral symmetry we assumed, taking into account that particles have a mass. Other questions involve the dynamics. How does this one approximately decouple the spin and momentum properties of the underlying chiral structure from those of the particle as a whole? How do particles get mass when the above chiral symmetry could be realized by a pair of massless particles of opposite helicities (like for a Majorana neutrino). A further but different-type question concerns the equation that could play here the role of the Dirac equation. In the picture developed above, particles and antiparticles are somewhat on the same footing, a unique combined parity operation, $\mathcal{P}$ ($= PC$), transforming one into the other. In the context of the Dirac equation, antiparticles appear as holes in a Fermi sea filled by negative-energy states.

5 A toy model

In order to get some insight into the answers to questions raised in the previous section, but also to emphasize possible problems, we consider a particular model that is inspired by the Nambu-Jona-Lasinio model [7], but with an essential difference [8]. As we want to associate "charges" with some chiral structure, we change the role of the spin, whose projection along some direction was assumed to be conserved by elementary excitations in this model, with the helicity one.

In a mean-field approximation, it is conceivable that the resulting Lagrangian density reads:

$$\mathcal{L}(x) = \frac{1}{2}\left(i\,\bar{\psi}(x)\gamma^{\mu}\partial_{\mu}\psi(x) - m\,\bar{\psi}(x)\gamma\cdot\epsilon\,\gamma_5\psi(x)\right), \quad (2)$$

where ϵ^{μ} (with $\epsilon^2 = -1$), which represents the polarization carried by the underlying chiral structure, could play a role similar to the deformation in the case of intrinsically non-spherical nuclei. It is easily checked that the above Lagrangian exhibits some chiral symmetry and that the associated axial current and chiral charge (the helicity number in the present case) read

$$J^A_{\mu}(x) = \frac{1}{2}\,\bar{\psi}(x)\gamma_{\mu}\gamma_5\psi(x), \qquad Q^A = \int \mathrm{d}\boldsymbol{x}\, J^A_0(x). \quad (3)$$

In momentum space, it leads to the equation

$$(\gamma\cdot p - m\,\gamma\cdot\epsilon\gamma_5)\,\psi_{\epsilon}(p) = 0. \quad (4)$$

Solutions to this last equation are obtained under the condition $\epsilon\cdot p = 0$, which corresponds to a decoupling between the spin carried by the underlying chiral structure and the momentum of the system as a whole. They can be expressed in terms of the standard solutions of the Dirac equation:

$$\begin{aligned}\psi_{\epsilon}(p) &\propto (1+\gamma_5)\,(1-\gamma\cdot\epsilon)\,u(p)\,,\\ \psi_{\epsilon}(p) &\propto (1-\gamma_5)\,(1-\gamma\cdot\epsilon)\,v(p)\,. \end{aligned} \quad (5)$$

The appearance of the solutions of the Dirac equation for particles and antiparticles is important as it allows one to make some relationship with its achievements. The presence of the front factors, $(1\pm\gamma_5)$, is essential with respect to the present developments. It shows that parity violation could be built into the description of the particles themselves. The standard model of electro-weak interactions would thus be the effective one accounting for the above feature.

The toy model considered in this section casts also some light on the currents and the appearance of the charge conjugation as an internal parity operation. Considering matrix elements of the axial current with solutions for particles and antiparticles, eq. (5), and integrating over ϵ^{μ} together with the further assumption $\epsilon\cdot w = 0$ (decoupling of the spin carried by the chiral structure and the spin of the particle as a whole, w), one gets for a particle and its antiparticle:

$$\begin{aligned}\langle S=\tfrac{1}{2},\,p\,|J^A_{\mu}(0)|\,S=\tfrac{1}{2},\,p\rangle \;&=\; \bar{u}(p)\gamma_{\mu}u(p)\,,\\ \text{and} \;&\; -\bar{v}(p)\gamma_{\mu}v(p)\,. \end{aligned} \quad (6)$$

The apparent change in the parity between the l.h.s. and r.h.s. could be disturbing but it is noticed that the current at the r.h.s. changes sign under a charge operation so that the behavior of both sides under the combined parity operation, $\mathcal{P} = PC$, is the same. This is a consequence of a charge conjugation operation generated dynamically as a parity operation on an internal chiral substructure. On the other hand, the appearance of a vector current at the r.h.s. solves the major problem of the current conservation that an axial current would raise unavoidably.

6 Conclusion and outlook

We presented a scheme in which particles and antiparticles could be associated to the parity partners of a chiral symmetry realized in the Wigner-Weyl mode. This provides a natural explanation for parity violation in physics. We describe how charge conjugation, which is no more than a parity operation on a substructure with a chiral character, emerges in this picture. Some of the questions and problems that could be raised have been outlined.

The present sketch is far from a realistic theory and it is not clear whether one can deal with some constraints on combining space-time and internal symmetries [9]. We nevertheless believe that some of the ideas considered here could cast some new light in developing theories. As examples, we notice that there is no point to restore the usual parity symmetry, its violation being intrinsically tied to the chiral substructure underlying some "charges". There is no need for introducing right-handed currents in the simplest case (see remark on $SU(2)_R$), and the difference between V and A currents could vanish at some point. Despite their speculative character, we thought it was appropriate to revive these ideas in this year where one celebrates 50 years of parity violation.

References

1. T.D. Lee, C.N. Yang, Phys. Rev. **104**, 254 (1956).
2. C.S. Wu *et al.*, Phys. Rev. **105**, 1403 (1957).
3. R.N. Mohapatra, J.C. Pati, Phys. Rev. D **11**, 2558 (1975).
4. L. Landau, Nucl. Phys. **3**, 127 (1957).
5. T.D. Lee, G.C. Wick, Phys. Rev. **148**, 1385 (1966).
6. J.H. Christensen *et al.*, Phys. Rev. Lett. **13**, 138 (1964).
7. Y. Nambu, G. Jona-Lasinio, Phys. Rev. **122**, 345 (1961).
8. B. Desplanques, AIP Conf. Proc. **539**, 130 (2000).
9. S. Coleman, J. Mandula, Phys. Rev. **159**, 1251 (1967).

Parity-violating DIS at 12 GeV

Prospects for a high-luminosity solenoidal spectrometer

P.A. Souder

Syracuse University, Syracuse, NY 13152, USA

Abstract. The advent of the 12-GeV upgrade at JLab will present an excellent opportunity to study parity violation in deep inelastic scattering at high values of x. The physics issues in this domain include charge symmetry violation, quark-quark correlations in the nucleon, and tests of the Standard Model. This program will require a high-luminosity detector with high acceptance for scattering angles up to about 35°. A possible design for a solenoidal spectrometer that meets this requirement is suggested.

PACS. 24.80+y Nuclear tests of fundamental interactions and symmetries – 24.85+p Quarks, gluons, and QCD in nuclei and nuclear processes – 11.30Er Charge conjugation, parity, time reversal, and other discrete symmetries – 13.60Hb Total and inclusive cross sections (including deep-inelastic processes)

1 Introduction

The first experiment to observe a nonzero parity-violating asymmetry

$$A^{PV} = (\sigma_R - \sigma_L)/(\sigma_R + \sigma_L)$$

in the scattering of polarized electrons was published by Prescott, et al. in 1978 [1]. Recent advances in the technology of performing parity-violation experiments allow us to consider the possibility of repeating this work with a precision improved by a factor of more than ten. Such high precision data obtained over a wide kinematic range can serve as a new test of the Standard Model and also provide information on new features in hadronic structure including charge symmetry violation (CSV) and correlations between quarks. This paper describes both this new physics as well as the apparatus required to make the measurement.

2 Cross sections and asymmetries

In deep-inelastic scattering (DIS), the kinematic variables are $\nu = E - E'$, $Q^2 = 4EE' \sin^2[\theta/2]$, $x = Q^2/2M\nu$, and $y = \nu/E$. Here M is the nucleon mass. The cross section may be expressed as

$$\frac{d\sigma}{d\Omega dE'} = \frac{\alpha^2}{4E^2}\frac{\cos^2(\theta/2)}{\sin^4(\theta/2)}[W_2 + 2W_1 \tan^2(\theta/2)].$$

The structure functions W_1 and W_2 usually expressed in terms of $F_2(x,Q^2)$ and $R(x,Q^2)$ where

$$\nu W_2 = F_2(x,Q^2), \quad MW_1 = F_1(x,Q^2)$$

and

$$R(x,Q^2) = \frac{\sigma_L}{\sigma_T} = \left[1 + \frac{Q^2}{\nu^2}\right]\frac{F_2(x,Q^2)}{2xF_1(x,q^2)} - 1.$$

For DIS[2],

$$A^{PV} = Q^2\frac{G_F}{2\sqrt{2}\pi\alpha}\left[a(x) + f(y)b(x)\right], \tag{1}$$

where in the quark-parton model (QPM)

$$a(x) = \sum_i f_i(x)C_{1i}Q_i/\sum_i f_i(x)Q_i^2,$$

$$b(x) = \sum_i f_i(x)C_{2i}Q_i/\sum_i f_i(x)Q_i^2.$$

and

$$f(y) = \frac{1-(1-y)^2}{2-2y+y^2(1+R)}.$$

Here the parton distribution function (PDF) $f_i(x)$ is the probability of the struck parton with flavor i having a fraction x of the momentum of the nucleon. Also Q_i is the electromagnetic charge and $C_{1i}(C_{2i})$ are the weak vector (axial vector) charges for the ith flavor. With the assumption of charge symmetry, the expression simplifies for an isoscalar target such as deuterium;

$$a(x) = \frac{6}{5}\left[\left(C_{1u} - \frac{1}{2}C_{1d}\right) + \frac{2}{15}\frac{s(x)+\bar{s}(x)-c(x)-\bar{c}(x)}{D}\right]$$

$$b(x) = \frac{6}{5}\left[\left(C_{2u} - \frac{1}{2}C_{2d}\right)\frac{q(x)-\bar{q}(x)}{D} + \text{heavy quark terms}\right].$$

Here $D(u) = u(x) + d(x) + \overline{u}(x) + \overline{d}(x) + \frac{2}{5}[s(x) + \overline{s}(x)] + \frac{8}{5}[c(x) + \overline{c}(x)]$ and $q(x) = u(x) + d(x) + s(x) + c(x)$. The $a(x)$ term is largest since $C_{2u} - C_2d/2 \sim -0.36$ whereas $C_{2u} - C_{2d}/2 \sim 0.06$.

3 Theoretical issues

There are a number of theoretical issues relevant to parity violation in deep-inelastic scattering (PV DIS). Some of the issues represent opportunities to learn about either the fundamental interactions or alternatively to learn new features of nucleon structure. Other issues reflect uncertainties that will cloud the interpretation of the interesting physics.

3.1 Testing the Standard Model

As pointed out by Musolf,et al., [3] a new measurement of PV DIS can provide an important test of the Standard model. In addition to being sensitive to new particles such as leptoquarks, a measurement might tell us something about supersymmetry.

However, a presision of better than 1% on a measurement of PV DIS is required to achieve the above goal. Thus as a prerequisite, any hadronic physics that might introduce uncertainty in the calculation of the asymmetry must be understood at the sub-percent level. A related measurement, that of the Paschos-Wolfenstein ratio R^{PW} for neutrino scattering, has been published by the NuTeV collaboration[4] with a precision of 0.5%. Although the authors suggested that the result gave evidence for a violation of the Standard Model, issues of hadronic physics have clouded the interpretation. Any program in PV DIS must be designed to directly address any similar issues.

The Paschos-Wolfenstein ratio

$$R^{PW} = \frac{\sigma\langle \nu N \to \nu X\rangle - \sigma\langle \overline{\nu} N \to \overline{\nu} X\rangle}{\sigma\langle \nu N \to \mu X\rangle - \sigma\langle \overline{\nu} N \to \overline{\mu} X\rangle} \sim \frac{1}{2} - \sin^2\theta_W$$

has similar hadronic physics to $a(x)$, as will be discussed in the next section. In terms of the Standard Model, however, the physics is somewhat different from that of parity violation. In particular, R^{PW} is the ratio of a neutral weak currents to a *charged weak current*, whereas A^{PV} is the ratio of a neutral weak current to an *electromagnetic* current.

3.2 Charge symmetry violation

Charge symmetry is the key assumption that leads to the cancellation of the PDF's in the expressions for $a(x)$ and R^{PW}. In the literature, CSV has been discussed as one of the prime candidates for hadron physics that can explain the NuTeV anomaly[5]. Possible CSV effects are now being included in global fits[6]. If charge symmetry is correct, then

$$u^p = d^n; \quad d^p = u^n.$$

Table 1. Prediction of $R^{CSV} \equiv (\delta u - \delta d)/(u + v)$ versus x.

x	$R^{CSV}(\%)$
0.45	6
0.55	9
0.65	14
0.75	26

We can define CSV structure functions by

$$\delta u(x) = u^p(x) - d^n(x); \quad \delta d(x) = d^p(x) - u^n(x)$$

and the ratio

$$R^{CSV} \equiv \frac{\delta u - \delta d}{u + v}.$$

The effect of the CSV parameters on R^{PW} is:

$$\frac{\delta R^{PW}}{R^{PW}} \sim 0.85 R^{CSV}.$$

whereas for parity, we have

$$\frac{\delta A^{PV}}{A^{PV}} \sim= 0.28 R^{CSV}.$$

How large might R^{CSV} be? Table 1 gives an optimistic estimate of R^{CVS} as a function of x which is based on effects discussed in the literature[7]. This amount of CSV in Table 1 could explain the NuTeV result and also could readily be observed in a PV DIS experiment. A particularly interesting feature of this prediction, which may well be valid in general, is that the CSV effect increases at large values of x. This feature can provide an additional signature for CSV for experiments that can provide precise data as a function of x.

3.3 Beyond the QPM

What other hadronic physics might influence the asymmetry? This subject was addressed by Bjorken [8], Wolfenstein [9], and Derman [10] shortly after the data of Prescott, et al. were published. As given by Bjorken, the $a(x)$ term in Eq. 1 can be written in terms of the fundamental currents:

$$a(x) \propto \frac{l_{\mu\nu}\int\langle D|j^\mu(x)J^\nu(0) + J^\mu(x)j^\nu(0)|D\rangle e^{iq\cdot x}}{l_{\mu\nu}\int\langle D|j^\mu(x)j^\nu(0)|D\rangle e^{iq\cdot x}},$$

where $|D\rangle$ is the deuteron state, j^μ is the electromagnetic current, and J^μ is the weak vector current. The parton model has not been used. We can decompose the currents in terms of isospin using

$$V_\mu = (\overline{u}\gamma_\mu u - \overline{d}\gamma_\mu d); \quad S_\mu = (\overline{u}\gamma_\mu u + \overline{d}\gamma_\mu d)$$

and define

$$\langle VV\rangle = l_{\mu\nu}\int\langle D|V^\mu(x)V^\nu(0)|D\rangle e^{iq\cdot x}$$

with similar expressions for $\langle SS \rangle$ and $\langle SV \rangle$ Then we have

$$a(x) \propto \frac{(C_{1u} - C_{1d})\langle VV \rangle + \frac{1}{3}(C_{1u} + C_{1d})\langle SS \rangle}{\langle VV \rangle + \frac{1}{3}\langle SS \rangle}$$

The key here is that the $\langle SV \rangle$ term vanishes in the absence of CSV. Strange quarks have also been neglected [9]. If $\langle VV \rangle = \langle SS \rangle$, the hadronic structure completely cancels. The difference between $\langle VV \rangle$ and $\langle SS \rangle$ can be written

$$\langle VV \rangle - \langle SS \rangle = \langle (V - S)(V + S) \rangle \propto$$

$$l_{\mu\nu} \int \langle D|\overline{u}(0)\gamma^\mu u(x)\overline{d}(0)\gamma^\nu d(0)|D\rangle e^{iq\cdot x}.$$

The expression on the right is a correlation between up and down quarks in the nucleon [11]. This correlation vanishes in the parton model. In QCD, it appears as a higher twist term that causes a nonvanishing Q^2 dependence in A^{PV}. The observation of extra Q^2-dependence in PV DIS would give us a unique handle on diquarks in the nucleon. All other higher twist terms cancel in the ratio.

3.4 Higher twist in DIS

A remarkable feature of DIS behavior is that higher twist effects for data where the mass of the final state $W > 2$ GeV are found to be small. For example, the higher twist terms have been determined recently for the measured e-p DIS structure functions $F_2(x, Q^2)$ [12] after the DGLAP evolution is removed. The ansatz is

$$F_2(x, Q^2) = F_2(x)(1 + D(x)/Q^2)$$

It turns out that the values of the $D(x)$ depend upon how many orders of α_s is taken in the DGLAP evolution of the PDF's. At leading order (LO), the higher twist contributions are significant and similar to the results of older analyses. [13,14]. However, as higher orders are taken, NLO, NNLO, and NNNLO, $D(x)$ becomes quite small, especially for $x < 0.4$. The values of D_i for both LO and NNNLO are summarized in Table 2.

To interpret the size of higher twist terms at large values of x, one must take into account the relationship between W, Q^2 and x:

$$Q^2 = (W^2 - M^2)/(1/x - 1).$$

If $W = 2$ is taken as the threshold for DIS behavior, then there is a threshold Q^2 denoted Q_t^2. Values for Q_t^2 are also given in Table 2. The maximum size of the higher twist effect that can be measured is thus $D(x)/Q_t^2$, which is also given in Table 2 as a fraction of $F_2(x)$. This fraction is large enough to motivate a measurement only at high x.

We can include higher twist terms in $a(x)$ by defining

$$a(x, Q^2) = a(x)(1 + C(x)/Q^2).$$

Only the quark-quark correlation that contributes to $C(x)$, whereas many possible higher-twist operators might

Table 2. Higher twist coefficients $D(x)$ from Ref. [12].

x	$D(x)$ (LO)	$D(x)$ (N^3LO)	Q_t^2	$D(x)/Q_t^2$ (%) (LO)	$D(x)/Q_t^2$ (%) (N^3LO)
0.15	-0.07	0.001	0.5	-14	2
0.25	-0.11	0.000	1.0	-11	0
0.35	-0.06	-0.001	1.7	-3.5	-0.5
0.45	0.22	.11	2.6	8	4
0.55	0.85	0.39	3.8	22	10
0.65	2.6	1.4	5.8	45	24
0.75	7.3	4.4	9.4	78	47

contribute to $D(x)$. Hence it is plausible that $C(x) \leq D(x)$. Based on this assumption and Table 2, higher twist effects in $C(x)$ are probably impractical to isolate in PV DIS for $x < 0.4$. However, for $0.5 < x < 0.7$, it is possible that these effects could be observed cleanly. Moreover, since the effects of the DGLAP evolution cancel in the ratio $a(x)$, there is no problem with the order to which the evolution is performed.

3.5 The axial hadronic term

The other term in Eq. 1, which involves $b(x)$, is proportional to the axial hadronic current. Typically

$$\frac{b(x)}{a(x)} \sim 13\%$$

This small ratio is due to the fact that the weak couplings C_{2i} are much smaller that the C_{1i}, not due to any cancellation of the PDF's. Thus the asymmetries are relatively insensitive to uncertain hadronic effects in $b(x)$. For practical kinematics with a beam energy of 11 GeV, the kinematic factor $f(y)$ appearing in Eq. 1 is between about 0.3 and 0.8.

It is the $b(x)$ term that is important for the Standard Model test, since the C_{1i} will be known from the planned Q_{weak} experiment at JLab and also published and planned atomic physics experiments. Indeed, PV DIS might be the first experiment that cleanly demonstrates that the C_{2i} are indeed nonzero.

Unfortunately, the uncertainty due to hadronic physics in $b(x)$ is larger than in $a(x)$. The arguments in Sect. 3.3, based on the conserved vector current, do not apply. There are two possible approaches to evaluating the hadronic structure in $b(x)$:

1. Use neutrino data on the structure function xF_3 as suggested by Wolfenstein [9].
2. Use the parton model and data on $R = \sigma_L/\sigma_T$.

The quality of data on xF_3 for $x \sim 0.5$ is good[15], and should suffice for a Standard Model test. For larger x, the second method may be required. The data on R is precise enough[16], but higher twist effects may be relevant if they are very large.

3.6 Measuring d/u at high x

Hydrogen is another useful target. Since it is not isoscalar, the structure functions do not cancel in the expression for $a(x)$. In particular,

$$a(x) \approx \frac{3}{4}\left[\frac{6C_{1u}u(x) - 3C_{1d}d(x)}{u(x) + \frac{1}{4}d(x)}\right]$$

$$\sim \left[\frac{u(x) + 0.912d(x)}{u(x) + 0.25d(x)}\right]$$

and we see that $a(x)$ is sensitive to the ratio d/u. The determination of this for the proton at high x is a topic of considerable interest at large values of x [17–20]. The ratio is difficult to determine experimentally because data from the neutron is required and at large x complicated nuclear physics effects become important.

If parity violation with hydrogen is used, there is no nuclear structure and, moreover, charge symmetry is not involved. Apart from the issue of higher twist, the measurement has a clean theoretical interpretation. Precision measurements in the range of x from 0.6 and 0.7 would be of great interest.

4 Experimental issues

Any experimental program in PV DIS must be able to untangle the various effects discussed above so that any unexpected result can definitely be attributed to, for example, CSV, instead of a violation of the Standard Model. Given sufficient kinematic coverage, the untangling of the physics is possible as indicated by Table 3. The issue is whether an apparatus can be constructed that will provide sufficient statistics for $0.5 < x < 0.7$ with a factor of two range in Q^2.

4.1 Basic requirements for the apparatus

With a beam energy of 11 GeV, a scattering angle of about 35° and scattered electron energies between 2-5 GeV are required to fully cover the range $0.5 < x < 0.6$. Sufficient rate can be obtained with a 60 cm LH_2 target and a beam current of 90 μA only if the azimuthal acceptance is about 50%. The resolution in energy and angle must be on the order of 1-2% in order to cleanly define the kinematics, especially since the cross section varies rapidly with x for large values of x.

The spectrometers planned for Hall C at JLab have good acceptance for smaller angles and also will detect higher energy scattered electrons. However, the reach in the high x region of these facilities is insufficient for the above program. Thus a spectrometer must be designed with large solid angle that can operate at the required high luminosity.

A large magnet is an obvious choice, both to shield the detectors from copious line of sight photons and to

Table 3. Dependence of the physics issues on the experimental conditions.

	x	y	Q^2	p	D
Higher Twist	Yes?	No	Yes		X
CSV	Yes?	No	No		X
d/u	Yes	No	No	X	
New Physics	No	Yes	No		X

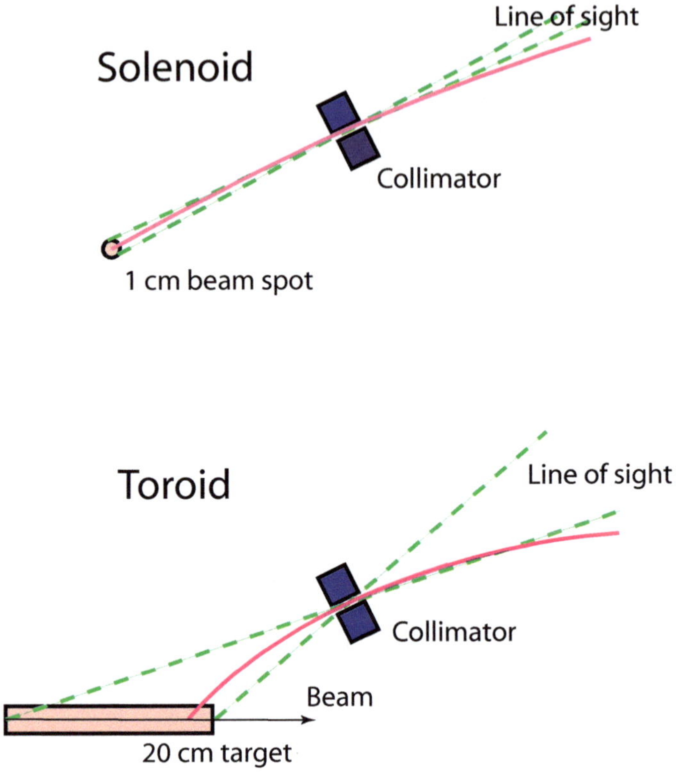

Fig. 1. Comparative geometries of the solenoidal and the toroidal spectrometers.

bend away low energy charged pions and Møller electrons. Two configurations have been studied, a toroid and a solenoid. A toroid usually provides more efficient use of the magnetic field and more bending. However for our target geometry, we have developed an efficient design using a solenoid. The idea behind a solenoid is illustrated in Fig. 1. The rastered beam spot at JLab is on the order of 1 cm, and if many collimators with small slits are used, the detectors may be shielded with little bending. For measuring momentum, the position of the beam is known within the raster pattern to about 0.1 mm, and again little bending is required to achieve excellent momentum resolution. With a toroid, the full length of the target must be shielded and the full target length enters into the uncertainty in the momentum measurement.

4.2 Possible design of a spectrometer

A sketch of a possible spectrometer design is given in Fig. 2. The scattered particles pass slits in carefully shaped

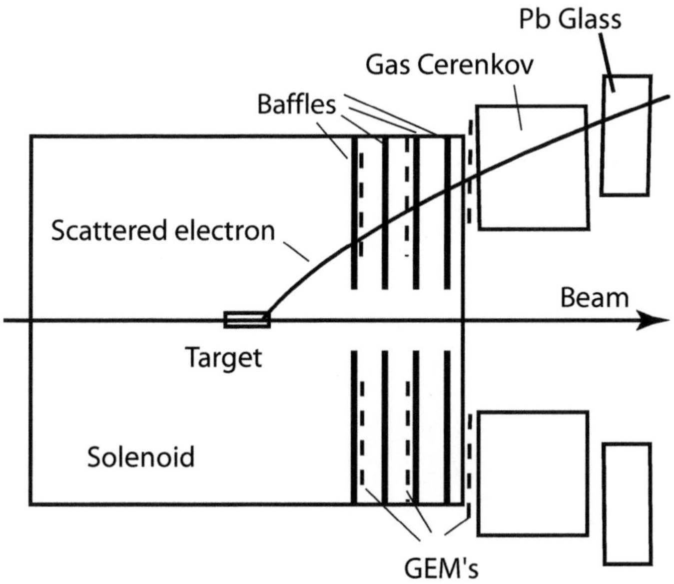

Fig. 2. Diagram of the solenoidal spectrometer

Table 4. Errors on data points that can be obtained in a 60-day runs at 8.8 and 11 GeV.

$\langle x \rangle$	$\langle Q^2 \rangle$	E_{beam}	$\delta A/A(\%)$
0.31	5.0	11	0.5
0.42	6.3	11	0.3
0.50	6.8	11	0.7
0.52	7.8	11	0.6
0.58	7.5	11	0.8
0.60	9.0	11	0.6
0.69	9.1	11	0.8
0.32	4.0	8.8	0.6
0.39	4.4	8.8	0.6
0.42	5.3	8.8	0.5
0.50	5.1	8.8	0.7
0.51	5.5	8.8	0.9
0.58	5.7	8.8	0.7
0.62	7.0	8.8	0.8

baffles, spiraling slightly in the solenoidal field. The baffles block the line of sight as well as low momentum particles. Cerenkov counters and lead glass total absorption counters are placed beyond the solenoid to serve as a trigger with modest rate and to provide for pion rejection. The tracking detectors will be subject to rather high rates. The recently developed GEM detector technology[22] has both the rate capability and the high spatial resolution required for this apparatus.

We have done a preliminary Monte Carlo study of a possible design. The projected statistical errors at the kinematic points within the acceptance are given in Table 4. In two 60-day runs at 11 GeV and 8.8 GeV, we obtain the points listed, which focus on the high Q^2 region. With a proton target, the statistical error in the asymmetry will be less that twice as large. These data can be augmented with either lower energy data or data from Hall C to complete the kinematic coverage.

5 Other applications of the spectrometer

In addition to PV DIS, the proposed spectrometer will be able to do numerous other experiments. One example is precision measurements of the asymmetry A_1^n that will serve as a major extension of work at JLab[23] at lower x and Q^2. A second example is seminclusive DIS.

I would like to thank E. Chudakov and K. Paschke for providing the results of a simulation of the solenoidal spectrometer. I would also like to thank D. Beck, P. Bosted, J. T. Londergan, K. Kumar, K. McFarland, F. Maas, Z. Meziani, W. Melnitchouk, M. J. Ramsey-Musolf, P. Reimer, A. Thomas, and X Zheng for useful discussions. This work is supported by the U. S. Department of Energy under Grant No. DE-FG020-84ER40146.

References

1. C. Y. Prescott, *et al., Phys. Lett.* B 77, (1978) 347; C. Y. Prescott, *et al., Phys. Lett.* B 84, (1979) 524.
2. A. Love, G. G. Ross and D. V. Nanopoulos, Nucl. Phys. B **49** (1972) 513.
3. A. Kurylov, M. J. Ramsey-Musolf and S. Su, Phys. Lett. B **582**, 222 (2004).
4. G. P. Zeller *et al.* [NuTeV Collaboration], Phys. Rev. Lett. **88**, (2002) 091802 [Erratum-ibid. **90**, (2003) 239902].
5. J. T. Londergan and A. W. Thomas, Phys. Lett. B **558** (2003) 132.
6. J. T. Londergan and A. W. Thomas, J. Phys. G **31** (2005) 1151.
7. M. Gluck, P. Jimenez-Delgado and E. Reya, Phys. Rev. Lett. **95** (2005) 022002.
8. J. D. Bjorken, *Phys. Rev.* D 18, (1978) 3239.
9. L. Wolfenstein, *Nucl. Phys.* B 146, (1978) 477.
10. E. Derman, *Phys. Rev.* D 19, (1979) 133.
11. S. Brodsky, in *"Proceedings from the JLab/Temple University HiX2000 Workshop,"*, 2000.
12. A. D. Martin, R. G. Roberts, W. J. Stirling and R. S. Thorne, Eur. Phys. J. C **35**, (2004) 325.
13. M. Virchenau *et al., Phys. Lett.* B 274, (1992) 221.
14. P. Amaudrez *et al., Nucl. Phys.* B 371, (1992) 3.
15. M. Tzanov *et al.* [NuTeV Collaboration], Phys. Rev. D **74** (2006) 012008.
16. U. K. Yang, A. Bodek and Q. Fan, arXiv:hep-ph/9806457.
17. W. Melnitchouk, I. R. Afnan, F. Bissey and A. W. Thomas, Phys. Rev. Lett. **84**, (2000) 5455.
18. W. Melnitchouk and A. W. Thomas, Phys. Lett. B **377**, (1996) 11.
19. S. I. Alekhin, Phys. Rev. D **63**, (2001) 094022.
20. S. Kuhlmann *et al., Phys. Lett.* B 476, (2000) 297.
21. M. J. Ramsey-Musolf, *Phys. Rev.* C 60, (1999) 015501.
22. B. Ketzer, Q. Weitzel, S. Paul, F. Sauli and L. Ropelewski, Nucl. Instrum. Meth. A **535** (2004) 314.
23. X. Zheng *et al.* [Jefferson Lab Hall A Collaboration], Phys. Rev. C **70** (2004) 065207.

Measuring $\sin^2\theta_W$ with parity violation in deep inelastic scattering with baseline spectrometers at JLab 12 GeV

P.E. Reimer

Physics Division, Argonne National Laboratory, Argonne, IL 60439, USA

Abstract. The couplings of leptons to quarks are fundamental parameters of the electroweak interaction. Within the framework of the Standard Model, these couplings can be related to $\sin^2\theta_W$. Parity violation (PV) in deep inelastic scattering (DIS) is proportional to these couplings and hence sensitive to $\sin^2\theta_W$. PV-DIS, first measured at SLAC in the mid-1970's, was used to establish the Standard Model. The high quality and intensity of the upgraded 11 GeV CEBAF beam at Jefferson Laboratory will make it an ideal tool for PV studies. In DIS the asymmetry from parity violation is large ($A_{PV} \approx 10^{-4}Q^2$), allowing precise measurements with modest beam time. This talk will explore a PV-DIS measurement which can be made using the baseline spectrometers that will exist as part of the 12 GeV JLab upgrade.

PACS. 12.60.-i Models beyond the Standard Model – 12.15.Mm Neutral currents – 12.15.-y Electroweak interactions – 11.30.Er Charge Conjugation, parity, time reversal and other discrete symmetries

1 The Standard Model and parity violation in deep inelastic scattering

The electroweak Standard Model has been remarkably successful in describing the unification of the electromagnetic and weak interaction at the level of a few parts per thousand. Even with this success, however, there are experimental hints, such as the *three standard deviation* difference between purely leptonic and semi-leptonic measurement of $\sin^2\theta_W$ [1,2], that the Standard Model may be part of a larger framework. Precision, low-energy experiments are an important element in the effort to elucidate this larger framework [3]. The couplings between leptons and quarks may be sensitive to these extensions [4]. Within the Standard Model, these couplings are vary linearly with $\sin^2\theta_W$. If their measured values deviate from Standard Model predictions, this will be a clear indication of processes not included in the Standard Model. The pattern of these deviations will help to clarify the larger framework governing the Standard Model.

Recent advances in the quality of polarized electron beams have made the measurement of PV asymmetries in electron scattering an important probe of the weak force and the structure of the nucleon. Recently, the HAPPEX II [5] and G0 [6] experiments at Thomas Jefferson National Accelerator Facility (JLab) have used this technique to explore the effects of intrinsic strangeness in the proton. In this article a measurement of parity violation in deep inelastic scattering (DIS) is outlined that will greatly improve the knowledge of the vector-electron × axial-quark couplings. This experiment can be realized at JLab in Hall C with the baseline 12 GeV equipment [7].

The asymmetry in polarized electron scattering on unpolarized deuterium in DIS kinematics, $A^d_{\rm DIS}$, was first measured by Prescott *et al.* at SLAC [8,9] and served to the establish what is now known as *the* Standard Model of the electroweak interaction. This asymmetry can be expressed in terms of quark distribution functions of the target, and the couplings C_{1q} (axial electron × vector quark) and C_{2q} (vector electron × axial quark) which, within the context of the Standard Model, depend on $\sin^2\theta_W$ [10, 11]. The C_{iq} coupling constants are, assuming the context of the Standard Model,

$$C_{1u} = g^e_A g^u_V = -\frac{1}{2} + \frac{4}{3}\sin^2\theta_W, \tag{1}$$

$$C_{1d} = g^e_A g^d_V = \frac{1}{2} - \frac{2}{3}\sin^2\theta_W, \tag{2}$$

$$C_{2u} = g^e_V g^u_A = -\frac{1}{2} + 2\sin^2\theta_W \text{ and} \tag{3}$$

$$C_{2d} = g^e_V g^d_A = \frac{1}{2} - 2\sin^2\theta_W. \tag{4}$$

Ignoring the intrinsic charmed quark contribution, on an isoscalar target such as deuterium, this asymmetry, expressed in terms of these couplings, is

$$A^d_{\rm DIS} = -\left(\frac{3G_F Q^2}{2\sqrt{2}\pi\alpha}\right) \times \frac{(2C_{1u} - C_{1d})\,[1 + R_s(x)] + Y\,(2C_{2u} - C_{2d})\,R_v(x)}{5 + R_v(x)}, \tag{5}$$

where the kinematical factor Y is a function of the incident and scattered electron energy and $R_v(x)$ and $R_s(x)$ depend solely on the quark distributions.

In the near future, the C_{1q} coefficients will be well determined by the JLab QWeak [12] experiment combined with existing measurements of atomic parity violation in cesium [13]. The C_{2q} coefficients, however, are poorly known and a measurement of parity non-conservation in DIS appears to be the best way to determine them. A new measurement of the C_{2q} coupling constants will be done by PV-DIS (JLab E05-007) [14,15] with limited statistical precision (Phase I approved) using a 6 GeV beam which restricts the available DIS kinematics. At lower Q^2, the interpretation of $A^d_{\rm DIS}$ in terms of C_{iq} may be clouded by uncertainties from higher-twist contributions.

2 The 12 GeV baseline DIS-parity experiment

The experiment will use the Hall C HMS and SHMS spectrometers to measure $A^d_{\rm DIS}$ for an overall statistical precision of $\delta A^d_{\rm DIS}/A^d_{\rm DIS} = 0.5\%$. This level of sensitivity can be reached with approximately 33 days of beam time using an 11 GeV, 85 μA beam with 80% polarization on a 40 cm long liquid deuterium target.

The dependence on the individual parton distribution of $A^d_{\rm DIS}$ shown in Eq. 5 through $R_v(x)$ and $R_s(x)$ emphasizes that this measurement must be made in DIS kinematics where the scattering takes place from the individual quarks. Fortunately, the 12 GeV JLab upgrade will open up a much larger kinematic space for DIS experiments. To optimize the sensitivity to the electron-quark couplings and to minimize possible hadronic effects, including higher-twist effects, the spectrometer settings were chosen to keep both Q^2 and W^2 as large as possible, while at the same time keeping the DIS rate sufficiently high. The optimized spectrometer settings and kinematics are given in Tab. 1. At these kinematics $A^d_{\rm DIS} \approx 280\times10^{-6}$, which is relatively large when compared to other completed (*e.g.* HAPPEX II, $A \approx 1\times10^{-6}$ [5]) and proposed (*e.g.* PREX, $A \approx 0.5\times10^{-6}$ [16]) experiments.

Because of this relatively large asymmetry, many of the systematic effects, especially those related to beam-induced false asymmetries, may be easily controlled to the required levels simply by using now standard techniques. A list of the main sources of systematic uncertainty is given in Tab. 2. The dominant systematic uncertainty in the measurement of $A^d_{\rm DIS}$ and the dominant technical challenge for this experiment is the measurement of the beam polarization. Here a 0.5% measurement was *assumed.* The planned JLab Hall C Compton polarimeter is expected to have an uncertainty of < 1%. If this experiment is approved, a significant effort will need to be devoted to improving this uncertainty. The second largest systematic uncertainty is the determination of Q^2, which is a combination of measurements of the incoming and scattered electron's momenta and the spectrometer angle. Based on existing measurements, an uncertainty of $\delta Q^2/Q^2 = 3.9\times10^{-3}$ should be achievable.

Table 1. This table lists the spectrometer settings, kinematic variables (and their ranges) as well as expected rates for both spectrometers [17,18]. The combined statistical uncertainty on A_d is 0.05%.

	HMS		SHMS	
	Average	Range	Average	Range
Cent. Angle	13.5°	-	13.5°	-
Momentum (GeV)	6.0	5.4-6.6 $\pm10\%$	5.8	4.9-6.7 $\pm15\%^a$
$\delta\Omega$ (msr)	-	6.8	-	3.8^b
Q^2 (GeV^2)	3.3	2.6-4.0^c	3.2	2.6-3.8
W^2 (GeV^2)	7.1	6.1-8.1	7.6	6.3-8.8
x	0.35	0.27-0.43	0.33	0.25-0.41
π/e ratio^d	0.3	0.1-0.6	0.45	0.1-1.0
$A_d\times10^{6e}$	285	220-350	280	210-340
$\delta A_d/A_d(\%)^f$	0.66		0.75	

^{a}While the SHMS momentum acceptance is nominally quoted as -15%/+20%, the high end of this acceptance is not complete. Since details of the high momentum acceptance are not final, a cut off at +15% has been used for the purpose of this rate estimate.
^{b}The spectrometer is still being optimized and it is possible that this will increase to > 4.5 msr [18].
^{c}The range for these kinematic variables describes the coverage of the RMS width.
^{d}The range of π/e depends on the difference between the π or e momentum and the central spectrometer momentum.
^{e}The range gives the variation in the asymmetry over each spectrometers acceptance.
^{f}Assuming 567 hrs of beam time.

The interpretation of $A^d_{\rm DIS}$ in terms of the C_{iq} couplings given by Eq. 5 is based on an assumption of quark scattering. With very high Q^2 and W^2, this assumption should be valid. As Q^2 decreases, however, higher-twist effects may also contribute to the observed asymmetry. The small amount of theoretical work which has been completed has shown that the effects of higher twist on $A^d_{\rm DIS}$ are small, but *possibly* not negligible [19,20]. At the same time, there is currently no experimental information on the contribution of higher twist to PV observables. In the near future, the 6 GeV PV-DIS experiment (JLab E05-007) [14,15] will look for these effects in the Q^2 de-

Table 2. This table lists the systematic uncertainties in the measurement of A_d. The contributions to the uncertainty of the measured asymmetry in terms of $\sin^2\theta_W$ are also shown.

Source	$\frac{\delta A_d}{A_d}$	$\frac{\delta\sin^2\theta_W}{\sin^2\theta_W}$
Polarization measurement	5.0×10^{-3}	2.5×10^{-3}
Determination of Q^2	3.9×10^{-3}	2.0×10^{-3}
Target Endcaps	0.5×10^{-3}	0.28×10^{-3}
Target Purity	0.4×10^{-3}	0.22×10^{-3}
Re-scatter background	0.2×10^{-3}	0.11×10^{-3}
π^- contamination	0.01×10^{-3}	0.006×10^{-3}
Total	6.4×10^{-3}	3.2×10^{-3}

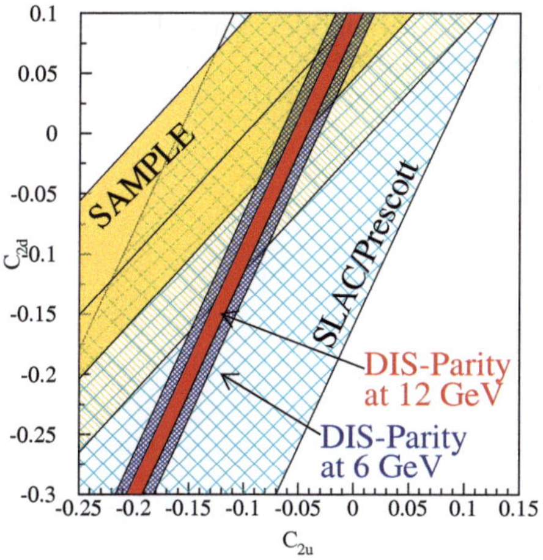

Fig. 1. This figure illustrates the current and proposed measurements of the C_{2u} and C_{2d} coupling constants, including data from SLAC [8,9] and the SAMPLE experiment [22]. The 6 GeV PV-DIS band assumes that both phases of the experiment will take place [14].

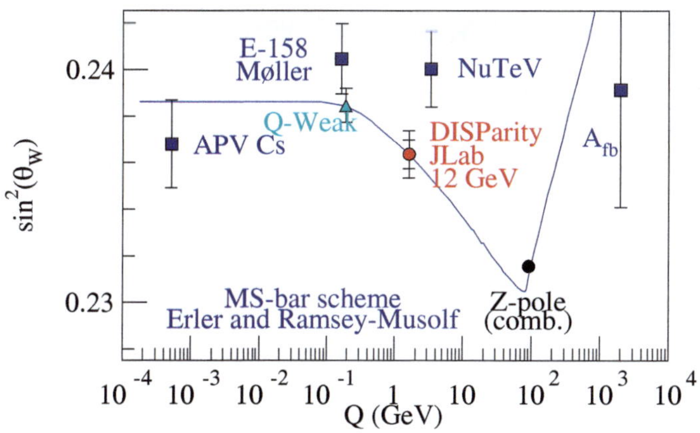

Fig. 2. This figure illustrates the running of $\sin^2\theta_W$ as a function of Q^2 [23,24] according to the Standard Model, together with the existing [13,24–27] and proposed measurements [12].

pendence of $A^d_{\rm DIS}$. Clearly, if a significant deviation from Standard Model-based predictions is measured, a systematic investigation of the dependence $A^d_{\rm DIS}$ on Q^2 and x will be necessary. This investigation could be part of a larger physics program based on a large solid-angle spectrometer under consideration [21].

3 Conclusions

$A^d_{\rm DIS}$ is sensitive to the poorly-measured C_{2q} couplings, and, within the context of the Standard Model, $\sin^2\theta_W$. A measurement of $A^d_{\rm DIS}$ has been outlined here using the JLab Hall C baseline 12 GeV upgrade equipment. Systematic effects are dominated by the measurement of the beam polarization, and this represents the major technical challenge of this experiment. The interpretation of this measurement of $A^d_{\rm DIS}$ in terms of C_{2q} depends on a theoretical and experimental understanding of the effects of higher twist on the measured asymmetry. While these effects are believed to be small, they have yet to be measured. A statistical precision of 0.5% can be quickly achieved for this measurement, because of the relatively large asymmetry.

The author would like to thank his co-spokespersons on the JLab 12 GeV DIS-Parity letter-of-intent, X. Zheng and K. Paschke. This work was in part supported by the U.S. Department of Energy, Office of Nuclear Physics, under Contract No. W-31-109-ENG-38.

References

1. M.S. Chanowitz, Phys. Rev. Lett. **87**, 231802 (2001), hep-ph/0104024
2. M.S. Chanowitz, Phys. Rev. **D66**, 073002 (2002), hep-ph/0207123
3. M.J. Ramsey-Musolf, Phys. Rev. **C60**, 015501 (1999), hep-ph/9903264
4. P. Langacker, M. Luo, A.K. Mann, Rev. Mod. Phys. **64**, 87 (1992)
5. K.A. Aniol et al. (HAPPEX), Phys. Rev. Lett. **96**, 022003 (2006), nucl-ex/0506010
6. D.S. Armstrong et al. (G0), Phys. Rev. Lett. **95**, 092001 (2005), nucl-ex/0506021
7. P. Reimer, K. Paschke, X. Zheng et al., *Precision measurement of the parity-violating asymmetry in deep inelstic scattering off deuterium using baseline 12 GeV equipment in Hall C* (2006), JLab LOI-12-06-111
8. C.Y. Prescott et al., Phys. Lett. **B77**, 347 (1978)
9. C.Y. Prescott et al., Phys. Lett. **B84**, 524 (1979)
10. R.N. Cahn, F.J. Gilman, Phys. Rev. **D17**, 1313 (1978)
11. J.D. Bjorken, Phys. Rev. **D18**, 3239 (1978)
12. R. Carlini, J.M. Finn, S. Kowalski, S. Page et al., *The Q_{weak} experiment: A search for new physics at the tev scale via a measurement of the proton's weak charge* (2001), JLab E02-020
13. S.C. Bennett, C.E. Wieman, Phys. Rev. Lett. **82**, 2484 (1999), hep-ex/9903022
14. X. Zheng, P. Reimer, R. Michaels et al., **e** $-^2H$ *parity violating deep inelastic scattering at CEBAF 6 GeV* (2004), JLab E05-007
15. X. Zheng, *Parity Violation in Deep Inelastic Scattering at JLab 6 GeV*, in *PAVI '06: From Parity Violation to Hadronic Structure and more... (Part III)* (Springer, Milos, Greece, 2006)
16. R. Michaels et al., *A clean measurement of the neutron skin of ^{208}Pb through parity violating electron scattering* (2000), JLab E00-003
17. J. Arrington et al. (Hall C), *Conceptual design report Hall C 12 GeV upgrade* (2002)
18. A. Bruell, P. Brindza, R. Ent, D. Potterveld, C. Yan (2006), Private communication
19. P. Castorina, P.J. Mulders, Phys. Rev. **D31**, 2760 (1985)
20. S. Fajfer, R.J. Oakes, Phys. Rev. **D30**, 1585 (1984)
21. P. Souder, *DIS-Parity at 12 GeV*, in *PAVI '06: From Parity Violation to Hadronic Structure and more... (Part III)* (Springer, Milos, Greece, 2006)
22. E.J. Beise, M.L. Pitt, D.T. Spayde, Prog. Part. Nucl. Phys. **54**, 289 (2005), nucl-ex/0412054

23. J. Erler, M.J. Ramsey-Musolf, Phys. Rev. **D72**, 073003 (2005), hep-ph/0409169
24. J. Erler (2006), Private communication
25. P.L. Anthony et al. (SLAC E158), Phys. Rev. Lett. **95**, 081601 (2005), hep-ex/0504049
26. G.P. Zeller et al. (NuTeV), Phys. Rev. Lett. **88**, 091802 (2002), hep-ex/0110059
27. W.M. Yao et al. (Particle Data Group), J. Phys. **G33**, 1 (2006)

Parity violation in deep inelastic scattering at JLab 6 GeV

X. Zheng[a]

For the JLab E05-007 Collaboration
Lab for Nuclear Science, Massachusetts Institute of Technology, Cambridge, MA 02139, USA

Abstract. The parity-violating asymmetry in $\mathbf{e}-{}^2$H deep inelastic scattering (DIS) can be used to extract the weak neutral-current coupling constants C_{2q}. A measurement of this asymmetry at two Q^2 values is planned at Jefferson Lab. Results from this experiment will provide a value of $2C_{2u} - C_{2d}$ to a precision of ± 0.03, a factor of eight improvement over our current knowledge. If all hadronic effects can be understood, this result will provide information on possible extensions of the Standard Model, complementary to other experiments dedicated to new physics searches. Presented here are the physics motivation, experimental setup, potential hadronic effects and their implications, and the future of PV DIS at Jefferson Lab.

PACS. 13.60.-r Photon and charged-lepton interactions with hadrons – 12.15.-y Electroweak interactions – 12.15.Mm Neutral currents – 12.60.-i Models beyond the standard model

1 Introduction

A measurement of the parity-violating (PV) asymmetry of $\mathbf{e}-{}^2$H deep inelastic scattering (DIS) is planned [1] at the Thomas Jefferson National Accelerator Facility (JLab) in Virginia, USA. The measurement will be performed at two Q^2 values of 1.1 and 1.9 GeV^2 and a fixed $x = 0.3$. The high Q^2 measurement will be used to extract the effective weak coupling constants $(2C_{2u} - C_{2d})$. The low Q^2 measurement may provide the first observation of the hadronic higher-twist contribution to the PV DIS asymmetry, which at JLab energies is the most likely source of a large deviation from the Standard Model prediction. Using the Standard Model value of $2C_{1u} - C_{1d}$ which will be tested by combining the results from Cs atomic parity violation (APV) and the future Qweak experiment, the expected uncertainty on $(2C_{2u} - C_{2d})$ is ± 0.03. This result will improve the current knowledge on this quantity by a factor of eight. It will help to extract the couplings C_{3q} from high-energy data, and might reveal possible physics beyond the Standard Model.

Results from this experiment (E05-007) will provide an important guidance for the future DIS-parity program, of which the ultimate goals are two-fold: by choosing a kinematics where hadronic effects are negligible, we can extract $\sin^2\theta_W$ and study possible extensions of the Standard Model to a high precision. By choosing kinematics where hadronic effects are expected to be large, various interesting phenomena can be studied, such as higher-twist effects, charge-symmetry violation (CSV), and the parton distribution function ratio d/u. These results may also have implications for the interpretation of existing neutrino scattering data.

2 Parity violation in deep inelastic scattering off a deuterium target

Historically, PV DIS was one of the first tests of the Standard Model and an early measurement of the PV DIS asymmetry off a deuterium target [2,3] served to establish the value of $\sin^2\theta_W$ at $\sin^2\theta_W \approx 1/4$. Since this ground-breaking experiment, parity violation has become an important tool not only for probing the Standard Model [4–6] but also for probing the structure of the nucleon [7–9].

2.1 PV DIS asymmetry

In electron scattering, the weak neutral current can be accessed by measuring a parity-violating asymmetry caused by the interference term between weak and electromagnetic scattering amplitudes [10]. The scattering amplitude, $\mathcal{M}$, for the scattering process is a product of the current for the electron with the photon or the Z^0 propagator and the hadron current: $\mathcal{M}_\gamma = j_\mu(1/q^2)J^\mu$ and $\mathcal{M}_Z = j_\mu(1/M_Z^2)J^\mu$. The cross section for scattering right- and left-handed electrons off an unpolarized target is proportional to the square of the total amplitudes: $\sigma^{l,r} \propto (\mathcal{M}_\gamma + \mathcal{M}_Z^{l,r})^2$, where only a longitudinally polarized electron beam was considered and $\mathcal{M}_Z^r$ and $\mathcal{M}_Z^l$ represent the incident right- and left-handed electrons,

[a] *Present address:* Department of Physics, University of Virginia, Charlottesville, VA 22904, USA.

respectively. The parity-violating asymmetry may be expressed as [10]

$$A_{LR} \equiv \frac{\sigma^r - \sigma^l}{\sigma^r + \sigma^l} \approx \frac{\mathcal{M}_Z^r - \mathcal{M}_Z^l}{\mathcal{M}_\gamma}. \quad (1)$$

Thus, measuring the parity-violating asymmetry gives access to the weak neutral current in a ratio of amplitudes rather than the square of this ratio, greatly enhancing its relative contribution. The size of the asymmetry can be *estimated* based on the ratio of the propagators:

$$A_{LR} \approx \frac{Q^2}{M_Z^2} \approx 120 \text{ ppm at } \langle Q^2 \rangle = 1 \text{ GeV}^2 \quad (2)$$

with $M_Z = 91.2$ GeV [11].

Following this formalism, the parity-violating asymmetry for scattering longitudinally polarized electrons from an unpolarized isoscalar target such as deuterium, assuming isospin symmetry, is given by [10,12]

$$A_d = \frac{\sigma_L - \sigma_R}{\sigma_L + \sigma_R} = -\left(\frac{3G_F Q^2}{\pi\alpha 2\sqrt{2}}\right) \times \frac{(2C_{1u} - C_{1d})\,[1 + R_s] + Y(2C_{2u} - C_{2d})R_v}{5 + R_s}. \quad (3)$$

Here, the kinematic variable Y is defined as

$$Y = \frac{1 - (1-y)^2}{1 + (1-y)^2 + f(y, R_{LT})} \quad (4)$$

with $y = \nu/E$ and $\nu = E - E'$ is the energy lost by an incident electron of energy E scattering to an electron of energy E'. The ratio $R_{\text{LT}}(x, Q^2) = \sigma_L/\sigma_T \approx 0.2$ measures the virtual photon-absorption cross-section ratio from longitudinally and transversely polarized photons, and $f(y, R_{\text{LT}})$ is a function describing the effect of longitudinally polarized photons to the asymmetry and vanishes when $R_{\text{LT}} = 0$. In a work by Blumlein *et al.* [13] it was derived that $f(y, R_{\text{LT}}) = -y^2 R_{\text{LT}}/(1 + R_{\text{LT}})$ when $\nu^2 \ll Q^2$. At JLab energies both Eqs. (3) and (4) need to be modified to take into account terms proportional to ν^2/Q^2. The ratios R_s and R_v depend on the parton distribution functions:

$$R_s(x) = \frac{s(x) + \bar{s}(x)}{u(x) + \bar{u}(x) + d(x) + \bar{d}(x)} \quad \text{and} \quad (5)$$

$$R_v(x) = \frac{u_v(x) + d_v(x)}{u(x) + \bar{u}(x) + d(x) + \bar{d}(x)}, \quad (6)$$

where we have neglected contributions from the charm quark. The weak couplings C_{1q} and C_{2q} with $q \in u, d$ are defined as:

$$C_{1u} = g_A^e g_V^u = -\frac{1}{2} + \frac{4}{3}\sin^2(\theta_W), \quad (7)$$

$$C_{1d} = g_A^e g_V^d = \frac{1}{2} - \frac{2}{3}\sin^2(\theta_W), \quad (8)$$

$$C_{2u} = g_V^e g_A^u = -\frac{1}{2} + 2\sin^2(\theta_W), \quad (9)$$

$$C_{2d} = g_V^e g_A^d = \frac{1}{2} - 2\sin^2(\theta_W), \quad (10)$$

where the second equality is valid at the tree-level of the Standard Model. These coupling constants, along with the Weinberg angle θ_W itself, are fundamental parameters of the Standard Model.

In an approximation of moderately large x where sea quark contributions vanish, $R_v \approx 1$ and $R_s \approx 0$. The uncertainty in $2C_{2u} - C_{2d}$ extracted from measured asymmetry is then approximately:

$$\frac{\delta(2C_{2u} - C_{2d})}{2C_{2u} - C_{2d}} \approx \frac{\delta A_d}{A_d}\left[1 - \frac{1}{Y}\frac{2C_{1u} - C_{1d}}{2C_{2u} - C_{2d}}\right] \quad (11)$$

2.2 Exploring new physics beyond the Standard Model from PV DIS

Although there exists a large amount of data confirming the electroweak sector of the Standard Model at the level of a few parts per thousand, there also exist strong conceptual reasons (*e.g.*, the so-called high-energy desert from $M_{\text{weak}} \approx 250$ GeV up to the Planck scale $M_P \approx 2.4 \times 10^{18}$ GeV) to believe that the Standard Model is only a piece of some larger framework [14]. This framework should provide answers to the conceptual puzzles of the Standard Model; but must also leave the $\text{SU}(3)_C \times \text{SU}(2)_L \times \text{U}(1)_Y$ symmetry of the Standard Model intact at $M_{\text{weak}} \approx 250$ GeV. Hence, there exists intense interest in the search for physics beyond the Standard Model.

The value of $\sin^2\theta_W$ at the Z-pole ($Q^2 = M_Z^2$) is measured to remarkable precision, $\sin^2\theta_W[M_Z]_{\overline{MS}} = 0.23120 \pm 0.00015$ [11]; however, a careful comparison of measurements involving purely leptonic and semi-leptonic electroweak currents shows a large inconsistency. This strongly suggests additional physics not included in the Standard Model or that one or more of the experiments has significantly understated its uncertainties [15,16]. Below the Z-pole, there are only three precise measurements: Atomic parity violation (APV) in Cs atoms [5] yields a result which, while in agreement with Standard Model predictions, has somewhat large uncertainties, and a difficult theoretical calculation is necessary to extract $\sin^2\theta_W$ from the measured asymmetry. The NuTeV experiment at Fermilab measured $\sin^2\theta_W$ through a careful comparison of neutrino and anti-neutrino deep inelastic scattering (DIS). Their result is approximately three standard deviations from Standard Model predictions [18]; however, the NuTeV result is not without considerable controversy. Most recently, the SLAC E-158 [4] experiment used the asymmetry in Møller scattering to determine a precise value of $\sin^2\theta_W$ that is consistent with the Standard Model prediction. A fourth measurement, the Qweak experiment [6], is planned at JLab, and will determine $\sin^2\theta_W$ to 0.3% by measuring the weak charge of the proton.

Among various experimental efforts to search for new physics, PV DIS involves the exchange of a Z^0 between electrons and quarks and thus is sensitive to physical processes that might not be seen in purely leptonic observables, such as the precision A_{LR} at SLC and A_{FB}^l at LEP.

The recent NuTeV [18] result on $\sin^2\theta_W$ at low Q^2 involves a particular set of semi-leptonic charged and neutral current reactions and disagrees with the Standard Model prediction by three standard deviations. A precision measurement of DIS-Parity will provide a clean semi-leptonic observable to the world data below the Z-pole and will provide essential clues as to the source of these discrepancies.

The values of C_{1q} have been determined to a reasonable precision [17]. However, our present knowledge of C_{2q} is poor: $\delta(2C_{2u} - C_{2d}) = \pm 0.24$. This also affect the extraction of $C_{3q} \equiv g_A^e g_A^q$ from neutrino scattering data. A precision measurement of C_{2q} is highly desirable to explore possible extensions of the Standard Model. PV DIS is a semi-leptonic process and is sensitive to the C_{2q}'s, therefore it is complementary to other Standard Model test experiments including the Qweak experiment – which studies semi-leptonic processes but is only sensitive to C_{1q}'s. For example, a large axial quark coupling could cause the NuTeV effect, but cannot be seen in C_{1q}. Quark and lepton compositeness is accessible only through C_{2q} but not C_{1q} if a particular symmetry, SU(12), is respected. A precision PV DIS measurement will significantly strengthen the constraints on these possible extensions to the Standard Model. Possible new physics that PV DIS may be sensitive to are Z' Searches, quark and lepton compositeness, leptoquarks and supersymmetry. We are currently working with theorists on an updated list of new physics limits achievable from the measurement described here [19,20]. The "mass scale" for which PV DIS is sensitive to physics beyond the Standard Model would be [20]

$$\frac{\Lambda}{g} = \frac{1}{\sqrt{2\sqrt{2}G_F\ \delta(2C_{2u} - C_{2d})}} \approx 1.0 \text{ TeV}. \qquad (12)$$

3 Experimental setup

3.1 Overview

The floor plan for our experimental setup is shown in Fig. 1. We will use an 85 μA polarized beam and a 25 cm long liquid deuterium target. The scattered electrons will be detected by the two standard Hall A High Resolution Spectrometers (HRS). In the following we will describe some details of the experimental setup, focusing on two major upgrades required by the measurement.

3.2 Beam line and the Compton polarimetry upgrade

We plan to use a 6.0 GeV 85 μA beam with an 85% polarization. To reduce the heat impact on the target, the beam will be circularly rastered such that the beam spot size at the target is $\sim$ 4mm in diameter. The beam energy can be measured to $\Delta E/E = 2 \times 10^{-4}$ using either the ARC or eP devices [21]. The helicity-dependent asymmetry of the electron beam will be controlled by a DAQ specially developed in Hall A for parity-violation experiments [7]. A luminosity monitor will be used downstream of the target to monitor helicity-dependent target-density fluctuations and other possible false asymmetries.

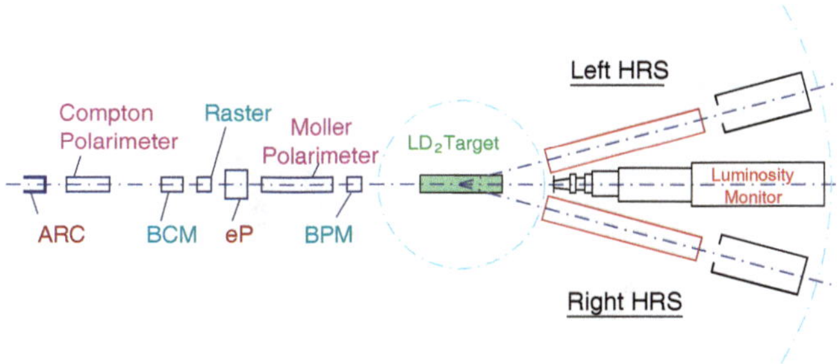

Fig. 1. Hall A floor plan for the proposed measurement.

We need 1% precision in the beam polarization measurement in order to achieve an acceptable systematic uncertainty on the final results. The current Compton polarimetry in Hall A measures the asymmetry of Compton scattering between the electron beam and a high-power IR laser (1 kW, 1064 nm) achieved by a Fabry-Perot cavity. The current systematic uncertainty of the Compton polarimeter is about 1.9% for a 6 GeV beam. In order to achieve a $\leqslant$ 1% precision, we will upgrade the current IR laser to a green laser (532 nm), and replace the current 600 μm micro-strips used in the electron detector by 300 μm strips. With these upgrades the total systematic uncertainty is expected to be reduced to $\leqslant$ 0.9%. These upgrades are presently being carried out at JLab and we expect to install and commission the new Compton polarimeter by late 2007.

3.3 Spectrometers and the DAQ upgrade

We will use the standard Hall A HRSs [21] to detect the scattered electrons. For each HRS the effective solid-angle acceptance for an extended target is 5.4 msr and the momentum acceptance is $\pm$4.5%. The HRS central momentum can be calculated from the dipole field magnitude to 5×10^{-4} and the central angle can be determined to $\pm$0.2 mrad, giving an uncertainty of $\pm 0.004/Q^2$ on $2C_{2u} - C_{2d}$ where Q^2 is in GeV^2. Particle identification (PID) in each HRS will be done with a CO_2 Čerenkov detector and a double-layered lead-glass shower detector. We expect to achieve a pion rejection factor of $\geq 10^4$ at an electron efficiency of $\geq$ 99%.

Because of the need to separate the pion background we must use a counting method instead of an integrating DAQ. The detector signals we will use include those from the two PID detectors and scintillators (for crude directional information). To process this information we are considering a Flash ADC (FADC) -based DAQ presently being designed by the Fast Electronics Group at JLab. This design will allow for the possibility of counting experiments at approximately 1 MHz with a low and precisely measurable dead time, *e.g.* a 1% dead time measured with a 0.3% absolute accuracy. The FADC fills an on-board memory at 250 MHz with $\sim$ 4 μs latency (buffer

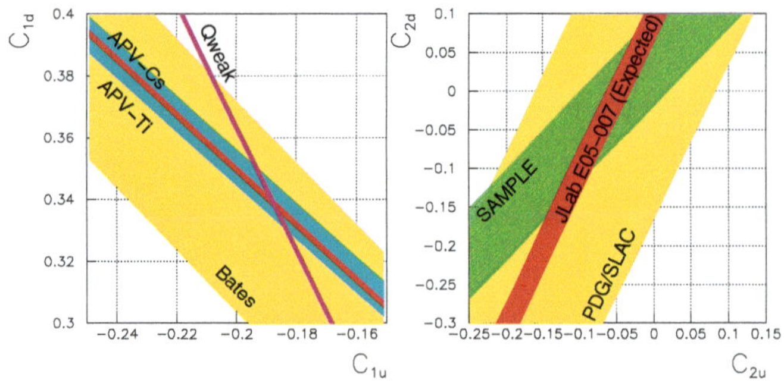

Fig. 2. The effective couplings C_{1q} (left) and C_{2q} (right) . The future Qweak experiment combined with the APV-Cs result will provide the most precise data and the best Standard Model test on C_{1q}. For C_{2q}, the SAMPLE result for $C_{2u} - C_{2d}$ at $Q^2 = 0.1\ \text{GeV}^2$ [22] and the current PDG value for $2C_{2u} - C_{2d}$ are shown. Assuming the SM prediction of $2C_{1u} - C_{1d}$, the value of $2C_{2u} - C_{2d}$ can be determined from this experiment to $\Delta(2C_{2u} - C_{2d}) = \pm 0.03$.

size). An on-board processor (FPGA) will analyze the digitized data and perform the PID. The DAQ will be flexible enough to accommodate a variety of experiments. A first version of the FADC is expected to be ready by 2007.

As an alternative, a scaler-based DAQ using fast NIM electronics is also being considered. We plan to build both systems such that the FADC-based DAQ can be cross-checked with the regular HRS counting DAQ at a low rate (<4 kHz) and with the scaler-based DAQ at a high rate (up to 1 MHz).

3.4 Expected results

The asymmetry A_d will be extracted from the measured raw counting asymmetry as

$$A_d = \frac{A_{raw}}{P_{beam}} + \Delta A_{EM}^{RC} \tag{13}$$

where P_{beam} is the beam polarization. A_{EM}^{RC} is the electromagnetic radiative correction and can be calculated to 0.2% (relative to A_d). Provided that all uncertainties related to hadronic structure are well under control, one can then extract $2C_{2u} - C_{2d}$ from A_d using Eq.(3). The expected total uncertainty in $2C_{2u} - C_{2d}$ is shown in Fig. 2 along with existing world data. For illustration, current knowledge of C_{1q}'s and their expected results from the Qweak experiments are also shown. We expect to run E05-007 in early 2009 and first results may become available within one year of its completion.

4 Interpretation of the data and discussions on hadronic effects

While we can extract $2C_{2u} - C_{2d}$ from the measured asymmetry, there are uncertainties coming from hadronic physics which may complicate the interpretation of this asymmetry in terms of Standard Model parameters. This section will describe some of these interesting issues.

4.1 Uncertainty from parton distributions and R_{LT}

The uncertainties due to R_s and R_v in Eq. (3) can be estimated using CTEQ [23,24] and MRST [25,26] PDF parameterizations. We find that our current knowledge of PDFs gives an uncertainty at the level of $\delta(2C_{2u} - C_{2d}) \leqslant \pm 0.0025$. The ratio $R_{LT} = \sigma_L/\sigma_T$ is taken from a global fit, R1998 [27]. Propagation of the uncertainty from the this fit yields an uncertainty of $\delta(2C_{2u} - C_{2d}) \leqslant \pm 0.0017$.

4.2 Higher-twist effects

Among all hadronic effects that could contribute to PV electron-scattering observables, higher-twist (HT) effects are expected to be the most probable for kinematics at JLab. Here higher-twist effects refer to the fact that the strong interactions between the quarks become observable at low Q^2 and the process cannot be described by the $\gamma(Z)$ exchange between the electron and a *single* quark (the leading-twist process). For electromagnetic scattering processes, these interactions introduce a scaling violation to the structure functions for $Q^2 < 1\ \text{GeV}^2$ that is stronger than the $\ln(Q^2)$-dependence of the DGLAP equations of perturbative QCD (pQCD). For PV $\mathbf{e}-{}^2\text{H}$ scattering, HT effects start from twist-four terms which diminish as $1/Q^2$. Hence, observation of any Q^2-dependent deviation from the expected asymmetry would strongly imply a contribution from HT.

The theory for HT is not well established. In a naive picture, HT is expected to vanish in the asymmetry A_d because the strong coupling between the struck quark and other spectator quarks (the origin of HT) should not depend on what type of the gauge boson (γ or Z^0) is probing the quark, hence the HT contributions cancel in the ratio. In less hand-waving pictures, most of existing HT models show that the HT contribution to A_d is below $1\%/Q^2$ level (see *e.g.* Ref. [1]), which is acceptable for interpreting our expected data in terms of Standard Model tests. However, these models were developed primarily during and after the previous PV DIS experiment in the 1970's. A modern calculation of the HT from QCD is needed and we hope such calculation will become available in the next 1-2 year(s) [19].

On the other hand, HT provides information on non-perturbative aspects of the strong interaction, which by itself has its own value. So far first-hand information on HT comes primarily from experimental data, *e.g.* by observing the non-ln Q^2 dependence of DIS structure functions. However, such an extraction is complicated by the determination of the leading-twist term, which has a large uncertainty from the truncation of higher order terms in α_S. By contrast, the prospects for observing HT contributions in PV-DIS are relatively uncomplicated because the QCD higher order terms are expected to cancel in the asymmetry. Thus our PV DIS experiment will provide valuable insights into the study of the non-perturbative side of strong interactions.

One interesting remark is that it has been shown that although the NuTeV measurement was performed at $\langle Q^2 \rangle = 20$ GeV^2, the HT contribution to the typically measured Paschos-Wolfenstein (P-W) ratio could be of the same magnitude as that of the PV DIS observable at $Q^2 \approx 2$ GeV^2 [28]. If the NuTeV deviation from the Standard Model is fully due to the HT, then this would imply a contribution to our expected results much above their uncertainties. Although this is a model-dependent calculation, it underscores the uncertainty in determination of HT effects in other contexts. The precision measurement proposed here would clearly be sensitive to this magnitude of the HT correction.

5 Future PV DIS program at JLab

More PV DIS measurements are being planned at the 12 GeV Upgrade of JLab [29] using the current high momentum spectrometer (HMS) and the planned baseline Super-HMS in Hall C. From this measurement we expect to extract the value of $\sin^2\theta_W$ to a precision of $\delta\sin^2\theta_W/\sin^2\theta_W = \pm 0.0025(stat.) \pm 0.0032(syst.)$ and the weak coupling C_{2q}'s to $\delta(2C_{2u} - C_{2d}) = \pm 0.015$. In addition, measurements of PV DIS asymmetries over a wide range of x and Q^2 are possible, provided a large-acceptance solenoid spectrometer [30] is built. If such a device is possible, we can measure HT effects to great precision. By measuring the PV DIS asymmetry on the proton, we can extract the unpolarized PDF ratio d/u at large x without contamination of the deuteron nuclear (EMC) effect and thus provide a clean test on various predictions of this ratio from *e.g.* pQCD and relativistic constituent quark models. By comparing the PV DIS asymmetry on the deuteron with that on the proton, we will be sensitive to charge-symmetry violation which implies that the $u(d)$ distributions in the proton differ from the $d(u)$ distributions in the neutron. This effect not only violates one of the known discrete fundamental symmetries (isospin symmetry) and has profound implications for our understanding of hadronic structure and the strong interaction, but is also one of the leading candidates for the cause of the NuTeV anomaly [31] and hence the picture of Standard Model test below the Z-pole will change dramatically should this effect be large.

6 Summary

We are planning to measure the parity-violation asymmetry A_d in $\mathbf{e}-{}^2$H deep inelastic scattering at $Q^2 = 1.1$ and 1.9 GeV^2 and similar $x(= 0.3)$ using a 85 μA polarized beam and a 25 cm long liquid deuterium target in Hall A of JLab. Two major instrumentational upgrades are required by this experiment and both are being developed at JLab. Assuming an 85% beam polarization and 46 days of beam time, we expect to extract the weak coupling constant combination to $\Delta(2C_{2u} - C_{2d}) = \pm 0.03$ from A_d at 1.9 GeV^2, a factor of eight improvement compared to our current knowledge of this quantity. This result will help to extract C_{3q} from high-energy data, and has the potential to reveal possible new physics beyond the Standard Model. The additional point at 1.1 GeV^2 will help to investigate the non-perturbative hadronic physics contribution to PV DIS asymmetry and may provide the first precision observation of this effect. This result may have implications for the interpretation of existing data from other Standard Model test experiments using semi-leptonic processes. This experiment will provide the first crucial guidance for the future PV DIS program with the JLab 12 GeV Upgrade.

The author would like to thank the PAVI06 organizers for their invitation to the beautiful Milos Island. The work was developed when the author was working for the Medium Energy Physics Group at Argonne National Lab, Argonne, IL 60439, USA, under the U. S. Department of Energy contract no. W-31-109-ENG-38. The Jefferson Science Associates, LLC operates the Thomas Jefferson National Accelerator Facility for the United States Department of Energy under contract DE-AC05-06OR23177.

References

1. X. Zheng, P.E. Reimer, R. Michales *et al.*, JLab Experiment Proposal 05-007 (2005).
2. C.Y. Prescott *et al.*, Phys. Lett. B **77** (1978) 347.
3. C.Y. Prescott *et al.*, Phys. Lett. B **84** (1979) 524.
4. P.L. Anthony *et al.*, Phys. Rev. Lett. **95** (2005) 081601.
5. S.C. Bennett, C.E. Wieman, Phys. Rev. Lett. **82** (1999) 2484.
6. D. Armstrong *et al.*, The $Q_{\rm weak}$ Experiment, JLab E02-020 (2001).
7. K.A. Aniol *et al.*, Phys. Rev. Lett. **96** (2006) 022003.
8. D.S. Armstrong *et al.*, Phys. Rev. Lett. **95** (2005) 092001.
9. R. Hasty *et al.*, Science **290** (2000) 2117.
10. R.N. Cahn, F.J. Gilman, Phys. Rev. D **17**, (1978) 1313.
11. S. Eidelman *et al.*, Phys. Lett. B **592** (2004) 1.
12. R. Arnold *et al.*, SLAC-PROPOSAL-E-149 (1993).
13. J. Blumlein, M. Klein, T. Naumann, T. Riemann, DESY HERA Workshop 1987:0067 (QCD161:H15:1987).
14. M.J. Ramsey-Musolf, Phys. Rev. C **60** (1999) 015501.
15. M.S. Chanowitz, Phys. Rev. Lett. **87** (2001) 231802.
16. M.S. Chanowitz, Phys. Rev. D **66** (2002) 073002.
17. J. Erler, M.J. Ramsey-Musolf, Prog. Part. Nucl. Phys. **54** (2005) 351.
18. G.P. Zeller *et al.*, Phys. Rev. Lett. **88** (2002) 091802.
19. M.J. Ramsey-Musolf, private communication.
20. J. Erler, private communication.
21. J. Alcorn *et al.*, Nucl. Instrum. Meth. A **522** (2004) 294.
22. E.J. Beise, M.L. Pitt, D.T. Spayde, Prog. Part. Nucl. Phys. **54** (2005) 289.
23. D. Stump *et al.*, Phys. Rev. D **65** (2002) 014012.
24. J. Pumplin *et al.*, Phys. Rev. D **65** (2002) 014013.
25. A.D. Martin, R.G. Roberts, W.J. Stirling, R.S. Thorne, Eur. Phys. J. C **28** (2003) 455.
26. A.D. Martin, R.G. Roberts, W.J. Stirling, R.S. Thorne, Eur. Phys. J. C **35** (2004) 325.
27. K. Abe *et al.*, Phys. Lett. B **452** (1999) 194.
28. M. Gluck, E. Reya, Phys. Rev. Lett. **47** (1981) 1104.
29. X. Zheng, P.E. Reimer, K. Paschke *et al.*, JLab 12 GeV Letter-of-Intent (2006)

30. *Initiatives of a large acceptance solenoidal spectrometer for Hall A at the JLab 12 GeV Upgrade for studying parity and spin*, informal discussions (2006).
31. J.T. Londergan, D.P. Murdock, A.W. and Thomas, Phys. Rev. D **72** (2005) 036010.

Res-Parity: Parity violation in inelastic scattering at low Q^2

P.E. Reimer[1], P.E. Bosted[2], J. Arrington[1], H. Mkrtchyan[3], and X. Zheng[4]

[1] Physics Division, Argonne National Laboratory, Argonne, IL 60439 USA
[2] Thomas Jefferson National Accelerator Facility, Newport News, VA 23606, USA
[3] Yerevan Physics Institute, Yerevan, Armenia
[4] Massachusetts Institute of Technology, Cambridge, MA 02139, USA

Abstract. Parity violating electron scattering has become a well established tool which has been used, for example, to probe the Standard Model and the strange-quark contribution to the nucleon. While much of this work has focused on elastic scattering, the RES-Parity experiment, which has been proposed to take place at Jefferson Laboratory, would focus on inelastic scattering in the low-Q^2, low-W domain. RES-Parity would search for evidence of quark-hadron duality and resonance structure with parity violation in the resonance region. In terms of parity violation, this region is essentially unexplored, but the interpretation of other high-precision electron scattering experiments will rely on a reasonable understanding of scattering at lower energy and low-W through the effects of radiative corrections. RES-Parity would also study nuclear effects with the weak current. Because of the intrinsic broad band energy spectrum of neutrino beams, neutrino experiments are necessarily dependent on an untested, implicit assumption that these effects are identical to electromagnetic nuclear effects. RES-Parity is a relatively straight forward experiment. With a large expected asymmetry ($\approx 0.5 \times 10^{-4}$) these studies may be completed with in a relatively brief period.

PACS. 13.60-r Photon and charged-lepton interactions with hadrons – 11.30.Er Charge conjugation, parity, time reversal, and other discrete symmetries – 24.30.-v Resonance reactions – 12.15.Mm Neutral Currents – 12.15.-y Electroweak interactions

1 Introduction

Until recently, the electromagnetic interaction has been *the* fundamental probe of the structure of the nucleon. It has been used in many kinematic domains from elastic scattering through deep inelastic scattering (DIS). In between these two extremes lies the nucleon resonance region. This region has also been well studied with the electromagnetic interaction with many interesting results, including quark-hadron duality in which the structure of the nucleon resonances remarkably averages to be the same as the DIS structure, indicating a cancellation of higher twist terms [1].

The weak interaction has also been used with high energies with neutrino beams to study the quark structure of the proton in DIS kinematics, but with reasonable precision possible only on nuclear targets. With recent advances in polarized electron parity violating experiments, it has become possible to use the weak force as a probe of nucleon structure in other kinematic regimes as well. Several experiments, most recently the HAPPEX [2] and G0 [3] experiments at Thomas Jefferson National Accelerator Facility (JLab), have used this technique to explore the effects of intrinsic strangeness in the proton through parity violation in elastic scattering.

Building on the recent success of experiments measuring parity violation at JLab and elsewhere, the Res-Parity experiment [4] has been proposed to explore parity violation in the resonance region ($Q^2 < 1$ GeV^2 and $M^2_{\text{proton}} < W^2 < 4$ GeV^2). These studies would not only examine the nucleon's parity violating response in the resonance region and study quark-hadron duality, but also study the nuclear dependence of parity violation with deuterium and carbon targets.

2 Parity violation and the resonance region

Lepton-nucleon scattering takes place either through the exchange of a photon or a Z^0. At energies well below the mass of the Z^0 boson, the electromagnetic interaction is dominant and the Z^0-exchange and interference terms can easily be neglected in the unpolarized cross section. The electromagnetic part is parity conserving, however, so that in a measurement of parity violating effects, the normally dominate electromagnetic term would not contribute to the measured asymmetry.

This asymmetry was first measured in the DIS region by Prescott *et al.* at SLAC [5,6] and served the establish the Standard Model of the electroweak interaction. In the DIS region, the asymmetry can be expressed in terms of the Standard Model couplings C_{1q} (axial electron × vector quark) and C_{2q} (vector electron × axial quark) which depend on $\sin^2\theta_W$ and the quark distribution functions of the target [7,8]. Ignoring the strange and charmed quark contribution, on an isoscalar target, such as deuterium or carbon, this asymmetry is

$$A^D_{\mathrm{DIS}} = -\left(\frac{3G_F Q^2}{2\sqrt{2}\pi\alpha}\right)\frac{(2C_{1u}-C_{1d})+Y\,(2C_{2u}-C_{2d})\,R_v}{5+R_v}, \tag{1}$$

where Y is a kinematical factor, and R_v is the ratio of valence up and down quark distributions to the total up and down quark and antiquark distributions. Assuming that quark-hadron duality (see Sec. 2.1) holds, this would give an asymmetry of $A^H/Q^2 \approx 50\times10^{-6}$ for hydrogen and $A^{D,C}/Q^2 \approx 65\times10^{-6}$ for deuterium and carbon.

The asymmetry in the resonance region, A_{Res}, can be described in terms of response functions to specific final states, $R(q,\omega)$, combined with a non-resonant background as [9–12]

$$A_{\mathrm{Res}} = A_0 Q^2 \frac{v_L R_L^{AV}(q,\omega) + v_T R_T^{AV}(q,\omega) v_{T'} R_{T'}^{VA}(q,\omega)}{v_L R_L(q,\omega) + v_T R_T(q,\omega)}, \tag{2}$$

with $A_0 \approx 6.5\times10^{-4}$. Here v is a lepton kinematic factor and the subscripts L, T and T' denote longitudinal, transverse and axial, respectively. The superscript AV (VA) represent the interference between axial-vector (vector) leptonic currents and the vector (axial-vector) hadronic currents. This general formalism has only been worked out in detail for elastic scattering [7] and for the $\mathrm{N}\to\Delta(1232)$ transition [7,13,14]

A recent calculation of parity violating electron scattering in the $\mathrm{N}\to\Delta$ transition has been made Matsui, Sato and Lee [15]. These predictions are based on a dynamical model of pion electroweak pion production reactions [16]. In this framework, the asymmetry may be expressed as

$$A^{\Delta(1232)}_{\mathrm{Res}} = -\frac{Q^2 G_F}{4\sqrt{2}\pi\alpha}\left(2-4\sin^2\theta_W+\Delta_V+\Delta_A\right) \tag{3}$$

$$\approx -Q^2\left(8.99\times10^{-5}\right)\left(1.075+\Delta_V+\Delta_A\right), \tag{4}$$

where Q^2 is in units of GeV^2. Here, Δ_A depends on the W_3 structure function with contributions only from the axial vector part of the neutral current and Δ_V depends on the $W_{1,2}$ structure functions with contributions from the vector part of the neutral current. Near the $\Delta(1232)$ resonance, the Δ_A term is dominant over the Δ_V term. The non-resonant contribution to the Δ_A term is minimal. This allows for the extraction of the axial form factors of the $\mathrm{N}\to\Delta$ transition from A_{Res}.

In studying the resonance region, Res-Parity would explore quark-hadron duality and the EMC effect with parity violation. It is important to remember that essentially no data exists to guide predictions, and so it is quite probable that significant surprises exist.

2.1 Duality and higher twist

Electron scattering experiments have shown that the unpolarized structure functions, when measured in the resonance region give on average the same result when measured at significantly higher Q^2 in DIS kinematics if the proper scaling variable is chosen [17]. This phenomena is known as quark-hadron duality and has also been shown to hold in separated longitudinal and transverse structure function and spin structure functions [1]. Quark-hadron duality can be described in terms of an operator product expansion (OPE) [18,1] of the moments of the structure functions. The leading twist terms in this expansion represent scattering from individual quarks in the target, while higher twist terms take into account quark-gluon correlations. In order for duality to hold, these higher twist terms must either cancel or be negligible. The model of Matsui, Sato and Lee, discussed above, predicts a 40% or greater departure from duality for the this parity violating asymmetry, while predicting duality for the unpolarized structure functions, in agreement with data for the $\mathrm{N}\to\Delta$ transition. This prediction is show in Fig. 1 as a function of the Nachman scaling variable,

$$\xi = \frac{2x}{1+\sqrt{1+4x^2m_N^2/Q^2}}, \tag{5}$$

where x is the Bjorken scaling variable and m_N is the nucleon mass.

2.2 Nuclear ratios and the EMC effect

The nuclear EMC effect in the resonance region would also be studied in this experiment through parity-violation. In addition to hydrogen, the experiment plans to collect data with deuterium and carbon targets. The parity violating asymmetry has very different sensitivity to the underlying quark structure than the traditional electromagnetic probes that have been used to study the EMC effect in the past. Electron scattering is proportional to the parton distributions weighted by their charge squared, thus emphasizing the effects of the u-quarks and antiquarks over the d-quarks and antiquarks. The weak probe has a different weighting. If the EMC effect is the result of identical modifications of both quark valence and sea and antiquark distributions, then there would be an identical effect for both the photon and Z^0 exchange, yielding no observable effect in the ratios of the parity violating asymmetries. Any observed effect would then necessarily point to something beyond simple rescaling of the quark distributions, for example a flavor dependence, additional nuclear dependencies related to Z^0 exchange, or a difference in the valence-sea dependence. With respect to the latter, it is interesting to note that the Fermilab E772 Drell-Yan experiment, which is sensitive to the sea quark distributions, searched for but did not observe any large nuclear effects aside from shadowing in the sea quark distributions [19].

3 Impact on other measurements

The measurement proposed by the Res-Parity experiment would also be important to other experiments in controlling their systematic uncertainties. Three such experiments at JLab are DIS parity violation family of experiments: the 6 GeV PV-DIS [20,21], the 12 GeV DIS-Parity [22] and the 12 GeV Large Solid Angle Spectrometer program under discussion for Hall A [23]. One of the largest uncertainties in interpreting the measured asymmetries from these experiments is likely to be from the effects of higher twist contributions. By measuring parity violation in the resonance region and testing duality in parity violation, Res-Parity would be able to limit or measure these contributions. Because of beam energy and rate limitations, all of these experiments would have data very near the resonance region. Radiative corrections to the DIS measurements would depend on input from these proposed measurements of parity violation in the resonance region by the Res-Parity experiment.

The proposed measurement would also aid in understanding the backgrounds the recently completed SLAC E158 Møller experiment [24]. This experiment is one of the most precise tests of the Standard Model below the Z-pole. The inelastic electron-proton scattering background was both the largest correction to the Møller asymmetry and had the largest systematic uncertainty. Direct measurements by Res-Parity would help understand this correction and reduce the systematic uncertainty assigned to it. This would also be extremely important for any future Møller scattering parity violation experiments at, for example, the 12 GeV JLab facility [25,26].

Because of the small weak cross sections, neutrino scattering experiments have, for the most part, depended on heavy nuclear target to achieve luminosities sufficient for their measurements. The weak-interaction nuclear effects in these targets is at best poorly understood and generally treated as identical to the electromagnetic nuclear effects. The Res-Parity experiment would directly measure the nuclear effects with the weak interaction. While the data would be in the resonance region rather than in DIS kinematics, this is precisely the data which the current generation of neutrino scattering experiment needs. The kinematics of these experiments is dictated by the neutrino oscillation parameters given by nature. This, and the broadband nature of most neutrino beams puts much of their kinematical coverage within the resonance region.

4 Res-Parity implementation in Jefferson Lab Hall A

The Res-Parity experiment would take advantage of the significant work which has gone into previous parity violation experiments at JLab. The experiment would use essentially the same experimental configuration and fast data acquisition as the 6 GeV PV-DIS E04-012 experiment [20,21]. The experiment is proposed to take place in JLab Hall A using both HRS spectrometers [27] positioned at 12.5° (the smallest angle available without septum magnets) with a 85 μA 4.8 GeV electron beam polarized at 80% incident on liquid hydrogen, liquid deuterium and carbon targets. The incident beam energy and spectrometer angle were chosen to optimize the coverage of the resonance region without the need for additional electron beam energies or spectrometer angle changes. For each target, the measured kinematics as well as the expected statistical uncertainties are shown in Tab. 1. The expected results and statistical uncertainties are plotted in Fig. 1. To obtain this kinematic coverage, the central momenta of the spectrometers would be varied over a range of 3.0 to 4.0 GeV. To obtain the desired statistical uncertainties and to allow for studies of systematic effects, 30 days of beam time (9 days for hydrogen, 8 for deuterium, 12 for carbon and 1 day for configuration changes and systematic effect studies) are required.

Table 1. This table lists the proposed kinematics for the measurement on each target as well as the expected counting rate (based on a DIS calculation) the asymmetry and the statistical uncertainty which would be achieved by the proposed experiment.

W (GeV)	x	Q^2 (GeV^2)	rate (MHz)	A (ppm)	dA/A (stat %)
		hydrogen			
2.00	0.17	0.64	0.5	46	7.0
1.78	0.24	0.73	0.5	51	6.5
1.52	0.36	0.82	0.6	53	4.3
1.21	0.61	0.91	0.8	52	4.2
		deuterium			
2.00	0.17	0.64	0.8	61	4.5
1.78	0.24	0.73	0.9	64	3.6
1.52	0.36	0.82	0.9	67	3.1
1.21	0.61	0.91	1.3	71	2.4
		carbon			
2.00	0.17	0.64	0.5	61	4.6
1.78	0.24	0.73	0.5	64	3.8
1.52	0.36	0.82	0.5	67	3.4
1.21	0.61	0.91	0.9	71	2.5

Depending on the kinematic bin, the pion to electron ratio may be up to 0.6:1 (based on pion photoproduction data [28]) so contamination could represent a significant background. Pions would be identified by using a CO_2 gas threshold Cherenkov detector and double-layered lead-glass shower counters in each HRS spectrometer. The combination of these two devices should provide a pion rejection of better than 10^{-4} with an electron identification efficiency of better than 99%. In addition, both the FADC DAQ and scaler DAQ described below would measure the pion asymmetry with precision roughly equal to the electron asymmetry so that corrections may be applied if necessary.

The need to separate electron signals from pionic background forces this to be a counting experiment. Event rates of up to 1.3 MHz are expected in this experiment. Two methods are being explored to handle these rates.

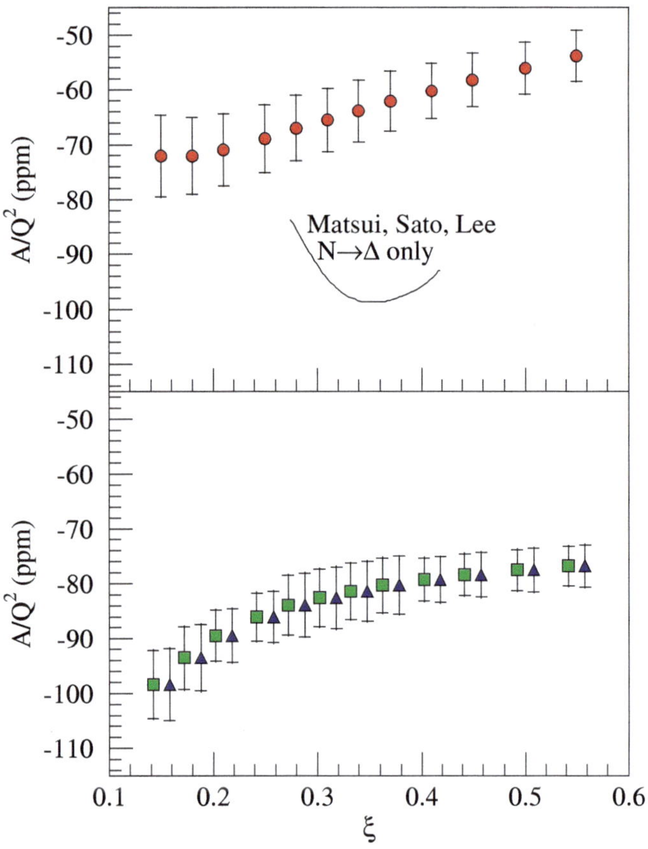

Fig. 1. The expected statistical uncertainties are shown in this figure for hydrogen (circles–top plot), deuterium (squares–bottom plot) and carbon (triangles–bottom plot) as a function of the Nachman scaling variable ξ. The points are plotted along the prediction of a simplified (purely $I = 1$ transition) model of the resonance region [4] for the asymmetry –with no nuclear effects. Also shown as a solid curve is the prediction of Matsui, Sato and Lee [15] for the $N \to \Delta$ transition in hydrogen at an incident beam energy of 4 GeV at $Q^2 = 0.5$ GeV^2.

The first uses a flash ADC (FADC) system to digitize the signals from the PID detectors and scintillators and passes them to an on board FPGA for processing. The logic in the FPGA would identify electrons, pions and associated pileup and count these in local memory which is read out to a VME bus once every helicity pulse. Alternatively a NIM based logic with a scaler counting algorithm can also be used to implement a counting scheme, at the expense of greater pileup and dead time due to the signal width needed for the scalers. More details may be found in Ref. [4,20].

As with any measurement of parity violation, careful control over systematic effects must be maintained. This is still true, even though the asymmetry measured in this experiment is relatively large ($A \approx 50 \times 10^{-6}$) when compared with those of the completed HAPPEX II [29] ($A \approx 1 \times 10^{-6}$) and proposed PREX [30] ($A \approx 0.5 \times 10^{-6}$) experiments. The known sources of systematic uncertainties are listed in Tab. 2. The most significant of these uncertainties are discussed below.

Table 2. Sources of experimental systematic uncertainties and their estimated size. In comparing the ratios of the asymmetry from different targets, many of these uncertainties would cancel either partly or completely. For the target ratios, an overall systematic uncertainty of $\delta A/A \approx 0.010$ is estimated.

Source	$\delta A/A$
Beam polarization	0.012
Kinematic determination of Q^2	0.009
Electromagnetic radiative corrections	0.005
Beam asymmetries	0.005
Pion Contamination	0.005
Dead time corrections	0.003
Pair symmetric background	0.002
Target purity and density fluctuations	0.002
Pole tip background	0.001
Total (in quadrature)	0.018
Total on target ratios	0.010

The beam polarization would be measured with a combination of a Compton and a Møller polarimeter. An upgrade of the JLab Hall A Compton polarimeter is underway and needed for both the PREX and 6 GeV PV-DIS experiments. Both the scattered photons and scattered electrons would be detected and the data may be analyzed independently or in coincidence. This upgrade would be able to achieve a combined statistical and systematic 1.2% measurement of the polarization, at the experiment's production beam current. This upgrade is expected to be complete in 2006. As a cross check, with lower beam current, the Hall A Møller polarimeter would have 1% accuracy.

The experimental asymmetry, $A_{\rm Res}$, is expected to be linearly proportional to $Q^2 = 4EE'\sin^2(\theta/2)$. In Hall A at JLab, the beam energy, E, can be determined to better than $\delta E < 0.02\%$ [27]. The scattered electron energy, E' as measured by either HRS spectrometer is known to $\delta E' = 0.1\%$ and previous experiments have acheived a pointing resolution of $\delta\theta \leq 0.5$ mr. When combined, the contribution of δQ^2 to the uncertainty of the asymmetry is expected to be $\delta A/A = 0.9\%$.

5 Conclusion

The Res-Parity experiment outlined in this talk would be the first measurement of parity violating electron scattering from hydrogen and deuterium targets covering the resonance region. The experiment has a large potential for discoveries in a kinematic region which is poorly explored with a weak probe. The experiment would test global quark-hadron duality in parity violation to 3% and local duality to better than 5%. In addition to providing interesting data on the resonance region, Res-Parity would be able to constrain higher twist contributions to the 6 GeV PV-DIS and 12 GeV DIS-Parity experiments and would be important inputs to the radiative corrections which would be necessary for those experiments. Res-Parity would place constraints on the parity violating background in Møller scattering experiments both at

SLAC [24] and at JLab [25,26]. The carbon and deuterium target data collected in this experiment would be used to study nuclear effects with the weak interaction. These data are especially interesting to the current generation of neutrino experiments which depend on nuclear targets for luminosity and scattering in and near the resonance region to search for neutrino flavor oscillations. Res-Parity is a relatively simple measurement of parity violation due to the large asymmetry, $A_{\text{Res}} \approx 50 \times 10^{-6}$, and it is a natural experiment to be completed in association with the PV-DIS experiment (JLab E05-007)[20,21], using the same experimental equipment and technique.

This work was supported in part by the U.S. Department of Energy, Office of Nuclear Physics, under Contract Nos. DE-AC05-84150 and W-31-109-ENG-38.

References

1. W. Melnitchouk, R. Ent, C. Keppel, Phys. Rept. **406**, 127 (2005), `hep-ph/0501217`.
2. K.A. Aniol et al. (HAPPEX), Phys. Rev. Lett. **96**, 022003 (2006), `nucl-ex/0506010`.
3. D.S. Armstrong et al. (G0), Phys. Rev. Lett. **95**, 092001 (2005), `nucl-ex/0506021`.
4. P. Bosted, J. Arrington, V. Dharmawardane, H. Mkrtchyan, X. Zheng et al., *Parity violating electron scattering in the resonance region (Res-Parity)* (2005), JLab P06-005.
5. C.Y. Prescott et al., Phys. Lett. **B77**, 347 (1978)/
6. C.Y. Prescott et al., Phys. Lett. **B84**, 524 (1979).
7. R.N. Cahn, F.J. Gilman, Phys. Rev. **D17**, 1313 (1978).
8. J.D. Bjorken, Phys. Rev. **D18**, 3239 (1978).
9. J.E. Amaro, M.B. Barbaro, J.A. Caballero, T.W. Donnelly, A. Molinari, Phys. Rept. **368**, 317 (2002), `nucl-th/0204001`.
10. M.J. Musolf et al., Phys. Rept. **239**, 1 (1994).
11. M.B. Barbaro, J.A. Caballero, T.W. Donnelly, C. Maieron, Phys. Rev. **C69**, 035502 (2004), `nucl-th/0311088`.
12. A. De Pace, M. Nardi, W.M. Alberico, T.W. Donnelly, A. Molinari, Nucl. Phys. **A741**, 249 (2004), `nucl-th/0403023`.
13. D.R.T. Jones, S.T. Petcov, Phys. Lett. **B91**, 137 (1980).
14. L.M. Nath, K. Schilcher, M. Kretzschmar, Phys. Rev. **D25**, 2300 (1982).
15. K. Matsui, T. Sato, T.S.H. Lee, Phys. Rev. **C72**, 025204 (2005), `nucl-th/0504051`.
16. T. Sato, T.S.H. Lee, Phys. Rev. **C54**, 2660 (1996), `nucl-th/9606009`.
17. E.D. Bloom, F.J. Gilman, Phys. Rev. Lett. **25**, 1140 (1970).
18. A. De Rujula, H. Georgi, H.D. Politzer, Ann. Phys. **103**, 315 (1977).
19. D.M. Alde et al., Phys. Rev. Lett. **64**, 2479 (1990).
20. X. Zheng, P. Reimer, R. Michaels et al., $\mathbf{e}-^2H$ *parity violating deep inelastic scattering at CEBAF 6 GeV* (2004), JLab E05-007.
21. X. Zheng, *Parity violation in deep inelastic scattering at JLab 6 GeV*, this conference.
22. P. Reimer, K. Paschke, X. Zheng et al., *Precision measurement of the parity-violating asymmetry in deep inelstic scattering off deuterium using baseline 12 GeV equipment in Hall C* (2006), JLab LOI-12-06-111.
23. P. Souder, *DIS-Parity at 12 GeV*, this conference.
24. P.L. Anthony et al. (SLAC E158), Phys. Rev. Lett. **95**, 081601 (2005), `hep-ex/0504049`.
25. D. Mack, *Outlook of an improved measurement of parity violation in Moeller scattering at JLab (e2ePV)*, this conference.
26. J. Arrington et al., eds., *Pre-Conceptual Design Report (pCDR) for The Science and Experimental Equipment for The 12 GeV Upgrade of CEBAF* (Thomas Jefferson National Accelerator Facility, 2004).
27. J. Alcorn et al., Nucl. Instrum. Meth. **A522**, 294 (2004).
28. D.E. Wiser, Ph.D. thesis, University of Wisconsin (1977).
29. K.A. Aniol et al. (HAPPEX), Phys. Lett. **B635**, 275 (2006), `nucl-ex/0506011`.
30. R. Michaels et al., *A clean measurement of the neutron skin of* ^{208}*Pb through parity violating electron scattering* (2000), JLab E00-003.

Parity violation, the neutron radius of lead, and neutron stars

J. Piekarewicz[a]

Department of Physics, Florida State University, Tallahassee, FL 32306, USA

Abstract. The neutron radius of a heavy nucleus is a fundamental nuclear-structure observable that remains elusive. Progress in this arena has been limited by the exclusive use of hadronic probes that are hindered by large and controversial uncertainties in the reaction mechanism. The parity radius experiment at the Jefferson Laboratory offers an attractive electro-weak alternative to the hadronic program and promises to measure the neutron radius of ^{208}Pb accurately and model independently via parity-violating electron scattering. In this contribution we examine the far-reaching implications that such a determination will have in areas as diverse as nuclear structure, atomic parity violation, and astrophysics.

PACS. 21.10.Gv Mass and neutron distributions – 26.60.+c Nuclear matter aspects of neutron stars

1 Introduction

The nucleus of ^{208}Pb is 18 order of magnitudes smaller and 55 orders of magnitude lighter than a neutron star. Yet remarkably, both the neutron radius of ^{208}Pb as well as the radius of a neutron star depend critically on our (incomplete) knowledge of the equation of state of neutron-rich matter. The emergence of such a correlation among objects of such a disparate size is not difficult to understand. Heavy nuclei develop a neutron-rich skin as a result of a large neutron excess and a large Coulomb barrier that reduces the proton density at the surface of the nucleus. Thus the thickness of the neutron skin depends on the pressure that pushes neutrons out against surface tension [1]. It is this same pressure that supports a neutron star against gravitational collapse [2,3]. Thus models with thicker neutron skins often produce neutron stars with larger radii [4].

Attempts at mapping the neutron distribution of ^{208}Pb, or of any other heavy nucleus, using hadronic probes have been met with limited success. Although highly mature and successful, the hadronic program will never attain the precision status that the electro-weak program enjoys. This is due to the large and controversial uncertainties in the reaction mechanism [5,6]. A particularly illustrative example of this situation is provided by the proton and neutron radii of ^{208}Pb. While elastic electron scattering experiments have determined the charge radius of ^{208}Pb to better than 0.001 fm [7], realistic estimates place the uncertainty in the neutron radius at about 0.2 fm [8].

[a] e-mail: jorgep@csit.fsu.edu

The enormously successful parity-violating program at the Jefferson Laboratory [9,10] provides an attractive electro-weak alternative to the hadronic program. Indeed, the Parity Radius Experiment (PREX) at the Jefferson Laboratory aims to measure the neutron radius of ^{208}Pb accurately (to within 0.05 fm) and model independently via parity-violating electron scattering [8]. Parity violation at low momentum transfers is particularly sensitive to the neutron density because the Z^0-boson couples primarily to neutrons. Moreover, the parity-violating asymmetry, while small, can be interpreted with as much confidence as conventional electromagnetic scattering experiments. PREX will provide a unique observational constraint on the thickness of the neutron skin of a heavy nucleus. We note that since first proposed in 1999, many of the technical difficulties intrinsic to such a challenging experiment have been met [11]. For further details on the status of the experiment, see the contribution by Robert Michaels to this conference.

2 Formalism

The starting point for the calculation of the properties of finite nuclei, their self-consistent linear response, and the structure and dynamics of neutron stars is based on a relativistic density functional. The underlying Lagrangian density includes an isodoublet nucleon field interacting via the exchange of two isoscalar mesons —a scalar and a vector— one isovector meson, and the photon. Details of this model may be found in refs. [12,13]. In addition to meson-nucleon interactions, the Lagrangian density must be supplemented by nonlinear meson interactions that are responsible for a softening of the equation of state of symmetric nuclear matter at both normal and high densities [14]. Of particular relevance to the present contribution is an effective coupling constant (denoted by $\Lambda_{\rm v}$) that induces isoscalar-isovector mixing and has been added to tune the poorly known density dependence of the symmetry energy [4,15]. As a result of the strong correlation between the neutron radius of heavy nuclei and the pressure of neutron-rich matter [1,16], the neutron skin of a heavy nucleus is also highly sensitive to changes in $\Lambda_{\rm v}$. Hence, the interacting Lagrangian density of ref. [14] has been supplemented by the following term:

$$\mathcal{L}_{\rm int} = \Lambda_{\rm v}\left(g_\rho^2\,\mathbf{b}_\mu\cdot\mathbf{b}^\mu\right)\left(g_{\rm v}^2 V_\mu V^\mu\right), \qquad (1)$$

where V^μ and $\mathbf{b}^\mu$ denote isoscalar and isovector vector fields, respectively.

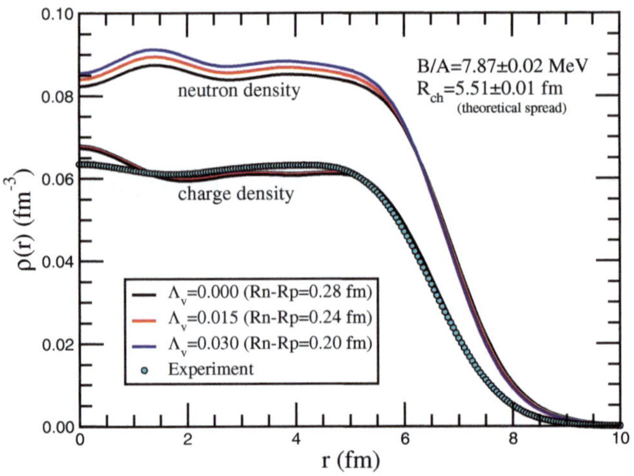

Fig. 1. Proton (charge) and neutron (point) densities for ^{208}Pb using a variety of values for the isoscalar-isovector coupling constant $\Lambda_{\rm v}$.

3 Results

We start this section by displaying in fig. 1 the impact of the addition of the empirical coupling constant $\Lambda_{\rm v}$ on both the charge and point-neutron densities of ^{208}Pb. An important constraint that must be satisfied by the addition of any new coupling constant into the Lagrangian is that the success of the effective model in reproducing well-determined ground-state observables is maintained. Figure 1 indicates that this is indeed the case for the charge density —an observable largely insensitive to the value of $\Lambda_{\rm v}$. In contrast, the changes in $\Lambda_{\rm v}$ depicted in the figure yield a significant reduction in the value of neutron skin of ^{208}Pb: from $R_n - R_p = 0.28\,$fm to $R_n - R_p = 0.20\,$fm. Note that the neutron skin $R_n - R_p$ is defined as the difference between the (point) root-mean-square neutron and proton radii.

3.1 Atomic parity violation

Due to the widespread interest of the *PAVI 2006* audience on atomic parity violation, we add a brief discussion on the fact that the impact of a 1% measurement of the neutron radius in ^{208}Pb could have on the neutron radius of those heavy nuclei that have been identified as promising candidates to the atomic parity violation program. These include barium, dysprosium, ytterbium, and francium. For more in-depth discussions on atomic parity violation, see the contributions to this conference by Profs. Derevianko, Lintz, Budker, Tsigutkin, Gwinner, and Sanguinetti.

In large part the choice of suitable atomic systems is the existence of very close (nearly degenerate) levels of opposite parity that enhance significantly the parity-violating amplitudes. Unfortunately, parity-violating matrix elements are contaminated by uncertainties in both atomic and nuclear structure. A fruitful experimental strategy for removing the sensitivity to the atomic theory is to measure ratios of parity violation observables along an isotopic chain. This leaves nuclear-structure uncertainties, in the form of differences in neutron radii, as the limiting factor in the search for physics beyond the standard model [17–19]. All three elements, barium, dysprosium, and ytterbium, have long chains of naturally occurring isotopes. While the experimental strategy demands a precise knowledge of neutron radii along the complete isotopic chain, we only correlate here (as means of illustration) the neutron radius of ^{208}Pb to the neutron radius of a member of the isotopic chain having a closed neutron shell (or subshell). That is, we focus exclusively on ^{138}Ba($Z=56; N=82$), ^{158}Dy($Z=66; N=92$), and ^{176}Yb($Z=70; N=106$).

The neutron skins of ^{138}Ba, ^{158}Dy, and ^{176}Yb, are correlated to the corresponding neutron skin of ^{208}Pb in the three panels of fig. 2. We observe a tight linear correlation that is largely model independent. The linear regression coefficients (slope m and intercept b) have been enclosed in parenthesis [20]. A theoretical spread of approximately 0.2 to 0.3 fm in the neutron radius of ^{208}Pb was estimated in refs. [16,17]. Most of this spread is driven by the difference between relativistic and nonrelativistic models, which has recently been attributed to the poorly known density dependence of the symmetry energy [21,22]. With the culmination of the parity radius experiment at the Jefferson Laboratory [11], the theoretical spread will be replaced by a genuine experimental error that is five times smaller, that is, $\Delta R_n(^{208}\text{Pb}) \simeq 0.056\,$fm. This 1% measurement of the neutron radius in ^{208}Pb translates into a neutron radius uncertainty of $\Delta R_n(^{138}\text{Ba}) \simeq 0.045\,$fm, $\Delta R_n(^{158}\text{Dy}) \simeq 0.034\,$fm, and $\Delta R_n(^{176}\text{Yb}) \simeq 0.052\,$fm, respectively. The results presented above employ a nuclear-structure model that lacks both deformation and pairing correlations —effects that may be important for the nuclei considered in atomic parity violation experiments. Thus, the present calculations were limited to the study of that single member of each isotopic chain having a closed neutron shell. This shortcoming was overcome in ref. [23] through the inclusion of both nuclear deformation and pairing correlations. Further, in anticipation of future experiments on high-Z atoms —where the accuracy in the

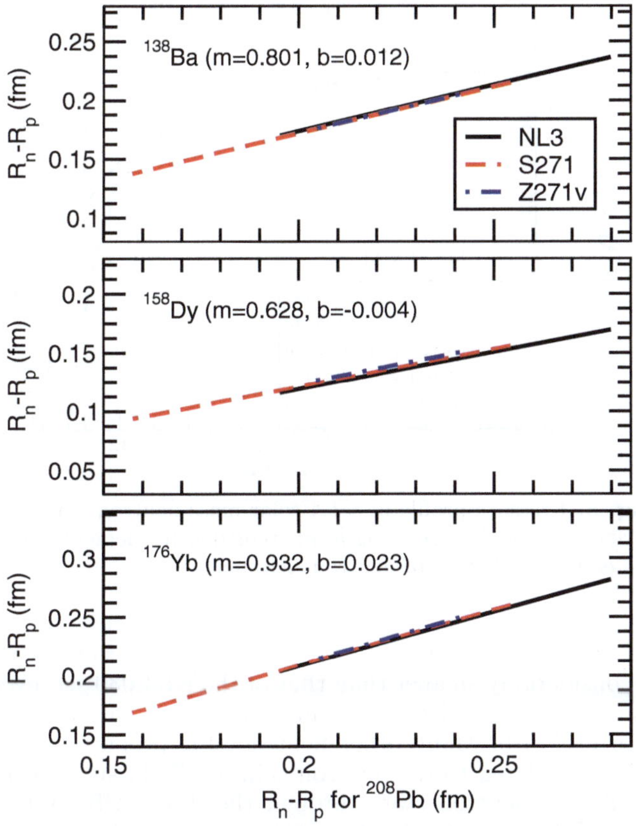

Fig. 2. Skin-skin correlations for three heavy nuclei of possible relevance to atomic parity violation (^{138}Ba, ^{158}Dy, and ^{176}Yb) as a function of the neutron skin of ^{208}Pb for three models predicting a different density dependence for the symmetry energy. Quantities in parenthesis represent linear regression coefficients (slope and intercept).

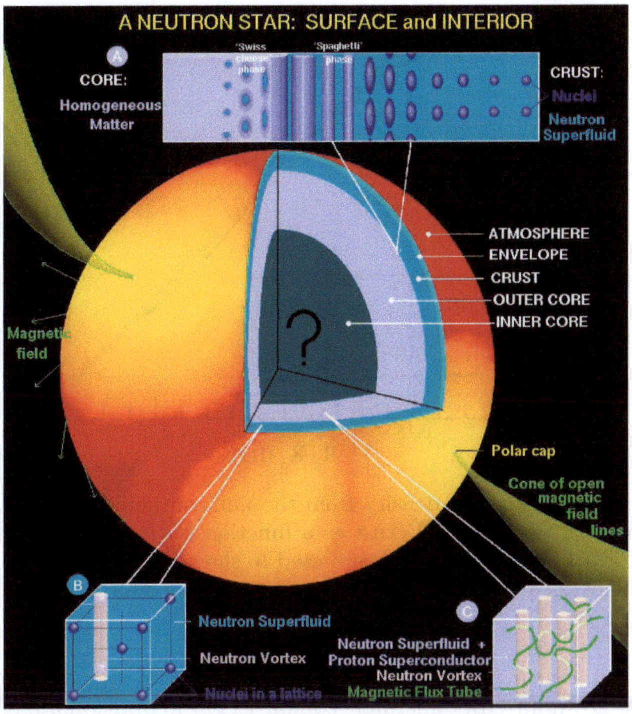

Fig. 3. State-of-the-art rendition of the structure of a neutron star (courtesy of Dany Page).

measurements of atomic parity-violating effects may be significantly improved— nuclear-structure corrections to the weak charge in francium isotopes were also computed. Insofar as the neutron radius of heavy nuclei may be regarded as a robust bulk property, the results presented in ref. [23] confirmed the results presented here in fig. 2.

3.2 Neutron star structure

Neutron stars contain a non-uniform crust above a uniform liquid mantle (or outer core). See fig. 3, courtesy of Dany Page, for an accurate rendition of the expected structure of a neutron star. To compute various properties of neutron stars, the equation of state for the uniform liquid phase is assumed to consist of neutrons, protons, electrons, and muons in beta equilibrium. Further, it is assumed that this description remains valid in the high-density inner core. Thus, the very interesting possibility of transitions to various exotic phases, such as meson condensates, hyperonic matter, and/or quark matter, are not considered here. In the opposite domain, namely, at the lower densities of the crust, the uniform system becomes unstable against density fluctuations. That is, at these densities it becomes energetically favorable for the system to separate into regions of high- and low-density matter. In this non-uniform region the system is speculated to consist of a variety of complex and exotic structures, such as spheres, cylinders, rods, plates, *etc.* —collectively dubbed as *nuclear pasta* [24,25]. While microscopic calculations of the nuclear pasta are now becoming available [26–29], it is premature to incorporate them in our calculation. Hence, after determining the transition density from the uniform liquid mantle to the non-uniform solid crust via an RPA stability analysis [15], a simple polytropic equation of state is used to interpolate between the outer crust [30] and the uniform liquid [31].

Results for the transition density from the uniform liquid mantle to the non-uniform solid crust as a function of the neutron skin in ^{208}Pb are displayed in fig. 4. Various models are used to show the nearly model-independent relation between these two seemingly distinct observables. The figure displays an inverse correlation between the neutron skin and the transition density found in ref. [15]. This correlation suggests that models with a stiff equation of state predict a low transition density, as it becomes energetically unfavorable to separate nuclear matter into regions of high and low densities. Finally, this "data-to-data" relation illustrates how an accurate and model-independent determination of the neutron skin of ^{208}Pb at the Jefferson Laboratory —assumed here purely on theoretical biases to be $R_n - R_p = 0.20$ fm— would determine an important neutron star observable.

Having constructed several accurately calibrated equation of states, such as NL3 [32] —together with its softer versions [15]— and FSUGold [33], we now examine their predictions for the structure of a "canonical" 1.4 solar-

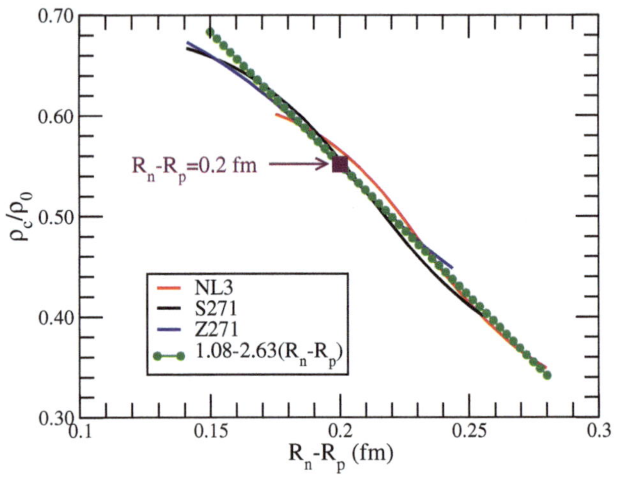

Fig. 4. Transition density from the uniform liquid mantle to the non-uniform solid crust as a function of the neutron skin of ^{208}Pb. Different models are used to show the largely model-independent relation between these two observables.

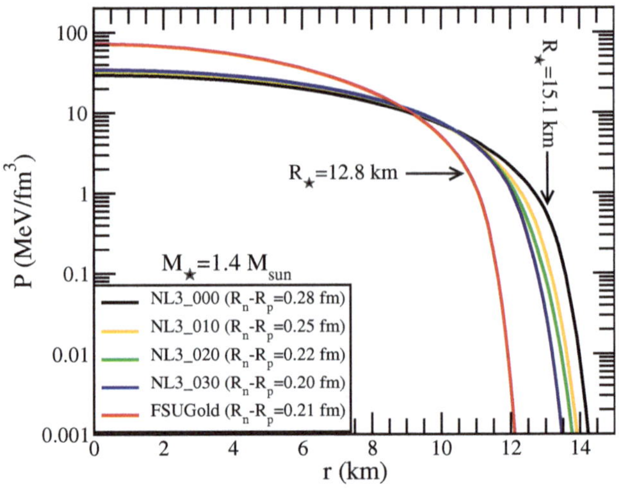

Fig. 5. Pressure profile of a 1.4 solar-mass neutron star for a variety of models predicting a different density dependence of the equation of state of neutron-rich matter.

mass neutron star ($M = 1.4\,M_\odot$). To do so, we solve the Tolman-Oppenheimer-Volkoff (TOV) equations, a set of equations appropriate for the structure of spherically symmetric neutron stars in hydrostatic equilibrium. These equations are a generalization of Newton's equation for hydrostatic equilibrium supplemented by three corrections from general relativity [34]. Incorporating these correction terms is critical, as typical escape velocities from neutron stars are of the order of half the speed of light. The TOV equations with their associated boundary conditions are still incomplete without the provision of an equation of state, *i.e.*, a relation between the pressure and the energy density. Indeed, if general relativity is assumed valid —a very modest and safe assumption— the only physics that the structure of neutron stars is sensitive to is the equation of state of neutron-rich matter in beta equilibrium.

In fig. 5 the pressure profile of a $M = 1.4\,M_\odot$ neutron star is displayed for various equations of states. The curves labeled as NL3 employ the original NL3 set of Lalazissis and collaborators [32] (denoted here as NL3_000) plus variants of it labeled by its value for the isoscalar-isovector mixing $\Lambda_{\rm v}$ (see eq. (1)). The addition of $\Lambda_{\rm v}$ allows one to tune the neutron radius of ^{208}Pb without sacrificing the success of the model in reproducing a variety of well-known ground-state observables, such as binding energies and charge radii. The figure clearly indicates the robust correlation between the neutron skin of ^{208}Pb (enclosed in parenthesis) and the radius of the neutron star: models with a softer symmetry energy yield both neutron skins and neutron star radii smaller relative to the stiffer NL3 set. Yet the correlation, as opposed to the one displayed in fig. 4, is not model independent. The radius of the neutron star depends on both the low- and high-density dependence of the equation of state, while the neutron skin of ^{208}Pb depends only on the former. Indeed, the FSUGold set, with a considerably softer high-density component of the equation of state, requires higher pressures —and thus higher densities— to prevent the gravitational collapse of the star. As a result, FSUGold predicts a star radius that is considerably smaller than that of the NL3_030 set, even though they predict very similar values for the neutron skin of ^{208}Pb. Many other neutron star properties have been correlated to the neutron skin of ^{208}Pb, such as the cooling of neutron stars through the direct URCA process. Unfortunately, due to space limitations they are not being addressed here. The interested reader is referred to ref. [33] for further details.

4 Conclusions

The neutron radius of a heavy nucleus is a fundamental nuclear-structure observable that remains elusive, mainly due to our inability to use electro-weak probes to sample the neutron distribution. While a highly mature hadronic program has been used to map the neutron distribution, the clean extraction of the neutron radius has been marred by controversial uncertainties in the reaction mechanism. The established and successful parity-violating program at the Jefferson Laboratory provides an attractive electroweak alternative to the hadronic program. The parity radius experiment at the Jefferson Laboratory will take advantage of the strong coupling of the Z^0-boson to neutrons to measure the neutron radius of ^{208}Pb accurately and model independently. While the intrinsic achievement of this experiment is undeniable, in this contribution we have examined the far-reaching consequences that such a measurement could have over fields as diverse as atomic parity violation and astrophysics. First, a tight correlation was found between the neutron skin of ^{208}Pb (the aim of the PREX experiment) and the neutron radius of a variety of elements (barium, dysprosium, and ytterbium) of relevance to the atomic parity violation program. Although the calculations presented here neglect both deformation and pairing correlations, we have argued —based on more sophisticated calculations— that being a bulk nuclear property, the neutron radius remains largely insensitive to these effects. Second, we demonstrated that PREX will also have a strong impact on various astrophysical observables. Indeed, a model-independent (or "data-to-data") relation between the neutron skin of ^{208}Pb and the transition density from the uniform liquid mantle to the non-uniform solid crust was established. This correlation emerged as a result of the similar composition of the neutron skin of a heavy nucleus and the crust of a neutron star: neutron-rich matter at similar densities. Further, we showed how the measurement of the neutron skin in ^{208}Pb is strongly correlated to the radius of a $M = 1.4\,M_{\odot}$ neutron star. This result emerges as a direct consequence of the density dependence of the symmetry energy, as the same pressure that pushes neutrons out against surface tension in the nucleus of ^{208}Pb is responsible for pushing neutrons out against gravity in a neutron star. Yet in contrast to the "data-to-data" relation described above, this correlation is model dependent —as the radius of the neutron star depends on both the low- and high-density component of the equation of state. Thus, we eagerly await observational results to constrain the high-density component of the equation of state. Fortunately, earth- and space-based telescopes have started to place important constrains on the high-density component of the equation of state. New telescopes operating at a variety of wavelengths are turning neutron stars from theoretical curiosities into powerful diagnostic tools. Significant advances in observational astronomy will soon yield the first combined measurement of mass-radius relations for a variety of neutron stars. These results —combined with the parity radius experiment at the Jefferson Laboratory— will provide the most complete information to date on the long-sought equation of state of neutron-rich matter.

The author wants to acknowledge the many colleagues that have contributed to this work —especially Prof. C.J. Horowitz and Dr. Bonnie Todd-Rutel. The author is very grateful to the organizers of the *PAVI 2006* conference for their kind invitation. This work was supported in part by DOE grant DE-FG05-92ER40750.

References

1. B.A. Brown, Phys. Rev. Lett. **85**, 5296 (2000).
2. J.M. Lattimer, M. Prakash, Astrophys. J. **550**, 426 (2001) astro-ph/0002232.
3. A.W. Steiner *et al.*, Phys. Rep. **411**, 325 (2005) nucl-th/0410066.
4. C.J. Horowitz, J. Piekarewicz, Phys. Rev. C **64**, 062802 (2001) nucl-th/0108036.
5. L. Ray, G.W. Hoffmann, Phys. Rev. C **31**, 538 (1985).
6. L. Ray, G.W. Hoffmann, W.R. Coker, Phys. Rep. **212**, 223 (1992).
7. G. Fricke *et al.*, At. Data Nucl. Data Tables **60**, 177 (1995).
8. C.J. Horowitz *et al.*, Phys. Rev. C **63**, 025501 (2001) nucl-th/9912038.
9. HAPPEX Collaboration (K.A. Aniol *et al.*), Phys. Rev. Lett. **96**, 022003 (2006) nucl-ex/0506010.
10. HAPPEX Collaboration (K.A. Aniol *et al.*), Phys. Lett. B **635**, 275 (2006) nucl-ex/0506011.
11. R. Michaels, P.A. Souder, G.M. Urciuoli, `http://hallaweb.jlab.org/parity/prex`.
12. B.D. Serot, J.D. Walecka, Adv. Nucl. Phys. **16**, 1 (1986).
13. B.D. Serot, J.D. Walecka, Int. J. Mod. Phys. E **6**, 515 (1997) nucl-th/9701058.
14. H. Mueller, B.D. Serot, Nucl. Phys. A **606**, 508 (1996) nucl-th/9603037.
15. C.J. Horowitz, J. Piekarewicz, Phys. Rev. Lett. **86**, 5647 (2001) astro-ph/0010227.
16. R.J. Furnstahl, Nucl. Phys. A **706**, 85 (2002).
17. S.J. Pollock, E.N. Fortson, L. Wilets, Phys. Rev. C **46**, 2587 (1992) nucl-th/9211004.
18. B.Q. Chen, P. Vogel, Phys. Rev. C **48**, 1392 (1993) nucl-th/9303003.
19. M.J. Ramsey-Musolf, Phys. Rev. C **60**, 015501 (1999) hep-ph/9903264.
20. B.G. Todd, J. Piekarewicz, Phys. Rev. C **67**, 044317 (2003) nucl-th/0301092.
21. J. Piekarewicz, Phys. Rev. C **66**, 034305 (2002).
22. J. Piekarewicz, Phys. Rev. C **69**, 041301 (2004) nucl-th/0312020.
23. T. Sil *et al.*, Phys. Rev. C **71**, 045502 (2005) nucl-th/0501014.
24. D.G. Ravenhall, C.J. Pethick, J.R. Wilson, Phys. Rev. Lett. **50**, 2066 (1983).
25. M. Hashimoto, H. Seki, M. Yamada, Prog. Theor. Phys. **71**, 320 (1984).
26. G. Watanabe, H. Sonoda, *Soft Condensed Matter: New Research* (Nova Science, 2006) cond-mat/0502515, and references therein.
27. C.J. Horowitz, M.A. Perez-Garcia, J. Piekarewicz, Phys. Rev. C **69**, 045804 (2004) astro-ph/0401079.
28. C.J. Horowitz *et al.*, Phys. Rev. C **70**, 065806 (2004) astro-ph/0409296.
29. C.J. Horowitz *et al.*, Phys. Rev. C **72**, 035801 (2005) nucl-th/0508044.
30. G. Baym, C. Pethick, P. Sutherland, Astrophys. J. **170**, 299 (1971).
31. J. Carriere, C.J. Horowitz, J. Piekarewicz, Astrophys. J. **593**, 463 (2003) nucl-th/0211015.
32. G.A. Lalazissis, J. Konig, P. Ring, Phys. Rev. C **55**, 540 (1997) nucl-th/9607039.
33. B.G. Todd-Rutel, J. Piekarewicz, Phys. Rev. Lett. **95**, 122501 (2005) nucl-th/0504034.
34. S. Weinberg, *Gravitation and Cosmology* (John Wiley & Sons, New York, 1972).

The lead radius experiment PREX

R. Michaels

Thomas Jefferson National Accelerator Facility 12000 Jefferson Ave Newport News, VA, USA

Abstract. The proposed PREX experiment at Jefferson Lab will measure the parity-violating electroweak asymmetry in the elastic scattering of polarized electrons at an energy of 850 MeV and a scattering angle of 6°. Since the Z_0 boson couples mainly to neutrons, this asymmetry provides a clean measurement of R_n with a projected experimental precision of ± 1 %. In addition to being a fundamental test of nuclear theory, a precise measurement of R_n pins down the density dependence of the symmetry energy of neutron rich nuclear matter which has impacts on neutron star structure, heavy ion collisions, and atomic parity violation experiments.

PACS. 20.10.Gv Mass and neutron distributions – 25.30.Bf Elastic electron scattering – 11.30.Er Parity Violation

1 Introduction

In a heavy nucleus like ^{208}Pb the difference between the neutron radius R_n and the proton radius R_p is believed to be several percent. This neutron skin has not been well established experimentally in stable nuclei. We plan to measure the neutron charge radius R_n (i.e. the RMS radius of neutrons in a nucleus) in a clean and model independent way analogous to the classic measurements[1] of the proton radius R_p and with unprecedented accuracy as suggested originally by Donnelly, Dubach, and Sick[2].

The experiment measures the parity violating asymmetry in elastic scattering $A = (\sigma_R - \sigma_L)/(\sigma_R + \sigma_L)$. This asymmetry arises due to the the interference of the Z^0 boson amplitude of the weak neutral interaction with the photon amplitude. The asymmetry is sensitive mainly to the neutron radius R_n because the weak charge of the neutron is much larger than that of the proton. In PWIA, the relationship between the asymmetry and the neutron form factor is given by equation (1)

$$A_{LR} = \frac{G_F Q^2}{4\pi\alpha\sqrt{2}} \left[1 - 4\sin^2\theta_W - \frac{F_n(Q^2)}{F_p(Q^2)} \right] \quad (1)$$

where G_F is the Fermi constant, $\alpha = \frac{1}{137}$ is the fine structure constant, θ_W is the Weinberg angle, and $F_n(Q^2)$ and $F_p(Q^2)$ are the neutron and proton form factor of the nucleus. Thus A_{LR} is approximately proportional to the ratio of neutron to proton form factors. In the above we used PWIA to illustrate. To achieve 1% accuracy requires corrections for Coulomb distortions, which have been calculated by Horowitz[4].

2 Physics impact and interpretation

It is remarkable that a single measurement of R_n with 1% accuracy can have such a broad impact on several areas of physics, including neutron star formation and structure[5], atomic parity violation[6, 7], nuclear theory[3], and heavy ion collisions[8].

From a theoretical perspective, the difference between our proposed measurement of R_n and the well-measured proton radius R_p is determined by fundamental parameters of bulk nuclear matter called the symmetry energy S_ν and the density dependence of the symmetry energy S'_ν. These parameters are important for other phenomena involving nuclear matter, such as heavy ion collisions. Today, perhaps the most important application is the field of supernovae and the properties of neutron stars. The proposed measurement of R_n to 1% will provide the best determination of the S_ν and S'_ν at the density of ordinary nuclear mater. This will provide vital input to the dynamics of stellar explosions and thus have implications for our understanding of how the heavy elements in our world were produced.

The relevant part of the energy of nuclear matter of density n and ratio x of protons to neutrons may be written

$$E(n, x) = E(n, x = 1/2) + S_\nu(n)(1 - 2x^2).$$

The connection between $S_\nu(n_0)$, where n_0 is the density of Pb, our measurement, and R_n is shown by Furnstahl[9] within the context of realistic nuclear calculations. The model dependence of the relations is small compared to our required sensitivity.

The importance of the symmetry energy S_ν for the formation and structure of neutron stars is given in several review articles[10,11]. A good introduction for the non-specialist by Lattimer and Prakish[12] appeared in Science. Neutron stars are giant nuclei ~13 km in diameter, fascinating objects for which astronomical data are rapidly improving. Data on neutron stars include radii, masses, luminosity, temperature, and cooling rates [13]. In addition to their intrinsic interest as newly explored natural phenomena, these stars provide "laboratories" to understand the nature of extremely dense matter. For example, do collapsed stars form "exotic" phases of matter, e.g. strange stars or quark stars ?

A precise measurement of R_n provides a calibration of the equation of state (pressure versus density) of neutron rich nuclear matter, which is an important input for calculating the structure of neutron stars[5]. To rule in or out possible exotic phases of dense matter one needs to combine the high density measurements of neutron stars with low density precision measurements of R_n in nuclei. R_n also has implications for the crust thickness because the transition density for the liquid-solid phase transition depends sensitively on the neutron skin of ^{208}Pb[5]. The proton fraction of neutron rich matter in beta equilibrium depends on the symmetry energy, which is calibrated by R_n. A large symmetry energy favors more protons, and if the proton fraction is high enough then the following "URCA" process can cool neutron stars $n \rightarrow p + e^- + \bar{\nu}_e$; $p + e^- \rightarrow n + \nu_e$ where the $\nu_e, \bar{\nu}_e$ carry off energy. URCA cooling might explain recent Chandra observations of the neutron star 3C58, a remnant of the supernova seen in the year 1186 that appears to be unexpectedly cold[13]. A neutron skin larger than about 0.2 fm may imply that URCA cooling is possible, while a smaller skin implies it is probably not possible[14].

Heavy ion reactions[15,16,8], including rare isotope beam experiments anticipated from RIA[17], can also pin down the density dependence of the symmetry energy by probing conditions at neutron densities up to 3 times normal nuclear matter density. A combination of our measurement at sub-nuclear densities and the RIA results will provide useful constraints on nuclear theory.

The impact of an accurate R_n measurement on atomic parity violation (APV) experiments has been analyzed by Pollock *et.al* [3],[7]. Knowledge of R_n at the 1% level is needed for interpreting atomic physics measurements of the Weinberg angle at the level of the Standard Model weak radiative corrections. The most accurate (1% in Q_{weak}) measurement of APV was carried out by Wieman and co-workers in atomic cesium ^{133}Cs[18]. Such low energy experiments provide powerful constraints on the standard model [19]. More independent tests of APV are needed at the accuracy of the Cesium experiment. An alternative approach which largely cancels the uncertainties due to difficult atomic structure calculations is to measure ratios of APV amplitudes between isotopes [20]. This is being exploited for example in the Berkeley atomic Yb experiment [21]. However, Fortson *et.al.* [22] pointed out that this technique has an enhanced sensitivity to uncertainties in R_n. A recent analysis[6] has emphasized the importance of this experiment on the field.

There have been renewed attempts to obtain R_n from hadronic data. A major effort has been devoted to the analysis of proton scattering [23–26]. Clark and Kerr [25] arrive at the smallest claimed error on R_n, less than 1%. Piekarewicz[23] suggests that the error should be at least 3%. Karataglidis et al.[24] suggest that the proton and proposed data are complementary and that both are important. We note that recent data on pion photoproduction from lead and other nuclei [27] have yielded matter radii that are less than the charge radii, implying a *negative* neutron skin in ^{208}Pb.

The physics interpretation of the experiment can be summarized as follows. From the measured asymmetry one may deduce the weak form factor, which is the Fourier transform of the weak charge density at the momentum transfer of the experiment. One must correct for Coulomb distortions, which has been done accurately by Horowitz[4] and others[28]. The weak charge density can be compared directly to theoretical calculations and this will constrain the density dependence of the symmetry energy. The weak density can be directly applied to atomic parity violation because the observables have approximately the same dependence on nuclear shape. From the weak charge density one can also deduce a neutron density at one Q^2 by making small corrections for known nucleon form factors. The uncertainty in these corrections for a realistic experiment have been estimated and are small[3].

Finally from a low Q^2 measurement of the point neutron density one can deduce R_n. This requires knowledge of the surface thickness to about 25% to extract R_n to 1%. The spread in surface thickness among successful mean field models is much less than 25%, hence we can extract R_n with the desired accuracy as shown by Furnstahl[9]. In summary, the physics results of the experiment are the weak charge density, the point neutron density and R_n.

3 Experiment

3.1 Experimental overview

The ^{208}Pb experiment was proposed for a beam energy of 850 MeV and a 6° scattering angle in Hall A at Jefferson Lab. The two Hall A 3.7 msr spectrometer systems supplemented by septum magnets focus elastically scattered electrons onto total-absorption detectors in their focal planes. A 50μA, 80% polarized beam with a 30 Hz helicity reversal will scatter from a foil of lead which is sandwiched between sheets of diamond to improve the thermal characteristics. Ratios of detected flux to beam current integrated in the helicity period are formed, and the parity–violating asymmetry in these ratios computed from the helicity–correlated difference divided by the sum: A = $(\sigma_R - \sigma_L)$ / $(\sigma_R + \sigma_L)$, where $\sigma_{R(L)}$ is the ratio for right(R) an left(L) handed electrons. Separate studies at lower rates are required to measure backgrounds, acceptance, and Q^2. Polarization is measured once a day by a Møller polarimeter, and monitored continuously with the Compton polarimeter.

Table 1. Rate and Asymmetry for ^{208}Pb.

Measured Asymmetry (p_e A)	0.51 ppm
Beam Energy	850 MeV
Beam Current	50μA
Required Statistical Accuracy	3%
Energy Cut (due to detector)	4 MeV
Detected Rate (ea spectrometer)	860 MHz
Running Time	680 hours

Table 1 shows the rates, asymmetries, and running times for the ^{208}Pb parity experiment proposal, the results averaged over the acceptance.

3.2 Technical issues

For the ^{208}Pb experiment the asymmetry of 0.5 ppm must be measured to 3% accuracy. Both the absolute error (15 ppb) and the relative accuracy (3%) are challenging to achieve. The main issues affecting the absolute error are the control of false asymmetries associated with helicity correlations in beam parameters such as intensity, energy, and position. The main issues affecting the relative error are the beam polarimetry and measurement of Q^2. In addition, we need a lead target that can withstand a high beam current.

3.3 Helicity correlated systematics

To evaluate the systematic error we need to know the helicity correlated differences as well as our sensitivity to these which are measured online by modulation of the parameters. During the 2005 HAPPEX run we have made tremendous progress in controlling helicity correlated systematics, see fig 1 which shows 1-month averages of ~1 nanometer in helicity correlated position differences. This accomplishment is due to a long-term effort to improve the setup of the polarized source and accelerator. For the polarized source we have a well-developed model for controlling the laser systematics, which allows us to minimize helicity correlations in the laser beam used to produce polarized electrons. For the accelerator, progress in understanding the betatron matching has helped achieve maximum dampening of position differences, while improved understanding of beam tuning has provided both excellent signal-to-noise ratios in the Compton polarimeter and good phase advances along our beamline that permit us to simultaneously have good measurements to the sensitivities of each of the independent beam parameters. For ^{208}Pb we will maintain position differences to less than 1 nm with an accuracy of 0.1 nm averaged over a 1 month run. The charge asymmetry will be maintained to less than 100 ppb with an accuracy of 10 ppb.

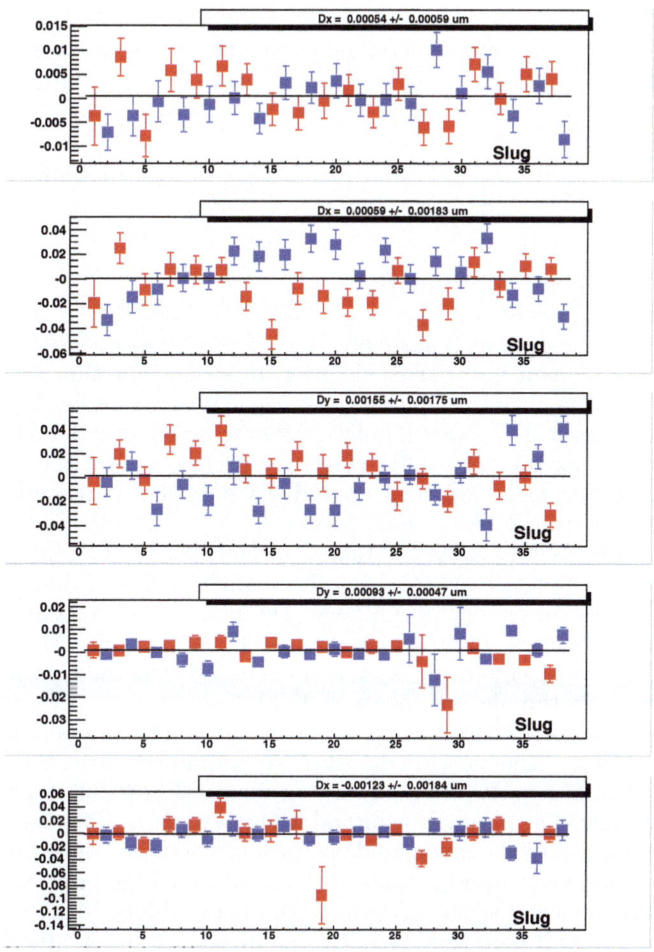

Fig. 1. Helicity correlated beam monitor differences (μm) versus slug number (1 "slug" ~ 1 day of running) for HAPPEX-2. The top four plots are X and Y monitors near the target, and the bottom plot is a monitor in the dispersive section of the magnet ARC leading into the hall, providing a relative energy monitor. The averages over a month of running are ~ 1 nanometer.

3.4 Lead target tests

We have built a high power lead target which will be stable at 40 Watt for a 50 μA beam. Successful beam tests at 80μA show that we have a good target design. Recently we have performed low-intensity tests at 1.1 GeV, which is approximately the energy of this proposal. These tests checked the rates, backgrounds, resolution, sensitivity to beam parameters (slopes), and width of asymmetry

3.5 Noise

The pulse-to-pulse noise in the lead experiment is 140 ppm. All other noises must be small compared to 140 ppm in order for counting statistics to dominate our errors and to keep the running time minimal. It will be important to avoid long cable runs for our detectors, and therefore we have installed a distributed DAQ system during

HAPPEX-2, in which the DAQ crates are situated near the detectors. For understanding the noise, beam systematics, and possible target boiling effects, we have implemented a new luminosity monitor which was used during HAPPEX. New 18-bit ADCs are being fabricated at JLab for use in upcoming parity experiments which will improve our pedestal noise.

3.6 Precision polarimetry

Improvements in polarimetry are of vital importance for the Jefferson Lab parity violation program. For this proposal, the polarization must be measured to 1% preferably, or at least 2%. With a polarization accuracy of 1% (2%) we can extract R_n to 1% (1.2%) respectively. Our plan is to use a combination of upgraded Compton polarimeter and upgraded Møller polarimeter.

The best accuracy obtained so far with the Compton polarimeter is 1.5% total relative error within 40 minutes for a beam energy of 4.5 GeV. We plan to implement a green laser to improve the figure of merit at 850 MeV. This brings the mean asymmetry to 0.65%. At the Compton edge the photon energy is 26 MeV and the associated scattered electron is 6 mm above the primary beam at the location of the electron detector. Assuming the same laser power of 1.5 kW at the Compton interaction point, a 1% statistical accuracy is achieved within 16 hours. The laser and much of the associated optics equipment has been obtained and is undergoing test. In addition to the laser, we plan on using a new crystal for the photon detector.

We will compare Møller to Compton at 5 μA and 6 GeV to give a clear comparison at the 1% level.

4 Summary and outlook

Measurement of R_n will have a fundamental impact on nuclear physics and its applications to neutron rich matter. The neutron skin is a qualitative feature of heavy nuclei which has never been cleanly observed. These experiments require state-of-the-art control of systematic errors. The technological developments will help drive other facets of the Jefferson Lab parity program, such as Q_{weak} which has similar demands.

The HAPPEX run in 2004 and 2005 have demonstrated excellent control of helicity correlated beam parameters. An upgrade of the Compton polarimeter, necessary to achieve 1% accuracy at 850 MeV, is underway.

References

1. H. de Vries, C. W. de Jager, and C. de Vries, Atomic and Nuclear Data Tables, **36**, 495 (1987).
2. T.W. Donnelly, J. Dubach, and I. Sick, Nucl. Phys. **A503**, 589 (1989).
3. C. J. Horowitz, S. J. Pollock, P. A. Souder, and R. W. Michaels, Phys. Rev. C **63**, 025501, (2001).
4. C. J. Horowitz, Phys. Rev. C **57**, 3430 (1998).
5. C. J. Horowitz, J. Piekarweicz, Phys. Rev. Lett **86**, 5647 (2001). C. J. Horowitz, J. Piekarewicz, Phys. Rev. C**64**, 062802 (2001). C. J. Horowitz, J. Piekarewicz, atro-ph/0201113 (2002). J. Carriere, C.J. Horowitz, J. Piekarewicz nucl-th/0211015 (2002).
6. T. Sil, M. Centelles, X. Vinas and J. Piekarewicz, Phys. Rev. C **71**, 045502 (2005) [arXiv:nucl-th/0501014].
7. S.J. Pollock, E.N. Fortson, and L. Wilets Phys. Rev. C **46**, 2587 (1992).
8. A. W. Steiner and B. A. Li, Phys. Rev. C **72**, 041601 (2005) [arXiv:nucl-th/0505051].
9. R. J. Furnstahl, Nucl. Phys. **A706**, 85 (2002).
10. A. W. Steiner, M. Prakash, J. M. Lattimer and P. J. Ellis, Phys. Rept. **411**, 325 (2005) [arXiv:nucl-th/0410066].
11. J. M. Lattimer and M. Prakash, Phys. Rept. **333**, 121 (2000) [arXiv:astro-ph/0002203].
12. J.M. Lattimer, M. Prakash, Science 304 (2004) 536.
13. New York Times, April 11, 2002 edition; P. Slane, D. Helfand, S. Murray, astro-ph/0204151; G.G. Pavolv, O.Y. Kargaltzev, D. Sanwal, and G.P. Garmire, ApJ **554**, L189 (2001); J.P. Halpern and F.Y.H. Wang, ApJ. **447**, 905 (1997); M. Colpi, U. Geppert, D. Page, and A. Possenti, ApJ. **553**, 382 (2001); D. G. Yakovlev *et.al.* astro-ph/0204233.
14. J. Piekarewicz, nucl-th/0207067 (2002).
15. A. Ono, P. Danielewicz, W. A. Friedman, W. G. Lynch and M. B. Tsang, Phys. Rev. C **70**, 041604 (2004) [arXiv:nucl-th/0409027].
16. D. V. Shetty *et al.*, Phys. Rev. C **70**, 011601 (2004) [arXiv:nucl-ex/0406008].
17. B. A. Li, Phys. Rev. C **69**, 034614 (2004) [arXiv:nucl-th/0312025].
18. C. S. Wood *et.al* Science **275**, 1759, (1997) S.C. Bennet and C.E. Wieman, Phys. Rev. Lett. **82**, 2484 (1999).
19. M.-A. Bouchiat and C. Bouchiat, Rep. Prog. Phys. **60**, 1351 (1997).
20. A. Derevianko and S.G. Porsev, physics/0112035 (submitted to Phys. Rev. A) (2002).
21. D. DeMille, Phys. Rev. Lett **74**, 4165 (1995), C.J. Bowers, *et.al* Phys. Rev. **A53**,(5), 3103 (1996); Phys. Rev. A **59**, 5 (1999)
22. E.N. Fortson, Y. Pang, and L. Wilets, Phys. Rev. Lett. **65**, 2857 (1990).
23. J. Piekarewicz and S. P. Weppner, arXiv:nucl-th/0509019.
24. S. Karataglidis, K. Amos, B. A. Brown and P. K. Deb, Phys. Rev. C **65**, 044306 (2002) [arXiv:nucl-th/0111020].
25. B. C. Clark, L. J. Kerr and S. Hama, Phys. Rev. C **67**, 054605 (2003) [arXiv:nucl-th/0209052].
26. M. Dupuis, S. Karataglidis, E. Bauge, J. P. Delaroche and D. Gogny, arXiv:nucl-th/0506077.
27. B. Krusche, arXiv:nucl-ex/0509003 Sept 2005
28. See refs in [3], Relativistic Optical Code RUNT, E.D. Cooper, Los Alamos preprint nucl-th/9911024, D. Vretenar *et al.*, and calculations by B.C. Clark *et al.*.
29. K. A. Aniol *et al.*, Phys. Rev. Lett. **82** 1096 (1999). K. A. Aniol, *et al.*, Phys. Rev. C 69, 065501 (2004). K. A. Aniol, *et al.*, nucl-ex/0506010. K. A. Aniol, *et al.*, nucl-ex/0506011.
30. P.L Anthony, *et.al.* Phys.Rev.Lett.95:081601,2005.
31. T. Holmstrom, R. Feuerbach, R. Michaels, "Q^2 for HAPPEX-2" technical report http://hallaweb.jlab.org/experiment/HAPPEX/docs/qsq_2004.ps
32. M. Hauger *et.al.* Nucl. Instrum. Meth. A**462** 382 (2001).

Helicity correlated asymmetries caused by optical imperfections

K. Aulenbacher[a]

For the A4 Collaboration
Institut für Kernphysik, Johannes-Gutenberg Universität Mainz, D-55099 Mainz, Germany

Abstract. False asymmetries in electron scattering experiments utilizing photoelectron sources are generated by imperfections of the light polarization optics. The false asymmetries are amplified by an analyzing power of the photocathodes. We have demonstrated compensation techniques which allow to approach zero intensity asymmetry with good stability. Even in the absence of analyzing powers fluctuations may be generated by more subtle effects like varying interference patterns of the exciting laser beam.

PACS. 42.25.Ja Polarization – 78.20.Jq Electrooptical effects

1 Introduction

Scattering experiments may attempt to measure the parity-violating (PV) asymmetry A_z

$$A_z = \frac{\sigma^+ - \sigma^-}{\sigma^+ + \sigma^-}\,. \tag{1}$$

Here $\sigma^\pm$ denotes the cross-section for a positive or negative helicity of the particle beam which moves along the z-direction onto an unpolarized target. The $\sigma^\pm$ are usually differential cross-sections which are determined from the rates $R^\pm$ measured by a detection system which is placed under a suitable angle with respect to the z-direction. In a first approximation the rates are proportional to the cross-section $R^\pm = f(\mathbf{x}^\pm)\sigma^\pm$. The function $f(\mathbf{x}^\pm)$ depends on a vector $\mathbf{x}^\pm = x_1^\pm, \ldots, x_N^\pm$ with N parameters. These in particular include the beam parameters (intensity I, position on target (x, y), angle at target $(x'y')\ldots$). If all parameters are independent of the helicity state $(x_1^+, \ldots, x_N^+) = (x_1^-, \ldots, x_N^-)$ it follows that the asymmetry A_z is equal to the measured rate asymmetry $A_R = (R^+ - R^-)/(R^+ + R^-)$ since the function $f(\mathbf{x}^\pm)$ cancels out. If this is not the case A_R is approximately given by

$$A_R \approx A_z + \sum_i \chi_i \Delta x_i = A_z + A_{false}\,. \tag{2}$$

$\Delta x_i = x_i^+ - x_i^-$ is the helicity-correlated (HC) fluctuation of the i-th parameter and $\chi_i = (\mathrm{d}f(\mathbf{x})/\mathrm{d}x_i)|_{\mathbf{x}}/f(\mathbf{x})$ represents the sensitivity of the rate asymmetry on the fluctuation of the i-th parameter. Since A_z has to be extracted from A_R the sum over the HC fluctuations has to be controlled in such a manner that the uncertainty of A_{false} is within the desired error budget. It should be kept in mind that A_z is typically of the order of 10^{-6} so that an accuracy goal of, *e.g.*, 10% leads to the requirement that the error in the determination of A_{false} must be below 100 ppb. To achieve this several issues emerge from an inspection of eq. (2). First, the sensitivities have to be known. Second, because the χ_i can only be determined with limited accuracy, the Δx_i must be kept small and, in addition, they themselves have to be measured with sufficient accuracy.

In all present day electron scattering PV experiments helicity switching is done by generating and reversing the circular polarization of a laser beam by a Pockels cell. The circularly polarized light is then used to produce electrons of alternating spin polarization (beam intensity $I^\pm$) via photoeffect from suitable semiconductor heterostructures like strained GaAs [1,2] or superlattices [3]. All types of photocathodes presently utilized exhibit a non-vanishing analyzing power with respect to linear light polarization components [4,5]. On the other hand, HC linear polarization fluctuations are virtually unavoidable because of optical imperfections. This is the dominant reason for the appearance of a beam intensity asymmetry

$$A_I = \frac{I^+ - I^-}{I^+ + I^-}\,. \tag{3}$$

It is the main contribution to A_{false} if no precautions are taken.

In the following section A_I is analyzed for the case of the PVA4 experiment at the MAMI accelerator in Mainz. It will be shown that the observed intensity asymmetry can be reliably compensated and its dependence on the parameters of the optical system is reasonably well under-

[a] e-mail: aulenbac@kph.uni-mainz.de

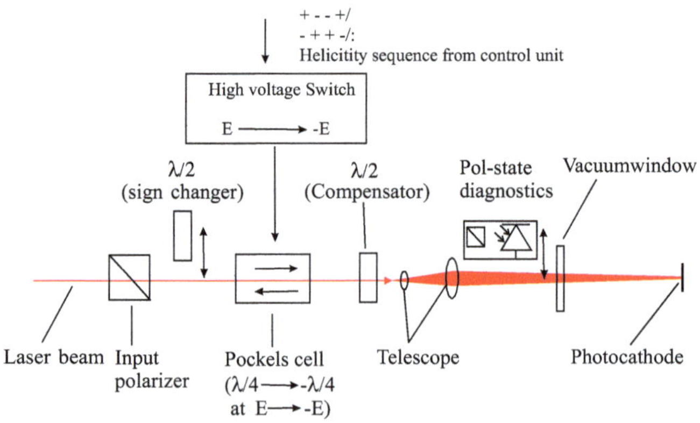

Fig. 1. Polarization optics for PVA4 (schematic).

stood. Section 3 provides additional investigations which for example demonstrate the existence of HC beam position fluctuations even in the absence of an analyzing power of the photocathode. The acceleration process may then map this variation onto HC fluctuations of all other beam parameters. Finally, the last section summarizes the present accuracy of the determination of false asymmetries for the PVA4 experiment at Mainz.

2 Optical imperfections and HC asymmetries

2.1 Polarization optics for the PVA4 experiment

The laser system of the Mainz polarized source has been described elsewhere [6]. An almost perfect linear polarization is achieved from the collimated laser beam after passing the input polarizer (fig. 1). The Pockels cell is oriented in a way that its electrically controlled optical axis is at 45 degree with respect to the linear polarization. If the phase shift of the cell is adjusted to $\pi/2$ ($-\pi/2$) positive (negative) circular polarization is generated. The phase shift ϕ is proportional to the applied voltage, $\phi^{\pm} = \pm\pi/2$ is achieved at ± 2800 V. The beam diameter in the cell is about 2 mm.

A control device generates a continuous sequence of $\pm$ state signals which is transferred to corresponding circular light polarization by the Pockels cell and its high-voltage switch (fig. 1). The pattern for four 20 ms long states is either $(- + + -)$ or $(+ - - +)$. After a four pattern is completed the first sign of the next pattern is determined at random. The state sequence is also transmitted to the PVA4 data acquisition. The measurement of all parameters contributing to eq. (2) is synchronized to the state sequence thus allowing to measure the HC fluctuations at a rate of 25 Hz.

An insertable half-wave plate ("sign changer") is used to rotate the input polarization by $\pi/2$, which changes the sign of helicity with respect to the control sequence. The physical asymmetry (A_z) then changes sign, whereas for example the false asymmetry resulting from an unequal length of the $+/-$ state will not. The observation of a sign change is therefore an important experimental check.

Behind the Pockels cell the beam passes another half-wave plate which is rotatable around the beam axis. It serves as a compensator for A_I (see sect. 2.3). The following telescope focuses the beam onto the photocathode to a spot of 0.3 mm diameter. Before hitting the cathode the laser beam has to pass a vacuum window which is located about 1.5 m in front of the photocathode.

Between telescope and vacuum window an insertable rotating polarizer/photodetector combination allows to determine the degree of circular polarization. A circular polarization of $P_{\sigma}^{\pm} \geq \pm 0.998$ is routinely achieved.

2.2 Residual linear polarization and compensation

Residual linear polarization components may perform HC variations by 90 degree if the asymmetric phase shift of the Pockels cell is non-zero

$$\Delta\phi_{PC}^{a} = (\phi^{+} + \phi^{-})/2 \neq 0. \tag{4}$$

As an extreme example one can imagine to arrange for phase shifts of the Pockels cell which are $\phi^{+} = 0$ and $\phi^{-} = -\pi$. In this case the linear input polarization is not changed at all in the first case and it is rotated by 90 degree in the second. In a real experiment the asymmetry of the phase shift will of course be minimized but the residual linear polarization will always contribute. An upper limit for residual linear polarization is given by the relation

$$P_{\sigma}^{2} + P_{lin}^{2} \leq 1. \tag{5}$$

For our case ($P_{\sigma} > 0.998$) P_{lin} may be as large as 0.06.

Only a small fraction of the light intensity is absorbed in the active regions of the photocathodes presently in use for PV experiments [1]. The absorption coefficient of these photocathodes varies with the orientation of the linear polarization as in an imperfect dichroic polarizer [4]. Since the emission current is proportional to the absorbed light intensity a corresponding modulation of the emission current with respect to a rotation of polarization will be observed. The amplitude of this modulation (for completely polarized light) is called the analyzing power. Photocathodes used for PVA4 exhibited a typical analyzing power of $\kappa = 3.5\%$. Since helicity switching generates a 90 degree rotation of the residual linear components a non-zero HC beam intensity asymmetry (A_I) of the order of $\kappa P_{lin} \approx 2000$ ppm is typically observed which is far from being acceptable.

The role of the compensator half-wave plate in fig. 1 is in principle the following: If the orientation of its fast axis with respect to the residual linear polarization components is given by an angle β it will rotate them by 2β (its action on the circular components is an additional sign change). Since the cathode "polarizer" has a 180 degree periodicity it follows that the beam intensity gets modulated around an average value $\overline{I}$ like $I^{\pm} = \overline{I}(1 \pm \kappa P_{lin}\cos(4\beta))$. A_I will then vanish for example at $\beta = \pi/8$.

2.3 Quantitative analysis of A_I

A_I can be calculated by using polarization transfer matrices that take into account realistic (*i.e.* imperfect) optical elements. A detailed description of this method is provided in a paper by Humensky *et al.* [7]. In this article we restrict ourselves to a brief outline and a presentation of the results at PVA4.

The polarization transfer matrices may act on the Stokes vector (Mueller matrices) or on the field amplitudes (Jones matrices). In our case the optical elements are either retarders (*e.g.*, Pockels cell) or polarizers (cathode). It is assumed that the optical elements between the input polarizer and the cathode are birefringent with retardations $\phi_{element}$ and have an orientation of their optical axis with respect to the input polarization given by angles $\theta_{element}$. The only imperfections taken into account are deviations from the nominal phase shifts $\Delta\phi_{element}$ and misalignments of the optical axis $\Delta\theta_{element}$. Stokes vectors behind the Pockels cell are helicity dependent, so that they carry the index $\pm$.

It is then possible to calculate the Stokes vector behind the active region of the cathode ($S^{\pm}_{cath}$) by consecutive application of the individual element matrices to the input Stokes vector S_{in} in front of the Pockels cell. The first vector component $S^{\pm}_{cath,0}$ corresponds to the transmitted intensity —which of course also defines the absorbed intensity and therefore the emission current. Using the matrices given in [8] we can therefore obtain the intensity asymmetry as a function of the matrix elements. The elements are approximated up to the first order with respect to the imperfections. A_I also depends on two additional variables β and θ_C which are the compensator angle and orientation of the cathode taken with respect to the input polarization. The resulting asymmetry is a special case of a more general form presented in [7] and can be written as

$$A_I = \kappa\big[\Delta\phi^a_{PC}\cos(2\theta_c - 4\beta) - \Delta\phi_C\sin(2\theta_c - 2\beta) + \Delta\phi_{optic}\sin(2\theta_c + 2\Delta\theta_{optic})\big]. \tag{6}$$

$\Delta\phi^a_{PC}$ is the asymmetric phase shift of the cell (eq. (4)). $\Delta\phi_C$ is the deviation of the compensator phase shift from its nominal half-wave retardation. The effective birefringence of telescope and vacuum window is taken into account by $\Delta\phi_{optic}$ and $\Delta\theta_{optic}$ (the nominal value would be $\phi_{optic} = 0, \theta_{optic} = 0$). The summands of eq. (6) are named according to their β-dependence as the "4β", "2β"and "offset" (β-independent) terms with their corresponding amplitudes $\kappa\Delta\phi^a_{PC}$, $\kappa\Delta\phi_C$ and $\kappa\Delta\phi_{optic}$.

2.4 Experimental result for A_I

Figure 2 shows a measurement of A_I with respect to a variation of the compensator angle at fixed cathode orientation. The fit function has two additional free phases since the orientation of the cathode "polarizer" and of the half-wave plate with respect to input polarization were unknown. From the amplitudes determined by the fit we find the following values for the imperfections.

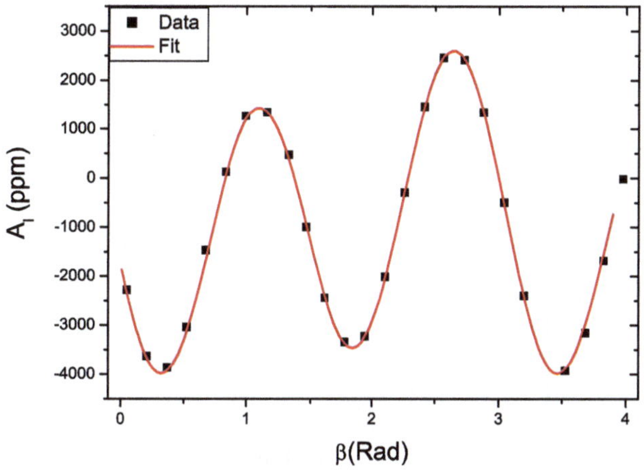

Fig. 2. A_I measurement at PVA4 target with fit.

- The 2β summand contributes with an amplitude of ≈ 640 ppm yielding $\Delta\phi_C = 18$ mrad, which is within the specifications of the half-wave plate.
- The 4β amplitude is approximately 2860 ppm, yielding $\Delta\phi^a_{PC} = 81$ mrad. The large value has been created mostly on purpose because no zero crossing of the A_I-curve may be obtained if $\kappa\Delta^a_{PC}$ is too small with respect to the offset term. The Pockels cell was operated with asymmetric voltages (+2900/−2700 V) which will result in $\Delta\phi^a_{PC} = 58$ mrad. The remaining deviation from the observation (23 mrad) is probably generated by a residual phase offset, *e.g.* strain-induced birefringence in the Pockels crystal.
- The offset term contributes about −830 ppm in this measurement. Since the angle $2\theta_c + 2\Delta\theta_{optic}$ was unknown $\Delta\phi_{optic}$ had to be determined by a separate measurement and is estimated to be ≈ 70 mrad (see sect. 2.5). A major contribution to this birefringence comes from the vacuum window.

Rotating the compensator to a zero crossing of A_I defines a condition which is suitable for data taking during PVA4. Its noteworthy that A_I does not (in first order) depend on the symmetric deviation of the phase shift ($\Delta\phi^s_{PC} = (\phi^+ - \phi^- - \pi)/2$). An example for such a symmetric deviation is the thermal phase drift of our KD*P Pockels cell which is $\approx -3\%/K$. Its influence on the temporal stability of the compensation is therefore suppressed.

2.5 Offset compensation

It is necessary to have non-zero errors $\Delta\Phi^a_{PC}$ and $\Delta\Phi_C$ to compensate for the offset term. The slope of the zero crossing determines the stability and accuracy of setting A_I to zero. Therefore it is desirable to minimize the offset term which then in turn would allow to minimize the other errors. Since the offset is $\propto \sin(2\theta_c + 2\Delta\theta_{optics})$ it can be compensated by changing θ_c which is possible in our set up because the cathode can be rotated under vacuum. We have therefore performed an experiment during which

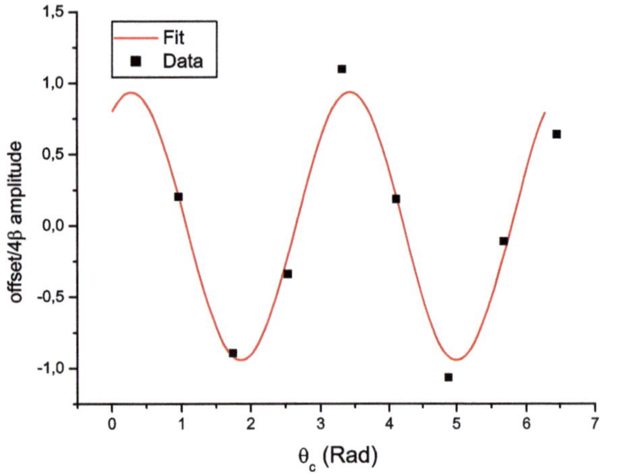

Fig. 3. Measurement of the offset/4β ratio as a function of θ_c.

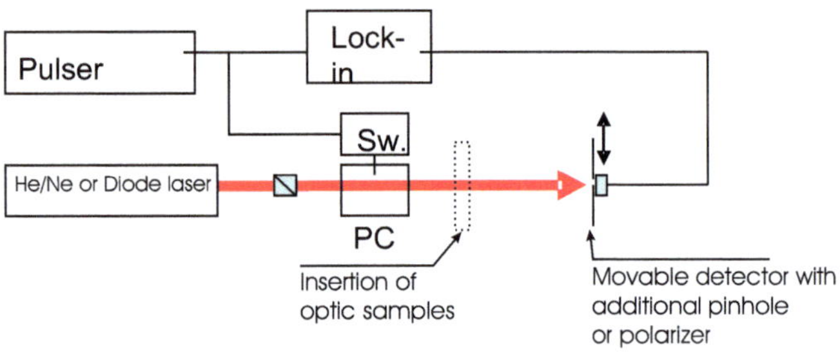

Fig. 4. Testbench for optical components.

$A_I(\beta)$ was measured for various angles θ_c. The summands entering in eq. (6) were determined for each angle. Figure 3 shows the ratio of the offset term and the 4β amplitude as a function of θ_c. The expected $2\theta_c$ variation and the expected zero crossings is clearly demonstrated.

From the maximum offset contribution observed in this measurement (2500 ppm) we can infer the total birefringent phase shift of the optics between Pockels cell and cathode to $\Delta\phi_{optic} = 70$ mrad.

3 Other HC asymmetries

Several observations which are relevant for false asymmetries were made with the help of the set-up shown in fig. 4. A high-voltage switch is triggered by a square wave signal generated by pulse generator. The switch forces the Pockels cell to change circular polarization synchronous to the square wave. A HC photodetector current is then phase locked to the pulser signal and detected by a lock in the amplifier. $A_I < 10^{-4}$ can be reliably measured within a few seconds, thus allowing for fast variation of parameters. In all cases the cell was carefully adjusted and the beam was centered in the cell with an accuracy of ≈ 0.2 mm.

3.1 A_I with minimized analyzing power

A_I was measured without elements between the Pockels cell and the detector. The analyzing power of the detector

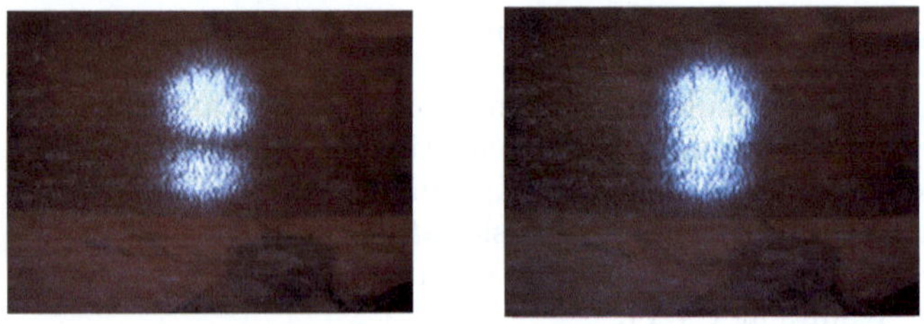

Fig. 5. Backreflection for the different helicity states.

was measured to be $< 10^{-2}$, leading to the expectation of a very small A_I. The experiment is therefore less sensitive to changes of the polarization state but detects HC changes of the light intensity. This situation is similar to early PV electron scattering experiments which were carried out with bulk-GaAs photocathodes. Bulk-GaAs also has a low analyzing power [5]. To promote this approach a HeNe laser was used to obtain better beam quality. Surprisingly an enormous asymmetry $A_I = 1\%$ was observed. An inspection of the backreflections from the cell windows revealed that their modulation was even larger. Since the windows are tilted by a small angle with respect to the Pockels crystal[1] the reflections can be separated from the main beam (fig. 5). The average intensity in fig. 5 is 10% of the transmitted intensity and A_I^R (R for reflected) is also 10%, leading to $A_I^T = -0.01$ for the transmitted beam in agreement with the observation. If an infrared (IR) laser diode is used instead of the HeNe —in this case the reflected intensity is $< 1\%$ because the Pockels cell is antireflection coated for IR— the asymmetry shrinks to $A_I^T < 100$ ppm.

3.2 Investigation of birefringence of inserted elements

In this case a polarizer was mounted in front of the detector in order to maximize the sensitivity according to eq. (6) ($\kappa = 1$) and rotated to obtain the parameter $\Delta\phi_{optic}$. The total $\Delta\phi_{optic}$ of a PVA4-type telescope and vacuum window was measured to 25 mrad. The largest contribution is caused by the vacuum window. We did not test the window under vacuum conditions but it seems probable that its effect will increase because of strain-induced birefringence. This may explain the difference to the 70 mrad observed at the PVA4 set-up.

3.3 Spatially resolved HC fluctuations

The detector was covered by a 25 μm pinhole and was moved transverse to the direction of the laser beam (x-direction). No optical elements were introduced between Pockels cell and the pinhole therefore again minimizing the analyzing power of the system behind the Pockels cell. Possible reasons for HC beam movements in the presence of an analyzing power are discussed in [7].

Figure 6 shows the resulting distributions for intensity $I(x)$ and $A_I(x)$. An IR laser diode was used simulating

[1] A precaution of the manufacturer against optical feedback.

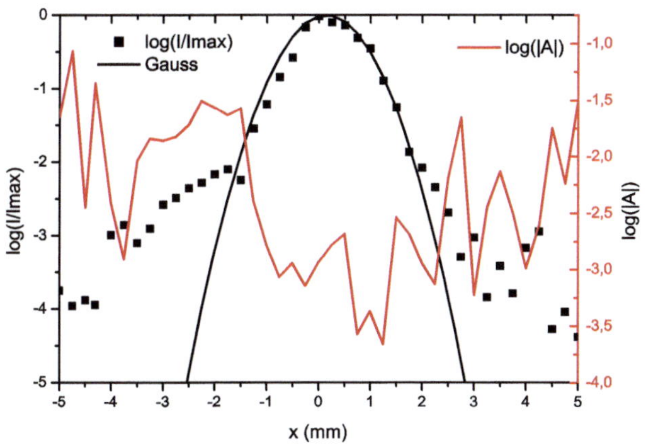

Fig. 6. Measurement of $A_I(x)$.

realistic PVA4 operating conditions. It can be seen that $I(x)$ in the beam center is approximately Gaussian with an $1/e^2$ diameter of 2.4 mm. At radii > 2 mm the intensity distribution is dominated by a superimposed halo. Since the distribution $A_I(x)$ is asymmetric (the largest values are found in the halo to the left) a HC movement of the beam centroid $\langle x \rangle$ is observed, for the data in fig. 6 we find $\Delta\langle x \rangle = 1\,\mu$m.

3.4 Interpretation of the effects

The HC asymmetries presented in 3.1 and 3.3 are caused by a modulation of the light intensity at the detector.

The first effect could be explained by interference between the beams reflected from the front and the backwindow of the cell. Since these are separated longitudinally the temporal coherence length of the laser device is important (see also [7]). The HeNe laser has a much larger coherence length if compared to the IR diode. Therefore its not surprising that the reduction of A_I observed when changing from HeNe to IR diode ($\approx 1/100$) is larger than the reduction of reflected intensity ($\approx 1/10$).

The spatial modulation of intensity with helicity within the halo of the laser beam (fig. 6) may also be a (transverse) coherence effect. It seems probable that polarization states of the halo have much larger linear components with 90 degree flip than the main beam. Since the interference pattern can even disappear completely if one of two interfering beams rotates its (complete) linear polarization by 90 degree (see [8], p. 256) it seems possible that large A_I values in the beam halo are generated by this mechanism. HC beam steering is probably not the reason because of the missing symmetry in the data.

3.5 Couplings

The results presented in fig. 6 lead to a HC movement of the beam section of 1 μm. Taking into account a possible averaging out of the effect for the whole two-dimensional distribution and the strong demagnification of the beam size by the telescope (fig. 1) we estimate the HC movement on the cathode to be about one order of magnitude smaller, *i.e.* 100 nm. Such a movement couples to angular variations due to the focusing elements of the electron optics. With a focusing strength of 1 m^{-1} (typical strength in the 100 keV injection system of the polarized source at MAMI) this transforms into 100 nrad HC angular variation. In addition, an angular variation can lead to path length differences which change the acceleration in the (*e.g.*) third microtron stage of MAMI and therefore lead to a HC energy variation which is estimated to be $\approx 5 \cdot 10^{-9}$. Therefore all beam parameters at the target may show HC fluctuations what is indeed observed at the PVA4 target with the expected order of magnitude.

Table 1. Correction for false asymmetries and resulting errors for PVA4 measurement at $Q^2 = 0.108$ (GeV/c)^2 [9].

Parameter	Correction (ppb)	Error (ppb)
Current A_I	330	40
Position $A_{x,y}$	10	90
Angle $A_{x',y'}$	10	90
Energy A_E	30	30

4 Conclusion: false asymmetries at PVA4

Table 1 shows final values of false asymmetries obtained during the 570 MeV PVA4 experiment [9]. A_I was minimized to 330 ppb with the technique described in sect. 2.3, for the other parameters a decreasing effect of the stabilization systems which work on the uncorrelated parameter fluctuations proved sufficient. It can be seen that angular and position fluctuations do both contribute 90 ppb to the error, which is small against the statistical error (290 ppb). Nevertheless, we envisage an active HC feedback system to compensate for the beam movement on the photocathode.

This work was supported by the Deutsche Forschungsgemeinschaft (DFG) within the framework of the Sonderforschungsbereich SFB-443.

References

1. T. Maruyama *et al.*, Phys. Rev. Lett. **66**, 2376 (1991).
2. T. Nakanishi *et al.*, Phys. Lett. A **158**, 345 (1991).
3. T. Omori *et al.*, Phys. Rev. Lett. **67**, 3294 (1991).
4. R.A. Mair *et al.*, Phys. Lett. A **212**, 231 (1996).
5. K. Aulenbacher *et al.*, *Proceedings of the Low Energy Polarized Electron Workshop, LE98* (SPES-Lab-publishing, St. Petersburg, 1998) p. 87.
6. K. Aulenbacher *et al.*, *Proceedings of the 6th European Particle Accelerator Conference, Stockholm 1998* (Institute of Physics Publishing, Bristol, 1998) p. 1388.
7. T.B. Humensky *et al.*, Nucl. Instrum. Methods A **512**, 261 (2004).
8. E. Collett, *Polarized light* (Dekker, New York, 1993).
9. F.E. Maas *et al.*, Phys. Rev. Lett. **94**, 152001 (2005).

Controlling helicity-correlated beam asymmetries in a polarized electron source

K.D. Paschke[a]

University of Virginia, VA, USA

Abstract. The control of helicity-correlated changes in the electron beam is a critical issue for the next generation of parity-violating electron scattering measurements. The underlying causes and methods for controlling these changes are reviewed with reference to recent operational experience at Jefferson Lab.

PACS. 29.25.Bx Electron sources – 29.27.Hj Polarized beams – 42.25.Ja Polarization – 11.30.Er Charge conjugation, parity, time reversal, and other discrete symmetries

1 Introduction

High-flux measurements in the next generation of parity-violating electron-scattering (PVES) experiments planned at Jefferson Lab aim to achieve a precision on the parity-violating asymmetry ($A_{\rm PV}$) in the range of 5–10 parts per billion (ppb). These goals represent an order-of-magnitude improvement in statistical uncertainty compared to the recently reported HAPPEX-II measurement of $A_{\rm PV}$ from hydrogen, with a precision of ≈ 100 ppb. Systematic uncertainties must also be reduced by a similar factor. Without further improvement, a leading source of systematic uncertainty in these experiments may be helicity-correlated beam asymmetries (HCBA).

The term HCBA generically refers to any difference in the beam between the right- and left-handed (R and L) longitudinal-polarization states. An asymmetry in the total charge delivered to the target for each helicity state (the charge asymmetry $A_{\rm Q}$) will result in a detected asymmetry in the scattered flux. The correction for $A_{\rm Q}$ is generally limited to $\sim 1\%$ by the accuracy of the measurement of $A_{\rm Q}$ or some non-linearity in the experimental apparatus. Future experiments will seek to hold $A_{\rm Q}$ to the level of 0.1 ppm; in the absence of any effort to control it, $A_{\rm Q}$ would typically be larger by four orders of magnitude.

Because the differential cross-section is a sensitive function of energy and angle, any helicity-correlated change in average position, angle, or energy of the beam, is reflected in the detected scattering rate asymmetry. The accuracy of corrections for beam position differences are typically limited to $\sim 10\%$ due to uncertainties in the measurement of the sensitivity of detected rates to the beam parameters.

[a] e-mail: paschke@jlab.org

The effect of position differences is suppressed by the symmetric acceptance of the spectrometers used for these measurements. However, this symmetry does not suppress sensitivity to spot size changes, or other beam shape changes. While effects due to beam spot size or shape changes have been assumed to be negligible in most recent experiments, simulations have suggested that such effects may be important in planned future experiments.

Asymmetries in the time domain may also be significant due to non-linearity of the experimental apparatus. One such example would be a significant non-linearity in position sensitivity. At Jefferson Lab, a strong time dependence of the HCBA has been observed at the few millisecond time scale, even when the average HCBA was small when integrated over the full 33.3 ms helicity windows. Although the net effect was negligible for the HAPPEX-II measurements, it may prove to be important for future measurement goals. The underlying mechanisms for $A_{\rm Q}$ and position differences will also produce spot size, shape, and time domain differences. Previous publications have detailed many of these mechanisms [1–3]; here the topic is reviewed and related to recent operational experience at Jefferson Lab.

2 Identifying the causes of HCBA

2.1 The basic features of the electron source

The polarized electron source used for PVES measurements is based on the use of a doped GaAs photocathode. The dopant induces a strain in the photocathode crystal which breaks a degeneracy in a valance band, allowing spin-selective promotion of electrons to the conduction band using circularly polarized laser light. Once in the

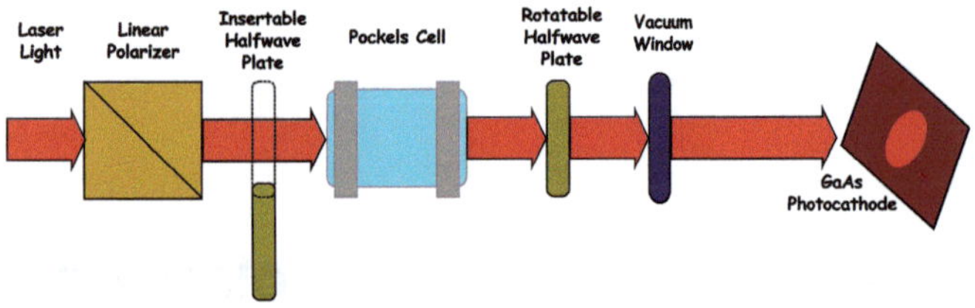

Fig. 1. Schematic of the optics used to prepare the circularly polarized laser light in the electron source.

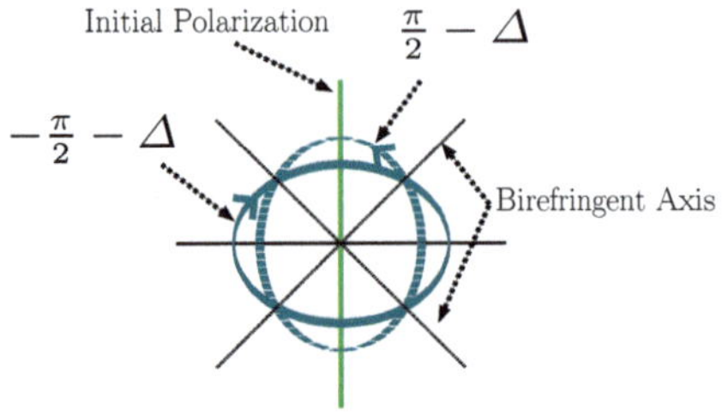

Fig. 2. Polarization ellipse, representing the shape circumscribed by the electric field vector, for each polarization state. An asymmetric phase shift Δ creates a small residual linear polarization of opposite sign for the two helicity states.

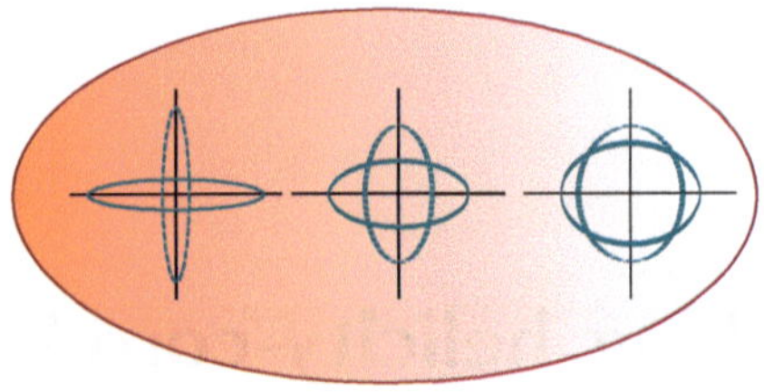

Fig. 3. Illustration of the effect of a phase gradient across a beam spot. A large A_Q will be produced on the left side of the beam spot due to the higher degree of linear polarization, leading to a helicity-correlated shift in the beam centroid.

conduction band, a negative work-function from the cesiated surface allows these polarized electrons to exit the cathode and be accelerated in an electrostatic field. The laser light polarization state is prepared using a Pockels cell, an electro-optic device in which the birefringence is proportional to the applied electric field. A schematic of the optics used to prepare the laser light is shown in fig. 1.

The laser light is initially linearly polarized in the vertical direction. The optic axis of the Pockels cell is aligned along the beam, with the birefringent axes at $\pm 45°$ to this initial-state polarization. High voltage is applied to the Pockels cell appropriate to create a $\pm\pi/2$ phase-shift between the birefringent axes, thus creating left- or right-handed circular polarization. The beam then enters a high-vacuum system through a transparent window and illuminates the GaAs photocathode. The two half-wave plates shown in fig. 1 are not essential for the production of circularly polarized light; their function in controlling HCBA will be described below.

2.2 Effect of imperfect circular polarization

The most troublesome source of HCBA is the imperfect circular polarization of the laser light. Following the prescription from [1], it is convenient to write the applied phase shift from the Pockels cell as:

$$\delta^{\mathrm{R}} = -\left(\frac{\pi}{2} + \alpha\right) - \Delta; \qquad \delta^{\mathrm{L}} = +\left(\frac{\pi}{2} + \alpha\right) - \Delta, \quad (1)$$

where R and L represent the sign of the produced circular polarization, and α and Δ parameterize the difference between the applied phase shift and a perfect $\pi/2$ phase shift. Non-zero α implies that both states are symmetrically under or over phase-shifted, so that both the R and L states have a residual linear polarization with the same sign. Non-zero Δ implies that the magnitude of the phase shift is different between the L and the R states so that one is over-shifted and the other under-shifted, resulting in the two states possessing oppositely signed residual linear-polarization, as illustrated in fig. 2.

The strain on the photocathode, which is critical for producing the highly polarized electron beam, induces an analyzing power for linear polarization; one linear polarization state produces more current than an equal intensity of the opposite linear polarization. For historical reasons, the interaction of linear polarization with the analyzing power of the photocathode is referred to as the "PITA" effect [1]. The PITA effect, in various forms and from various sources, is responsible for most (but not all) of HCBA.

It should be noted that the Δ phase-shift contributes with the same sign to each helicity state, that is to say, it is a static effect uncorrelated to the voltage applied to the Pockels cell. Thus any static birefringent optics element contributes to this Δ phase shift. The Pockels cell itself possesses residual birefringence, due to latent stress on the crystal. The vacuum window also displays stress-induced birefringence, which introduces an additional source of Δ phase-shift, at an arbitrary axis with respect to the Pockels cell axes.

2.3 Phase gradients

Section 2.2 describes how a charge asymmetry is created by a non-zero Δ phase-shift. In general, the source of Δ is not uniform, which implies that A_Q can vary across the beam spot. Figure 3 illustrates this effect: A_Q is large at the left of the beam spot and small at the right of the spot, implying that the centroid of the charge distribution is shifted between helicity states. The shift in the beam centroid can be thought of as the first moment of the intensity asymmetry; higher-order moments are also possible. In particular, while a Δ-phase gradient will lead to a position difference, the second derivative of the Δ-phase will create a spot size asymmetry.

While the above-argument applies to a non-uniform Δ-phase, a similar effect is seen from a non-uniform analyzing power at the photocathode. A gradient (or second derivative) of the analyzing power will combine with a non-zero Δ to create position and spot size differences. One clear difference between the two effects is that position differences due to an analyzing-power gradient are proportional to the average Δ, while those from a Δ-phase gradient are the same regardless of the average Δ phase-shift. This difference is a useful diagnostic tool for distinguishing the two effects.

2.4 Role of beam divergence

It has long been recognized that a well-aligned Pockels cell is key to producing highly circularly polarized laser light. If the direction of beam propagation is not aligned to the Pockels cell optical axis, an offset birefringence is introduced. In studies leading up to the installation of the HAPPEX-II experiment at Jefferson Lab, it was recognized that the divergence of the laser beam produces a coupling of position and phase-shift. Consider a simple ray-trace view of a diverging (far-field) beam. In a far-field beam spot, one can imagine diverging optic rays with a strong correlation between a position in the beam spot and the angle of the associated optic ray. Because the Δ phase-shift due to the Pockels cell varies with the angle of the optic ray, this introduces a correlation between the residual linear polarization and the position in the beam spot.

For a well-aligned Pockels cell, the leading-order effect will be a spot shape asymmetry. However, a slight misalignment of the Pockels cell breaks the first-moment symmetry of this effect, and the beam divergence and misalignment conspire to create A_Q and position differences. While the Pockels cell is typically aligned to the beam within a few milliradians, tests at Jefferson Lab reveal a sensitivity to this effect of $\sim 500\,\text{nm/mrad}$. Uncorrected, this effect would create unacceptable position differences. Assuming other sources of position differences are under control, the Pockels cell alignment can be tuned to zero measured position differences. Such a procedure uses the divergence effect to counteract contributions from other sources of HCBA.

2.5 Other effects

As described in the above section, residual linear polarization and the analyzing power of the photocathode combine to create the majority of HCBA in a well-tuned source. Additional examples of effects seen to be important at Jefferson Lab, other than the various sources of residual linear polarization, are described below.

The Pockels cell is piezo-electric as well as electro-optic, and studies have shown that it deflects the beam as if it were a lens with a helicity-correlated focal length. In practice, carefully centering the beam spot on the crystal minimizes this effect.

A detail of the Jefferson Lab source not shown in fig. 1 is that the laser reflects from several mirrors between the last polarizing element and the Pockels cell. These additional optical elements can degrade the initial polarization state, creating a gradient in the linear polarization of the incident beam and therefore in the residual linear polarization at the cathode.

Table 1. Average beam asymmetries under polarization reversal in intensity and energy and beam differences in horizontal and vertical position and angle for G0 (2004) [4] and the HAPPEX-II experiments (2005).

	G0		HAPPEX-He		HAPPEX-II	
A_Q	−0.28	ppm	−0.38	ppm	0.41	ppm
A_E	19	ppb	3	ppb	0.2	ppb
Δx	6	nm	−0.2	nm	0.5	nm
$\Delta x'$	2	nrad	4.4	nrad	−0.2	nrad
Δy	8	nm	−26	nm	1.7	nm
$\Delta y'$	3	nrad	−4.4	nrad	0.2	nrad

In general, great care is taken to electrically isolate the helicity signal which triggers voltage changes on the Pockels cell or feedback systems. During the 2005 HAPPEX-He measurement, this signal was not well isolated; leakage of the helicity signal was sufficient to induce large position differences in the 100 keV injector. The DAQ electronics were not affected. This large effect dominated other sources of position differences for this run.

There is cross-talk between the helicity-correlated properties of the three independent beams which run simultaneously for delivery to three separate experimental halls. The physical mechanism of this beam cross-talk at the photocathode is not understood, but has proved to be significant at Jefferson Lab. The experiments which have achieved a very high level of control of position differences each ran as the only high-current beam.

Apertures in the injector are seen to clip the beam in a helicity-correlated manner. In most cases, this effect can be reduced with careful configuration of the beam transport. A_Q is monitored throughout the injector to guide the beam optics configuration.

3 Techniques for reducing HCBA

Recent experiments at Jefferson Lab have successfully controlled position differences at a high level. In 2004, the G0 experiment used helicity-correlated feedback in the laser source optics to meet goals on intensity and position HCBA. In 2005, improved results were achieved by the HAPPEX Collaboration with relatively small feedback corrections on intensity asymmetry and no feedback loop acting on position differences. A careful configuration of the source laser optics and electron beam transport optics was sufficient for the control of position differences, averaged over the run, at the level of 1 nm. Results for those experiments, including results for the 2005 HAPPEX-He run in which leakage of the real-time helicity-signal drove large beam-orbit deflections, are shown in table 1.

3.1 Controlling the "PITA" effect

As described in sect. 2.2, the helicity-correlated difference in the residual linear polarization (arising from the Δ

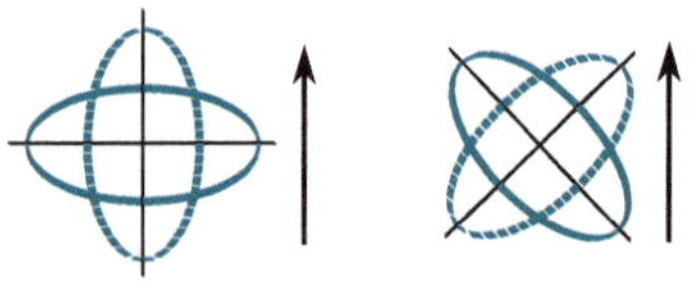

Fig. 4. Illustration of the effect of rotating the polarization ellipses. (a) Aligning the ellipses with the photocathode analyzing power maximizes A_Q, while (b) an angle of 45° with respect to the analyzing power axis minimizes A_Q.

phase-shift) couples with the analyzing power of the photocathode to create a charge asymmetry A_Q. This dominant source of A_Q can be simply tuned out using an asymmetric offset to the voltages applied to the Pockels cell for each helicity state. In fact, a residual linear polarization can be added to the beam to cancel contributions from other birefringent elements, such as the vacuum window. In practice, this global correction is easy to apply: A_Q is measured over a range of asymmetry voltage offsets ("PITA voltage") to determine the slope (A_Q *vs.* PITA voltage), then the PITA voltage is chosen to zero A_Q. This correction results in a minimization of asymmetric linear polarization along the cathode analyzing-power axis.

An independent way of controlling A_Q is by rotating the residual linear polarization to be 45° to the photocathode analyzing-power axis, as illustrated in fig. 4. This is accomplished by rotating a half-wave plate (the RHWP), which is located after the Pockels cell as shown in fig. 1. In practice, this wave plate is oriented to provide a small, but non-zero, sensitivity to the cathode analyzing power; this allows a lever arm for a moderate PITA voltage offset to cancel the residual linear polarization induced by other components, such as the vacuum window, while still reducing sensitivity to imperfections in the Pockels cell birefringence.

Using the RHWP to leave only a small sensitivity to the Pockels cell birefringence similarly reduces position differences induced by birefringence gradients of the Pockels cell. The sensitivity is tuned to be non-zero so that the average residual linear polarization due to vacuum window birefringence can be tuned out using the PITA voltage; doing so zeroes the effect of analyzing-power gradients. In this way, these dominant sources of position differences are both highly suppressed.

One significant source which cannot be controlled using the above prescription is the birefringence *gradient* of the vacuum window. A new polarized gun to be installed at Jefferson Lab will allow for the rotation of the photocathode. When this is installed, the axis of stress-induced birefringence in the vacuum window will be the only arbitrary axis remaining among dominant contributors to position differences. A rotating photocathode analyzing-power axis can be used to zero effects from the vacuum window. This would eliminate the last-known major source of PITA-induced position differences which is presently uncontrolled.

3.2 Reversals

Suppression of HCBA may also be provided by slow reversals, in which the sign of the electron beam helicity is reversed relative to the sign of all electronic signals of the helicity (including the Pockels cell voltage). One obvious method for such a reversal would be a Wien polarization filter; however, in practice such devices typically lack sufficient dynamic range. Another method for accomplishing a similar goal is the use of an insertable half-wave plate (IHWP) in the source optics, just upstream of the Pockels cell as shown in fig. 1. By reversing the sense of incident linear polarization, this wave plate changes the sign of the phase-shift induced by the Pockels cell and therefore the helicity of the electron beam. This reversal provides an excellent method for canceling asymmetries rooted in electronics pickup of the helicity signal.

However, not all HCBA will cancel under IHWP reversal. In particular, any effect directly related to polarization of the laser light, including all PITA-type effects, change sign along with the polarization and therefore do not cancel. On the other hand, steering/lensing does cancel since this effect is related only to the Pockels cell voltage and not to the polarization of the laser light.

Another type of reversal, which was successfully employed for the SLAC E-158 experiment, is a change of beam energy to create a spin reversal using the $g-2$ precession. The $g-2$ reversal is entirely independent of the electronic helicity signal or dynamics of the polarized source. While this reversal is too inefficient to be done frequently and therefore any cancellation will be reduced by slow drifts, it nonetheless remains a full reversal of a large class of effects and is a valuable alternative diagnostic for the integrity of the asymmetry measurement.

3.3 Feedback systems

The use of helicity-correlated correctors in the source optics has become standard for the control of A_Q and position differences. While such feedback systems play an important role, they can mask the underlying problems. As one example, consider the Intensity Attenuator (IA) system often used for feedback on A_Q. The IA consists of a Pockels cell between two linear polarizers, configured so that it can be used as a variable electro-optic shutter. Using such a system, A_Q can be easily controlled, but if large intensity corrections are needed, then underlying problems are only being masked. For example, large A_Q would imply a large average Δ phase shift which would also create large position differences due to analyzing-power gradients. Another correction mechanism for A_Q is the PITA voltage set point described above. Such a mechanism may be preferred as it is directly related to the most likely underlying cause of charge asymmetry. In any case, a reasonable guiding principle would be to use feedback only for fine control, after first optimizing the system to reduce the underlying sources of HCBA.

Feedback systems for position differences are more problematic. Typically, the laser position is modulated using a piezo-electric actuated mirror or prism. Of course, the general principle still applies: this tool, which does nothing to control the root causes of HCBA, should not be used to hide larger problems. A mirror was successfully used for the G0 experiment in 2004. For reasons which have not been fully determined, neither the IA nor the piezo-electric actuated mirrors operated on charge or position alone; each device changed both charge asymmetry and position differences. Periodic recalibration of the response matrix of the mirror was required, as well as occasional adjustments to the electron beam transport optics to create a non-singular response matrix [5]. Position feedback might also be performed on the electron beam using magnets, which would remove complicating effects from the photocathode. Such a system has recently been implemented in the injector at Jefferson Lab.

3.4 Adiabatic damping

The problem of helicity-correlated position differences can be reduced by maximizing the benefit of the reduction of beam emittance during acceleration. The x/x' phase space is reduced by a factor of p_0/p as the beam is adiabatically accelerated from momentum p_0 to p, which implies that helicity-correlated differences in position and angle should each be reduced by a factor of $\sqrt{p_0/p}$. For acceleration from the 100 keV injector ($p = 355$ keV) to 3 GeV in the experimental Halls, one expects position differences to be reduced by a factor of ~ 95.

It is difficult to realize the full benefit of this phase-space reduction for position differences. In transport of the electron beam, correlations develop between x, x', y, and y' so that the phase-space becomes a rotated ellipsoid. In such a case, while the total phase-space volume remains small, the projection of the phase-space into any coordinate can still be large, so this phase-space coupling still allows helicity-correlated changes in beam position and angle to be large.

While the accelerator optics were originally designed to keep such couplings small, in practice the transport optics are often far from design specifications due to small deviations from specification of beam-line elements or initial conditions. The coupling develops from large-amplitude oscillations around the central trajectory, exposing the beam to aberrations or increased skew contributions from both magnetic and RF acceleration elements. The best strategy for avoiding problematic phase-space coupling is to tune the transport optics back to the original machine design, a process referred to at Jefferson Lab as "matching" the optics. Such matching assures a controlled beam envelope and reduced coupling, and thus maximizes the benefits of adiabatic damping for HCBA.

Major work was invested towards an algorithmic approach to matching transport optics to design specification [6,7]. Important diagnostic tools were developed, including a tool for measuring differential orbit launched by a 30 Hz oscillation of a piezo-electric actuated mirror in the source optics. For the HAPPEX-II hydrogen run in 2005, the transport optics were well matched from the 5 MeV region of the injector through the high-energy region. The expected emittance reduction was verified to within a few percent. The fact that HCBA were very small complicates estimates of the adiabatic damping effect on beam asymmetries, but the reduction in position difference was at least a factor of 5 and, in some estimates, may have been as large as a factor of 30.

Significant benefits may also be realized from further improvements, in particular, improved control of coupling from the region of the injector before any RF acceleration, where the beam energy is only 100 keV. Progress here will be challenging; very small fringe fields can have a large effect on the properties of such a soft beam, and the present configuration of correctors and beam monitors in the region does not support a clean determination of beam parameters.

4 Conclusions

Recent progress on controlling HCBA has allowed sub-ppm control of charge asymmetry and ~ 1 nm control of position differences. The design and alignment of the polarized-source laser optics is now guided by an improved understanding of the underlying causes of HCBA, and careful configuration of the electron-beam transport optics has been demonstrated to provide significant benefits in the reduction of HCBA. While work in these areas must continue in order to meet the demanding specifications of the next-generation experiments, particularly with respect to beam spot size asymmetries, this recent success suggests that these future challenges can be met.

This report details work by members of the HAPPEX Collaboration, the Electron Gun Group, and the Center for Advanced Study of Accelerators at Jefferson Lab. I must specifically acknowledge contributions from G.D. Cates, Y-C. Chao, T.B. Humensky, L.J. Kaufman, and R. Snyder although many others have also contributed significantly to these efforts. This work was supported by the Department of Energy and by The Southeastern Universities Research Association, Inc. under U.S. DOE Contract No. DE-AC05-84150.

References

1. G.D. Cates *et al.*, Nucl. Instrum. Methods Phys. Res. A **278**, 293 (1989).
2. T.B. Humensky *et al.*, Nucl. Instrum. Methods Phys. Res. A **521**, 261 (2004).
3. G.D. Cates, Eur. Phys. J. A **24**, s2, 109 (2005).
4. K. Nakahara, Eur. Phys. J. A **24**, s2, 119 (2005).
5. D.H. Beck, M.L. Pitt, Eur. Phys. J. A **24**, s2, 115 (2005).
6. Y.-C. Chao, *Proceedings of the Particle Accelerator Conference 2003*, JLAB-ACP-03-97, `http://accelconf.web.cern.ch/AccelConf/p03/PAPERS/TOPB011.PDF`.
7. Y.-C. Chao *et al.*, *Proceedings of the European Particle Accelerator Conference 2004*, JLAB-ACP-04-252 and EPAC-2004-TUPLT163, `http://accelconf.web.cern.ch/AccelConf/e04/PAPERS/TUPLT163.PDF`.

Laser Compton polarimetry at JLab and MAMI

A status report

J. Diefenbach[1,a], Y. Imai[1], J. Han Lee[1], F. Maas[1,b], and S. Taylor[2,c]

[1] Institut für Kernphysik, Johannes Gutenberg-Universität Mainz, Joh.-Joachim-Becherweg 45, D-55128 Mainz, Germany
[2] Laboratory for Nuclear Science, Massachussetts Institute of Technology, Cambridge, MA 02139, USA

Abstract. For modern parity violation experiments it is crucial to measure and monitor the electron beam polarization continuously. In the recent years different high-luminosity concepts, for precision Compton backscattering polarimetry, have been developed, to be used at modern CW electron beam accelerator facilities. As Compton backscattering polarimetry is free of intrinsic systematic uncertainties, it can be a superior alternative to other polarimetry techniques such as Møller and Mott scattering. State-of-the-art high-luminosity Compton backscattering designs currently in use and under development at JLab and Mainz are compared to each other. The latest results from the Mainz A4 Compton polarimeter are presented.

PACS. 29.27.Hj Polarized beams – 41.85.Qg Beam analyzers, beam monitors, and Faraday cups

1 Introduction

The most important methods for polarimetry of high-energy electron beams are Mott scattering, Møller scattering, and laser Compton backscattering. While Mott scattering polarimeters operate at low beam energy (only a few MeV) and Møller polarimeters can —so far— not handle beam currents beyond some hundred nanoamperes or about ten microamps (depending on the design), Compton backscattering provides polarization measurements at the same beam energy and current as the main experiment while not derating the beam properties as the solid targets for Møller and Mott scattering do.

The lower beam current needed by a Møller target introduces uncertainties, because today's polarized electron sources produce bunches whose polarization is a non-trivial function of time.

The low beam energy required by Mott polarimeters leads to systematic uncertainties, because the polarization and spin transportation through the accelerator needs to be determined precisely. The measurement cannot be performed during the main experiment as the Mott target derates beam quality too much.

Compton backscattering polarimetry provides a destruction-free tool for online monitoring of the electron beam polarization. One more advantage is that the cross-section for Compton scattering can be calculated to arbitrary precision from QED. Possible systematic effects from the uncertainty of the polarization of the target can be controlled very well as the target is a polarized laser beam whose polarization can be determined by measurement during operation of the polarimeter.

2 Principle of Compton backscattering polarimetry

In Compton backscattering polarimetry one uses a colliding beam geometry where a laser beam is shot through the electron beam whose polarization is to be measured. Most of the backscattered photons are radiated into a cone of full opening angle $2/\gamma$ (where γ is the relativistic factor of the electron beam). For usual electron beam energies of at least several hundred MeV the photon beam is thus strongly collimated into the forward direction of the electron beam. Therefore electrons and photons need to be separated by a magnetic field. Then the number of backscattered photons and also their energy spectrum can be obtained.

2.1 Compton cross-section

The cross-section for Compton scattering of circularly polarized photons off polarized electrons consists of three

[a] e-mail: defi@kph.uni-mainz.de

[b] *Present address:* IPN Orsay, 91406 Orsay, France.

[c] *Present address:* Ohio University, Department of Physics and Astronomy, Athens, OH 45701, USA.

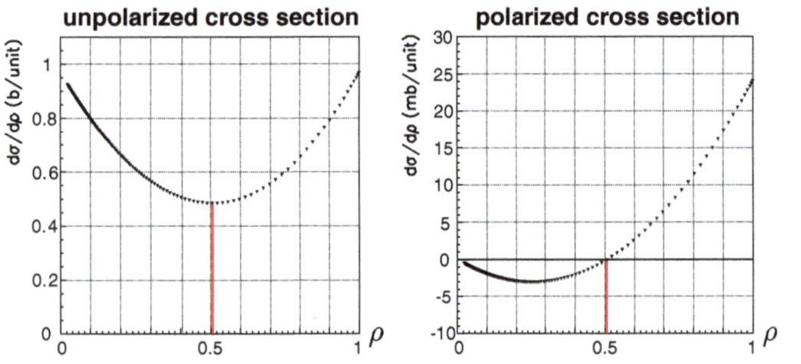

Fig. 1. Left, unpolarized part (σ_0) and right, longitudinal-polarization-dependent part (σ_p) of the Compton cross-section. The x-axis shows the energy of backscatterd photons in terms of the reduced energy ρ, *i.e.* normalized to the maximum backscattered photon energy.

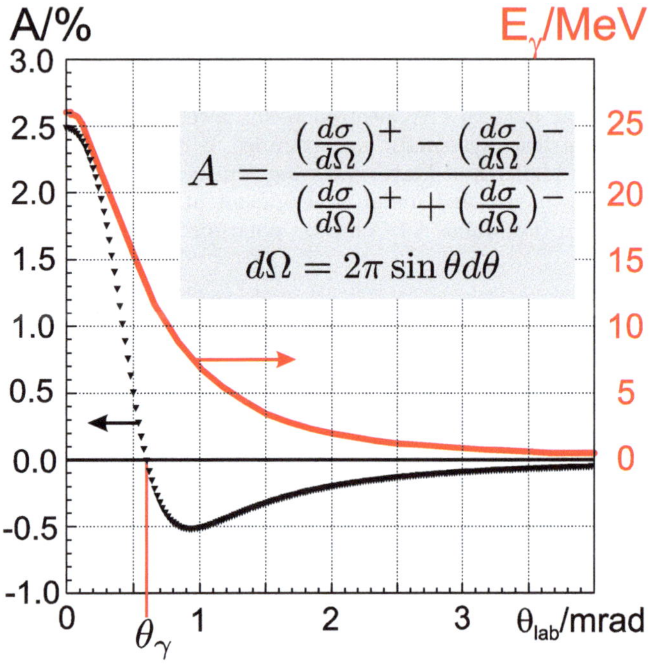

Fig. 2. Energy spectrum and asymmetry calculated from the Compton cross-section for a beam energy of 855 MeV and green laser light (514.5 nm) as used in the A4 Compton polarimeter. The left axis shows the Compton asymmetry in percent, the right axis the backscattered photon energy in the lab frame in MeV.

contributions:

$$\sigma = \sigma_0 + P_e^{long} P_\gamma \sigma_p + P_e^{trans} P_\gamma \sigma_a \cos\theta'_f \,. \qquad (1)$$

For the measurement of the transverse polarization of the electron P_e^{trans} a position-sensitive photon detector would be needed, as can be seen from the dependence on the cosine of the photon angle in the lab frame. The cross-sections for the longitudinal case discussed here are shown in fig. 1. Figure 2 shows the energy spectrum and the resulting asymmetry of the backscattered photons for the special case of the A4 polarimeter that will be discussed in detail in sect. 4.

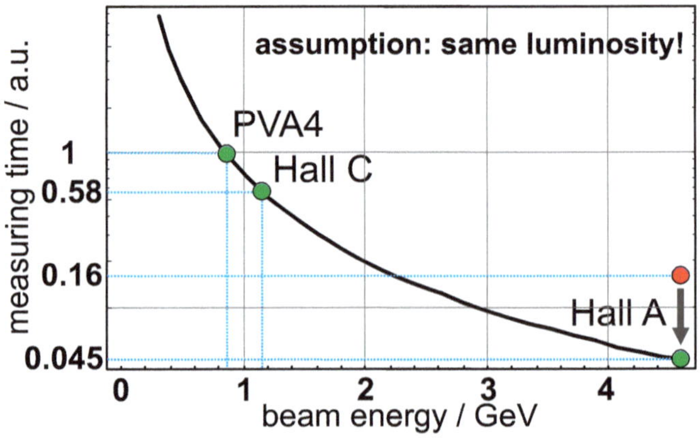

Fig. 3. Comparison of the measuring time (in arbitrary units) needed to obtain a polarization measurement with the Compton backscattering polarimeters at JLab and MAMI, assuming same luminosity and absence of *e.g.* bremsstrahlung background. The JLab Hall-C polarimeter is currently in the design phase. The MAMI polarimeter is realised and the systematic effects are currently beeing studied. For the Hall-A case two points are given: the upper one for the current polarimeter using infrared laser light and the lower one demonstrating the performance improvement of the green upgrade currently under development.

2.2 Determination of the Compton asymmetry

The analyzing power increases with increasing beam energy and decreasing wavelength of the laser light. Therefore Compton backscattering polarimetry becomes more demanding at lower electron beam energies. Assuming the same luminosity and that there is no dilution from bremsstrahlung background, one obtains the time needed for a polarization measurement for the beam energies of some typical experiments as shown in fig. 3. As the polarization-dependent part of the cross-section has a zero crossing at about half the maximum backscattered photon energy ($\rho = 0.508$ for a beam energy of 855 MeV and green argon light), the asymmetry, when switching either the laser or the electron helicity, will also have a zero crossing at this point. The method to obtain the asymmetry from the backscattered photons depends on the conceptual design of the polarimeter and the detectors.

When doing an *energy-weighted* measurement, the amount of energy deposited in the photon detector for each polarization state is recorded. This means, that the detector response as well as the background conditions need to be very well known and have to be stable during the measuring time for a data sample. Otherwise effects like *e.g.* changing background dilution would introduce false asymmetry contributions. Another drawback is that even backscattered photons of lower energy will contribute to the signal, thus diluting the Compton asymmetry, because of the zero crossing of the asymmetry. Therefore the time for a polarization measurement will be proportional to the square of the mean Compton asymetry $\langle A\rangle^2$.

The *differential* method measures the energy spectrum of backscattered photons. Then the asymmetry of a certain part of the spectrum is calculated from the data.

This method provides an energy spectrum for every data sample, so that *e.g.* changes in background conditions can be identified and data can be corrected for. Also bremsstrahlung background subtraction can be done in a very straightforward manner. As the full information from the energy spectrum is available, the flip of the sign of the asymmetry can also be considered to improve the statistical accuracy. The measurement time will be proportional to $\langle A^2 \rangle$, but the range for averaging A^2 can be determined in the analysis. Using the zero crossing of the asymmetry together with the high-energy edge of the Compton spectrum for an intrinsic energy calibration is highly desirable to keep systematic effects as low as possible.

When using the *counting* method, only the number of photons for each polarization state is obtained, so the same precautions against systematic and apparative errors have to be taken into account. As low-energy photons have the same weight as high-energy photons, the measuring time to obtain a certain statistical accuracy is longer than for the integral method.

3 High-luminosity designs for Compton polarimetry

In modern parity violation experiments beam currents are usually limited to some 10 up to 100 μA. In contrast to storage ring experiments with currents of several mA, the usual commercial laser systems with output powers of less than 100 W do not provide high enough luminosity at CW accelerators. Therefore special, high-luminosity Compton polarimeter designs are needed for high-accuracy polarization survey.

3.1 Fabry-Pérot or external-cavity design

At JLab Hall A an external-cavity polarimeter [1] is used for beam polarization measurement. The uncertainty for a single polarization measurement has been reported to be less than 1.5% at 40 μA beam current. Within 25 minutes a statistical accuracy of 1% can be reached [2]. The polarimeter consists of a magnetic chicane, an optical resonator that is pumped by a solid-state laser, a calorimeter to detect the backscattered photons and a silicon microstrip detector for the electrons.

The optical cavity provides a very high gain of the order of 3000. Therefore the available laser power is up to about 1700 W. The optical cavity is set up on the electron beamline, such that both beams will cross. As a drawback due to the sensitive laser optics a crossing angle of 20 mrad between electron and laser beam cannot be avoided, thus reducing the luminosity. Furthermore, the frequency of the pump laser needs to be stabilized accurately to match the cavity as the bandwidth of such a high-gain resonator is only a few kHz.

3.2 Internal-cavity design

Instead of pumping an external cavity one can also make use of the high internal power inside the cavity of a laser itself. A gas laser for example consists of the plasma tube (the laser medium) and two mirrors that form an optical cavity. One of these mirrors is usually highly reflective, the other one has a reflectivity of only about 80 up to 90%. For an argon-ion laser with an output power of 10 W (single-line) and an output coupler of about 90% reflectivity, the internal power is about 100 W.

By extending the optical cavity and guiding the electron beam directly through the cavity one can make use of this high-power density without the need for frequency stabilization of the laser light. Making both beams overlap within two or three meters, the luminosity can reach the same level as for an external-cavity design with a crossing angle between the beams. Still, because of the length of such a cavity (several meters) one needs to design the optical cavity carefully, taking care of low sensitivity to mechanical vibrations and misalignment of the optics. Such a polarimeter has been set up for the A4 experiment at the MAMI accelerator facility in Mainz.

3.3 Pulsed laser

For the upcoming experiments at JLab Hall C a Compton polarimeter with a pulsed laser as light source is currently under development [3]. The mean laser power will be about 90 W at 527 nm. The signal-to-background ratio will be greatly reduced by enabling the data acquisition only during the laser pulses and inhibiting it during the rest of the time. To obtain the Compton asymmetry the integral method will be used. The polarimeter will also make use of an electron detector, either a silicon microstrip as in Hall A or a fiber detector as in the A4 design.

4 The internal-cavity Compton polarimeter of the A4 experiment at MAMI

At the MAMI accelerator facility in Mainz, the A4 Collaboration has set up an internal-cavity Compton polarimeter. Figure 4 shows a schematic overview. We are using a commercial argon-ion laser as light source. So far, we have achieved a maximum intra-cavity power of 83 W with 90% circularly polarized light.

4.1 Providing beam overlap

To get overlap between electron and laser beam, we determine the positions of both beams simultaneously with three wire scanners, so that the axes of both beams and their intersection can be calculated. Then the electron beam can be tilted and moved in the horizontal plane using shunts connected in parallel to the dipole magnets of the chicane. The laser beam can be tilted and moved in

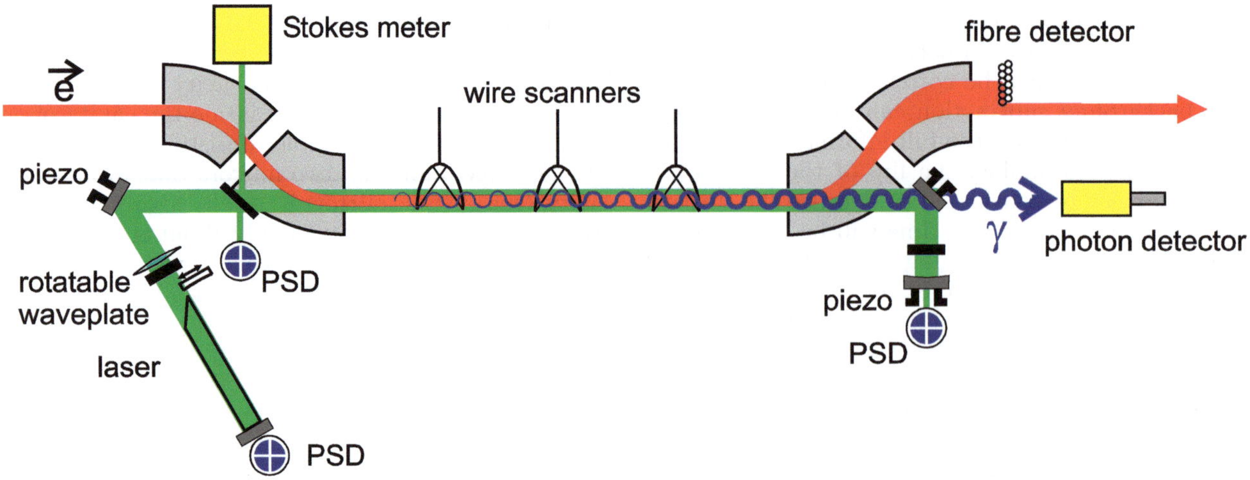

Fig. 4. Schematic view of the A4 Compton backscattering polarimeter. The electron beam is bent by a magnetic chicane consisting of four dipole magnets and two quadrupoles (not shown). The electron beam is guided through the 7.8 m long laser cavity. The plasma tube of an argon-ion laser serves as the light source. The polarization state of the light can be switched between left- and right-handed circular using a rotatable half-waveplate. The current polarization state is determined with a Stokes meter. The position-sensitive detectors (PSD) and piezo platforms are used to stabilize the laser cavity. Backscattered photons are detected in a calorimeter and the corresponding electrons in a fiber detector.

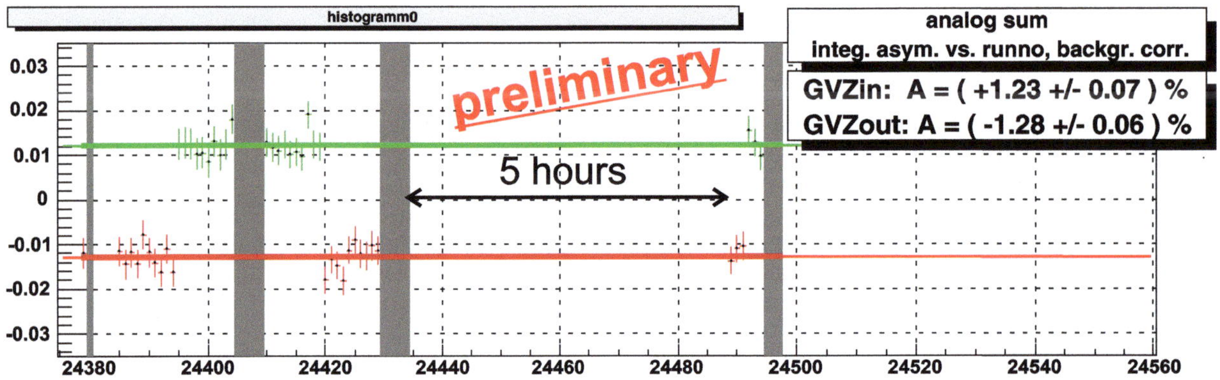

Fig. 5. Compton asymmetry data obtained in August 2005. The laser power was about 34 W and the degree of circular polarization of the laser light was 92%. Grey bands stand for background runs without laser light to correct data for bremsstrahlung background. During a five hours long interval without asymmetry data, tests of the trigger system were performed. The errors shown are from statistics only.

both, the horizontal and vertical planes. Once overlap is established, a stabilization system [4] keeps the position and orientation of the laser beam fixed.

4.2 Detectors

We are using a bundle of 24 scintillating fibers to detect the electrons that have interacted with the laser light. The maximum energy for the backscattered photons is 26.2 MeV for a beam energy of 855 MeV and green argon light (514.5 nm). The energy of the backscattered photons is currently measured in a small BaF_2 calorimeter. We have so far performed tests with NaI(Tl) and $PbWO_4$ and are considering to use a LYSO calorimeter. First tests with LYSO will be completed until the end of 2006. Very dense materials with fast time constants of the scintillation light, such as LYSO or cooled $PbWO_4$ have the advantage of making very compact calorimeters. They are easy to shield against background from the bypassing beamline and provide virtually no pile-up.

The electron detector can be position-calibrated with respect to the electron beam. From the calculated dispersion of the third and fourth dipole magnet of the chicane we know the corresponding photon energy range for every fiber. By triggering on coincidences between electron detector and photon detector, we can then make use of the tagging technique to calibrate our photon detector.

The polarization state of the laser light is measured using a Stokes meter [5]. The polarization state can be switched between left- and right-handed using a rotatable $\lambda/4$-waveplate.

4.3 Analysis

For Compton asymmetry measurements we are taking data with and without laser light to be able to correct our spectra for bremsstrahlung background. We are using the differential method to obtain the Compton asymmetry from the measured energy spectra of backscattered photons. Calculating the weighted mean of the asymmetry of all bins between the zero crossing and $\rho = 1$ of a background-corrected photon energy histogram, we obtain the asymmetry for that run. Combining several runs, one can compare the measured asymmetry with the Compton asymmetry calculated from the Compton cross-sections, taking the measurements for the degree of circular polarization of the laser light into account.

4.4 First beam polarization measurement

In August 2005 we took Compton asymmetry data with 92% circularly polarized light. The measured Compton asymmetries are shown in fig. 5. The weighted mean of the Compton asymmetry for all runs is shown with a one sigma error band (statistical error only). One can clearly see that even after a five hours long interruption for trigger tests the asymmetries are consistent with the measurements done before, proving the stability and reliability of the whole system.

The electron beam polarization obtained from this measurement was $(78 \pm 4_{stat})\%$. The systematic errors treated so far give about 2%. Not yet included are effects from the trigger efficiency of our electron detector and uncertainties from the laser polarization measurement.

4.5 Outlook

In summer 2006 we were able to improve our photon and electron detector performance. By shielding the photon detector we reduced the coincident background in our detectors coming from shower development in parts of the beamline elements so far by a factor of five. Due to standard procedures performing automatic setting of discriminator thresholds and rate measurements for all electron detector channels, our tagging efficiency can be determined now. For further improvement, polarization-dependent measurement of the count rates of our fibers is under development. Deriving the Compton asymmetry independently from such a measurement would uncover possible systematic effects in the photon detector calibration.

Further investigation of *e.g.* charged events and other possible background contributions will be done before the end of this year, together with first tests of a new LYSO photon calorimeter. We aim at a statistical accuracy of 1% within 24 hours.

5 Conclusions

Compton backscattering polarimetry provides a powerful tool for the measurement and monitoring of high-energy electron beam polarization for today's and future parity violation experiments at electron accelerators. The main advantage is that polarization measurements can be performed destruction-free. Therefore they can be carried out in parallel to the main experiment. Systematic effects can be controlled by monitoring of critical parameters such as the polarization state of the laser light during the measurement. Even rather low beam currents of less than 100 μA provide enough luminosity for fast and accurate measurements, when using the high-luminosity designs developed in the recent years.

We are indebted to Prof. Stanley Kowalski for his support and many useful discussions. This work is supported by the Deutsche Forschungsgesellschaft (DFG) in the framework of the SFB 443 and by the U.S. Department of Energy. This article comprises parts of the author's PhD Thesis.

References

1. G. Bardin *et al.*, DAPNIA-SPhN 96-14, 1996, unpublished.
2. Escoffier *et al.*, Nucl. Instrum. Methods A **551**, 563 (2005).
3. D. Gaskell, private communication.
4. J. Diefenbach, Eur. Phys. J. A **24**, s2, 131 (2005).
5. Y. Imai, this conference.

New methods for precision Møller polarimetry*

D. Gaskell[a], D.G. Meekins, and C. Yan

TJNAF, Newport News, VA, 23606, USA

Abstract. Precision electron beam polarimetry is becoming increasingly important as parity violation experiments attempt to probe the frontiers of the standard model. In the few GeV regime, Møller polarimetry is well suited to high-precision measurements, however is generally limited to use at relatively low beam currents ($< 10\,\mu$A). We present a novel technique that will enable precision Møller polarimetry at very large currents, up to $100\,\mu$A.

PACS. 29.27.Hj Polarized beams – 29.27.Fh Beam characteristics

1 Introduction

Until recently, electron beam polarimetry with 2-3% precision has been considered adequate for most electron scattering experiments. However, the next generation of parity violation experiments require ever increasing precision for electron beam polarization measurements. The approved Q_{Weak} [1] and lead-parity [2] experiments at Jefferson Lab require $dP/P \approx 1\%$. Møller scattering and parity-violating deep-inelastic scattering (DIS-parity) standard model tests envisioned for a 12 GeV Jefferson Lab require polarimetry at the 1.0% and 0.5% level of precision, respectively.

The Basel Møller polarimeter in Hall C [3] marked a significant improvement in Møller polarimetry techniques by reducing the dominant systematic uncertainty, for a given target polarization. The Hall-C Møller is in principle able to make polarization measurements with systematic uncertainties on the order of 0.5%. However, this precision is confined to rather low beam currents (a few μA) due to target heating and subsequent depolarization at higher beam currents. We describe here a novel beam-kicker-based system that will mitigate these effects and allow beam polarization measurements that will achieve systematic uncertainties on the order of 1% at very high beam currents ($\approx 100\,\mu$A).

2 The Basel Møller polarimeter in Hall C

Møller polarimeters can be used to extract an electron beam polarization via polarized electron-electron scattering. The electron beam polarization is extracted via,

$$A_{meas} = P_{beam} P_{target} A_{th},$$

where A_{meas} is the experimental asymmetry, P_{beam} and P_{target} are the beam and target electron polarizations, and A_{th} is the acceptance averaged theoretical Møller scattering asymmetry. This last quantity is known from QED and for longitudinally polarized electrons (beam and target) the asymmetry is $-7/9$ at 90 degrees in the center of mass. In the end, the dominant systematic uncertainty for most Møller polarimeters is influenced by the target electron polarization (2–3%).

2.1 Polarimeter overview

The Basel Møller polarimeter improves on earlier polarimeters by dramatically increasing the precision with which one knows the target polarization. In the past, polarized target electrons have been created using iron-alloy foils in a relatively small magnetic field (of the order 100 gauss) aligned along the plane of the foil. The degree of target polarization then had to be carefully determined by measuring the magnetic flux with and without the foil present. In the end, systematic errors smaller than 2–3% on the target polarization are difficult to achieve.

The Basel Møller polarimeter instead uses a pure iron foil magnetized to saturation by a 3–4 tesla superconducting solenoid [4]. In this case, the solenoid field is aligned along the beam direction and the foil is magnetized out of plane. Since the saturation magnetization of pure iron is extremely well known, the net electron polarization can be known to better than 0.3%.

The Basel polarimeter also improves on earlier Møller polarimeters via the use of coincidence electron detection

* Talk presented by D.J. Mack, TJNAF.

[a] e-mail: gaskelld@jlab.org

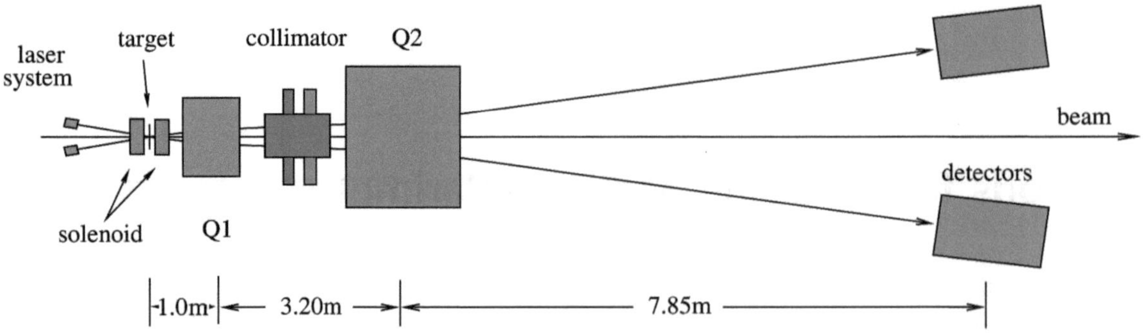

Fig. 1. Layout of the Basel Møller polarimeter in Hall C. A superconducting solenoid magnetizes a pure iron foil out of plane. Møller-scattered electrons are transported using a 2-quadrupole spectrometer to total absorption lead-glass detectors 11 m from the target.

Table 1. Dominant contributions to the systematic uncertainty associated with the Basel Møller polarimeter.

Source	Systematic uncertainty
Levchuk effect	0.30%
Spin polarization in Fe	0.25%
Beam position	0.16%
Multiple scattering	0.12%
Quadrupole settings	0.12%
Total	0.47%

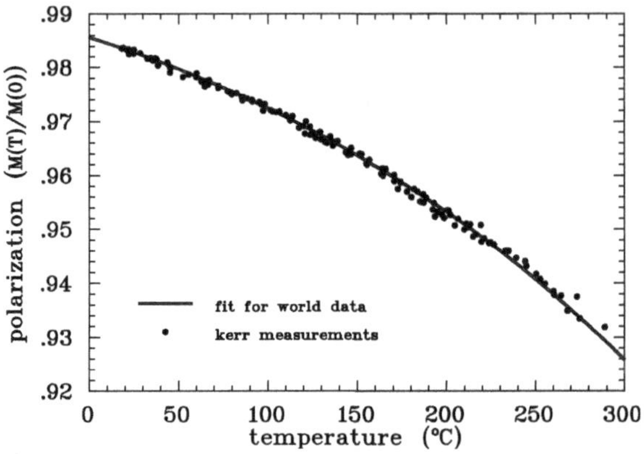

Fig. 2. Relative magnetization *vs.* temperature for a pure iron foil driven to magnetic saturation (from [4]). The iron foil used in the Basel Møller polarimeter is normally at room temperature with no active cooling.

to suppress background from Mott scattering and the use of large-acceptance detectors, which reduces sensitivity to the orbital motion of electrons bound in the iron atom (the Levchuk effect [5]). The polarimeter layout is shown in fig. 1 with the dominant systematic uncertainties associated with the system given in table 1.

2.2 High-current operation

The systematic uncertainties listed in table 1 assume minimal depolarization of the iron target due to beam heating effects. The effect of foil heating on the iron magnetization is shown in fig. 2. A temperature increase of 60–70 °C results in a $\approx 1\%$ reduction in foil magnetization. In order to keep heating effects small enough to have negligible impact on the total systematic error of the measurement, most polarimeter data are taken at rather low beam currents, 1–2 μA. Higher currents can be achieved by rastering the beam over a larger area of the iron foil, but for reasonable raster sizes, the maximum beam current is still on the order of 10 μA.

While delivering low current for operation of the Hall-C Møller polarimeter is straightforward, one must make the assumption that the beam polarization is the same at low currents (1–2 μA) as it is at high currents ($\approx$ 100 μA). It has been suggested that the polarization of the electron beam may change at higher currents due to effects at the polarized source or due to changes in accelerator loading properties. It is desirable, therefore, to measure the beam polarization at currents comparable to the running conditions of the main experiment. As an example, the recent polarimetry uncertainty for the forward-angle portion of the G0 experiment was $dP/P = 1.2\%$ [6], and this was dominated by the (conservative) 1% uncertainty assigned due to the fact that the experiment ran at 40 μA while most polarization measurements were made at 2 μA. Higher-current tests were performed but only extended to a maximum current of 10 μA.

In the next section, we describe a novel approach for mitigating these target heating issues.

3 Kicker magnet for high-current Møller polarimetry

Target foil heating effects can be overcome by quickly scanning the electron beam over a wire or strip target. By minimizing the time on the target, the energy deposited in the foil is kept small and by allowing sufficient time between scans, the foil has a chance to cool before the next scan. A series of "kicker" magnets have been designed and built for this application. In general, these magnets share the following properties: 1) a fast kick that results in a total time on the polarimeter foil on the order of microseconds and, 2) a relatively small duty cycle to

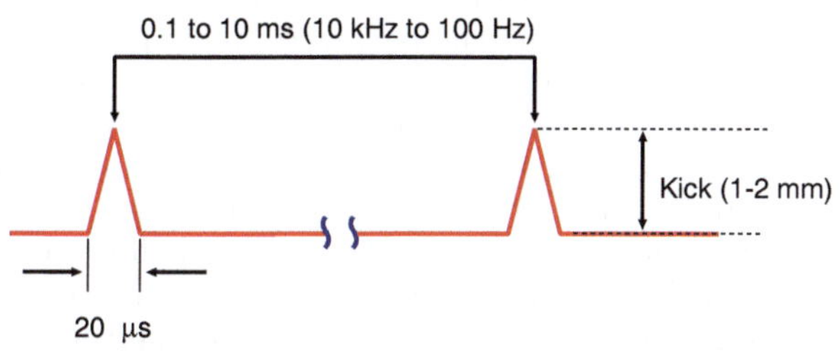

Fig. 3. Graphical representation of the original kicker magnet operating parameters. The kicker magnet kicks the beam 1–2 mm (at the Møller target position) at a frequency between 100 Hz to 10 kHz. During the "off" period, the beam proceeds unimpeded along the normal beam path to the Hall-C dump. During the 20 μs kick period, the beam is deflected across either an iron wire or thin-foil strip target.

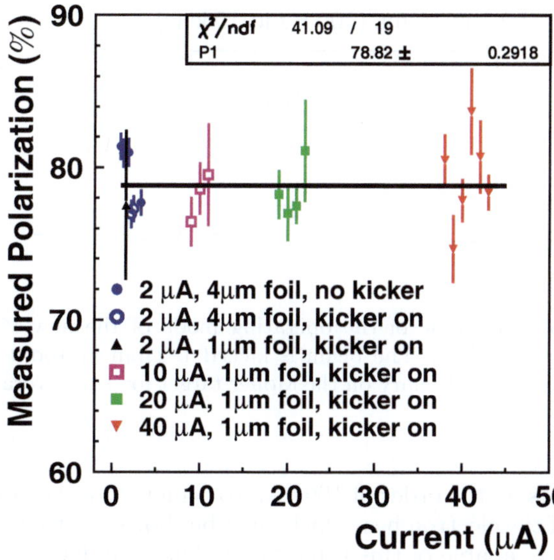

Fig. 4. Results of polarization measurements taken in December 2004 using the second-generation kicker magnet impinging on a 1 μm thick iron foil strip target. Measurements were made at beam currents up to 40 μA. Higher currents were not accessible to due to beam losses from the deflected electron beam. Control measurements at 2 μA suggest instability at the polarized source, so these measurements cannot be used to prove 1% precision at high currents. However, the relative consistency of the results suggest the method should be feasible.

allow the target the opportunity to cool. The magnet itself is a very low inductance, air core magnet installed in the Hall-C beamline several meters upstream of the Møller polarimeter.

3.1 Kicker prototype tests

A first generation kicker was installed in 2003 and tested during the forward-angle phase of the G0 experiment. In that case, the kick duration was 20 μs with an operating frequency between 100 Hz and 10 kHz (see fig. 3). This first-generation kicker was used in conjunction a 20 μm diameter iron wire target. While polarization measurements were made with this device up to 20 μA, the uncertainties at the largest currents were quite large. At high currents, it was necessary to reduce the kick duty factor substantially to keep heating effects small. This lead to quite small average luminosities, which required extremely long measurement times. In addition, since the beam size was comparable to the iron wire diameter, the instantaneous luminosity was quite large when the beam was scanning the target, leading to large random coincidence rates in the polarimeter.

To overcome the two problems encountered above, the target design was reconsidered and a 1 micron thick, strip target was installed. This target had the advantage that during the beam "kick", the entire beam was incident on the target for most of the kick duration (in contrast to the wire target in which one could conceivably be measuring the polarization of the "beam tails" as the beam scans the wire). In addition, the thin foil greatly reduces the instantaneous rates and keeps the random coincidences at a manageable level.

This new target design was tested in December 2004 with very promising results. In this case we were able to make measurements of the beam polarization up to beam currents of 40 μA with no noticeable effects from beam heating (see fig. 4). A slightly improved kicker was also used for this test, the waveform very similar to that shown in fig. 3 but with kick duration a factor of two smaller (10 μs).

During early stages of the 2004 test, it was extremely difficult to transport the electron beam without tripping machine protection systems. Naively, since the kicker was tens of meters upstream of the Møller target, and the kick amplitude was quite small (on the order of 1 mm), the beam should have easily propagated to the beam dump in Hall C. However, the focusing elements downstream of the Møller actually conspired to amplify the kick and resulted in the electron beam scraping various apertures in the downstream beamline. With support from the accelerator division scientists, we were able to set up the beamline optics appropriately to suppress the kick amplitude downstream of the Møller target. Even with this configuration however, we were unable to successfully transport the beam at currents higher than 40 μA. Undoubtedly, the improved beam optics still resulted in the electron beam coming quite close to some aperture in the beamline.

Additionally, we should note that while very promising, the tests performed in 2004 cannot be used to verify the Hall-C Møller operation at 40 μA with 1% precision. Unfortunately, our control measurements with the polarimeter in its nominal configuration were not stable. Measurements made at 2 μA using our standard 4 μm foil yielded results that fluctuated between $\approx$ 78% and $\approx$ 81%. This fluctuation can likely be attributed to non-optimal polarized source conditions.

3.2 Future plans

The eventual goal of this project is to make high-precision measurements using the Basel Møller polarimeter at

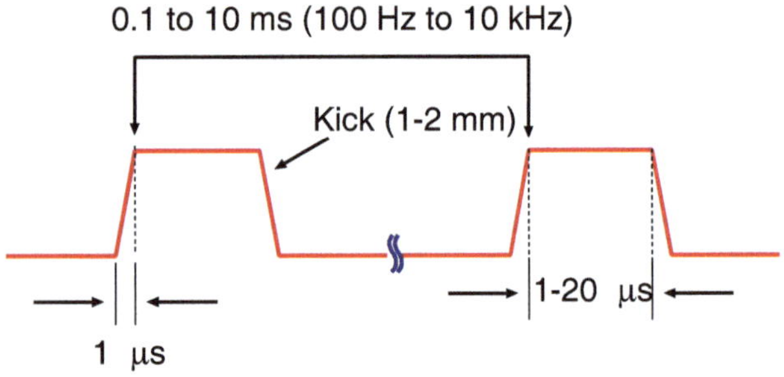

Fig. 5. Schematic of the operating mode of the new kicker magnet waveform. The beam is kicked 1–2 mm in about 1 μs and remains stationary on the Møller target for 1 to several μs.

currents on the order of 100 μA. To achieve this, the duration of the electron beam kick must be shortened to 2 μs. A magnet capable of this rapid beam deflection has already been developed. In addition, this new kicker magnet has the capability to allow the beam to "dwell" on the target at a fixed position for a certain amount of time (see fig. 5). This modified kick cycle will allow us to form a gate such that data is only acquired when the beam is fully "on target" and is not transitioning over a foil edge which may have irregularities and non-optimal magnetization.

In addition we plan to improve the target design. The target used in the December 2004 tests suffered from wrinkles in the foil that had the potential to negatively impact the uniformity of the target magnetization. We are designing a new mounting mechanism to avoid these problems. We are also looking at alternate designs to improve the cooling and heat transport in the foil.

Initially, there were plans to conduct further tests of the kicker system during the backward-angle phase of the G0 experiment. However, the beam energies used were quite low (687 MeV and 362 MeV). Below 1 GeV, the polarimeter becomes extremely sensitive to beam position and trajectory at the target (small position and angle offsets induce larger changes in acceptance and effective analyzing power than at higher energies) so that the measured polarization with and without the kicker magnet would undoubtedly be different. In addition, steering effects from the fixed field 3 tesla solenoid are significant at these low energies and difficult to overcome. Our next opportunity to perform further tests will likely be in the spring and fall of 2007, when higher-energy polarized beams will be available in Hall C.

References

1. R. Carlini *et al.*, *The Q_{Weak} Experiment: A Search for Physics at the TeV Scale via a Measurement of the Proton's Weak Charge*, Jefferson Lab Experiment 05-008, `http://www.jlab.org/exp_prog/proposals/05/PR05-008.pdf`.
2. R. Michaels *et al.*, *A Clean Measurement of the Neutron Skin of* ^{208}Pb *Through Parity Violating electron Scattering*, Jefferson Lab Experiment 06-002, `http://www.jlab.org/exp_prog/proposals/06/PR06-002.pdf`.
3. M. Hauger *et al.*, Nucl. Instrum. Methods A **462**, 382 (2001).
4. L.V. de Bever *et al.*, Nucl. Instrum. Methods A **400**, 379 (1997).
5. L.G. Levchuk, Nucl. Instrum. Methods A **345**, 496 (1994).
6. D.S. Armstrong *et al.*, Phys. Rev. Lett. **95**, 092001 (2005).

Rearrangement of the A4 calorimeter for the backward angle measurement

B. Gläser

For the A4 Collaboration
Institut für Kernphysik, J.J. Becherweg 45, 55099 Mainz, Germany

Abstract. The A4 collaboration investigates the contribution of strange quarks to the vector form factors of the nucleon at the electron accelerator facility MAMI in Mainz by measuring the parity violating asymmetry in elastic electron proton scattering at forward and backward angles. A 1022 channel PbF_2-Calorimeter with a weight of 5 tons has been designed for detecting the scattered elastic electrons under forward angles. For the backward angle measurements the detector support had been redesigned and had been put on a rotatable platform. Changing from forward to backward kinematics is possible with little effort. The PbF_2-detector had been operational after the reconstruction in April 2005. First measurements at backward angles showed, that photons from π^0 decay in the energy range of the elastic scattered electrons dominate the counting rate. An additional system of 72 scintillation counters has been developed and installed in order to separate elastic scattered electrons from background. The system is operational and has been used with electron beam on target for the parity violation experiment during more that 1500 h. The new experimental setup and first measurements will be presented.

PACS. 07.05.Fb Design of experiments – 13.40.Gp Electromagnetic form factors – 25.30.Bf Elastic electron scattering

1 Introduction

The A4 collaboration investigates the contribution of the strange quarks to the vector form factors of the nucleon at the electron accelerator facility MAMI in Mainz by measuring the parity violating asymmetry in elastic electron proton scattering at forward and backward angles [1,2]. For this purpose we use a 1022 channel PbF_2-Calorimeter with a weight of 5 tons to detect the scattered elastic electrons. The calorimeter has been designed to detect the electrons under forward angles. For backward angle measurement the detector suppurt had been redesigned. The calorimeter had been put on a rotatable platform, so that changing from forward to backward kinematics is possible with little effort.

First measurements at backward angles with the redesigned detector showed, that the counting rate in the energy range of the elastic scattered electrons is dominated by photons from π^0 decay. To avoid this, an additional system of 72 scintillation counters has been designed and installed which tags elastic scattered electrons and thus separates them from background.

2 The new calorimeter support

Before the rearrangement, the A4 calorimeter was mounted on a linear moveable platform. The support was used to adjust the focus of the crystal calorimeter on the target center. A special requirement for a possible backward angle setup came through the fact, that not only the detector but also the cryogenic hydrogen target with its scattering chamber and the luminosity monitors have to be adjusted. A new support for the calorimeter and the scattering chamber was needed.

2.1 The scattering chamber and the luminosity monitors

Fig. 1 shows the solid angle of the calorimeter crystals and of the luminosity monitors. An elongation chamber for the main scattering chamber was installed, so that the luminosity monitors have the same solid angle with respect to the target center for forward and backward scattering geometry (Fig. 1).

2.2 The new experimental setup

We have designed a new, common support structure so that the calorimeter and the scattering chamber are supported on a rotatable platform.

To ensure a vibration free, shock free and easy positioning of the platform, it is supported on three hydraulic oil plunger cylinders. By its small stroke (1/10

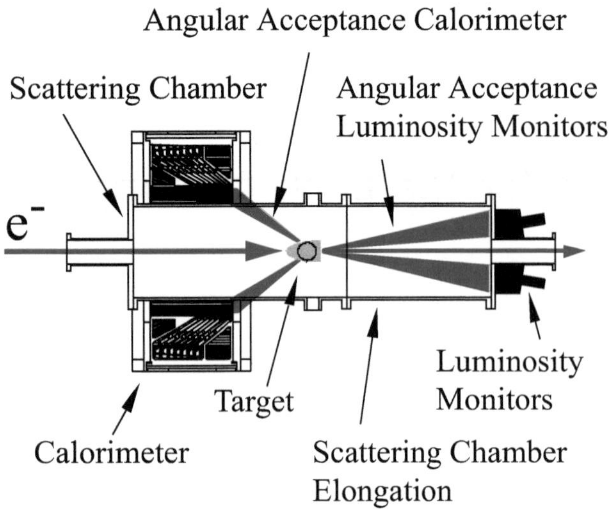

Fig. 1. New scattering chamber. The electron beam comes from the left. The scattering chamber elongation has been installed, so that the luminosity monitors have the same solid angle with respect to the target center for forward and backward scattering geometry.

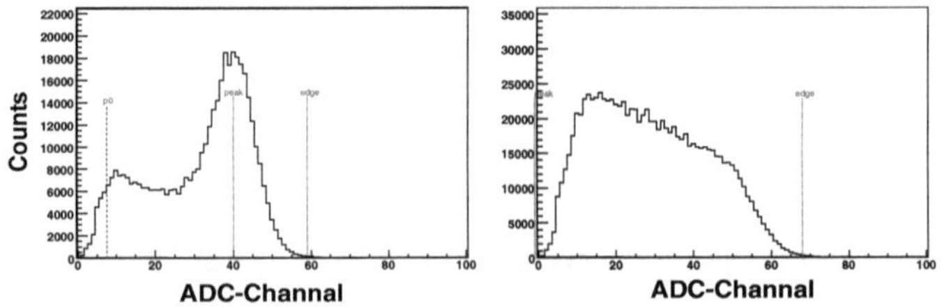

Fig. 2. Energy spectra at 225 MeV (left) and 315 MeV (right). In the 315 MeV spectrum the counting rate in the energy range of the elastic scatterd electrons is dominated from photons from π^0 decay.

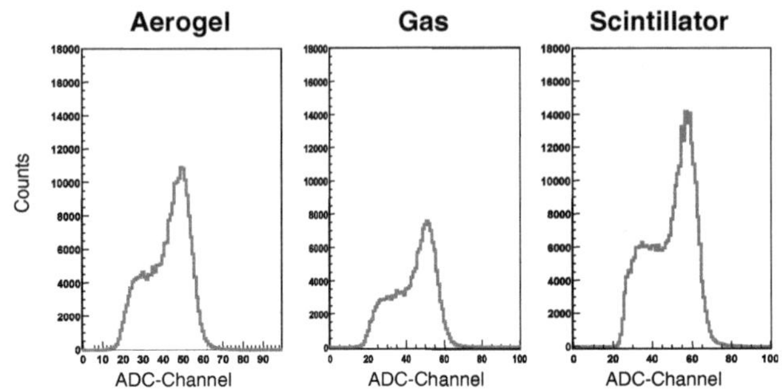

Fig. 3. Tagged electron energy spectra at a beam energy of 315 MeV with the three different tagging detectors.

mm) contrary to air cushions (stroke some mm), an easy adjustment without sudden strokes during set off is ensured without vibrations which could damage the fragile crystals of the calorimeter.

The support platform, the scattering chamber elongation and the hydraulic oil plunger cylinders are already installed, and the experiment was operational for measurements at backward angles since March 2005.

3 First measurements at backward angles

After the rearrangement of the experimental setup first measurements at backward angles (May 2005) have shown, that the counting rate in the energy range of the elastic scattered electrons at 315 MeV beam energy is dominated by a large background (Fig. 2). Simulations have shown that this background is produced by photons from π^0 decay in the target. Fig. 2 left shows clearly the peak of the elastic scattered electrons at 225 MeV beam energy. At a beam energys of 315 Mev (Fig. 2 right) or higher one can only imagine the peak. An additional electron tagging detector system was developed and integrated in the experimental setup to separate photons from electrons.

4 The electron tagging detector

For the electron tagging detector we developed tree prototype detector modules. The first detector type employed a diffusely reflective aerogel Cherenkov radiator from Novosibirsk with a refractive index of 1.05. The walls of the light tight box of the detector had been covered with the diffuse reflector Millipore. With this setup we expected about 5 photoelectrons per incoming electron. The next detector prototype was a gas Cherenkov counter. We used the same light tight box geometry as for the aerogel Cherenkov counter, filled with isobutane gas. The walls were covered with aluminized Mylar as reflector material. With a refractive index of the gas of n=0.019 we expected about 4 photoelectrons per tagged electron. The third prototype was a 2 cm thick plastic scintillator. The detector has to be positioned between the scattering chamber and the crystals of the calorimeter, so that it can measure the scatterd electrons in coincidence with the calorimeter. Because of the low counting rate of elastic scattered electrons per calorimeter crystal it is not necessary to build an individual electron tagging detector for each crystal. We decided to cover 14 calorimeter crystals (two frames) with one tagging detector, which corresponds to a counting rate of the elastic scattered electrons of about 15 kHz.

Fig. 3 shows the coincidence energy spectra of a single calorimeter cystal with the tested electron tagging detectors at 315 MeV electron beam energy and 5 minutes measurement time (The spectra have all the same scaling). One can see the peak of the elastic scattered electrons cleary with all tagging detectors and most of the background is removed. The detection efficiency for elastic scattered electrons is best with the scintillation counter. The detection efficiency for elastic scattered electrons corresponds to 80% with the scintillation counter, to about about 70% with the Aerogel cherenkov counter and to about 50% with the gas cherenkov counter. We decided to build the whole tagging system with scintillation detectors.

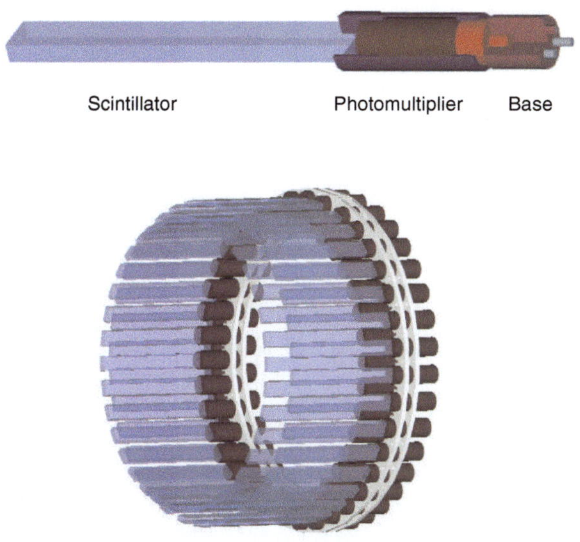

Fig. 4. Electron Tagging Detector. Upper Picture: One Detector Module, lower Picture: Double ring of 72 Modules.

Fig. 5. Rearranged A4-Calorimeter with integrated Electron Tagging Detektor.

4.1 Electron tagging detector setup

Fig. 4 shows the setup of the electron tagging detector. The upper drawing shows one detector module. It consists of plastic scintillator (type EJ-204 from Eljen Tecnologies) with a size of 40cm x 5cm x 2cm, a housing for the photomultiplier with a XP 2262B tube from Photonis employing a actively stabilized volatage devider which we designed ourselves. The lower part of Fig. 4 shows the whole electron tagging system. We built a double ring of 72 scintillator modules, each ring with 36 modules, to cover all calorimeter crystals with a little overlap between the scintillator modules. The whole electron tagging system is shown in Fig. 5 integrated in the rearranged A4 calorimeter with its new support platform and scattering chamber elongation.

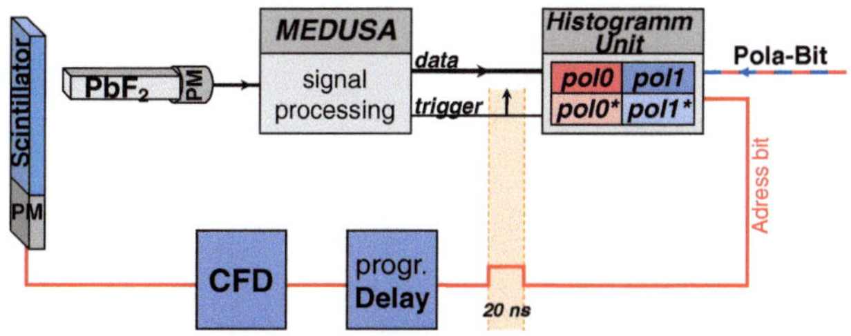

Fig. 6. Electronics for the electron tagging detector.

4.2 Electronics

The electronics for the electron tagging detector and the A4 calorimeter is shown on fig: 6. The main electronics of the A4 calorimeter has two main parts. First, the analog electronics called MEDUSA, which converts the incoming signals into digital data and produces a trigger. The second part is a histogramming unit, where the energy of the accepted events is histogrammed directly into hardware memory. Depending on the electron helicity the data is written into different memory areas, so that we receive two spectra for each helicity state. The scintillator tagging signal is used as an adress bit during histogramming. We obtain in total 4 spectra: 2 energy spectra (tagged and untagged events) for each helicity state. The main parts of the new electronics are a constant fraction discriminator to discriminate the signal at a certain threshold and get a stable timing. A following programmable delay allows us to tune the timing with the calorimeter trigger.

4.3 Conclusion

The system is operational and has been used with electron beam on target for the parity violation experiment during more that 1500 h. It allows an accurate separation of elastic scattered electrons from photons from pion-production. Those photons would otherwise constitute a unwanted background for the parity vioaltion experiment since their parity violating asymmetry is not know. See the contributions of S. Baunack and J. Diefenbach in this volume.

References

1. F.Maas *et al.*, Phys. Rev. Lett. **93**, 022002 (2004).
2. F.Maas *et al.*, Phys. Rev. Lett. **94**, 152001 (2005)

The optical system of the A4 laser Compton polarimeter

Y. Imai

Institut für Kernphysik, Johannes-Gutenberg-Universität, Joh.-Joachim-Becher-Weg 45, D-55128 Mainz, Germany

Abstract. Compton backscattering polarimeters for experiments at sub-mA accelerator facilities use optical cavities to increase the laser light intensity. Therefore, parasitic properties of the optical elements have an increased influence on the polarization state of the light and need to be analyzed carefully. This contribution shows examples of such difficulties and the possible solutions, found during the setup of the A4 Compton backscattering polarimeter at the MAMI accelerator facility of the university of Mainz.

PACS. 29.27.Hj Polarized beams – 42.60.Da Resonators, cavities, amplifiers, arrays, and rings

1 Introduction

Compton backscattering polarimeters make use of an asymmetry in the cross-section for photon-electron-scattering that arises when switching either electron or photon helicity. This asymmetry is proportional to the degree of circular polarization of the light, which therefore needs to be maximized. The analyzing power increases with shorter wavelength. Alas already in the visible spectrum, common lasers do not provide enough intensity to achieve the necessary statistical accuracy within reasonable time [1].

Different schemes have been proposed to overcome this difficulty. One possibility is to feed the output beam of a commercial laser into an external Fabry-Pérot cavity, where constructive interference of the incoming and recirculating light waves leads to massive intensity buildup. The electron beam is then guided through this cavity. Such a system has been successfully set up for the HAPPEx experiment at JLab [2]. However, a complex feedback system is needed to lock the light wavelength to the cavity length.

The Compton polarimeter built for the A4 experiment at the MAMI accelerator uses a different approach, in which the resonant cavity of the laser itself is used for the intra-cavity beam-beam-interaction (see also [3]). In this scheme, the lasing cavity of a commercial laser is extended, and the only partially reflecting output coupling mirror replaced by a high-reflective mirror. The laser medium adapts to fluctuations of the cavity length, and a stabilization is not necessary.

The polarimeter is shown schematically in fig. 1. It is based on an Ar-ion-laser designed for 10 W at 514.5 nm. A lens is used to focus the light in the interaction region for maximum luminosity while keeping the original beam profile in the laser medium. The electron beam is guided through the cavity in a collinear arrangement by using a magnetic chicane. After reflection at the curved end mirror, the photons undergo Compton scattering off the electrons. The backscattered photons are detected in a scintillating calorimeter (currently BaF_2). Fused silica windows are used at the interfaces between air and vacuum, and quartz waveplates are installed to manipulate the polarization state of the light (cf. sect. 2.2). The latter is measured below one of the vacuum windows, where a portion of the light is reflected out of the cavity. In order to establish overlap between the beams, their positions are measured simultaneously using wire scanners.

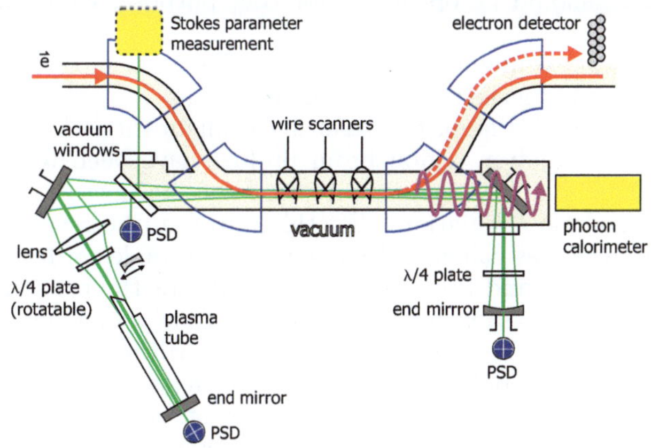

Fig. 1. Schematic layout of the polarimeter. The total resonator length is 7.8 m, the interaction region is 2.7 m long.

2 Design of the A4 polarimeter

2.1 Resonator design

The laser resonator has been designed in two steps. In a first step, general equations for the positions and re-

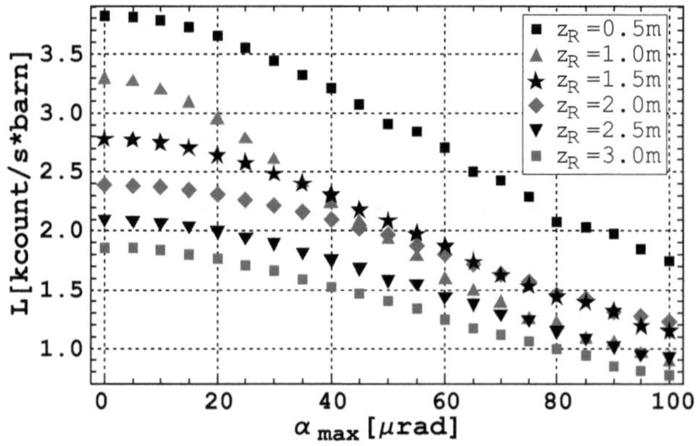

Fig. 2. Effective luminosity per 10 W laser power for different beam profiles, characterized by the Rayleigh length z_R. A current of 20 μA and z_R=10 m were assumed for the electron beam.

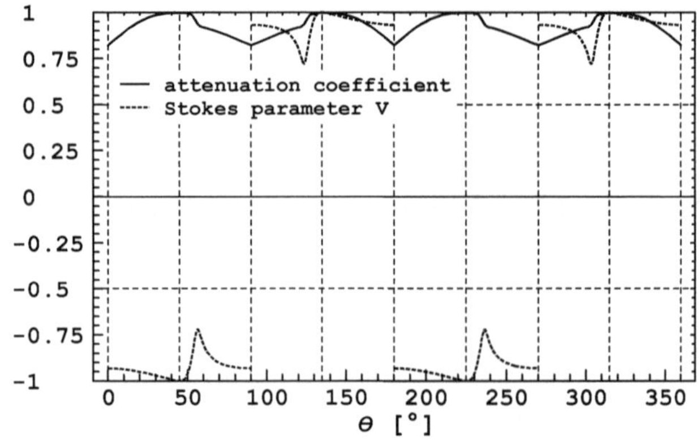

Fig. 3. Circular polarization V and attenuation coefficient (*i.e.* the modulus of the eigenvalue) as a function of the orientation angle of the first waveplate. This calculation accounts for the real mirror reflectivities (but not phase advance) and the fact that the Brewster's window consists of crystalline quartz.

fractive powers of the optical elements have been derived as a function of the desired beam profile using the matrix formalism for Gaussian beams. The beam profile in the interaction region determines the achievable luminosity and should be optimized for that purpose, while the beam profile in the laser medium should stay as close to the factory configuration as possible. The validity of these results was checked in a smaller test setup.

In a second step, the resulting configurations for different interaction region beam profiles have been tested for their stability against vibrations of the optics using Monte-Carlo simulations. For this purpose, the resulting beam axis was calculated for random tilt angles of lenses and mirrors, and the average luminosity for the resulting overlap volumes extracted (see fig. 2). Translational effects were found to be negligible compared to tilting.

For the A4 polarimeter a Rayleigh length of z_R=2.5 m has been chosen for the interaction region because of constraints arising from the chicane geometry.

2.2 Polarization of the laser light

If the light is purely circular, and the electrons only longitudinally polarized or the dependence on the azimuth angle is averaged out, then the Compton asymmetry arising from flipping the light helicity can be derived from the cross-section formula (*e.g.* [4]) to be

$$A = \frac{(d\sigma/d\Omega)_+ - (d\sigma/d\Omega)_-}{(d\sigma/d\Omega)_+ + (d\sigma/d\Omega)_-} = -|V|P_e^{long}\frac{d\sigma_2/d\Omega}{d\sigma_0/d\Omega} \quad (1)$$

Here, V denotes the Stokes parameter describing the circular polarization of the light and the index +,− the light helicity. It can be seen that V needs to be maximized (ideally $V = \pm 1$) and measured in order to extract P_e.

The challenge in this polarimeter setup is having all optics inside the resonator, and that in order to obtain laser emission the polarization state also needs to fulfill a resonance condition, *i.e.* after one round-trip through the resonator, the polarization state of the light must again be the initial state. The resulting polarization state of a cavity can be calculated using the Jones matrix formalism for polarized light. The resonator can then be described by one resulting matrix, and the polarization states possible are the eigenvectors of this matrix.

Using this technique, it was found that two quarter waveplates are necessary in the cavity in order to establish circular light. Furthermore, since one end of the plasma tube is a Brewster's window, any deviation from the ideal setting not only diminishes the degree of circular polarization, but also introduces losses into the cavity, thereby reducing the laser intensity (see fig. 3).

3 Operation experience

The optical system of the polarimeter is successfully operated since 2003. We have achieved powers of up to 83 W in the cavity at 91% left circular polarization. The maximum power so far was 113 W, which is already higher than the intra-cavity power of approx. 100 W in factory configuration. Standard procedures for alignment have been established, and overlap with the electron beam is achieved on a regular basis. We come, however, to the conclusion that these values can still be improved further.

3.1 Parasitic effects

The first Stokes parameter measurements showed that even without waveplates, the laser light had an elliptical polarization. It was found out that the pressure on the vacuum windows exerted by the metal sealings leads to stress birefringence. This effect was minimized by using rubber sealings and only the vacuum suction for holding the windows against the sealings.

After installation of the waveplates, it was possible to increase the laser power by up to 50% by optimizing the waveplate orientation. However, the settings for maximum power didn't provide the maximum degree of circular polarization, and vice versa. Even when both were optimized at the same time, the high degree of polarization could only be established for one helicity, not for the other. There are several possible reasons for this behaviour:

- Intrinsic birefringence in the optics from the HR/AR coatings or from stress in the substrates
- Anisotropy in reflectivity/transmission and phase advance of the AR and HR coatings at oblique incidence
- Birefringence due to external stress, *e.g.* from the optic mounts or thermal expansion

The first point can only be avoided by performing a rigorous quality control of all optics, which is not always possible. The second point is an issue in all folding mirrors, because the reflectivity and phase advance is never equal for the *s*- and *p*-components of the wave, even though coatings optimized for circular polarization have been chosen. Calculations show, however, that the anisotropic reflectivity has only minor impact on the polarization. The distortions arising from the 60° folding mirror may be eliminated by moving the first waveplate behind the folding point (but outside the vacuum). Then, the linear vertical polarization favoured by the Brewster's window would hit the mirror, and the anisotropy between the reflectivities would not become effective.

Heat-induced stress can arise in any mounted optical element, and also in unmounted optics if glued with strong adhesives. This is a problem in cavities with stabilized mirrors, since these are often glued to piezo-actuated platforms. During alignment of the A4 polarimeter cavity, it was often observed that when blocking the beam with an intra-cavity shutter, the laser wavelength had changed and drifted back visibly after opening the shutter again. The magnitude of the effect was dependent on the duration of the beam interruption. This is interpreted as absorption heating in the mirrors, since the other operating conditions were not changed during the process. The use of soft silicon glue or glueing only at one spot can improve the situation. It should be verified, however, that the bandwidth of the stabilization system is not affected.

To reduce stress applied to the optics by the mounting mechanics, new optics holders have been designed that need less holding force and better distribute it over the optical element. A compensator, *e.g.* of Bowen type, can be used to eliminate residual anisotropy.

3.2 Systematic effects

The Stokes parameters are measured by transmitting the light reflected off the 45° vacuum window through a rotating quarter waveplate and a linear polarizing prism (see fig. 4). The light is thereby intensity-modulated at multiples of the waveplate rotation frequency, with modulation amplitudes characteristic of the polarization state. This signal is normalized to the intra-cavity power to correct for fluctuations. The system works stable and reproducibly, but the analysis yields a considerable fraction (several percent) of unpolarized intensity in the measurements. A possible reason is the calibration of the device. The analysis depends on the exact knowledge of the angle between the waveplate fast axis and the polarizer transmission axis. In case of misalignment, unreasonable degrees of polarization may be obtained. A new design with improved electronics and high-resolution mechanics is being developed to overcome these problems. Furthermore, the reflection and subsequent transmission through a vacuum window might alter the measured polarization with respect to the intra-cavity polarization. The transfer properties of these elements will be measured in a dedicated test setup to eliminate this systematic error source.

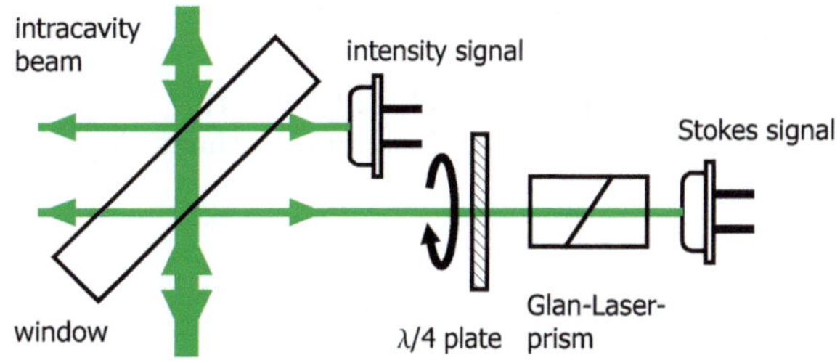

Fig. 4. Schematic view of the Stokes meter.

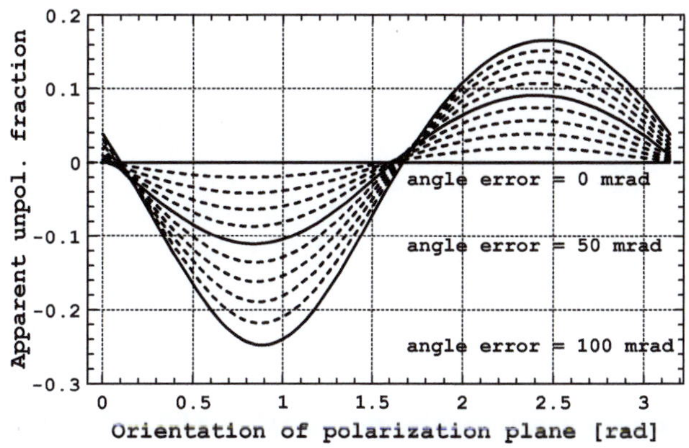

Fig. 5. Calculated unpolarized intensity for purely linear light of different orientations when analyzed with wrong zeroing.

4 Conclusion and outlook

The optical system of the A4 Compton backscattering polarimeter has been designed to provide optimum performance in light intensity and polarization. Reliable operation has proven possible, but unwanted properties of the components can have serious effects on these parameters, because all optics is installed inside the lasing cavity. Several sources of parasitic effects have been identified, and work is underway to further optimize the performance.

This work is supported by the Deutsche Forschungsgemeinschaft in the framework of SPP 1034 and SFB 443. This article comprises part of the author's PhD thesis work. The author is indebted to J. Diefenbach and F. Maas for many fruitful discussions and support during this work.

References

1. Y. Imai, Eur. Phys. J. A **24**, s2 (2005) 127.
2. N. Falletto *et al.*, Nucl. Instrum. Methods A **459**, 412 (2001).
3. J. Diefenbach *et al.*, this conference.
4. F. Lipps, H. Tolhoek, Physica **XX**, 395 (1954).

Conclusions and perspectives*

K.S. Kumar

University of Massachusetts, Amherst, MA 01003, USA

Abstract. The principal focus of the workshop, the study of neutral weak interactions at low energy ($Q^2 \ll M_Z^2$), is a mature field that has major impact on a variety of physics topics, ranging from atomic, nuclear and nucleon physics to probing the limits of the electroweak theory itself. We highlight some important developments in this article, much of which is elaborated on in detail in the rest of the proceedings.

PACS. 12.15.Mm Neutral Currents – 11.30.Er C, P, T and other discrete symmetries

1 Historical perspective

The notion of a weak *neutral* current dates back to the 1950s, when Yang-Mills theory was proposed. Zel'dovich was the first to point out the possibility of an experimental signature for an interaction involving the weak neutral current in the scattering of two charged particles. Weinberg introduced the Z boson as the mediator of a weak neutral current interaction that was an orthogonal admixture of the electromagnetic current in the $SU(2)_L \times U(1)_Y$ gauge model of electroweak interactions.

The 1970s saw the first experimental verification of the above-mentioned theoretical ideas, with the observation of anti-neutrino-electron scattering at Gargamelle and the observation of weak-electromagnetic interference by the E122 experiment at SLAC. The discoveries of W and Z bosons and the measurements of weak neutral current interactions in atomic parity violation experiments, deep-inelastic neutrino nucleon scattering experiments and further measurements of parity-violating electron scattering in the 1980s established a productive subfield of nuclear and particle physics.

Beginning in the late 1980s, the study of weak neutral current interactions at low energy ($Q^2 \ll M_Z^2$) has joined the ranks of high precision electroweak measurements, complementing the specatacular measurements of the Z factories LEP and SLC. Quite apart from testing the electroweak theory, low energy weak neutral current interactions touch on important problems in atomic, nuclear and nucleon physics.

In what follows, we attempt to provide a flavor of these topics, much of which is elaborated on in other articles in these proceedings. Wherever appropriate, we provide speakers' names in (*italics*) to guide the reader towards more detailed information in the proceedings.

2 Weak neutral currents and the atom

Significant progress has been made over the years in experimental techniques to observe small parity admixtures in atoms due to the electron-nucleus weak neutral current interaction, first suggested by Bouchiat and Bouchiat in 1974. The best published measurement is in the 6S-7S amplitude in ^{133}Cs, which yields a 0.35% measurement of the weak mixing angle $\sin^2\theta_W$. Significant progress in the theoretical prediction for this transition over the past few years (especially the Breit interaction, vacuum polarization and vertex/self-energy corrections) has resulted in impressive agreement between theory and experiment (*Derevianko*).

Improved measurements on many atoms and molecules are being pursued (*Lintz, Gwinner, Chardonnet, Budker, Sanguinetti, Tsigutkin*). Fr and Yb are especially good candidates not only to measure the parity-nonconserving transition, but also to measure transitions on a series of isotopes. This has the potential to measure a different linear combination of lepton-quark weak neutral current couplings, complementing lepton scattering measurements. In order to extract these couplings, information on the ground state neutron distribution is required. It may well be that future measurements in atoms and lepton scattering would be combined to obtain unique and important information on neutron skin thicknesses in heavy nuclei, which has important implications for nuclear astrophysics.

Future measurements on Cs and Yb will also focus on further measurements of the nuclear spin-dependent parity non-converving effect, which manifests itself as the so-called nuclear anapole moment. This is a T-even, P-odd observable that probes weak meson-nucleon coupling

* Research funded by US Department of Energy Grant No. DE-FG02-88R40415.A018.

constants in heavy nuclei. These future measurements will improve our undertanding of nucleon-nucleon weak neutral current interactions.

3 Weak neutral currents and the nucleus

3.1 Hadronic parity violation

Parity-violating hadronic interactions have been studied in nuclei for nearly fifty years (*Desplanques, Holstein, Hyun, Ramsey-Musolf, Schiavella*) . Over the past couple of decades, the nucleon-nucleon flavor-conserving weak neutral current interaction has been analyzed in a meson-exchange model with seven coupling constants. However, experimental measurements on light and heavy nuclei, such as photon circular polarization and decay asymmetries, polarization asymmetries in nucleon-nucleon and lepton-nucleon scattering and nuclear anapole moment measurements, have not been mutually consistent within this framework.

Recently, it has been proposed to analyze the parity-violating hadronic interaction in an Effective Field Theory framework. The analysis is confined to nuclei with $A \leq 4$ and the pionless model has five coupling constants that can be constrained by 5 measurements of parity-violating amplitudes. Dynamical pions could be introduced, adding two more coupling constants, if a mutually consistent data set of five measurements is not feasible.

For the abovementioned program, challenging new experiments (*Bowman, Snow*) are being designed and constructed and new few-body ab-initio calculations are being pursued. Once a successful framework is established to describe the parity-violating hadronic interaction in light nuclei, a systematic study of such effects in heavy nuclei can be carried out.

3.2 The neutron skin in a heavy spinless nucleus

In a heavy nucleus such as ^{208}Pb, the difference between the neutron and proton radii is believed to be several percent. A new experiment known as PREX is being planned at Jefferson Laboratory (*Michaels*), to measure the parity-violating asymmetry in the elastic scattering of 850 MeV longitudinally polarized electrons off ^{208}Pb. The measurement is proportional to the ratio of the neutron to proton form factors since the weak neutral current coupling to protons is much smaller than that to neutrons. The planned measurement would measure the proton-neutron radius difference to an accuracy of 1%.

The measurement can have impact on nuclear theory, neutron star structure and atomic parity violation experiments. Relativisitic vs non-relativistic models and the relative size of the symmetry energy affect the size of the predicted neutron skin in heavy nuclei. A precision measurement using an electroweak probe would significantly constrain nuclear models. In addition, models also relate the neutron skin thickness in heavy nuclei to the transition density from liquid to non-uniform matter in the core of a neutron star (*Piekerewicz*). Such an independent constraint on the critical density in a neutron star would provide new insights into neutron star size and composition.

4 Weak neutral currents and the nucleon

One important thrust of theoretical and experimental efforts in the study of the structure of the nucleon has been to explore the relationship between the Quark Model and QCD. In particular, the role of strange quarks in determining the static properties of the nucleon is a way to study the dynamics of low energy QCD (*Schaefer, Leader*).

4.1 Strangeness in nucleons

While studies of the pion-proton scattering and double-spin asymmetries in deep-inelastic lepton-nucleon scattering have yielded information on the contribution of strange quarks to nucleon mass and spin respectively, the relationship between experimental observables and the extracted quantities involve many theoretical uncertainties. In contrast, measurements of the parity-violating elastic electron-nucleon scattering amplitudes, coupled with measurements of electromagnetic nucleon form factors, provide information on the strangeness contribution to the nucleon charge and magnetization distributions with little theoretical uncertainty.

Measurements from the four experimental programs, A4, G0, HAPPEX and SAMPLE, are mutually consistent (*Armstrong, Real, Baunack, Kox, Glaeser, Moffit*). While no single measurement has yet to establish a statistically significant signal for strange form factors, there are two regions where there are hints of non-zero values, one at $Q^2 \sim 0.1$ GeV^2 and one a t $Q^2 \sim 0.6$ GeV^2, though the statistical significance is marginal. This is the focus of current and future measurements.

Most recently, the HAPPEX measurements at $Q^2 \sim 0.1$ GeV^2 have determined that both the electric and magnetic strange form factors are much smaller than early indications at low Q^2. Future measurements from A4, G0 and HAPPEX will allow both electric and magnetic strange form factors to be separately determined to high accuracy over a range of Q^2.

Theoretical calculations of the strangeness content of the nucleon have used several different approaches (*Bijker, Goeke, Riska, Toublan, Young, Zanotti*), including quark models, chiral quark soliton model, heavy quark effective theory, vector meson dominance models and lattice QCD. An important limitation of the technique in terms of obtaining information on the strange form factors is the effect of the assumption of charge symmetry (*Lewis*). As the experimental measurements get more precise, significant information about the strengths and limitations of various approaches is being obtained. Ultimately, the experimental measurements will be tested against unquenched lattice QCD calculations that use light, chiral quarks.

4.2 Electromagnetic and axial form factors

Extraction of the strange form factors requires precise knowledge of the proton and neutron electromagnetic form factors, especially at low Q^2. The BLAST experiment has recently carried out measurements of double-spin asymmetry measurements using polarized ^{1}H and ^{2}H targets (*Alarcon*). In particular, their measurements of the neutron electric form factor has a significant impact on reducing the systematic error in the extraction of the strange form factors (*Pacetti, Dubnicka*).

Another important issue, especially for parity-violation measurements at backward angles, is the axial form factor. For the first time, independent information on this form factor will be obtained from neutrino quasi-elastic scattering, planned for the MINERVA experiment (*Bodek, Schindler*) to be placed in the long baseline neutrino beam line at Fermilab. This experiment, as well as backward angle measurements off ^{2}H in parity-violating electron scattering, will yield precise information on the axial form factor, including its Q^2 dependence.

4.3 Beam-normal asymmetries

The advent of precision parity-violating electron scattering experiments has provided the opportunity to make beam-normal asymmetries measurements. The azimuthal dependence of elastic electron-proton scattering with the incident beam polarized normal to beam direction provides information on the imaginary part of the two-photon exchange amplitude. Depending on the incident beam energy, information on the virtual proton intermediate states over a wide W range is obtained (*Vanderhaeghen, Pasquini*).

The SAMPLE measurement was sensitive to W near pion threshold, while the A4 measurements was sensitive to the W range of several hundred MeV (*Capozza*), where the two-photon amplitude is related to multi-pion production amplitudes. The HAPPEX measurements are sensitive to W upto 2.5 GeV, where estimates of the beam normal asymmetry can be related to the virtual photo-absorption cross-sections (*Kaufman*).

In the future, it might be very interesting to consider beam-normal asymmetry measurements with a 11 GeV beam at an upgraded Jefferson Laboratory. If the measurement of the elastic electron-proton asymmetry can be made with $Q^2 \sim 1$ GeV^2, one can compare to parton-level calculations and obtain new insights into Generalized Parton Distributions. One can also make measurements in the deep inelastic scattering regime, where higher twist effects might be enhanced.

4.4 Gluon polarization

Apart from the strangeness content of the nucleon, studies of the role of gluons in determining nucleon properties is another way to learn about low energy QCD dynamics. With the advent of the RHIC-spin program, the PHENIX and STAR experiments are obtaining new measurements of high energy polarized proton-proton collisions. Information on gluon polarization obtained from asymmetries complement polarized deep-inelastic lepton-nucleon scattering experiments (*Leader, Proceurer*).

The first measurements have shown that the gluon polarization contribution to the nucleon spin is small, in agreement with lepton scattering measurements. With increased luminosity and improved detectors, the RHIC-spin program will provide the first precision measurements of the gluon polarization. Ultimately, in order to provide a model-independent estimate of the total contribution of gluons to the proton spin in the framework of perturbative QCD, new measurements of double-spin asymmetries at a lepton-ion collider will be required.

5 Weak neutral currents and the valence quark

The upgrade of Jefferson Laboratory to 11 GeV beams will allow precision measurements in parity-violating deep inelastic scattering (PV DIS) (*Reimer, Souder, Zheng*). For the first time, high statistics can be accumulated at high $x \sim 0.7$, where x is the fraction of the nucleon momentum carried by the struck quark. PV DIS provides access to novel aspects of nucleon structure, complementing and enhancing precision electromagnetic DIS studies.

The expression for the parity-violating asymmetry involves two terms. For an isoscalar target such as ^{2}H, the first term is a constant that depends purely on the quark weak-neutral current vector couplings, under the assumption of charge symmetry. The second term involves the poorly known axial-vector quark couplings, which can thus be determined to 10% accuracy with a sub-1% measurement of PV DIS off ^{2}H.

5.1 Charge symmetry violation

If parity-violating asymmetry off ^{2}H is measured with high precision over a range of x values, one can test the assumption of charge symmetry for the first time at the partonic level. While this has been tested at the few percent level for $x < 0.4$ from neutrino DIS measurements, calculations suggest that it might be significantly larger for $x \sim 0.6$ and higher. This is one of the likely explanations of the discrepancy between the theoretical predication and the measurement of the charged to neutral current ratio in deep-inelastic neutrino scattering in the NuTeV experiment (*Londergan*).

5.2 Higher twist effects

It has been empirically observed that higher twist effects in the unpolarized structure functions in lepton-nucleon deep inelastic scattering are small. In the case of PV DIS, it can be shown that all except one higher-twist term cancels in the asymmetry ratio. The remaining term involves a correlation between up and down quarks in the nucleon and could provide a unique handle on diquark configurations.

By making a series of precision measurements of the parity-violating asymmetry at high x and over a range of Q^2 values, it might be possible to cleanly observe this higher-twist effect and disentangle it from possible signatures of charge symmetry violation as well as new physics at the TeV scale.

5.3 d/u at high x

For PV DIS off the proton, the first term in the theoretical prediction for the parity-violating asymmetry is quite sensitive to the ratio $d(x)/u(x)$. The value of d/u as $x \to 1$ is a very important parameter to pin down in DIS physics. It is also required in order to properly constrain fits of parton distribution functions and impacts predictions for QCD processes at high energy colliders.

More importantly, d/u at high x provides new information on important pieces of the nucleon wave function. There is empirical evidence that the minority quark in the nucleon is suppressed at high x, an intuitive notion in terms of a hyperfine interaction. While the SU(6) wave function would predict $d/u \sim 0.5$, simple SU(6)-breaking arguments would predict $d/u \sim 0$. However, a perturbative QCD calculation predicts $d/u = 0.2$ as $x \to 1$.

Currently, the best estimates of d/u comes from ^{2}H DIS structure function data, but uncertainties in the ^{2}H wave-function limits the ability to discriminate between various predictions for d/u. There are plans to measure d/u via the ratio of ^{3}H and ^{3}He structure functions and also via measurements of deuteron structure functions with tagged slow recoiling protons.

A precise enough measurement of PV DIS for the proton at $x \sim 0.7$ would be able to distinguish between competing predictions for d/u as $x \to 1$. The advantage of this measurement over other methods is that there are no nuclear corrections since the PV DIS measurement would be made on a proton target.

5.4 Experimental equipment for PV DIS at high x

To comprehensively address the physics topics discussed above experimentally, a series of A_{PV} measurements in the range of 1 to 2% accuracy are required for the x range from 0.3 to 0.7, with a lever arm of a factor of 2 in Q^2 while keeping $W^2_{\text{min}} > 4$ and $Q^2_{\text{min}} > 1$. With the upgrade of Jefferson Laboratory, high luminosity with a beam energy of 11 GeV becomes possible for the first time. However, to achieve sufficient statistics at the highest possible Q^2, a spectrometer with at least 50% acceptance in the azimuth is required (*Souder*).

A large magnet is required to shield the detectors from target photons and to sweep out low energy charged particles. Recent studies have indicated that a solenoid would be ideal for the required geometry, mainly the long target length and the large laboratory scattering angles. Feasibility studies and detailed simulations studies have just begun. It might be possible to use one of the existing collider detector solenoids after the corresponding experiments have been decommissioned.

6 Weak neutral currents and TeV physics

The electroweak interaction has been tested to high precision at high energy colliders (*Erler*), especially by measurements of the properties of the W and Z bosons and their couplings to leptons and quarks. While all data are consistent with the standard model, experiments continue to probe for the indirect effects of new physics at the TeV scale by making more and more precise measurements of electroweak parameters (*Marciano*).

Weak neutral current interactions at $Q^2 \ll M_Z^2$ can probe for heavy Z$'$ bosons or leptoquarks whose effects might be highly suppressed in measurements on the Z pole. Since Z pole amplitudes are imaginary, there are no interference terms with new, real amplitudes. At low Q^2 on the other hand, interference effects might be measurable if sufficient accuracy is achieved. The goal of low energy neutral current measurements is to reach a sensitivity on contact interactions at the 10 TeV level, similar sensitivity to that of the highest energy colliders.

Three published measurements have achieved such sensitivity to date. The weak charge measurement with ^{133}Cs discussed earlier, the NuTeV neutrino deep-inelastic scattering measurement and the measurement of the parity-violating asymmetry in electron-electron (Møller) scattering at SLAC (*Kumar*). These measurements are used to set important new constraints on new parity-violating contact interactions at the 10 TeV scale.

6.1 Planned and future measurements at Jefferson Laboratory

At sufficiently forward angles and low Q^2, the hadronic structure undertainty in the WNC elastic electron-proton amplitude becomes small enough such that one can measure the underlying electron-quark amplitude combination to high precision. This combination is proportional to $1 - 4\sin^2\theta_W$, so that a 4% measurement of A_{PV} would achieve a precision of $\delta(\sin^2\theta_W) = 0.0007$. Such a measurement has been proposed by the Qweak collaboration in Hall C at Jefferson Lab (*Page*). Initial construction funds have been approved and the first data collection is projected for 2010.

With the upgrade of Jefferson Laboratory to 12 GeV, it is possible to achieve higher precision in the measurement of the parity-violating asymmetry in Møller scattering (*Mack*). It might be possible to measure the 40 ppb asymmetry to an accuracy of 2%, which in turn would lead to a weak mixing angle measurement with an uncertainty of 0.00025. At this level, the asymmetry competes directly with the best collider measurements and is sensitive to SUSY loops in the radiative corrections. In particular, the higher precision facilitates further stringent constraints on R-parity violating terms, with important implications for SUSY dark matter candidates.

The measurement of the PV DIS asymmetry to an accuracy of 0.5% at $Q^2 \sim 5$ GeV^2 and $x \sim 0.4$ for an isoscalar target such as deuterium will provide a precision constraint on the axial-vector weak neutral current quark couplings for the first time. This measurement would be robust only if the comprehensive PV DIS program described above is carried out, providing new constraints on TeV scale physics complementary to hadron collider measurements. The measurement is also interesting since it would test the WNC amplitude in the lepton-quark sector, where there is currently a 3 σ discrepancy in the NuTeV result.

6.2 Fixed target Møller scattering at a linear collider

One of the goals of precision electroweak physics measurements is to probe the scalar sector of the electroweak theory via precise measurements of fundamental electroweak parameters. For example, a measurement of $\sin^2\theta_W$ to a precision better than 0.00005 would measure the mass of the standard model Higgs boson to within 10% of itself. Such a measurement is extremely challenging but several methods are being explored in future high energy collider experiments.

Measurements of such precision are very complementary if they are done at different energy scales. One promising method to reach such sensitivity at $Q^2 \ll M_Z^2$ is with fixed target Møller scattering with the electron beam at a future linear collider (*Kumar*). One idea is to develop a test beam and fixed target facility at a linear collider that would use the "exhaust" electron beam after the primary interaction at the collision point with the positron beam. Thus, it would be possible to run fixed target experiments simultaneously with the electron-positron annihilation experiments. It is estimated the $\sin^2\theta_W$ would be measured with a precision of 0.00007 with physics runs of 10^7 seconds at 250 GeV/c^2 and 500 GeV/c^2.

7 Experimental challenges in parity-violating electron scattering

As described above, many of the physics topics discussed have triggered the design of new experiments using parity-violating electron scattering. They involve the measurements of asymmetries smaller than 0.1 ppm. In addition, the fractional accuracy required will be smaller than 1%.

A great deal of technical progress has been achieved in recent years in order to meet these exacting demands on systematic control. The properties of the polarized electron beam, which is generated by photoemission of a GaAs wafer, is now carefully monitored and controlled, such that the cumulative differences in the beam properties upon polarization reversal are the part per million level in intensity, part per billion level in energy and at the nanometer and nanoradian level for the beam trajectory on target. Significant progress in understanding the properties of the laser beam at SLAC, MIT-Bates, Mainz and Jefferson Lab polarized sources is responsible for this achievement (*Aulenbacher, Paschke*).

Cryogenic target technology has also made significant advances over the last few years. Density fluctuations are now routinely controlled at the 100 ppm level over the time-scale of polarization reversal for beam powers up to 1 kW. This control will have to be improved by an order of magnitude for future experiments. Ultra-quiet electronics have been developed to allow flux integration techniques to be used for scattered electron rates of more than 1 GHz. In addition, the G0 and A4 experiments have developed novel counting electronics that allow for online background rejection while counting at rates of 100 MHz.

Finally, significant progress has been made in controlling the systematic error in the knowledge of the electron beam polarization, with the development and improvements in Møller and Compton polarimeters (*Diefenbach, Mack*). While a systematic error of 0.5% was achieved with a Compton polarimeter using a 50 GeV beam, achieving the same level of control with few GeV beams is a significant challenge. Recently, devices have been operational at both Mainz and Jefferson Laboratory and systematic errors approaching 1% have been achieved. For the future program, it is desirable to achieve 0.5% accuracy. At that level, it will be crucial to have two devices that achieve sub-1% accuracy and the potential of using atomic Hydrogen Møller polarimetry will be explored for use with the future 12 GeV program at Jefferson Laboratory.

8 Summary

As seen by the excellent presentations given at the workshop, the field of low energy weak neutral current interactions studies provides access to fundamental measurements over a range of length scales and touches on important problems in atomic, nuclear and particle physics. It is likely that, before one can achieve a complete understanding of electroweak interactions, further important measurements in low energy neutral current phenomena will be required which will keep us busy for years to come!

It is a pleasure to thank the organizers for a stimulating workshop. Input, discussions and advice from all the speakers at the conference are gratefully acknowledged.

MIX
Papier aus verantwortungsvollen Quellen
Paper from responsible sources
FSC® C105338

If you have any concerns about our products,
you can contact us on
ProductSafety@springernature.com

In case Publisher is established outside the EU,
the EU authorized representative is:
Springer Nature Customer Service Center GmbH
Europaplatz 3, 69115 Heidelberg, Germany

Printed by Libri Plureos GmbH
in Hamburg, Germany